Superplasticity in Advanced Materials
ICSAM 2023

The 14th International Conference on Superplasticity in Advanced Materials, ICSAM 2023, held on July 10-12th, 2023 in Barcelona, Spain.

https://www.icsam2023.net/

Editor

José María Cabrera Marrero

Peer review statement

All papers published in this volume of "Materials Research Proceedings" have been peer reviewed. The process of peer review was initiated and overseen by the above proceedings editors. All reviews were conducted by expert referees in accordance to Materials Research Forum LLC high standards.

Published under License by **Materials Research Forum LLC**
Millersville, PA 17551, USA

Published as part of the proceedings series
Materials Research Proceedings
Volume 32 (2023)

ISSN 2474-3941 (Print)
ISSN 2474-395X (Online)

ISBN 978-1-64490-260-8 (Print)
ISBN 978-1-64490-261-5 (eBook)

Distributed worldwide by

Materials Research Forum LLC
105 Springdale Lane
Millersville, PA 17551
USA
https://www.mrforum.com

Manufactured in the United State of America
10 9 8 7 6 5 4 3 2 1

Table of Contents

Preface
Committees

Plenary Talks

New developments in the processing of metallic alloys for achieving exceptional superplastic properties
Chuan Ting Wang, Zheng Li, Jing Tao Wang, Terence G. Langdon 3

Achieving superplasticity in hard-to-deform metallic materials using high-pressure sliding (HPS) process
Zenji Horita, Yoichi Takizawa 15

Keynote Talks

In-situ neutron and synchrotron methods for the investigation of plastic deformation and annealing in metals
Klaus-Dieter Liss 25

Ultra-severe plastic deformation for room-temperature superplasticity and superfunctionality
Kaveh Edalati 41

Regular Talks

An investigation of forming of Ti-6Al-4V at lower temperatures
Hosam Elrakayby, Diego Gonzalez, Paranjayee Mandal, Paul Blackwell 55

Fabrication and superplastic forming property research of TiB whisker reinforced titanium alloy matrix composite sheet
Mingjie Fu, Fuxin Wang, Zhen Li, Yingying Liu 64

Superplasticity of Ti-6Al-7Nb alloys with different initial microstructures processed by high-pressure torsion
Kathy González Jiménez, Joaquín E. González Hernández, Jorge M. Cubero-Sesin, Jeremy Barrantes Rodríguez, Zenji Horita 72

Precise forming of Ti60 alloy by superplastic forming /diffusion bonding
Jiang Shaosong, Kang Liangwei, Li Yang 81

Numerical simulation and experimental trials on superplastic forming for thin-walled structure of Ti2AlNb alloy
Ning Zhang,Yaoqi Wang, Yanhong Mu, Zhen Li 87

Effect of friction stir forming parameters on mechanical properties and plastic flow of material in the mechanical interlocking of optical fiber and SP-700 superplastic titanium alloy
Hamed Mofidi Tabatabaei, Takahiro Ohashi, Tadashi Nishihara 95

Mechanical and microstructural analysis of Ti-6Al-4V material in a wide range of superplastic forming conditions
Alexander Sancho, Paranjayee Mandal, Craig Knowles, Diego Gonzalez 103

Effects of deformation conditions on the superplastic deformation behavior of LZ91 Mg-Li alloy under electric field
Bao Meng, Feng Pan, Jianqiang Yang, Dongsheng Li, Min Wan 111

Integration of HF and SPF technology in an efficient industrial production flow for high rates aeronautical parts
Yves Marcel, Jeremy Pion, Sharan Raj Rangasamy Mahendren 119

Investigation of the complexities inherent in manufacturing near-unconstrained superplastic parts by experiments and simulation
Bryan Ferguson, Eric Bol, D.G. Sanders, and M. Ramulu 127

Formation mechanism and elimination method for surface groove of the hollow lattice structure by superplastic forming
Yanhong Mu, Zhiqiang Li, Bing Zhao, Feng Yang, Ning Zhang, Lihua Du, Yong Li 141

Study on superplastic gas-bulging properties and microstructure evolution of as-rolled 2A97 Al-Li alloy
Zhang Yanling, Liu Jiajia, Xu Huiyuan, Du Lihua 149

Superplastic forming process research of front aluminum alloy opening and closing mechanism hatch for multiple units
Ranran Chen, Guangzhong He, Kefei Wang, Guofeng Wang 157

Low-temperature superplasticity of Si-added medium-Mn steel
Hyun-Bin Jeong, Seok-Won Choi, Seok-Hyeon Kang, Young-Kook Lee 167

Friction stir processing to improve grain refinement and superplasticity of Al-Mg-Mn-Cr alloy
Anna A. Kishchik, Ahmed O. Mosleh, Olga A. Yakovtseva, Anton D. Kotov, Anastasia V. Mikhaylovskaya 173

Effect of Co addition on the microstructure evolution and superplastic behavior of Ti-4Al-3Mo-1V-0.1B alloy
Anton D. Kotov, Maria N Postnikova, Ahmed O. Mosleh, Anastasia V. Mikhaylovskaya 181

Predicting ECAP misorientation evolution and its influence on superplasticity for an Al-Zn-Mg-Cu alloy
Fernando Carreño 189

Microstructural characteristics, mechanical and corrosion properties of a low-alloyed Mg alloy after different deformation processing
Zahra Abbasi, Jose Maria Cabrera, Ramin Ebrahimi, Erhard Schafler 197

Effect of severe plastic deformation on microstructure and properties of duplex stainless steel
Claudio Gennari, Luca Pezzato, Nuria Llorca Isern, Alisiya Biserova-Tahchieva, Irene Calliari 205

Consolidation of AZ31 magnesium chips using equal channel angular pressing
Majid Al-Maharbi, Almundher Al-Namani 213

Secondary phase precipitation in ultrafine-grained superduplex stainless steels
Alisiya Biserova-Tahchieva, Núria Llorca-Isern, Jose Maria Cabrera 220

Grain boundary sliding at low temperatures
Roberto B. Figueiredo 227

In-situ heating observations on microstructure relaxation of ultrafine-grained high-entropy alloys using neutron diffraction and laser-scanning confocal microscopy
Megumi Kawasaki, Jae-Kyung Han, Suk-Chun Moon, Klaus-Dieter Liss 235

Hot deformation of ultrafine-grained TiC-reinforced AlCrFeNbMoTiV refractory high entropy alloys
P. Martin, C. Aguilar, B. Ferrari, A.J. Sánchez-Herencia, J.M. Cabrera 244

Deformation behaviour and microstructural evolution of high-entropy CoFeMnNi alloy at hot deformation condition with low strain rate
Lihua Du, Jiaai Shi, Jiayu Liang, Kailun Zheng 253

Comparative studies of hyperbolic sine constitutive models for constant pressure superplastic tests
Luis García-Barrachina, Antonio J. Gámez 264

Warm tensile properties of low-alloy TRIP steel warm-rolled just prior to isothermal holding treatment
Katsumi Hattori, Genta Kojima, Junya Kobayashi, Goroh Itoh, Shigeru Kuramoto, Tomohiko Hojo 274

Mechanical behavior and microstructure evolution of Ti-6.5Al-2Zr-1Mo-1V/TiB metal-matrix composites during deformation
Maxim Ozerov, Vitaly Sokolovsky, Alexander Galtsev, Nikita Stepanov, Sergey Zherebtsov 280

Investigation of the influence of kinematic asymmetry on the properties of laminated materials
Olesya D. Biryukova, Alexander M. Pesin, Denis O. Pustovoytov 287

FEM simulation of dynamic recrystallization during asymmetric hot rolling of high-speed steel M2
Alexander Pesin, Denis Pustovoytov, Ilya Pesin, Hailiang Yu, Puneet Tandon, Harshal Y Shahare, Abhay Kumar Dubey, Leonid Nosov 294

Improvement in strength and ductility of flame-retardant magnesium alloy through uniaxial hot pressing
Koichi Kitazono, Kotaro Wada 303

Experimental study of the effect of increasing technological plasticity during asymmetric rolling of aluminum alloys
Anna Kozhemyakina, Alexander Pesin, Denis Pustovoytov, Leonid Nosov, Anna Baryshnikova, Natalia Lokotunina, Dmitry Grachev 309

Posters

Microstructural and corrosion study of a “non-comercial” high manganese steel
Marcos Natan da Silva Lima, Mauro Andres Cerra Florez, Jorge Luiz Cardoso,

João Vitor da Silva Pinto, Lucas Moura Farias de Medeiros, Walney Silva Araújo, Hamilton Ferreira Gomes de Abreu 317

Hot plastic behavior of an ultrafine-grained aluminum alloy fabricated by laser powder bed fusion and equal channel angular pressing
Jairo Alberto Muñoz, Louis Huvelle, Enrique Manuel Huerta, José María Cabrera 330

Improved stochastic dissimilar diffusion bonding model with experimental validation
Bryan Ferguson, Neha Kulkarni, D.G. Sanders, Eric Bol, M. Ramulu 338

Development of asymmetric cold rolling technology of high-strength steel grades in order to exclude intermediate annealing operations
Alexander Pesin, Georgy Raab, Alexey Sverchkov, Denis Pustovoytov, Gennady Kornilov, Alexey Bochkarev, Ilya Pesin, Leonid Nosov 355

Texture and strain rate sensitivity analysis of solid solution and precipitation hardening aluminum alloys processed by repetitive corrugation and straightening
Sergio Elizalde, Marco Ezequiel, Liliana Romero-Resendiz, Jose Maria Cabrera, Gonzalo González 362

Exploring the effects of repetitive corrugation and strengthening on alloy A5083 using X-ray diffraction techniques
Carlos Alberto De la Trinidad, Ignacio A. Figueroa, José María Cabrera, Gonzalo González 371

Keyword index
About the editor

Preface

This book brings together most of the presentations made at ICSAM2023, during the days 10 to 12 July 2023. After the St. Petersburg conference in 2018, and the subsequent cancellation due to the COVID 19 pandemic, the conference was resumed in Barcelona in person. Practical and theoretical works were presented on superplasticity in metals and ceramics, on fundamentals of deformation mechanisms, on processes to obtain large ultrafine-grained structures (especially by severe plastic deformation), on advanced characterization techniques and finally on hot deformation works of traditional and especially advanced materials. Very important contribution as plenary and keynote talks were given by Prof. Terence Langdon, Prof. Zenji Horita, Prof. Hidehiro Yoshida, Prof. Klaus-Dieter Liss and prof. Kaveh Edalati

The congress also paid tribute to the memory of Dr. Alexander Zhilyaev, a very active member of the ICSAM community, who passed away on November 20, 2020, at the age of 61. Dr Alexander Zhilyaev was principal researcher of the Institute for Metals Superplasticity Problems of the Russian Academy of Sciences (IMSP RAS). Alexander Zhilyaev was an internationally recognized scientist and made important contributions in the fields of physics of grain boundaries, superplasticity, severe plastic deformation and nanostructured materials. He worked at various leading research centers in Russia, Canada, USA, Spain and the UK. And above all, he was a very good friend of all of us.

Sponsorship
Organizer thanks the sponsorship provided by Group Lauak, France, as well as all the support provide by Universitat Politècnica de Catalunya (UPC), the Faculty of Engineering Barcelona Est (EEBE), and most specially by Fundacio Centre CIM.

Note on the image taken as conference logo.

"Trencadís": also known as pique assiette, broken tile mosaics, is a type of catalan mosaic (Gaudi's style) made from cemented-together tile shards and broken chinaware. It resembles the ultrafine grained structure needed to promote superplasticity.
Image taken from Pixabay (free use)

Committees

International Committee

Prof. Goroh Itoh, Chairman, Japan
Prof. Gèrard Bernhart, France
Dr. Vincent Velay, France
Dr. Vit Janik, UK
Mr. Larry D. Hefti, USA
Prof. Sergey Zherebtsov, Russia
Prof. Megumi Kawasaki, USA
Prof. Koichi Kitazono, Japan
Prof. Zhi Qiang Li, China
Dr. Koji Morita, Japan
Prof. Jose María Cabrera, Spain
Prof. Gilmar Ferreira Batalha, Brazil
Prof. Paul Blackwell, UK
Prof. Guofeng Wang, China
Prof. Gencage Purcek, Turkey

National Committee

Prof. José María Cabrera Marrero, Universitat Politècnica de Catalunya, Barcelona, Chairman
Dr. María Teresa Pérez Prado, IMDEA Materiales, Madrid
Prof. Nuria Llorca Isern, Universitat de Barcelona, Barcelona
Dr. Fernando Carreño Gorostiaga, CENIM-CSIC, Madrid
Prof. Jessica Calvo Muñoz, Universitat Politècnica de Catalunya, Barcelona
Prof. José Antonio Benito Páramo, Universitat Politècnica de Catalunya, Barcelona
Prof. Jairo Alberto Muñoz Bolaños, Universitat Politècnica de Catalunya, Barcelona
Prof. Pere Barriobero Vila, Universitat Politècnica de Catalunya, Barcelona

Plenary Talks

Materials Research Proceedings **32** (2023) 3-14 https://doi.org/10.21741/9781644902615-1

New developments in the processing of metallic alloys for achieving exceptional superplastic properties

Chuan Ting Wang[1,a], Zheng Li[2,b], Jing Tao Wang[2,c], Terence G. Langdon[3,d,*]

[1]School of Mechanical Engineering, Nanjing University of Science and Technology, Nanjing 210094, China

[2]School of Materials Science and Engineering, Nanjing University of Science and Technology, Nanjing 210014, China

[3]Materials Research Group, Department of Mechanical Engineering, University of Southampton, Southampton SO17 1BJ, U.K.

[a] ctwang@njust.edu.cn, [b] lizheng@njust.edu.cn, [c] jtwang@njust.edu.cn, [d] langdon@soton.ac.uk

Keywords: Equal-Channel Angular Pressing, High-Pressure Torsion, Severe Plastic Deformation, Tube High-Pressure Shearing, Ultrafine-Grained Materials

Abstract. The process of superplasticity has a long history dating back to the early experiments of Pearson conducted in the U.K. in 1934. Since that time, superplasticity has become of increasing importance because of the recognition that superplastic forming provides a simple procedure for the processing of complex and curved parts for use in a wide range of industrial applications. The fundamental requirement for superplastic flow is a small grain size typically smaller than ~10 μm. These fine grains were achieved traditionally through the use of appropriate thermo-mechanical processing which provided a procedure for developing microstructures having grain sizes of the order of a few micrometers. Over the last two decades the processing procedures have been further developed through the use of techniques based on the application of severe plastic deformation (SPD) where it is possible to achieve ultrafine-grained materials with grains sizes in the submicrometer or even the nanometer range. Early SPD experiments were conducted using the processes of equal-channel angular pressing or high-pressure torsion but more recently a new and improved technique was developed which is known as tube high-pressure shearing (t-HPS). Experiments show that t-HPS provides a capability of producing exceptional superplastic elongations with, for example, an elongation of ~2320% in a Bi-Sn alloy when tested at a strain rate of 10^{-4} s^{-1} at room temperature. This report examines these recent developments with an emphasis on the potential for improving the superplastic capabilities of metallic alloys.

Introduction

When metals are pulled in tension, they generally break at relatively low elongations of the order of ~50-100% or less. However, some interesting and unusual results started appearing at the beginning of the last century with a report in 1912 of an elongation of 163% in brass [1] and a later report in 1928 of elongations of ~300% in Cd-Zn and Pb-Sn alloys [2]. In order to firmly place these results in perspective, it is necessary to have a unique and unambiguous definition of the meaning of superplasticity. This definition was achieved by critically examining an early plot of the strain rate sensitivity, m, against the elongation to failure for a very large number of metals, where m is defined as $\partial \ln \sigma / \partial \ln \dot{\varepsilon}$ where σ is the applied stress in creep experiments or the measured flow stress in conventional tensile testing and $\dot{\varepsilon}$ is the measured steady-state strain rate in creep or the strain rate imposed on the sample in tension [3]. This plot was published in 1969 and it showed that the elongation increased with increasing values of m so that a high value of m is a critical requirement for high elongations in tension.

To definitively establish a value of m for superplastiuc flow, it is first necessary to examine the meaning of the strain rate sensitivity and this requires an examination of the different flow mechanisms that may occur during the plastic deformation of metals.

Flow mechanisms in metals during plastic deformation

When flow occurs in metals tested in either creep conditions of constant stress or in conventional tensile testing, the strain rate may be expressed by a relationship of the form [4,5]

$$\dot{\varepsilon} = \frac{ADG\mathbf{b}}{kT}\left(\frac{\mathbf{b}}{d}\right)^p\left(\frac{\sigma}{G}\right)^n \tag{1}$$

where D is the diffusion coefficient, G is the shear modulus, $\mathbf{b}$ is the Burgers vector, k is Boltzmann's constant, T is the absolute temperature, d is the grain size, A is a constant having a value of ~10, p is the exponent of the inverse grain aize and n is the stress exponet which is equal to $1/m$. Inspection shows that, under any selected testing conditions, the only variables in Eq. 1 are A, D, p and n. Therefore, in order to use Eq. 1 in superplasticity, it is necessary to conduct experiments to determine the values of these various terms.

As discussed in detail in other reports [6,7], the earliest descriptions of superplastic properties provided inconclusive results because tests were conducted in tension at a fixed strain rate at an elevated temperature in order to measure the flow stress and then the same sample was used to measure additional flow stresses at other strain rates without removing the sample from the testing facility. This procedure is expeditious but it fails to take account of the grain growth which will occur during the testing procedure. Accordingly, a set of careful experiments was conducted on the Zn-22% Al eutectoid alloy where each flow stress was measured on a different sample at the same tempetrature of 473 K but different strain rates and all samples had the same initial grain size of 2.5 μm [8]. Later experiments were conducted at the additional temperatures of 423 and 503 K and the full results are shown in Fig. 1 where each datum point was obtained from a different sample, ΔL is the total increase in length and L_o is the initial length of each sample [9].

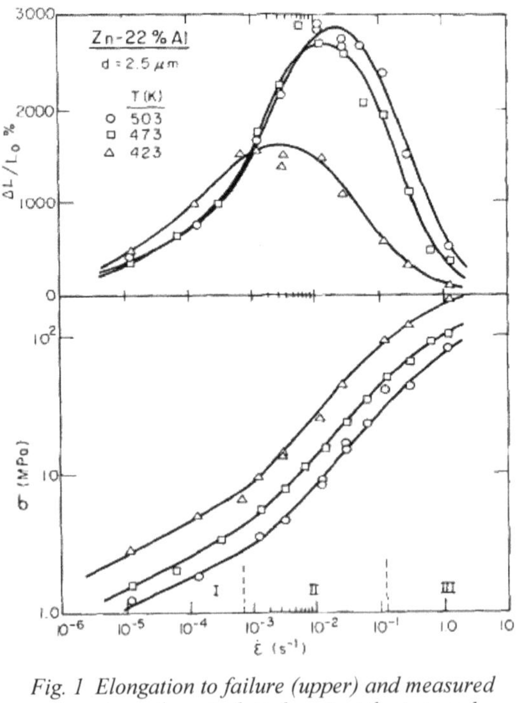

Fig. 1 Elongation to failure (upper) and measured flow stress (lower) plotted against the imposed strain rate for the Zn-22% Al eutectoid alloy [9].

It is readily apparent from inspection of Fig. 1 that all experimental points are mutually consistent. High superplastic elongations occur at intermediate

Materials Research Proceedings **32** (2023) 3-14 https://doi.org/10.21741/9781644902615-1

strain rates in the area labelled region II and there are lower elongations at both slower strain rates in region I and faster strain rates in region III. Typically, the high elongations occur over about two orders of magnitude of strain rate and the slopes of the lines in Fig. 1 indicate strain rate sensitivities of about ~0.2 in regions I and III and ~0.5 in region II.

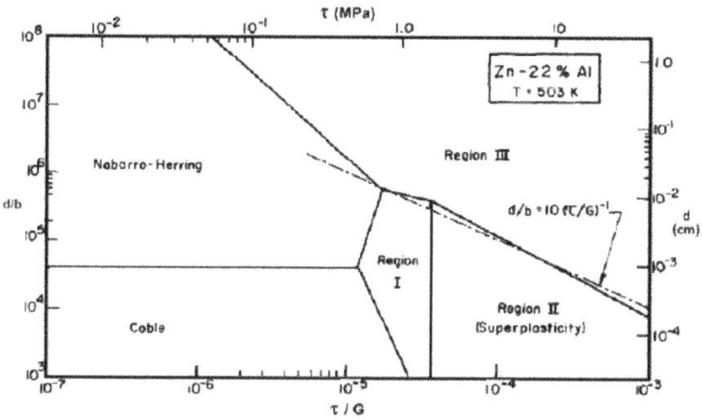

Fig. 2 Deformation mechanism map for Zn-22% Al alloy at 503 K showing the region of superplastic flow in grain size-stress space [10].

The optimum procedure for obtaining a simple visual presentation of these results is to use the data to plot a deformation mechanism map of the normalized grain size against the normalized stress. A map was constructed using the results obtained at a temperature of 503 K and it is shown in Fig. 2 where region II is the area of superplastic flow in grain size-stress space, regions I and III are the areas of lower elongations and τ is the shear stress [10]. The map in Fig. 2 also includes the theoretical predictions for diffusional creep where vacancies flow through the lattice in Nabarro-Herring creep [11,12] and along the grain boundaries in Coble creep [13]. It is apparent that at higher stresses there is no superplasticity when the grain size is large but as the grain size is reduced there is a sharp transition from the non-superplastic region III to the superplastic region II.

To obtain a better understanding of this transition, it is important to note that a very wide range of creep date, covering many different metals, shows that subgrains are formed within the grains under normal creep conditions and the average size of these subgrains, λ, varies inversely with the applied stress through a general relationship of the form [14]

$$\frac{\lambda}{\mathbf{b}} = B\left(\frac{\tau}{G}\right)^{-1} \qquad (2)$$

where B is a constant having a value of ~10. Similar relationships were also reported for ceramic [15] and geological [16] materials. Setting the grain size equal to the subgrain size so that $d = \lambda$, Eq. 2 is plotted as a broken line in Fig. 2 where this line is in remarkably good agreement with the boundary separating the fields for regions II and III. This plot shows unambiguously that superplastic flow requires a refined grain size that is sufficiently small that no subgrain boundaries are formed within the grains.

When a material pulls out to a very large superplastic elongation, the grains remain essentially equiaxed but they move with respect to each other so that the major flow process in superplasticity is grain boundary sliding [17]. Nevertheless, sliding cannot occur in isolation and it must be accommodated by some limited intragranular slip to prevent the opening of voids within the material. This suggests a mechanism for sliding as depicted schematically in Fig. 3 [18]. Thus, sliding under conventional creep conditions is shown in Fig. 3(a) where $d > \lambda$ so that sliding produces a stress concentration at the triple point A, dislocations move into the next grain and then pile up at the first subgrain boundary at B and climb into the boundary. By contrast, when $d < \lambda$ in superplasticity, the stress concentration at the triple point C generates dislocations in the next grain which cross the grain and then pile up and climb into the opposite grain boundary at D. By modelling these two processes it was shown that the mechanism of superplasticity leads to Eq. 1 with $A \approx 10$, $D = D_{gb}$ for grain boundary diffusion and p and n both equal to 2. Excellent experimental evidence is now available demonstrating the occurrence of an accommodating intragranular slip during superplastic flow and this provides direct support for this model [19-22].

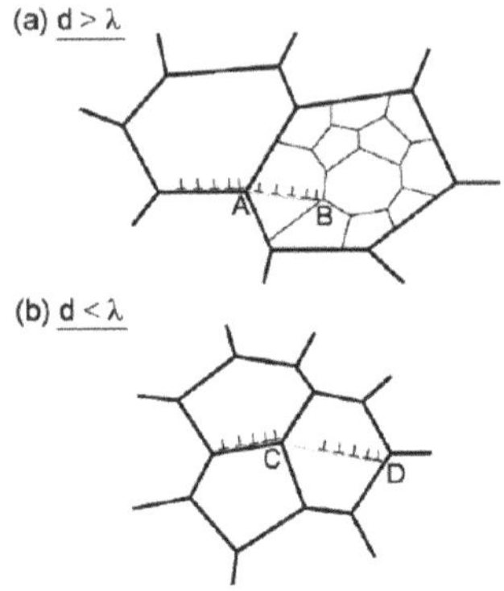

Fig. 3 Grain boundary sliding in (a) creep with $d > \lambda$ and (b) superplasticity with $d < \lambda$ [18].

In conventional creep, flow generally occurs by a dislocation climb process where the diffusion coefficient is the value for lattice self-diffusion, $n \approx 4.5$ and $p = 0$. However, in solid solution alloys the rate of glide may become slower than the rate of climb because of the segregation of solute atoms around the moving dislocations and in these conditions the diffusion coefficient is the value for interdiffusion of the solute with $n = 3$ and $p = 0$. This means that the strain rate sensitivity in creep controlled by dislocation glide is $m \approx 0.3$ and this is quite high so that, since the elongation increases with increasing values of m, dislocation glide will lead to some reasonably high elongations prior to failure. It is important, therefore, to establish a criterion for superplasticity that avoids any confusion with metals deforming by the viscous glide process. Accordingly, a careful analysis of published data was conducted for the elongations achieved in metals deforming by dislocation glide and it was found that these elongations may be as high as, and even exceed, a value of 300% [23]. For example, there is a report of an elongation of 325% in an Al-Mg alloy deforming by dislocation glide with a measured strain rate sensitivity of $m = 0.33$ [24]. Based on extensive analysis of published data, superplasticity was formally defined as a measured elongation of at least 400% and a strain rate sensitivity close to $m \approx 0.5$ [23].

Recent developments in the superplasticity of metals

Superplasticity has become an important flow process over the last two or three decades, mainly because of the development of industrial superplastic forming processes which provide the opportunity for fabricating relatively complex curved shapes from sheet metal. The potential uses of superplastic forming are very wide ranging from automotive applications to consumer products and various uses in architectural design. This processing is now employed annually to physically form many thousands of tons of sheet metals [25] and some reviews are available describing these developments as superplasticity has progressed from a laboratory curiosity to a viable and valuable industrial tool [26,27]. There are several recent developments within the field of superplasticity but four of these developments have important implications and they will be reviewed briefly in this report.

First, superplasticity requires a small grain size that is typically below ~10 μm and, because it is a diffusion-controlled process, the testing should be performed at an elevated temperature that is generally above ~$0.5T_m$ where T_m is the absolute melting temperature. These characteristics were discussed in detail in an early review [6]. Materials suitable for superplastic testing were traditionally fabricated using thermo-mechanical processing but this has a limitation because it is difficult or even impossible to achieve grain sizes smaller than ~2-5 μm. A classic publication in 1988 showed that it was possible to use a new processing technique, based on the application of severe plastic deformation (SPD), to achieve a submicrometer grain size in an Al-Cu-Zr alloy [28]. This led to widespread use of SPD in the processing of ultrafine-grained (UFG) metals and the various processing procedures for achieving significant grain refinement in bulk solids were described in a detailed review [29]. Basically, there are two major SPD procedures and these are equal-channel angular pressing (ECAP) where a rod or bar is pressed through a die constrained within a channel that is bent through a sharp angle within the die [30] and high-pressure torsion (HPT) where the sample, generally in the form of a thin disk, is subjected to an applied pressure and concomitant torsional straining [31]. In practice, processing by HPT has an advantage over ECAP because it produces materials having smaller grain sizes [32] and a higher fraction of high-angle grain boundaries which contributes to easier grain boundary sliding during plastic flow [33]. Finally, an advantage of producing UFG microstructures is that it leads to the occurrence of superplastic flow at high strain rates [34] and this is advantageous for use in superplastic forming operations [35].

Second, reports of superplasticity have been extended to new materials such as high-entropy alloys (HEAs). The first report was a superplastic elongation of 1240% in an AlCoCrCuFeNi alloy processed by multiaxial forging [36] and there were later reports of superplasticity in a CoCrFeNiMn HEA [37] and an elongation of up to 2000% in an $Al_9(CoCrFeNiMn)_{91}$ (at.%) alloy [38]. A comprehensive review is now available summarizing the properties of the traditional CoCrFeNiMn HEA [39]. More recently, results have become available showing the occurrence of superplastic elongations in multi-principal element alloys (MPEAs) [40,41].

Third, an important requirement in attaining superplastic results for any selected alloy at a fixed temperature is to summarize the data in a simple format so that predictions are then possible concerning the limitations of the superplastic regime in terms of stress and grain size. This is most easily accomplished by using the experimental data to construct a deformation mechanism map of the type shown earlier in Fig. 2. Within these maps, it is advantageous to include the predicted regimes for the Nabarro-Herring and Coble creep mechanisms since these processes are well-established with good theoretical relationships. The development of deformation mechanism maps, and the various types of maps that are available, were discussed in an earlier report [42]. Examples of this approach are given in Fig. 4 where results are plotted for a Zn-22% Al eutectoid alloy tested at a temperature of 473 K after processing by (a) ECAP for 8 passes at 473 K and (b) HPT for 1 to 5 turns at room temperature (RT) under an applied pressure of 6.0 GPa [42]. These maps show the lines for the subgrain size based on Eq. 2 and all experimental points lie within the correct fields.

Materials Research Forum LLC
https://doi.org/10.21741/9781644902615-1

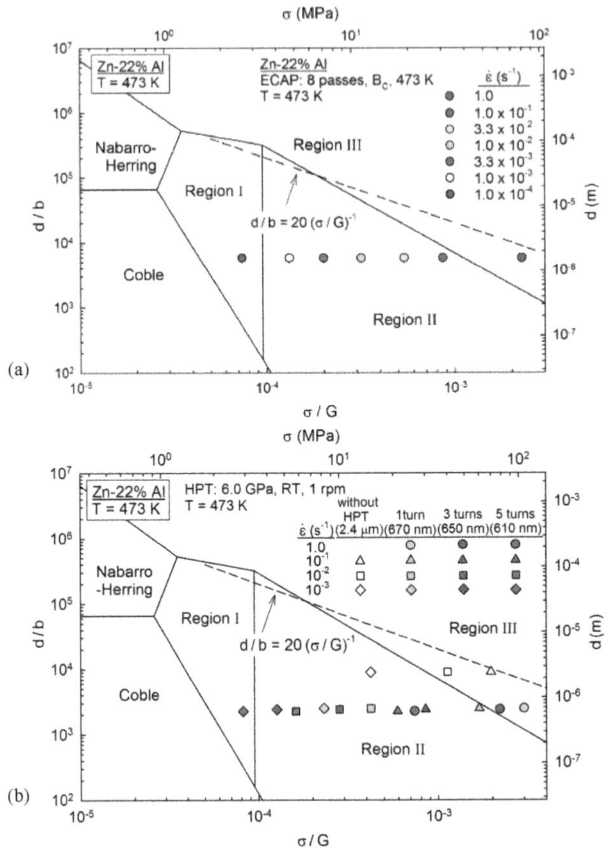

Fig. 4 Deformation mechanism maps for Zn-22% Al tested at 473 K after processing by (a) ECAP and (b) HPT [42].

Fourth, calculations show that the relationship developed earlier for superplastic flow [18], based on grain boundary sliding with intragranular accommodation, is in excellent agreement with experimental data available for both Al and Mg alloys tested after processing by either ECAP or HPT [43]. Specifically, in plots of the temperature and grain size compensated strain rate against the normalized stress, the experimental datum points lie close to, and scatter about, the predicted values for the strain rate in superplasticity. However, this is a direct test of the validity of this approach at elevated temperatures where the relationship incorporates the occurrence of rapid diffusion and a negligible supersaturation of vacancies. In order to extend this same analysis to flow at low temperatures, it is necessary to modify the relationship to permit a supersaturation of vacancies. This approach was used effectively in a recent model which led to good agreement with experimental data for several different metals [44] and, more recently, a high-entropy alloy [45]. A very recent comprehensive analysis demonstrates that this approach is in good agreement with more than 2300 individual datum points taken from the literature covering multiple materials over a very wide range of grain sizes, strain rates and temperatures [46].

The introduction of a new procedure for achieving grain refinement and superplasticity

The preceding analysis emphasized the importance of the SPD procedures of ECAP and HPT which are now widely used to achieve very significant grain refinement. Nevertheless, an alternative procedure was developed recently and preliminary experiments show this procedure has a capability of producing exceptionally high superplastic elongations.

This new procedure is tube high-pressure shearing (t-HPS) where a sample, in the form of a tube, is subjected to shearing under the action of a hydrostatic pressure. A detailed description of this procedure was presented earlier [47-49] and a schematic illustration is shown in Fig. 5 where (a) is conventional HPT for a solid disk, (b) is a torsion/twisting process for a tube and (c) is the new process of t-HPS as applied to a tube: in Fig. 5 the radius is r, the height is h and the angle of rotation is θ. Inspection shows that HPT and torsion/twisting are similar processes because in both cases a vertical line scribed parallel to the axis on the outer surface will be sheared into a curve as indicated by the broken line whereas a radial line on the upper surface will remain straight. Conversely, in t-HPS the outer surface of the tube is rotated so that a vertical line will remain parallel to the tube axis during deformation but a radial line will shear into a curve as shown by the broken line in Fig. 5(c).

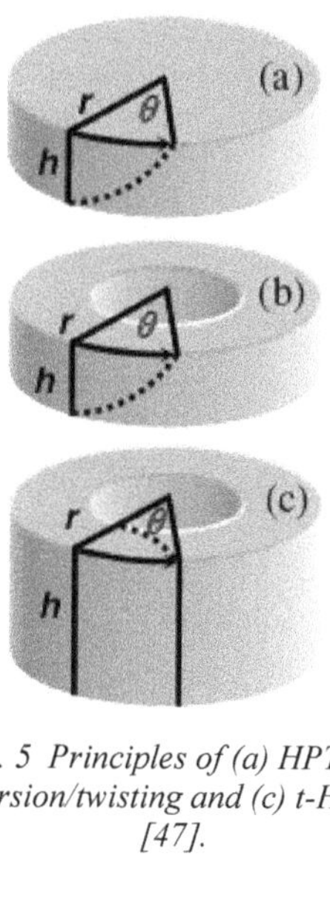

Fig. 5 Principles of (a) HPT, (b) torsion/twisting and (c) t-HPS [47].

Experiments were conducted where t-HPS was applied to alloys of Pb-40% Sn and Pb-62% Sn (wt. %) by constraining these alloys within an outer cylinder under a hydrostatic pressure of 1.0 GPa and with the outer cylinder rotated up to 40 or 50 turns for these two alloys, respectively; this gives equivalent true strains for these alloys of ~1600 and ~2000, respectively, thereby providing a good mechanical mixing between the Pb and Sn [50]. Full details of the grain sizes for the α and β phases and the measured α/β phase volume ratios were given for the alloys in the as-cast, rolled and t-HPS conditions in a tabulation presented earlier [50]. However, as an example, for as-cast, rolled and t-HPS the α phase sizes were ~2.3, ~2.0 and ~0.7 μm in the Pb-62% Sn alloy and ~1.0, ~1.8 and ~0.8 μm in the Pb-40% Sn alloy, respectively. Tensile specimens were

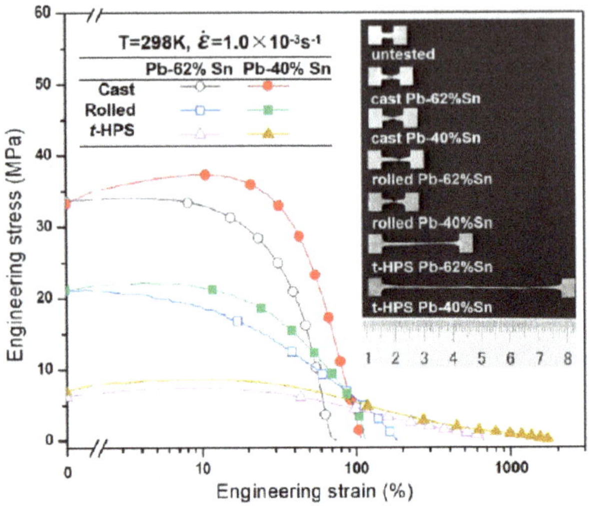

Fig. 6 Engineering stress-strain curves for two Pb-Sn alloys tested in tension at room temperature: the inset shows samples after pulling to failure [50].

machined for these three separate fabrication conditions and then the samples were pulled in tension to failure at room temperature (298 K) at a strain rate of 1.0×10^{-3} s^{-1}. The resultant engineering stress-strain curves are shown in Fig. 6 and the inset at upper right shows the appearance of the specimens after fracture. The results from these tests demonstrate that the as-cast condition gives elongations up to ~100% but this was increased to almost ~200% by rolling where this is not sufficient to demonstrate the occurrence of superplastic flow. These elongations are consistent with an earlier report for flow in a Pb-62% Sn alloy [51]. Nevertheless, processing by t-HPS gave good superplastic properties with an elongation of ~670% in the Pb-62% Sn alloy which represents the highest elongation achieved at this strain rate at room temperature when using an alloy prepared by casting and extrusion. Furthermore, there was an unprecedented elongation of ~1870% in the Pb-40% Sn alloy processed by t-HPS where this elongation exceeds all earlier reported values by a factor of at least three times. This unusual result is due to the presence of equiaxed grain after t-HPS, the grain size of ~0.7 μm and the mixing of the Pb and Sn domains in nearly equal proportions.

Very recently, tests were conducted by processing a cast Bi-43% Sn (wt. %) alloy by t-HPS using a procedure described earlier [52]. Samples were processed through 20 and 100 revolutions at room temperature and then tensile tests were undertaken at 298 K using different strain rates from 1.0×10^{-4} to 1.0×10^{-2} s^{-1}. The results are shown in Fig. 7 for the 20 turns condition where the maximum elongation was 1820% at a strain rate of 1.0×10^{-4} s^{-1}. Figure 8 gives results after 100 turns where the maximum elongation was 2320% at this strain rate. The occurrence of remarkable superplasticity at RT is evident from the absence of any localized necking within the gauge length [53].

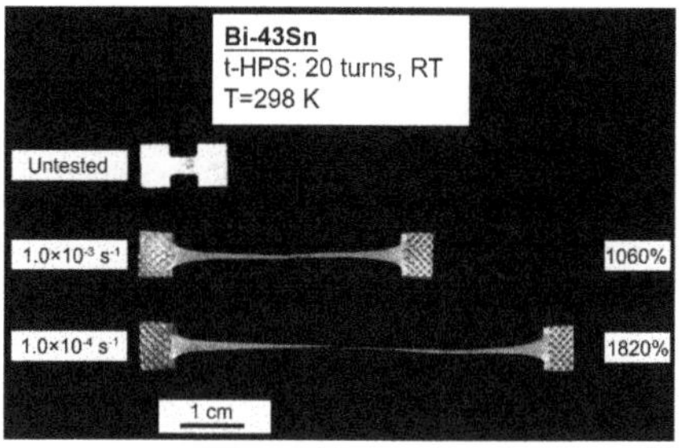

Fig. 7 Appearance of Bi-43% Sn samples after processing by t-HPS for 20 turns and pulling to failure at two different strain rates at 298 K.

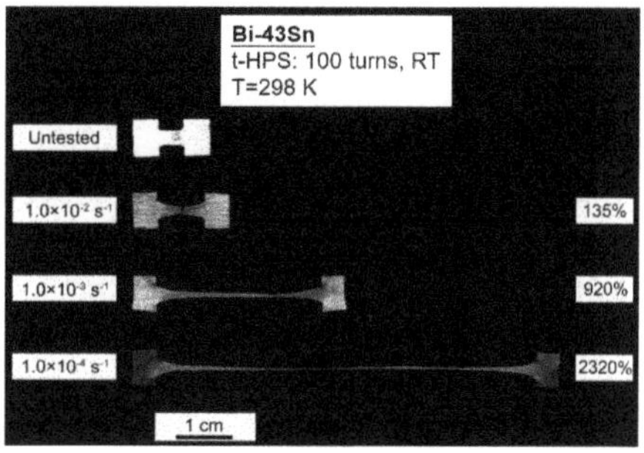

Fig. 8 Appearance of Bi-43% Sn samples after processing by t-HPS for 100 turns and pulling to failure at 298 K.

Summary and conclusions

1. Superplasticity occurs in metals when the grain size is equal to or smaller than the subgrain size. A model based on grain boundary sliding is in excellent agreement with the available data.
2. Recent developments include the use of severe plastic deformation to achieve exceptional grain refinement, the extension of superplasticity to new materials such as high-entropy alloys and the visual display of the superplastic regime through the use of deformation mechanism maps.
3. Tube high-pressure shearing is a new processing technique providing the capability of fabricating metals that exhibit exceptionally high superplastic elongations.

Acknowledgements

This work was supported in part by the National Natural Science Foundation of China (Grant No. 52074160) and in part by the European Research Council (ERC Grant Agreement No. 267464-SPDMETALS).

References

[1] G.D. Bengough, A study of the properties of alloys at high temperatures. J. Inst. Metals 7 (1912) 123-178.

[2] C.H.M. Jenkins, Strength of Cd-Zn and Sn-Pb alloy solder. J. Inst. Metals 40 (1928) 21-32.

[3] D.A. Woodford, Strain rate sensitivity as a measure of ductility, Trans. ASM 62 (1969) 291-293.

[4] T.G. Langdon, An analysis of flow mechanisms in high temperature creep and superplasticity, Mater. Trans. 46 (2005) 1951-1956. https://doi.org/10.2320/matertrans.46.1951

[5] T.G. Langdon, Identifying creep mechanisms in plastic flow, Z. Metallk. 96 (2005) 522-531. https://doi.org/10.3139/146.101066

[6] T.G. Langdon, The mechanical properties of superplastic materials, Metall. Trans. A 13A (1982) 689-701. https://doi.org/10.1007/BF02642383

[7] T.G. Langdon, Overview: Using severe plastic deformation in the processing of superplastic materials, Mater. Trans. (2023) in press. https://doi.org/10.2320/matertrans.MT-MF2022021 https://doi.org/10.2320/matertrans.MT-MF2022021

[8] H. Ishikawa, F.A. Mohamed, T.G. Langdon, The influence of strain rate on ductility in the superplastic Zn-22% Al eutectoid, Phil. Mag. 32 (1975) 1269-1271. https://doi.org/10.1080/14786437508228105

[9] F.A. Mohamed, M.M.I. Ahmed, T.G. Langdon, Factors influencing ductility in the superplastic Zn-22 Pct Al eutectoid, Metall. Trans. A 8A (1977) 933-938. https://doi.org/10.1007/BF02661575

[10] F.A. Mohamed, T.G. Langdon, Deformation mechanism maps for superplastic materials, Scripta Metall. 10 (1976) 759-762. https://doi.org/10.1016/0036-9748(76)90358-6

[11] F.R.N. Nabarro, Deformation of crystals by the motion of single ions, in: Report of a Conference on Strength of Solids, The Physical Society, London, U.K. (1984), pp. 75-90.

[12] C. Herring, Diffusional viscosity of a polycrystalline solid, J. Appl. Phys. 21 (1950) 437-445. https://doi.org/10.1063/1.1699681

[13] R.L. Coble, A model for boundary diffusion controlled creep in polycrystalline materials, J. Appl. Phys. 34 (1963) 1679-1682. https://doi.org/10.1063/1.1702656

[14] J.E. Bird, A.K. Mukherjee, J.E. Dorn, Correlations between high-temperature creep behavior and structure, in: (D.G. Brandon and A. Rosen, Eds.), Quantitative Relation Between Properties and Microstructure Israel Universities Press, Jerusalem, Israel (1959) pp. 255-342.

[15] W.R. Cannon, T.G. Langdon, Review: Creep of ceramics, Part 2 An examination of flow mechanisms, J. Mater. Sci. 23 (1988) 1-20. https://doi.org/10.1007/BF01174028

[16] R.J. Twiss, Theory and applicability of a recrystallized grain size paleopiezometer, Pure Appl. Geophys. 115 (1977) 227-244. https://doi.org/10.1007/BF01637105

[17] T.G. Langdon, An evaluation of the strain contributed by grain boundary sliding in superplasticity, Mater. Sci. Eng. A 174 (1994) 225-230. https://doi.org/10.1016/0921-5093(94)91092-8

[18] T.G. Langdon, A unified approach to grain boundary sliding in creep and superplasticity, Acta Metall. Mater. 42 (1994) 2437-2443. https://doi.org/10.1016/0956-7151(94)90322-0

[19] L.K.L. Falk, P.R. Howell, G.L. Dunlop, T.G. Langdon, The role of matrix dislocations in the superplastic deformation of a copper alloy, Acta Metall. 34 (1986) 1203-1214. https://doi.org/10.1016/0001-6160(86)90007-6

[20] R.Z. Valiev, T.G. Langdon, An investigation of the role of intragranular dislocation strain in the superplastic Pb-62% Sn eutectic alloy, Acta Metall. Mater. 41 (1993) 949-954. https://doi.org/10.1016/0956-7151(93)90029-R

[21] Y. Xun, F.A. Mohamed, Slip-accommodated superplastic flow in Zn-22wt% Al, Phil. Mag. 83 (2003) 2247-2266.

[22] Y. Xun, F.A. Mohamed, Superplastic behavior of Zn-22% Al containing nano-scale dispersion

particles, Acta Mater. 52 (2004) 4401-4412. https://doi.org/10.1016/j.actamat.2004.03.039

[23] T.G. Langdon, Seventy-five years of superplasticity: historic developments and new opportunities, J. Mater. Sci. 44 (2009) 5998-6010. https://doi.org/10.1007/s10853-009-3780-5

[24] E.M. Taleff, G.A. Henshall, T.G. Nieh, D.R. Lesuer, J. Wadsworth, Warm-temperature tensile ductility in Al-Mg alloys, Metall. Mater. Trans. A 29A (1998) 1081-1091. https://doi.org/10.1007/s11661-998-0300-1

[25] A.J. Barnes, Superplastic forming 40 years and still growing, J. Mater. Eng. Perform. 16 (2007) 440-454. https://doi.org/10.1007/s11665-007-9076-5

[26] T.G. Langdon, The background to superplastic forming and opportunities arising from new developments, Solid State Phenom. 306 (2020) 1-8. https://doi.org/10.4028/www.scientific.net/SSP.306.1

[27] J. Wongsa-Ngam, T.G. Langdon, Advances in superplasticity from a laboratory curiosity to the development of a superplastic forming industry, Metals 12 (2022) 1921. https://doi.org/10.3390/met12111921

[28] R.Z. Valiev, O.A. Kaibyshev, R.I. Kuznetsov, R.Sh. Musalimov, N.K. Tsenev, Low-temperature superplasticity of metallic materials, Dokl. Akad. Nauk SSSR 301 (1988) 864-866.

[29] R.Z. Valiev, Y. Estrin, Z. Horita, T.G. Langdon, M.J. Zehetbauer, Y.T. Zhu, Producing bulk ultrafine-grained materials by severe plastic deformation, JOM 58(4) (2006) 33-39. https://doi.org/10.1007/s11837-006-0213-7

[30] R.Z. Valiev, T.G. Langdon, Principles of equal-channel angular pressing as a processing tool for grain refinement, Prog. Mater. Sci. 51 (2006) 881-981. https://doi.org/10.1016/j.pmatsci.2006.02.003

[31] A.P. Zhilyaev, T.G. Langdon, Using high-pressure torsion for metal processing: Fundamentals and applications, Prog. Mater. Sci. 53 (2008) 893-979. https://doi.org/10.1016/j.pmatsci.2008.03.002

[32] A.P. Zhilyaev, B.K. Kim, G.V. Nurislamova, M.D. Baró, J.A. Szpunar, T.G. Langdon, Orientation imaging microscopy of ultrafine-grained nickel, Scripta Mater. 46 (2002) 575-580. https://doi.org/10.1016/S1359-6462(02)00018-0

[33] J. Wongsa-Ngam, M. Kawasaki, T.G. Langdon, A comparison of microstructures and mechanical properties in a Cu-Zr alloy processed using different SPD techniques, J. Mater. Sci. 48 (2013) 4653-4660. https://doi.org/10.1007/s10853-012-7072-0

[34] R.Z. Valiev, D.A. Salimonenko, N.K. Tsenev, P.B. Berbon, T.G. Langdon, Observations of high strain rate superplasticity in commercial aluminum alloys with ultrafine grains sizes, Scripta Mater. 37 (1997) 1943-1950. https://doi.org/10.1016/S1359-6462(97)00387-4

[35] Z. Horita, M. Furukawa, M. Nemoto, A.J. Barnes, T.G. Langdon, Superplastic forming at high strain rates after severe plastic deformation, Acta Mater. 48 (2000) 3633-3640. https://doi.org/10.1016/S1359-6454(00)00182-8

[36] D.G. Shaysultanov, N.D. Stepanov, A.V. Kuznetsov, G.A. Salishchev, O.N. Senkov, Phase composition and superplastic behavior of a wrought AlCoCrCuFeNi high-entropy alloy, JOM 35 (12) (2013) 1815-1828.

[37] H. Shahmir, J. He, Z. Lu, M. Kawasaki, T.G. Langdon, Evidence for superplasticity in a CoCrFeNiMn high-entropy alloy processed by high-pressure torsion, Mater. Sci. Eng. A685 (2017) 342-348. https://doi.org/10.1016/j.msea.2017.01.016

[38] N.T.C. Nguyen, P. Asghari-Rad, P. Sathiyamoorthi, A. Zargaran, C.S. Lee, H.S. Kim, Ultrahigh high-strain-rate superplasticity in a nanostructured high-entropy alloy, Nature Commun. 11 (2020) 2736. https://doi.org/10.1038/s41467-020-16601-1

[39] H. Shahmir, M.S. Mehranpour, S.A.A. Shams, T.G. Langdon, Twenty years of the CoCrFeNiMn high-entropy alloy: achieving exceptional mechanical properties through

microstructure engineering, J. Mater. Res. Tech. 23 (2023) 3362-3423. https://doi.org/10.1016/j.jmrt.2023.01.181

[40] P. Asghari-Rad, N.T.C. Nguyen, A. Zargaran, P. Sathiyamoorthi, H.S. Kim, Relation of phase fraction to superplastic behavior of multi-principal element alloy with multi-phase structure, Scripta Mater. 207 (2022) 114239. https://doi.org/10.1016/j.scriptamat.2021.114239

[41] N.T.C. Nguyen, P. Asghari-Rad, H. Park, H.S. Kim, Differential superplasticity in a multi-phase multi-principal element alloy by initial annealing, J. Mater. Sci. 57 (2022) 18154-18167. https://doi.org/10.1007/s10853-022-07616-8

[42] M. Kawasaki, T.G. Langdon, The many facets of deformation mechanism mapping and the application to nanostructured materials, J. Mater. Res. 28 (2013) 1827-1834. https://doi.org/10.1557/jmr.2013.55

[43] M. Kawasaki, T.G. Langdon, Review: achieving superplastic properties in ultrafine-grained materials at high temperatures, J. Mater. Sci. 51 (2016) 19-32. https://doi.org/10.1007/s10853-015-9176-9

[44] R.B. Figueiredo, T.G. Langdon, Deformation mechanisms in ultrafine-grained metals with an emphasis on the Hall-Petch relationship, J. Mater. Res. Tech. 14 (2021) 137-159. https://doi.org/10.1016/j.jmrt.2021.06.016

[45] R.B. Figueiredo, W. Wolf, T.G. Langdon, Effect of grain size on strength and strain rate sensitivity in the CrMnFeCoNi high-entropy alloy, J. Mater. Res. Tech. 20 (2022) 2358-2368. https://doi.org/10.1016/j.jmrt.2022.07.181

[46] R.B. Figueiredo, M. Kawasaki, T.G. Langdon, Seventy years of Hall-Petch, ninety years of superplasticity and a generalized approach to the effect of grain size on flow stress, Prog. Mater. Sci. 137 (2023) 101131. https://doi.org/10.1016/j.pmatsci.2023.101131

[47] J.T. Wang, Z. Li, J. Wang, T.G. Langdon, Principles of severe plastic deformation using tube high-pressure shearing, Scripta Mater. 67 (2012) 810-813. https://doi.org/10.1016/j.scriptamat.2012.07.028

[48] Z. Li, P.F. Zhang, H. Yuan, K. Lin, Y. Liu, D.L. Yin, J.T. Wang, T.G. Langdon, Principle of one-step synthesis for multi-layered structures using tube high-pressure shearing, Mater. Sci. Eng. A658 (2016) 367-375. https://doi.org/10.1016/j.msea.2016.02.024

[49] Z. Li, Y. Liu, J.T. Wang, T.G. Langdon, Tube high-pressure shearing: a simple shear path to unusual microstructures, Mater. Trans. (2023) (doi.org/10.2320/matertrans.MT-MF2022051).

[50] K. Lin, Z. Li, Y. Liu, E. Ma, J.T. Wang, T.G. Langdon, Exploiting tube high-pressure shearing to prepare a microstructure in Pb-Sn alloys for unprecedented superplasticity, Scripta Mater. 209 (2022) 114390. https://doi.org/10.1016/j.scriptamat.2021.114390

[51] M.M.I. Ahmed, T.G. Langdon, Ductility of the Pb-Sn eutectic at room temperature, J. Mater. Sci. Lett. 2 (1983) 59-62. https://doi.org/10.1007/BF00725431

[52] C.T. Wang, Z. Li, Y. He, J.T. Wang, T.G. Langdon, Microstructural evolution and tensile testing of a Bi-Sn (57/43) alloy processed by tube high-pressure shearing, Crystals 11 (2021) 1229. https://doi.org/10.3390/cryst11101229

[53] T.G. Langdon, Fracture processes in superplastic flow, Metal Sci. 16 (1982) 175-183. https://doi.org/10.1179/030634582790427208

Achieving superplasticity in hard-to-deform metallic materials using high-pressure sliding (HPS) process

Zenji Horita[1,2,3,a*] and Yoichi Takizawa[4,b]

[1]School of Engineering, Kyushu Institute of Technology, Kitakyushu, 804-8550, Japan

[2]Synchrotron Light Application Center, Saga University, Saga, 840-8602, Japan

[3]Magnesium Research Center, Kumamoto University, Kumamoto 860-8555, Japan

[4]Technology Department, Nagano Forging Co., Ltd., Nagano 381-0003, Japan

[a]horita.zenji.688@m.kyushu-u.ac.jp, [b]ytakizawa@nsc-com.co.jp

Keywords: Severe Plastic Deformation, High-Pressure Sliding, Superplasticity, Superalloy, Cup Forming

Abstract. This presentation shows that superplasticity is achieved in hard-to-deform materials when they are processed by severe plastic deformation (SPD) through high-pressure sliding (HPS). The HPS process is similar to the high-pressure torsion (HPT) process as the straining is made under high pressure in highly constrained conditions. It is applicable to a sheet form of samples while the HPT process uses disks or rings, and thus it has advantage that the sample size can be increased. In addition, when the HPS process is combined with a feeding process, the SPD-processed area can be further enlarged without increasing the machine capacity, of which process is called the incremental feeding HPS (IF-HPS). The HPS process is applied to a Ni-based superalloy (Inconel 718), a Ti-6Al-7Nb alloy (F1295) and a Mg-6Al-1Zn alloy (AZ61), and superplasticity is well attained in all the hard-to-deform alloys with total elongations more than 400%. It is also demonstrated that cup forming of the Inconel 718 with practical dimensions is realized by application of the IF-HPS process.

Introduction

It is well established that the fine-grained structure is an important prerequisite for the advent of superplasticity. Severe plastic deformation (SPD) is a useful technique to produce such ultrafine-grained structures in many metallic materials [1]. In particular, SPD under high pressure such as using a process of high-pressure torsion (HPT) is of great importance because grain refinement is feasible even in hard-to-deform materials [2]. High-pressure sliding (HPS) [3,4] is also an SPD process under high pressure, and it is similar to the HPT process [5] as the straining is made under highly constrained conditions [6]. While the HPT process uses disks or rings and the sample size is rather limited, the HPS process is applicable to sheets and it has then advantage over the HPT process that the sample size can be increased as discussed in recent overview papers [7,8]. Furthermore, the HPS process provides a homogeneous development of microstructure throughout the sample [4]. However, the microstructure development is rather heterogeneous in the HPT-processed disk because the strain is introduced in proportion to the distance from the disk center so that it is more in the outer areas but is less in the inner area and theoretically zero at the center of the disk [6]. It should be noted that, when the HPS process is combined with a sample feeding process, the SPD-processed area can be further increased without increasing the machine capacity. This combined process, called the incremental feeding HPS (IF-HPS) [9], has been applied to hard-to-deform metallic materials such as a Ni-based superalloy (Inconel 718) [9,10] and a Ti-6Al-7Nb alloy (F1295) [10] including conventional Al alloys [11].

This presentation first gives a brief introduction of the HPS process and the IF-HPS process, and thus it is shown that superplastic elongation more than 400% is well attained in the former two hard-to-deform alloys. It is also demonstrated that cup forming of the Inconel 718 with practical dimensions is realized by taking advantage of the superplasticiy through grain refinement by the IF-HPS process [9].

HPS
As illustrated in Fig. 1, sheet samples are placed between the anvils and the plunger. While applying a pressure on the anvils, the plunger was moved with respect to the anvils so that shear strain is introduced in the sample. Reciprocation of the plunger with respect to the anvils may be adopted to accumulate the strain [7,8]. It is noted that the HPS process is applicable not only to sheets [3,4] but also to rods [12] and pipes [13].

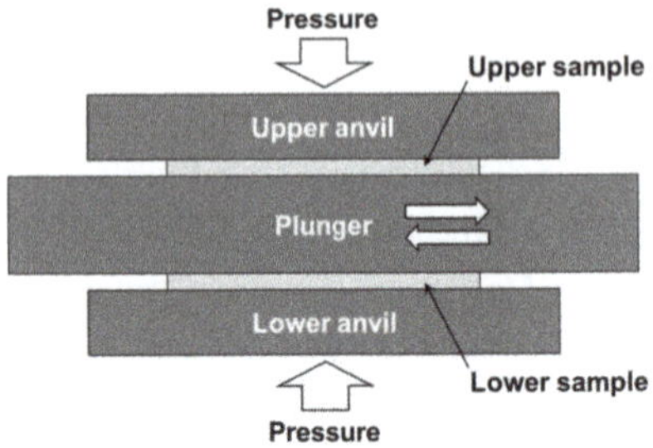

Fig. 1 Schematic illustrations of high-pressure sliding (HPS).[7]

IF-HPS
As illustrated in Fig.2, when a feeding process is incorporated in the HPS process, the SPD-processed area can be enlarged. This combined process, IF-HPS, has an advantage that upsizing is feasible without increasing the machine capacity. The conditions for optimal operation of the IF-HPS process was described in our recent overview paper [8] in terms of the following three points: (1) use of flat-type anvils to make sample feeding easier and sheet surface smoother, (2) control of the sliding mode determined by the sliding distance and the numbers of the reciprocation of the sliding direction, and (3) control of the feeding pattern determined by the feeding distance and the feeding direction.

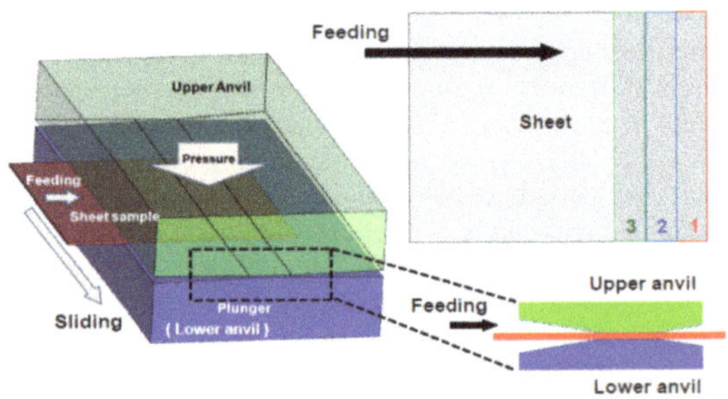

Fig. 2 Schematic illustration of incremental feeding high-pressure sliding (IF-HPS).[9]

Application to Ni-based superalloy (Inconel 718)
Figure 3 shows a microstructure observed by a transmission electron microscope (TEM) after processing by HPS for a sliding distance of 15 mm under 4 GPa at room temperature [4]. It is

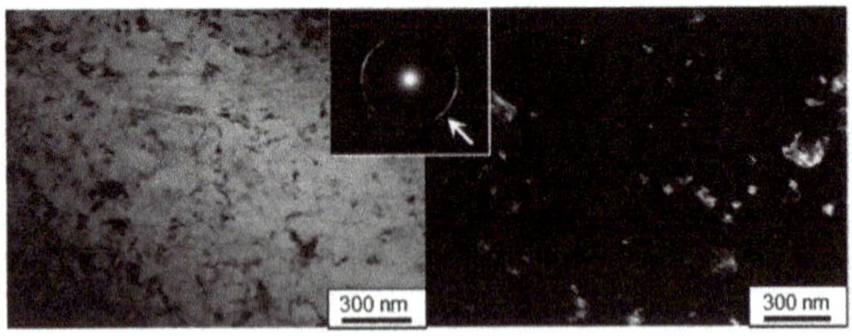

Fig.3 TEM bright-field image (left), dark-field image (right) and SAED pattern (center) for Inconel 718 alloy after processing by HPS for sliding distance of 15 mm under 4 GPa at room temperature.[4]

Materials Research Proceedings **32** (2023) 15-22 https://doi.org/10.21741/9781644902615-2

confirmed that the grain size is refined to ~100 nm from the dark-field image (on the right) taken by a diffracted beam indicated by the arrow in the selected area diffraction (SAED) pattern (at the center). Furthermore, the SAED pattern exhibits rings so that the structure consists of fine grains with high-angle boundaries. Nevertheless, the grain boundaries are ill-defined, and this is a typical feature observed in SPD-processed alloys [14].

Figure 4 shows results of tensile testing for HPS-processed specimens and those without processing, for comparison [15]. Here, the HPS-processing was made for an as-received state and an annealed state, while the underformed specimens include the as-received state, solution-treated state and solution-treated plus aged state. Tensile tests were conducted with an initial strain rate of 2×10^{-2} s^{-1} at 1073 K in all specimens for comparison. The unprocessed specimens were deformed only to 35% or less with maximum flow stresses exceeding 500 MPa, whereas supeplasticity was well attained to the total elongations more than 400% with the maximum flow stresses below 200 MPa in the HPS-processed specimens. It is clearly demonstrated that the HPS processing is effective to create a superplastic flow which gives rise to not only a significant reduction of the maximum flow stress but also to a significant increase in the total elongation.

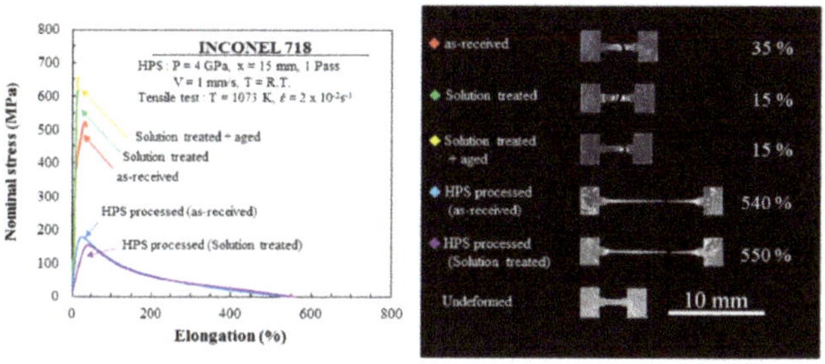

Fig.4. Stress-strain curves (left) and appearance of tensile specimens after deformation to failure (right) for as-received and solution-treated specimens with and without HPS processing including solution-treated plus aged specimen without HPS processing for comparison. All tensile specimens were deformed at 1073 K with initial strain rate of 2×10^{-2} s^{-1}.[15]

Tensile tests were further conducted at temperatures of 973, 1023, 1073, 1123 and 1173 K with initial strain rates in the range of 5.0×10^{-4} to 2×10^{-2} s^{-1} [15]. The total elongations are plotted against the initial strain rate in Fig.5. Here, the horizontal dotted line is drawn at the elongation of 400% above which is considered to be superplastic [16] and the vertical dotted line is drawn at the strain rate of 1x 10^{-2} s^{-1} beyond which the superplastic elongation is defined as the high-strain rate superplasiticity (HSRS) [17]. Inspection reveals that the HSRS was attained in the HPS-processed sample even though the temperature is as low as 1073 K, which is ~0.6 Tm where Tm is the melting temperature.

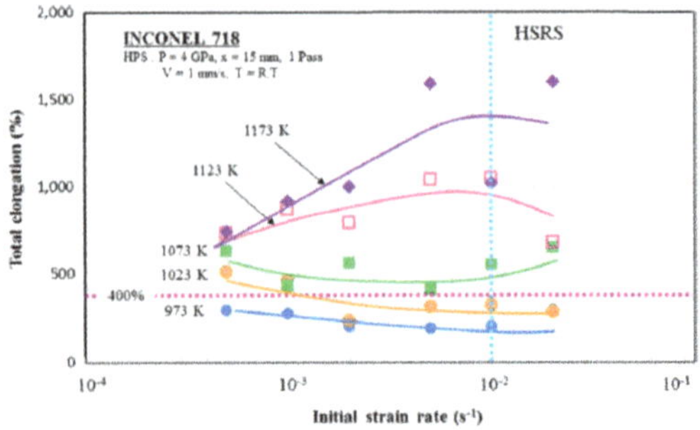

Fig. 5. Total elongation to failure plotted against initial strain rates for testing temperatures ranging from 973 K to 1173 K. Region of high strain rate superplasticity (HSRS) (> 400% and > 1×10^{-2} s^{-1}) is indicated by drawing horizontal and vertical dotted lines.[15]

The IF-HPS process was employed because the sample dimensions of 10 x 100 x 1 mm processed by a maximum capacity of the HPS machine is still insufficient. As shown in Fig.6, it was successful to cover a square area of a sheet with 100 x 100 mm without increasing the machine capacity [10]. Four tensile specimens were extracted from the sheet using an electrical discharge machine, where the traces of the extracted specimens are visible in the initial and ending sides as numbered 1 and 8, respectively. They were extracted from the shiny and rough surface areas. Here, the rough surface appeared because the anvil at the flat area was deliberately roughened to prevent slippage and the shiny surfaces arose because the sample was made contact with the smooth surfaces at both sides of the roughened parts of the anvil and plunger. All the tensile specimens exhibited superplastic elongations well more than 400% with maximum flow stresses reduced to ~ 200 MPa or less as shown in Fig.7.

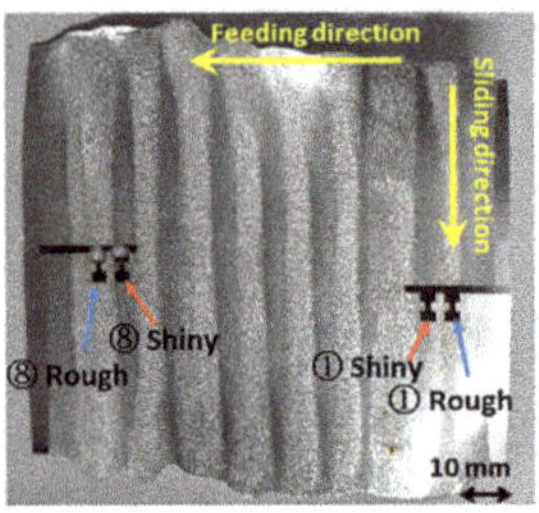

Fig. 6 Appearance of sheet sample after 8 passes of IF-HPS processing. [10]

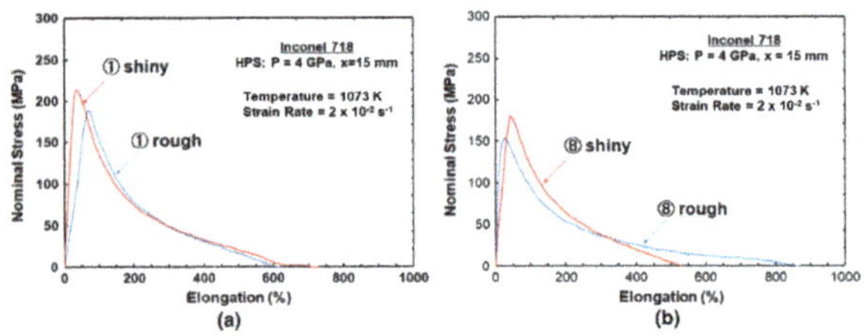

Fig. 7 Stress–strain curves at (a) initial and (b) ending sides of processed sheet. All tensile specimens were deformed at 1073 K with initial strain rate of 2×10^{-2} s^{-1}. [10]

Superplastic formability was further checked by extracting 30mm diameter disks from the IF-HPS processed sheet. The disk was then subjected to cup forming at 1073 K in air with a forming speed of 0.1mm/s. It was successful as shown in Fig. 8(a), whereas the disk without the IF-HPS processing was fractured at the flange as shown in Fig. 8(b) and many cracks appeared at the bottom of the cup even under the same forming condition.

Materials Research Proceedings **32** (2023) 15-22 https://doi.org/10.21741/9781644902615-2

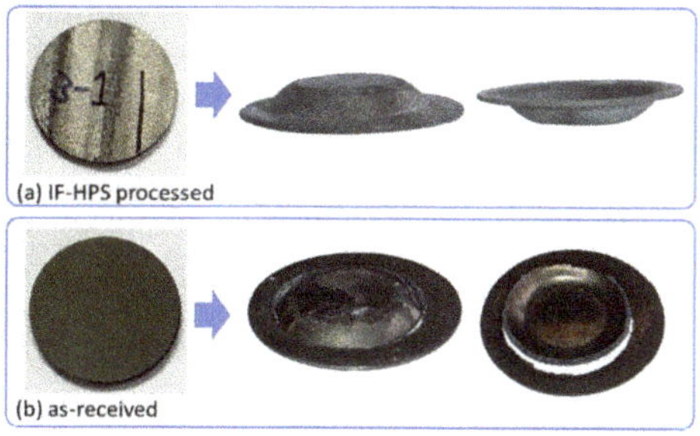

Fig. 8 Appearance of circular disks (left) extracted from IF-HPS processed sheet (upper) and as-received sheet (lower). Appearance after cup-shape forming (right) for IF-HPS-processed sheet (upper) and as-received sheet (lower). [8]

Figure 9 plots the total elongation to failure against the hardness measured on the Inconel 718 after processing by HPS including HPT from past studies [18]. There exists a linear relation, suggesting that it is possible to predict the advent of superplastic deformation by hardness measurement. Such a relation is reasonable because grain refinement promotes the total elongation at high temperatures due to enhanced contribution of grain boundary sliding to the total deformation [19]. The grain refinement can also enhance the room-temperature hardness due to the Hall-Petch strengthening mechanism [20,21].

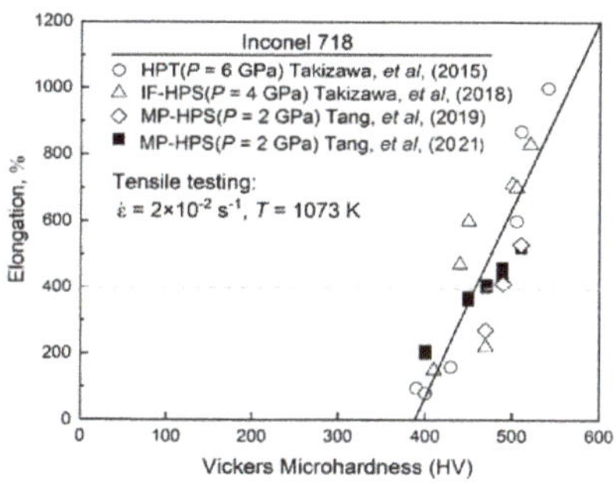

Fig. 9. Relation of total elongation plotted against hardness constructed using published data. Horizontal dotted line represents critical lower limit for superplasticity defined earlier [18].

Ti alloys

Figure 10 shows the stress-strain curves and the appearance of the tensile specimens of the F1295 alloy after processing through x = 10 mm under 3 GPa. The superplastic elongations more than 400% are confirmed. The tensile specimens are extracted from three different positions along the longitudinal axis of the sheet samples as shown in the inset. For comparison, the stress–strain curve and the appearance of a tensile specimen in the as-received state are included in Fig.10. While reduction in gage width is prominent in the as-received sample, more uniform deformation occurs throughout the gage length after the samples are processed by HPS.

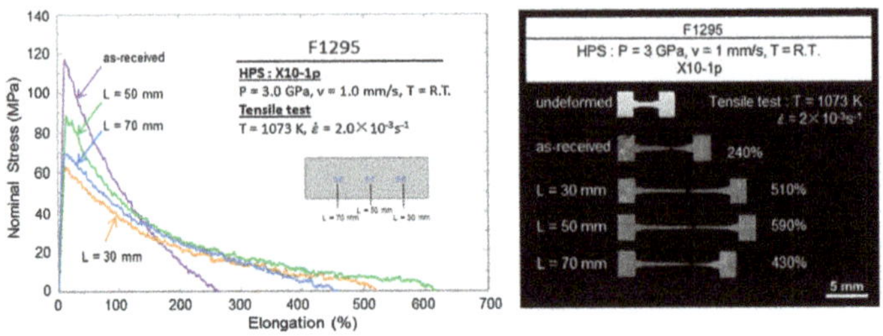

Fig. 10 Stress–strain curves (left) and appearance of tensile specimens after deformation to failure including undeformed specimen (right) for F1295 alloy. HPS was perfomed for sliding distance of 10 mm under 3 GPa at room

Mg alloys

Figure 11 shows the stress–strain curves and the appearance of the AZ61 samples processed for x = 10 mm at room temperature and at 403 K. The elongation to failure reached ~400% at both conditions with smooth and uniform elongations throughout the specimens. A close observation of microstructures after HPS processing at room temperature revealed that the grain size is refined to ~100 nm, but this fine-grained structure contains subgrains with low- angle boundaries. It is suggested that dynamic recrystallization should have occurred during the tensile testing because superplastic flow through grain boundary sliding hardly occurs in the subgrain structure but requires grain boundaries with high angles of misorientations.

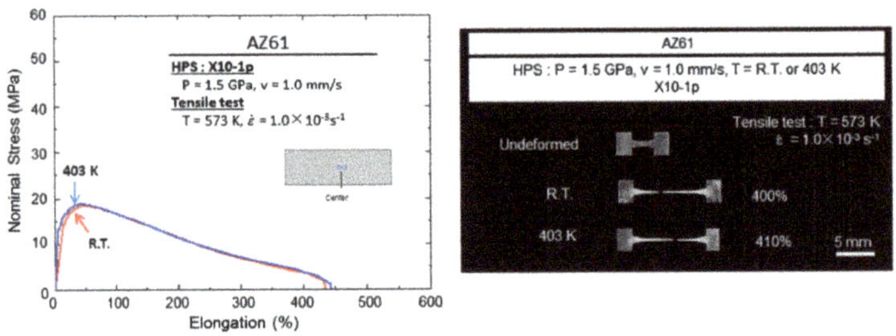

Fig. 11 Stress–strain curves (left) and appearance of tensile specimens after deformation to failure including undeformed specimen (right) for AZ61 alloy. HPS was performed for sliding distance of 10 mm under 1.5 GPa at room temperature and 403 K.

Summary and conclusions

Superplasticity successfully appeared in hard-to-deform metallic materials such as a Ni-based superalloy (Inconel 718), a Ti alloy (F1295) and a Mg alloy (AZ61) after processing by severe plastic deformation (SPD) though high-pressure sliding (HPS). In particular, high-strain rate superplasticity with a strain rate higher than 1×10^{-2} s^{-1} was attained in the Inconel 718 alloy. Incorporating a sample feeding process to the HPS process, called the incremental feeding HPS (IF-HPS), it was possible to enlarge the SPD-processed area as large as 100 x 100 mm in the Inconel 718 alloy without increasing the machine capacity. This enlargement thus enabled cup forming of the alloy without fracturing of the sample.

Acknowledgements
This study was conducted under a project subsidized by the New Energy and Industrial Technology Development Organization (NEDO).This study was also supported in part by a Grant-in-Aid for Scientific Research (A) from the MEXT, Japan (19H00830).

References

[1] R.Z. Valiev, Y. Estrin, Z. Horita, T.G. Langdon, M.J. Zehetbauer and Y.T. Zhu: JOM 58(4) (2006) 33-39. https://doi.org/10.1007/s11837-006-0213-7

[2] K. Edalati and Z. Horita: Mater. Sci. Eng. A 652 (2016) 325-352. https://doi.org/10.1016/j.msea.2015.11.074

[3] T. Fujioka and Z. Horita: Mater. Trans. 50 (2009) 930-933. https://doi.org/10.2320/matertrans.MRP2008445

[4] Y. Takizawa, T. Masuda, K. T. Fujimitsu, Kajita, K. Watanabe, M. Yumoto, Y. Otagiri and Z. Horita : Metall. Mater. Trans. A 47 (2016) 4669-4681. https://doi.org/10.1007/s11661-016-3623-3

[5] P.W. Bridgman: Phys. Rev. 48 (1935) 825-847. https://doi.org/10.1103/PhysRev.48.825

[6] A.P. Zhilyaev, T.G. Langdon: Prog. Mater. Sci. 53 (2008) 893-979. https://doi.org/10.1016/j.pmatsci.2008.03.002

[7] Z. Horita, Y. Tang, T. Masuda and Y. Takizawa : Mater. Trans. 61 (2020) 1177-1190. https://doi.org/10.2320/matertrans.MT-M2020074

[8] Y. Takizawa and Z. Horita: Mater. Trans. 64 (2023).https://doi.org/10.2320/matertrans.MT-MF2022025

[9] Y. Takizawa, K. Sumikazwa, K. Watanabe, T. Masuda, M. Yumoto, Y. Kanai, Y. Otagiri and Z. Horita : Metall. Mater. Trans. A 49 (2018) 1830-1840. https://doi.org/10.1007/s11661-018-4534-2

[10] Y. Takizawa, K.Watanabe, T. Kajita, K. Sumikawa, T. Masuda, M. Yumoto, Y. Otagiri and Z. Horita: J. Japan Inst. Met. Mater. 82 (2018) 25-31. https://doi.org/10.2320/jinstmet.J2017038

[11] T. Komatsu, T. Masuda, Y. Tang, I.F. Mohamed, M. Yumoto, Y. Takizawa and Z. Horita: Mater. Trans. 64 (2023) 436442. https://doi.org/10.2320/matertrans.MT-LA2022032

[12] T. Masuda, K. Fujimitsu, Y. Takizawa and Z. Horita: Lett. Mater. 5 (2015) 258263 https://doi.org/10.22226/2410-3535-2015-3-258-263

[13] Y. Tang, K. Matsuda, Y. Takizawa, M. Yumoto, Y. Otagiri and Z. Horita: Mater. Sci. Technol. 36 (2020) 877886. https://doi.org/10.1080/02670836.2020.1746538

[14] Z.Horita, D.J.Smith, M.Furukawa, M.Nemoto, R.Z.Valiev and T.G.Langdon: J. Mat. Res., 11 (1996) 1880-1890. https://doi.org/10.1557/JMR.1996.0239

[15] Y. Takizawa, T. Kajita, P. Kral, T. Masuda, K. Watanabe, M. Yumoto, Y. Otagiri, V. Sklenicka and Z. Horita: Mater. Sci. Eng. A 682 (2017) 603-612. https://doi.org/10.1016/j.msea.2016.11.081

[16] T.G. Langdon: J. Mater. Sci. 44 (2009) 5998-6010. https://doi.org/10.1007/s10853-009-3780-5

[17] K. Higashi, M. Mabuchi and T.G. Langdon: ISIJ Int. 36 (1996) 1423-1438. https://doi.org/10.2355/isijinternational.36.1423

[18] Y. Tang, K. Edalati, M. Masuda, Y. Takizawa, M. Yumoto, and Z. Horita: Mater. Lett. 300 (2021) 130144. https://doi.org/10.1016/j.matlet.2021.130144

[19] T.G. Langdon: Metall. Mater. Trans. A 13 (1982) 689-701. https://doi.org/10.1007/BF02642383

[20] E.O. Hall: Proc. Phys. Soc. B 64 (9) (1951) 747-753. https://doi.org/10.1088/0370-1301/64/9/303

[21] N.J. Petch: J. Iron. Steel Inst. 174 (1953) 25-28. https://doi.org/10.1111/j.1600-0447.1953.tb09440.x

Keynote Talks

Superplasticity in Advanced Materials - ICSAM 2023 Materials Research Forum LLC
Materials Research Proceedings **32** (2023) 25-40 https://doi.org/10.21741/9781644902615-3

In-situ neutron and synchrotron methods for the investigation of plastic deformation and annealing in metals

Klaus-Dieter Liss[1,a*]

[1]School of Mechanical, Materials, Mechatronic and Biomedical Engineering, Northfields Avenue, University of Wollongong, Wollongong, NSW 2522, Australia

[a]kdl@uow.edu.au, liss@kdliss.de

Keywords: Neutron Scattering, Synchrotron Radiation, Diffraction Theory, Reciprocal Space, Titanium Aluminides, High-Pressure Torsion, Phase Transformation, High-Temperature, In-Situ, Plastic Deformation, Thermo-Mechanical Processing, Dynamic Recrystallization

Abstract. Following a crash course in neutron and synchrotron diffraction standards, applications are demonstrated on selected metallic systems, comprising the atomic order in titanium aluminide intermetallics at thermal and mechanical processing. High pressure torsion processed specimens show heterogeneous structure and order. Upon heating, their nanostructure evolves revealing regimes of recovery, recrystallization and grain growth, which can be exploited for engineering designated microstructures with enhanced physical and mechanical properties. Advanced analysis of two-dimensional diffractograms by synchrotron radiation allows to distinguish microstructure transformations as well as deformation mechanisms in thermo-mechanical processing. The methods are applicable to a wide range of materials and processes allowing to speed up materials development by orders of magnitude.

Prolegomenon

Superplasticity is a phenomenon on large plastic deformation of materials occurring through many crystalline and amorphous classes of solid materials, such as metals, crystalline materials, glasses and polymers. While the phenomenon is defined by plastic strain surpassing 4, i.e., four times straining its original length in a tensile test, its micromechanical and microstructural mechanism and control are highly complex, depending on multiple internal and external parameters, such as composition, microstructure, temperature, pressure, and time. Such mechanisms in crystalline materials demand some self-healing processes counteracting work hardening, dislocation pile up, localization and involve processes as dynamic recovery, dynamic recrystallization, grain boundary gliding. In other words, diffusivity has to be high either in the bulk or though specific lower dimensional channels, such as grain boundaries and along dislocation. A study of the mechanisms ideally occurs in-situ, time-resolved and in the bulk of the material while external parameters are applied.

Neutron and high-energy synchrotron radiation are both bulk penetrating probes [1], delivered by large user-facilities of highest intensities and brilliance to allow for parametric studies, spatially and time resolved, in-situ under the applied conditions. A wide range of spectroscopic methods, imaging and diffraction are typically provided at such facilities for tackling specific aspects of the materials and their behavior. Scope of the present paper is an overview on the advanced diffraction techniques and presenting selected examples.

Neutron- and Synchrotron X-ray Diffraction

Both X-rays and neutrons are short-wavelength quantum beams with wavelengths λ of interest in the range or smaller than atomic distances in the crystalline or amorphous structure [2], subject to interferences of waves scattered by the individual atoms, presenting the effect of diffraction. By the quantum-mechanical wave-particle dualism, important kinematic quantities are the wave number k, or momentum p

$$k = 2\pi / \lambda \quad ; \quad \vec{p} = \hbar\, \vec{k} \tag{1}$$

with $\hbar = h/2\pi$ the Planck constant, and the energy

$$E = \hbar\, \omega \tag{2}$$

of the particle, which are globally conserved under any transformation. An important fundamental difference between X-rays and neutrons are the dispersion relations between E and k, namely

$$E_X = \hbar c k \text{ for X-rays, and} \tag{3$_x$}$$

$$E_n = \frac{\hbar^2 k^2}{2m} \text{ for neutrons.} \tag{3$_n$}$$

Herein c is the speed of light and m the mass of a neutron. Values for say $k = 3.49$ Å^{-1} result in wavelength $\lambda = 1.80$ Å with energies $E_X = 6.89$ keV and $E_n = 25.3$ meV, respectively, for X-rays and neutrons. These typical values can vary in a range spanned by an order of magnitude and the selected numbers correspond to the tabulated values at neutron velocity $v = 2200$ m/s [3] and are close to typical lab X-ray energies [4].

A propagating X-ray or neutron wave contains the term

$$\psi_o = e^{i(\vec{k}\vec{x} - \omega\, t)} = e^{i\vec{k}\vec{x}} e^{-i\omega\, t} \tag{4}$$

which is the oscillating function in space x and time t. If we focus on the $k\,x$ dependence, then we are talking of diffraction, while spectroscopy is defined by studying the $\omega\, t$ behavior. We also recognize in Eq. (4) symmetry between k and x, as well as for ω and t, demonstrating that momentum space and energy space are the Fourier transforms of position and time, respectively, further called reciprocal space and frequency domain.

Under pure diffraction conditions ω is constant, $\Delta\omega = 0$, and therefore by Eqs. (2) and (3), the length k of the wavevector is conserved by

$$|\vec{k}_f| = |\vec{k}_i| \tag{5}$$

while its direction can change in a scattering processes by the momentum transfer

$$\vec{q} = \vec{k}_f - \vec{k}_i\,, \tag{6}$$

also called the scattering vector. The Laue interference condition, or also the Ewald construction state, if

$$\vec{q} = \vec{G}\,, \text{ especially } \vec{G} = \vec{G}_{hkl}\,, \tag{7}$$

in which $\vec{G}_{hkl}$ is a reciprocal lattice vector of a crystal, then we obtain high intensity in the reflected beam direction $\vec{k}_f$. This vectorial Laue equation, displayed in Fig. 1 is a more elegant representation of the often cited Bragg equation where twice the Bragg angle θ equals the scattering angle 2θ between $\vec{k}_f$ and $\vec{k}_i$, and lattice spacing $d_{hkl} = 2\pi/G_{hkl}$.

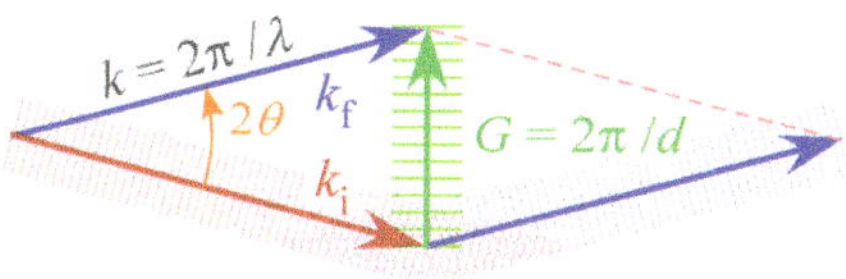

Figure 1: Laue-Bragg interference occurs when the Moiré pattern in the overlap of the incoming wave $\mathbf{k}_i$ *and the outgoing wave* $\mathbf{k}_f$ *matches the crystal lattice spacing d described by the reciprocal lattice vector* **G**, *so that* $\mathbf{G} = \mathbf{k}_f - \mathbf{k}_i$. *The wavelength is* λ *and the Bragg angle* θ, *while* $k = |\mathbf{k}_f| = |\mathbf{k}_i|$.

The intensity of the reflection with Miller indices *hkl* is proporional to the structure factor, which is the Fourier transform of the crystallographic unit cell, computing

$$F_{hkl} = \sum_{\vec{\rho}_j} l_j \, e^{-i\vec{G}_{hkl}\vec{\rho}_j} \quad (8)$$

by summing scattered waves from all *j* atom positions $\vec{\rho}_j$ in the unit cell. The latter are weighted each by their individual scattering amplitudes l_j, expressed by the coherent bound scattering length

$$l_j = b_{c.j} \quad (9_n)$$

for neutrons and

$$l_j = r_e \cdot f_j(q) \cdot P(\vec{q}) \quad (9_x)$$

for X-rays, where r_e is the classical electron radius, i.e., the scattering length for one electron, f_j the atomic form factor, i.e., the Fourier transform of the electron distribution, and *P* the polarization factor, which can be set equal to $P = 1$ for small scattering angles as for high-energy X-rays.

There exist basically two orthogonal kinds of scanning reciprocal space, namely angle dispersive along the orientation of $\vec{k}_f$, i.e., along 2θ, and wavenumber dispersive by keeping 2θ fixed and varying the length of *k*. Some modern instruments can combine both by positioning energy dispersive detectors simultaneously at various scattering angles, making maximal usage out of the scattered radiation.

Plotting and further evaluation of diffractograms are then plotted against *q* or $\vec{q}$ in one to three dimenstions, representing reciprocal space in a linear coordinate system describing the sample crystal lattice rather than the machine.

Practically, conversion of the abcissa from an angular or wavenumber dispersive scale, 2θ or *k*, transforms via

$$q = 2\,k \sin(\theta) = 4\pi/\lambda \sin(\theta) \quad (10)$$

by fixing either *k* or θ, respectively. Note, in the case of energy-dispersive X-ray diffraction, Eq. (3_x) needs to be solved for *k*. For neutrons, the wavenumber dispersion is often measured by the time of flight method, where a neutron pulse of kinetic energy in Eq. (3n) is measured by the neutron time of flight τ over a flight path of length *L*, the diffractometer length, delivering its velocity $v = p/m$. Putting all together, the neutron time-of-flight dispersive scale transforms through

Materials Research Forum LLC
https://doi.org/10.21741/9781644902615-3

$$q = \frac{2m}{\hbar} sin(\theta) \frac{L}{\tau} \tag{11}$$

to linear reciprocal space, q, which is directly comparable to any angle dispersive measurement.

Bulk Penetration of Neutron- and Synchrotron Radiation

The main advantage of neutron and Röntgen radiation is their penetration into the bulk of materials while probing their structure on the atomic length scale. On top of this, immense intensities or radiation fluxes delivered by modern sources, as well as advanced instrumentation, radiation optics and detection allow for most refined and time resolved studies, enabling the recording of minute structural changes and in-situ studies.

The penetration along a linear dimension x is characterized by the attenuation into a material following an exponential law

$$\frac{I(x)}{I_0} = e^{-\mu x} \tag{12}$$

where I is the intensity, $I_0 = I(0)$ and μ the attenuation coefficient. The attenuation length is defined by the depth x_{att} at which the intensity ratio drops to e^{-1} giving $x_{att} = 1/\mu$.

Figure 2 displays the attenuation lengths for representative metals covering the light metals Al and Mg, Fe for steels, advanced structural metals Ti and Zr, precious metals Cu, Ag and Au, and heavy metals Pb and U. The X-ray curves display three ranges, namely the typical steep increase of photo-absorption at lower energies of 1~70 keV, decorated by the element characteristic absorption edges, a gentle increase between 100~1000 keV dominated by Compton scattering and finally decrease by electron-positron pair production above their threshold of 1022 keV. At the lower energies, including most common laboratory X-rays around 8 keV, photo absorption varies strongly, roughly with the third power. Passing an absorption edge removes electrons from the atom and therefore increases absorption. Compton scattering is an inelastic scattering process occurring at any electron while pair production converts photon energy into matter and anti-matter.

Neutron absorption scales with k^{-1} and is tabulated for $k = 3.49$ Å^{-1} along the scattering cross sections [3], leading to the neutron attenuation curves in the upper left corner of Fig. 2. Apart from some strong absorbing exceptions, their penetration is by orders of magnitude larger than for lab X-rays, and well larger than for high-energy X-rays.

At laboratory energies of 8 keV with $k = 4.05$ Å^{-1}, X-rays penetrate 144 μm and 74 μm into Mg and Al, 4.2 μm into Fe or steel and even less into the heavy metals Pb, Au and U. Penetrations are increased by two orders or magnitude to 37 mm, 22 mm, 3.8 mm and hundreds of micrometers, respectively, at high-energy X-rays of 100 keV ($k = 50.7$ Å^{-1}) – and those still increase significantly beyond. Neutron attenuation lengths at $k = 4.05$ Å^{-1} evaluate to 1.7 m, 0.8 m and 45 mm for Mg, Al, Fe, respectively, and 2 mm into Au.

It shall be noted that depths of a few attenuation lengths can be probed regarding the high-flux X-ray synchrotron sources, so that bulk investigations by diffraction methods are routinely feasible on most materials in the high-energy range of [80 … 200] keV and beyond. High-energy X-ray diffraction studies of penetrating 25 mm of steel have been reported in a strain scanning experiment [5] and structures of heavy element containing, few millimeter thick $EuAs_3$ been solved [6],[7].

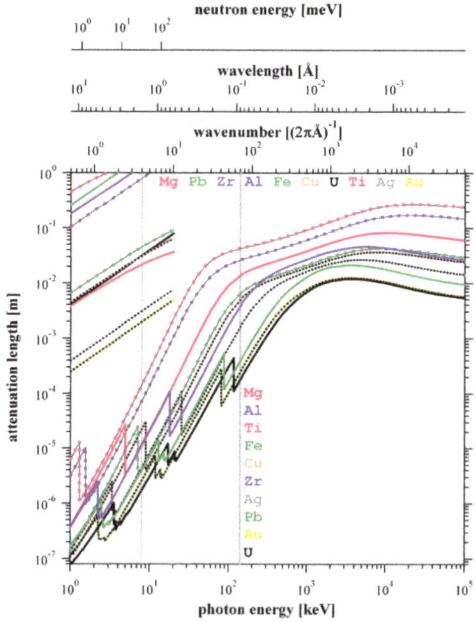

Figure 2: Attenuation lengths for typical selected elements through the periodic table. The upper left curves displayed up to 10 Å^{-1} are neutron data while the lower full-range data belong to X-rays. The colors and sequences of the element symbols correspond to their appearance from top to bottom. Cross section data taken in 1996 form the NNDC tables at NIST [8].

Usage of Neutron and Synchrotron Radiation

The applications of both kinds of quantum beams are most versatile covering all disciplines on condensed matter. Because of their deep penetration power into bulk materials, particularly high-energy synchrotron, as well as neutron radiation allow for easy sample environment, at which the beam may pass solid window materials, such as aluminum windows of a cryostat or cryo-magnet, niobium heat elements and heat shields of a high-temperature furnace, chemical reactor windows, gaskets of high-pressure cells, submerged environments, embedded sample systems, to mention but a few. Diffraction measurements range from structure determination, kinetics and dynamics of the sample system, microstructural, interfacial, and engineering applications. The salient differences for using neutrons or synchrotron radiation are compiled in Table 1.

For example, the *powder diffraction* method, as also applied to polycrystalline material demands a good statistical grain average with random orientation, in order to extract reliably peak intensities and positions, necessitating many thousand grains in the illuminated volume. As the grain size may be given by the processing of the material, neutron scattering may be advantageous over synchrotron high-energy X-ray diffraction illuminating a large gauge volume. Similar can be said for texture determination, which is the preferred crystallographic orientation averaged over a large ensemble of grains, use neutron diffraction if grains are large. On the other hand, small beam synchrotron techniques allow to probe for tiny volumes, locally, which may give information on individual or a small number of grains, even embedded in a polycrystalline matrix. When high-energy X-ray diffraction came up, *bad* grain statistics has been a problem and experimentalists spun and shifted specimens through the beam in order to get a better orientation powder average. Nowadays, we make use out of this local information, distinguished in both direct as in reciprocal space, as demonstrated in the second last chapter of this manuscript.

Table 1: Comparison of High-Energy X-rays with neutrons on some features.

feature	**HEX-rays**	**neutrons**
penetration	millimeters	meter
beam size / focus	micrometers - millimeters	centimeter
hydrogen sensitivity	zero	huge
light / heavy element detection	challenging to impossible	easy
magnetism	zero	sensitive
inelastic scattering	very difficult	easy
resolution	high	lower
timing	microseconds - milliseconds	seconds - hours

Another complement between the two kinds of quantum beams is the difference in scattering lengths for the various elements. For X-rays, the total scattering amplitude of the atom just increases with its atomic number *Z*, resulting that the heavy elements shine much stronger than the lighter ones. This is even dramatized as the intensity of a reflection scales with the square of the scattering length. For neutrons, the scattering length is more or less in the same order of magnitude for most of the isotopes, although unsystematic. This means, light elements can have similar scattering contribution than their heavy counterparts. Some isotopes even possess negative scattering lengths, meaning there is a 180° phase jump at those atoms in the structure. An extreme example is hydrogen, H, which scatters neutron strongest of all isotopes, unfortunately incoherently. However, the coherent scattering length of H is still strong, and in this case negative. Positive coherent scattering with much less incoherent background can be obtained through replacing the ^{1}H isotope by its heavier brother ^{2}H, also called deuterium, ^{2}D. By mixing H and D, the average scattering length can be adjusted, which is called contrast variation, allowing to match specific features in a structure (mostly in organic materials). For X-rays, however, scattering from H is practically zero, as they scatter from electrons and H usually donates its electron in a chemical compound. Thus, by comparing X-ray and neutron data, one can obtain the H-insensitive and H-sensitive part of a structure and refine the H positions. In metallurgy, H may play a role in titanium sintering, enhancing strongly the processing kinetics [9],[10].

There are other choices for neutrons prior to X-rays, which are not further emphasized here, such as sensitivity to magnetism, as the neutron itself is an elementary magnet. It allows to refine the magnetic structure and ordering in magnetic phases. Last not least, inelastic scattering is a vast field to study motions of atoms, like in diffusion, in phonons, and other collective excitations. Since thermal energies of 25 meV at room temperature are comparable with the energy of the neutron, this is the probe of choice rather than X-rays with 10^6 times larger probe energies.

An important aspect is the spatial and time resolution at which a diffractogram can be taken. This allows to follow the in situ processes in real time, while a phase transformation takes place, during the processing of a microstructure, operando while a machine is running. Selected examples will be given throughout the manuscript.

Complementary Neutron and X-ray Diffraction in Titanium Aluminides

The different atomic scattering lengths l_j between X-rays and neutrons give rise to different Laue-Bragg peak intensities through the structure factor, Eq. (8). As an example, Ti has a negative neutron scattering length, while most of the metals show positive values. For X-rays, the scattering lengths are always positive, Eq. (9_x). This feature can be applied to study the ordering in the intermetallics of the titanium-aluminide system. Figure 3 compares the diffractograms of a titanium aluminide of composition approximately Ti-45Al, comprising ordered α_2 and γ-phase. The positions of the reflections are identical on the scattering-vector scale, however, the intensities vary dramatically. The neutron scattering lengths b_c for Ti and Al are −3.438 fm and +3.449 fm, respectively [11], while their atomic numbers *Z* are 22 and 13. The strongest X-ray peak in Fig. 3 is a superposition of the α_2-002 and γ-111 reflections. For calculating their structure factors, the scattering lengths of all atoms simply add up, which leads to ~ |22+13| = 35, a large value for X-rays, which is taken to the square for obtaining the intensity. The neutron structure factor for those reflections contains ~ |−3.438+3.449| fm = 0.011 fm, leading to a tiny intensity only, as observed in Fig. 3. On the other hand, the superstructure reflections, e.g., the first 3 neutron peaks, contain the differences of the scattering lengths, which renders them relatively small with ~ |22-13| = 9 for X-rays and huge for neutrons, by ~ |−3.438−3.449| fm = 6.887 fm, to the square, delivering the large neutron intensities, whereas X-ray peaks a tiny. Moreover, the effect of positive and negative scattering lengths can be exploited to study the crystallographic disorder of the intermetallics, when Al atoms occupy Ti sites and vice versa. Such disordered structure is described by averaging the scattering lengths on each site, which approaches the concentrations of the composition when fully disordered. Coincidentally, this value again is close to zero for neutrons, meaning, for this composition, all Bragg peaks of the disordered phase will disappear, although the material is solid and crystalline. With knowledge of the phase composition, e.g. as determined by by high-energy X-ray diffraction, the intensity of the neutron superstructure peaks gives most sensitive insight into the crystallographic order. Rather than being concentrated in Laue-Bragg interferences, the scattered intensity goes between the Bragg peaks into diffuse scattering, allowing to determine short-range order in the critical fluctuations around a phase transition [12].

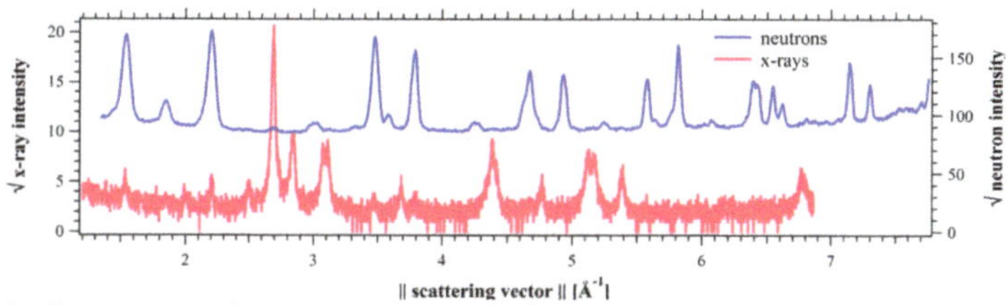

Figure 3: Comparison of neutron and X-ray diffractograms on the same scale of momentum transfer. Because of the negative scattering length of Ti, neutron peaks are strong where X-ray peaks are weak, and vice versa.

Titanium aluminides are a novel high-temperature structural material for aerospace and turbine applications. The phase composition and their crystallographic order in titanium aluminide based intermetallics is a critical parameter for their mechanical properties, influencing strongly the dislocation behavior and other plasticity aspects [13]. Quantitative thermal transformation details have been analyzed by Kabra et alii on Ti-44Al-3Mo and Ti-44Al-7Mo (in at.%) alloys [14], in which Mo may stabilize the ductile β-phase in an $\alpha+\beta$ two-phase field at high, while three ordered phases, α_2, γ and β_o exist at lower temperature, as extracted from Fig. 4. It demonstrates that the lines of all three ordered phases disappear upon heating while reappearing on cooling, revealing their individual transition temperatures and potential undercooling effects, due to nucleation. Basically, ordered, hexagonal α_2 disorders to *hcp* α in an eutectoid reaction, *ccp*-based tetragonal γ dissolves at the γ-transus, and β_o disorders to β of a ductile *bcc* structure. Similar kind of in-situ studies have been undertaken to address critical aspects in the Ti-Al-Nb phase diagram, revealing transition temperatures and phase sequences with gentle admixture of Nb by synchrotron studies [15]. The higher-Nb concentration range around the o (omicron) phase and its transformation to neighboring phases and their order clarified long-lasting discrepancies and misinterpretations in literature, particularly stating that there is only one independent kind of o phase while others can be understood as a continuous disorder transition, thriving to well-known α or α_2 in its asymptotic limit [16]. Moreover, ordered β_o phase plays a role, which disorders again at high temperature to β. Pioneering in-situ X-ray diffraction studies of the Ti-Al system go back to Shull et alii in 1990 [17], solving part of the binary phase diagram. Although an earlier, unpublished in-situ neutron work has been cited therein, only the more challenging X-ray experiment could solve e.g. the eutectoid temperature at which α_2 disorders to α and the region at which α-phase is stable. Further clarification of phase diagrams and adjustments of databases are based on in-situ studies by Chladil et alii [18],[15]. Yeoh et alii [19] undertook the first comprehensive, synchrotron high-energy X-ray analysis connecting crystallographic order to lattice parameters, which later has been more quantified by Li et. alii [20], including effects under high hydrostatic pressure. The latter, is another physical parameter to address phase changes and stabilize the ductile β-phase under hydrostatic pressure for a potential processing window, while it disappears under operating conditions, keeping the material better creep resistant, as studied by energy-dispersive synchrotron X-ray diffraction with conditions up to 9.6 GPa and1668 K [21]. Those in-situ high-pressure experiments not only determine pressure-induced phase transformations but also indicate how a counterintuitive transformation compromising volume increase and ordering energy modify the equation of state.

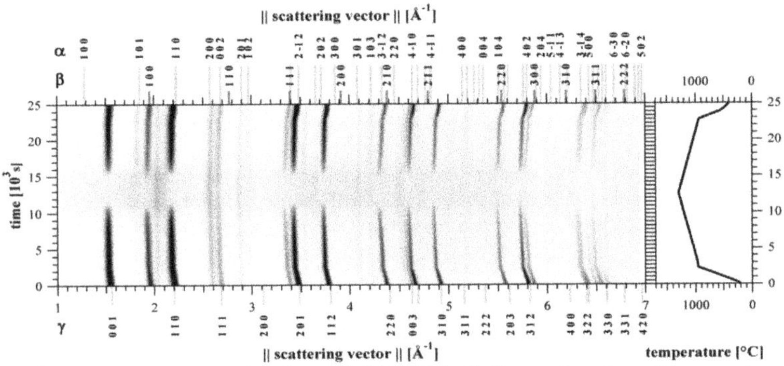

Figure 4: In-situ neutron scattering from Ti-44Al-3Mo (left) as a function of temperature (right) and time. Reproduced from Kabra et. alii [14].

Materials After Severe Plastic Deformation

Bulk nano-structured materials in metastable and heterogeneous states can be achieved by severe plastic deformation. They expose enhanced mechanical and strongly altered physical properties. For example, CoCrFeNi high entropy alloys processed by high-pressure torsion have been studied increasing Vickers hardness from 160 to 525 [22], while drastically changing ferromagnetic response to spin-clustering resulting in superparamagnetic or spinglass behavior with even inverted hysteresis [23]. Important aspects in characterizing and understanding those materials are both local volume probes for assessing heterogeneity and temperature dependent studies for revealing phase and microstructural changes.

Materials Research Proceedings **32** (2023) 25-40
https://doi.org/10.21741/9781644902615-3

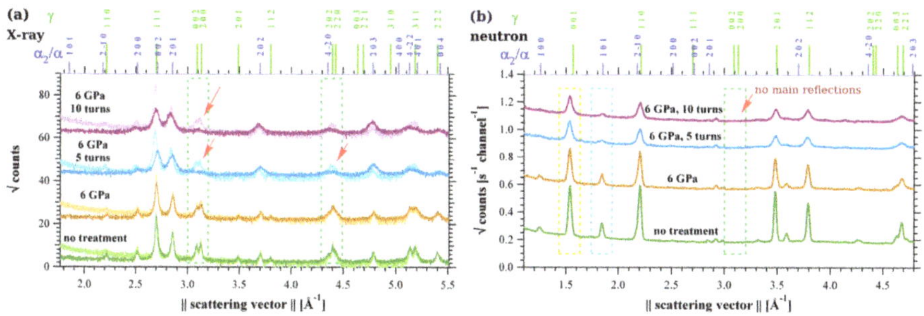

Figure 5: Lab X-ray diffraction (a) and neutron diffraction (b) of titanium aluminide Ti-45Al-7.5Nb as-received and processed by high-pressure-torsion under 6 GPa and given number of turns. The darker and lighter superimposed X-ray lines are taken at the specimen surface and mid-section layer of the disk, respectively, revealing strong heterogeneity, particular visible at the arrows. Bulk-integrated neutron diffractograms reveal decrease in intensity due to crystallographic disorder. Note the different scale ranges on the abcissa. Reproduced form Li et alii [26].

A comprehensive table has been published by Kawasaki et alii, listing both neutron and synchrotron studies on specimens after severe plastic deformation and their conditions of investigation [24]. An early study combining both synchrotron and neutron diffraction on high-pressure torsion processed material has been undertaken on a twinning-induced-plasticity steel, revealing the local and global stress induced martensitic transformation from α-Fe to ε-Fe phase, retained in the ultrafine-grained microstructure, alongside a texture analysis [25]. Titanium aluminides have been extensively investigated after high pressure torsion. Figure 5 compares laboratory Cu-K_α X-ray and neutron diffractograms by Li et alii [26]. Because X-rays around 8 keV penetrate only a few tens of micrometers, only surface-near regions are probed, first at the surface of the processed disk and then after grinding off the specimen to its mid-section. First of all, significant peak broadening can be recognized with increasing numbers of turns, i.e. shear strain, which is related to size-strain broadening as the grain size reduces and the microstrain, which is the stress-induced strain gradient between the grains of the microstructure, increases. Those X-ray data reveal that the microstructural and phase transformation at the surface is significantly different than in the mid-section, as seen e.g. at the γ-002/220 reflection pair. Upon high-pressure torsion, these peaks not only merge but almost disappear. The splitting of this couple stems from the tetragonal distortion, or the other way round, their merging reveals disorder by random anti-site occupation of Ti and Al, driving the structure to the *ccp* lattice. Such atomic disorder is valuably expressed in the neutron diffractograms by decrease of the strong superstructure reflections of both phases γ and α_2, in a similar way as discussed on the thermal driven phase transformation in Figs. 3 and 4. As the γ-phase exposes the highest ordering energy of all possible *ccp* and *hcp* based configurations [27], it rather transforms to α / α_2 and shrinks in amount. The latter has been observed already under hydrostatic pressure alone [21], while the large applied shear strain under high-pressure torsion leads to much larger disorder. Altogether, this X-ray and neutron combined study distinguishes local and global structural evolution, revealing that there exist higher plastically strained zones in the surface-near region of a high-pressure torsion specimen than in the bulk. Further inhomogeneities and their thermal evolution have been imaged by laser scanning confocal microscopy by Kawasaki et alii on high-entropy alloy specimens [24].

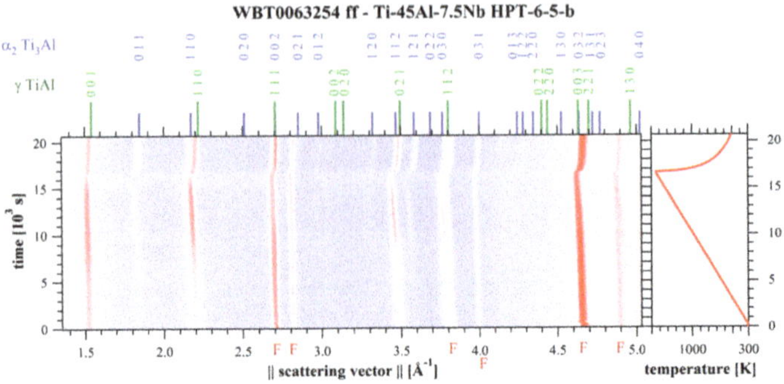

Figure 6: In-situ heating of the high-pressure torsion processed titanium aluminde under neutron diffraction from Wombat (left) undergoing a heating-cooling cycle (right). Reproduced form [28].

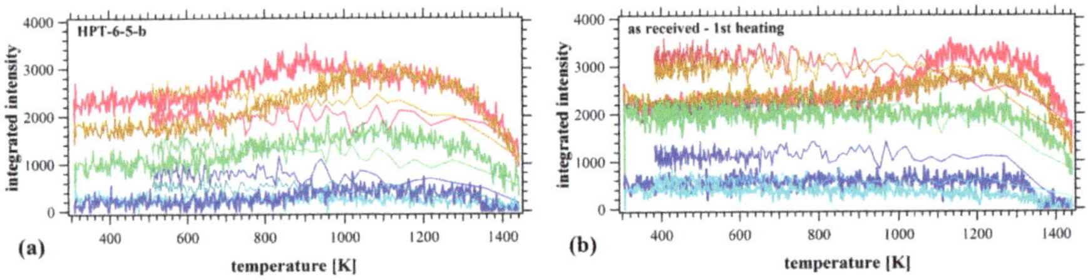

Figure 7: Extracted peak intensities from Fig. 6 on the high-pressure torsion processed (left) and as-received material (right). γ-001 (red), γ-110 (orange), γ-021 (green), α_2-011 (dark blue) and α_2-121 (light blue). Reproduced from [28].

Thermal stability and microstructural evolution of such high-pressure torsion processed specimens can be investigated by in-situ diffraction upon a heating ramp. Such neutron diffraction data is displayed in Fig. 6 showing various features: First, there are peak shifts to smaller values upon heating, due to thermal expansion of the lattice, revealing thermal expansion coefficients for the given material. Secondly, there are peak intensity and peak profile evolutions due to microstructural transformations. The salient observation in the present study, as displayed in Fig. 7a, are an intensity increase setting-in at 660 K, a relative low temperature, due to the reinstatement of crystallographic order, well before the equilibrium transformations become remarkable at about 1000 K in Fig b. As shown in the original manuscript [28], this recovery of order is consistent with afore-mentioned lattice parameter trends between fully disordered, hypothetical *ccp* and the ordered tetragonal γ-TiAl phase.

Further in-situ neutron diffraction investigations on high-pressure torsion processed CoCrFeNi high-entropy alloys are reported by Kawasaki et alii in this same journal issue [29], based on work by Liu et alii [30], which is not further reproduced here. The study employs high-resolution time-of-flight acquisition at the iMateria beamline [31] at the Materials and Life Science experimental facility [32] of the J-PARC neutron spallation source [33]. By single-peak fitting, parameters like peak positions, peak widths and integrated intensities are extracted and evaluated as a function of temperature. A minute evaluation of lattice parameters not only reveals thermal expansion and segregation of the elements, but furthermore, lattice widening due to a high concentration of vacancies has been observed. At high temperatures, these supersaturated defects recombine in a regime of recovery, followed by a sharp onset of recrystallization at which the peak widths shrink drastically, and grains grow. As a conventional Williamson-Hall analysis, plotting peak width against peak location behaves unsystematic, the modified Williamson Hall method taking into account the anisotropic strain fields around dislocations, has been applied for obtaining reliable values of dislocation densities [30]. Very similar studies have been reported by the same group [34] on high-pressure torsion processed stainless steel. Both materials show consistently equivalent microstructural behavior on heating, correlating the different temperature regimes with hardness after annealing, revealing that hardness reaches a maximum after recovery, just before recrystallization. Such data is most valuable for designing microstructures and thus mechanical properties through severe plastic deformation and well controlled subsequent annealing. As an outlook, such microstructural design based on in-situ diffraction studies establishes a wide playground for engineering their physical properties, as the aforementioned change in magnetic response [23].

Thermo-Mechanical Processing in a Synchrotron Beam

In previous chapters we have treated the importance of grain size for accurate *powder diffraction* averaging in polycrystalline materials, which is essential for crystallographic structure determination, strain analysis, and texture acquisition. Figure 8 shows a couple of high-energy diffraction images of the same specimen of Ti-45Al-10Nb, before and after undergoing a heating-cooling cycle. It is eye-striking that the smooth continuous rings have evolved into spotty patterns, each intensity dot representing a reflection from a particular crystallite. Apart from some oxide layer (fine rings), the bulk phase composition is qualitatively similar, while certain features can be recognized in the spotty pattern: (i) intensities are concentrated in 6 regions, representing texture, which is inherited from the as-produced material. (ii) the intensity grouping across different rings reveals orientation correlations, such as Burgers, Potters and Blackburn correlations between the various phases, as well as domain orientation correlations in the γ-phase. Details have been elaborated in the pioneering works in 2006 [35].

Moreover, such rich data can be acquired during physical thermo-mechanical simulation under plastic deformation at high temperature. Azimuth-time plots have been introduced by reproducing the intensity distribution of selected reflections along a straight line, the abscissa, while streaking time along the ordinate of a two-dimensional plot [36],[37], resumed and expanded in a classical paper [38]. Results of a Ti-43.5Al-4Nb-1Mo-0.1B titanium alloy undergoing heating to 1573 K for plastic deformation are presented in Fig. 9, showing the azimuth-time plots of a selected reflection of each phase, α-101 and β-200, published by Liss et alii in 2009 [39]. The inclined, elongated streaks on α-101 prove for a grain rotation around the incoming beam axis, while their azimuthal broadening is a signature of subgrain formation. Gentle fluctuations and narrowing of the timelines witness of recovery; when in competition with the broadening, dynamic recovery; and the appearance of new lines dynamic recrystallization. The time lines seen on β-200 are very short temporally, basically dots on the graph, revealing a highly fluctuating almost random orientation distribution, which is a continuous creation and disappearance of grains, in other words, dynamic recrystallization. This data showed for the first time in-situ the deformation mechanisms at such high temperature, in real time, in two co-existing phases, proving the high formability in the $\alpha+\beta$ phase field, at which the dynamic recrystallizing β-phase takes most of the plastic strain, being the grease of the system.

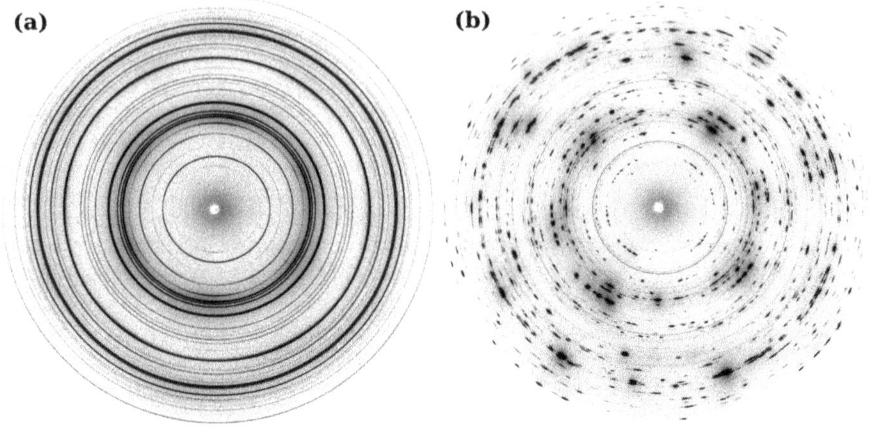

Figure 8: High-energy synchrotron X-ray diffraction from Ti-45Al-10Nb before and after heating into the high-temperature $\alpha+\beta$-phase regime. A fine microstructure has evolved into coarse (lamellar) grains.

Extensive examples have been published by the authors describing in detail the interpretation of diffraction phenomena as well as the sample systems on Cu, Zr alloys, Mg alloys, and twinning induced plasticity steels [38],[40],[41]. A study on magnesium alloys shows the transition from extension twinning as a plastic deformation mechanism to the activation of different slip systems, resulting in different textures. Moreover, fluctuating grain orientations at static high temperatures have been observed giving raise to ongoing investigations on abnormal grain growth.

Compendium

Neutron and high-energy synchrotron radiation are most powerful tools for the investigation of bulk materials, particularly in-situ, in real time or spatially resolved. Despite the different character of radiation, individual dispersion relations and various detection and scanning methods, transformation of data into linear reciprocal space units makes them directly comparable, which can be extended to other kinds of radiation, such as electron, light and phonon scattering.

Applications in the field of metals are vast and only a few examples can be given in this limited space, selecting titanium aluminides as a primary example. Therein, negative and positive scattering lengths of neutrons versus positive ones for X-rays allow uniquely to separate crystallographic order and disorder from other effects, such as phase composition upon a transformation. An important field of application are ultra-fine grained materials obtained by severe plastic deformation, especially high-pressure torsion. Combined X-ray and neutron studies reveal different states of disorder in surface-near and deep bulk layers, testifying a high degree of heterogeneity. High-temperature studies of such specimens reveal recovery, recrystallization and grain growth regimes, allowing to micro-tune and engineer the nanostructure for increased mechanical and physical properties.

Novel methods making usage out of spotty diffraction rings can reveal a universe of complementary information, such as grain orientation correlations, transitions between them, thermally activated microstructure evolution, and during high-temperature plastic deformation. Diffraction allows to focus on a selected phase in a multi-phase material, in parallel to the others, revealing deformation by slip, dynamic recovery and dynamic recrystallization.

The methods are still in their infancy. Ongoing work actually focuses on the investigation of nanostructures with microbeams to reveal similar micostructural evolution. More comprehensive studies allow to follow grain orientations in their two dimension, revealing full orientation rearrangements, which are relevant for grain boundary rearrangements, such as in abnormal grain growth and superplasticity.

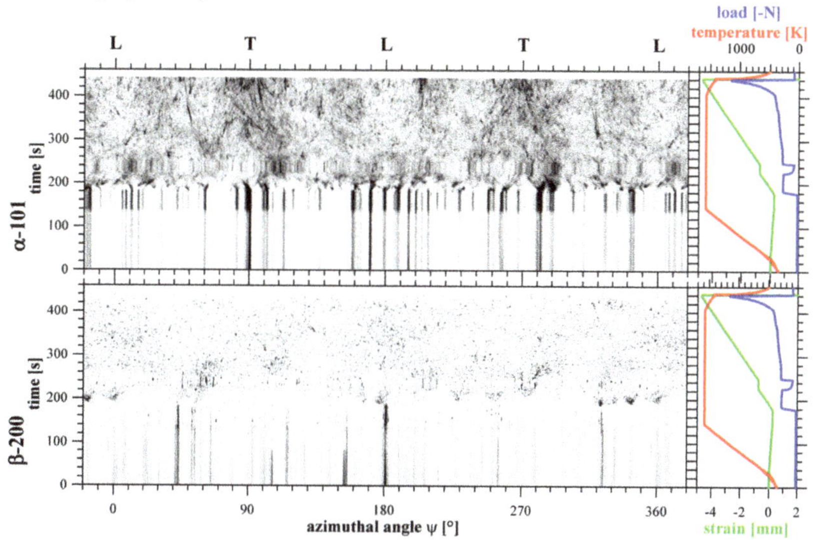

Figure 9: Azimuth-time plots of co-existing α and β phase in Ti-43.5Al-4Nb-1Mo-0.1B showing different reorientation behavior. While α deforms by crystallographic slip, some recovery and very sluggish recrystallization, the high-frequency fluctuations in the β phase reveal dynamic recrystallization. Reproduced base on [39].

Acknowledgments

The author likes to acknowledge the collaborators on the cited papers, in particular the groups around Prof. Rian Dippenaar, University of Wollongong, Prof. Megumi Kawasaki, Oregon State University, and Prof. Helmut Clemens, Montan Universität Leoben. I appeciate the many beamtimes, especially for the displayed data. At the Australian Nuclear Science and Technology Organisation ANSTO we used the WOMBAT diffractometer under proposal numbers P333 for Fig. 3, DB6022 for Fig. 4b, P6229 for Figs. 6 and 7, and time for Fig. 3. The beamline ID15B at the European Synchrotron Radiation Facility ESRF has been used for access to high-energy X-rays under proposal numbers MA-77 for Fig. 8 and MA-736 for Fig. 9.

References

[1] Liss, K.-D.; Bartels, A.; Schreyer, A.; Clemens, H. High-Energy X-Rays: A Tool for Advanced Bulk Investigations in Materials Science and Physics. *Textures Microstruct.* **2003**, *35*, 219–252. https://doi.org/10.1080/07303300310001634952

[2] Liss, K.-D. Quantum Beam Science—Applications to Probe or Influence Matter and Materials. *Quantum Beam Sci.* **2017**, *1*, 1. https://doi.org/10.3390/qubs1010001.

[3] Sears, V.F. Scattering Lengths for Neutrons. In *International Tables for Crystallography*; Vol. C, Sec. 4.4.4, pp. 444–454.

[4] Thompson, A.; Lindau, I.; Attwood, D.; Liu, Y.; Gullikson, E.; Pianetta, P.; Howells, M.; Robinson, A.; Kim, K.-J.; Scofield, J.; et al. *X-Ray Data Booklet*; 3rd Edition.; Lawrence Berkeley National Laboratory: Berkeley, 2009;

[5] Steuwer, A.; Santisteban, J.R.; Turski, M.; Withers, P.J.; Buslaps, T. High-Resolution Strain Mapping in Bulk Samples Using Full-Profile Analysis of Energy Dispersive Synchrotron X-Ray Diffraction Data. *Nucl. Instrum. Methods Phys. Res. Sect. B Beam Interact. Mater. At.* **2005**, *238*, 200–204. https://doi.org/10.1016/j.nimb.2005.06.049

[6] Tapan Chatterji; Liss, K.-D.; Tschentscher, T. Anomalous Thermal Expansion Due to Magnetism in EuAs3 and MnS2. In *ESRF Highlights 1996/1997*; 1997; p. 19.

[7] Chatterji, T.; Liss, K.D.; Tschentscher, T.; Janossy, B.; Strempfer, J.; Bruckel, T. High-Energy Non-Resonant X-Ray Magnetic Scattering from EuAS(3). *Solid State Commun.* **2004**, *131*, 713–717. https://doi.org/10.1016/j.ssc.2004.06.026

[8] Hubbell, J.H.; Seltzer, S.M. Tables of X-Ray Mass Attenuation Coefficients and Mass Energy-Absorption Coefficients 1 KeV to 20 MeV for Elements Z = 1 to 92 and 48 Additional Substances of Dosimetric Interest. *NIST* **1995**.

[9] Chen, G.; Liss, K.D.; Auchterlonie, G.; Tang, H.; Cao, P. Dehydrogenation and Sintering of TiH2: An In Situ Study. *Metall. Mater. Trans. A* **2017**, 1–11. https://doi.org/10.1007/s11661-017-4043-8.

[10] Liss, K.-D.; Whitfield, R.E.; Xu, W.; Buslaps, T.; Yeoh, L.A.; Wu, X.; Zhang, D.; Xia, K. In Situ Synchrotron High-Energy X-Ray Diffraction Analysis on Phase Transformations in Ti-Al Alloys Processed by Equal-Channel Angular Pressing. *J. Synchrotron Radiat.* **2009**, *16*, 825–834. https://doi.org/10.1107/S090904950903711X

[11] Sears, V.F. Neutron Scattering Lengths and Cross Sections. *Neutron News* **1992**, *3*, 26–37. https://doi.org/10.1080/10448639208218770

[12] Li, X.; Bhattacharyya, D.; Jin, H.; Reid, M.; Dippenaar, R.; Yang, R.; Liss, K.-D. In-Situ Studies of TiAl Polysynthetically Twinned Crystals: Critical Fluctuations and Microstructural Evolution. *J. Alloys Compd.* **2019**, 152454. https://doi.org/10.1016/j.jallcom.2019.152454

[13] Shechtman, D.; Blackburn, M.J.; Lipsitt, H.A. The Plastic Deformation of TiAl. *Metall. Trans.* **1974**, *5*, 1373–1381. https://doi.org/10.1007/BF02646623

[14] Kabra, S.; Yan, K.; Mayer, S.; Schmoelzer, T.; Reid, M.; Dippenaar, R.; Clemens, H.; Liss, K.-D. Phase Transition and Ordering Behavior of Ternary Ti-Al-Mo Alloys Using in-Situ Neutron Diffraction. *Int. J. Mater. Res.* **2011**, *102*, 697–702. https://doi.org/10.3139/146.110528

[15] Chladil, H.F.; Clemens, H.; Zickler, G.A.; Takeyama, M.; Kozeschnik, E.; Bartels, A.; Buslaps, T.; Gerling, R.; Kremmer, S.; Yeoh, L.; et al. Experimental Studies and Thermodynamic Simulation of Phase Transformations in High Nb Containing Gamma-TiAl Based Alloys. *Int. J. Mater. Res.* **2007**, *98*, 1131–1137. https://doi.org/10.3139/146.101569

[16] Xu, S.; Reid, M.; Lin, J.; Liang, Y.; Yang, L.; Zhang, J.; Liss, K.-D. The Crystal Structure and Transformations of the Omicron Phase O in the Ti-Al-Nb System and on the Ambiguity of Its Subvariants O1 and O2. *Scr. Mater.* **2022**, *219*, 114841. https://doi.org/10.1016/j.scriptamat.2022.114841

[17] Shull, R.D.; Cline, J.P. High Temperature X-Ray Diffractometry of Ti-Al Alloys. In *Materials Chemistry at High Temperatures*; Hastie, J.W., Ed.; Humana Press: Totowa, NJ, 1990; pp. 95–117 ISBN 978-1-4612-6781-2.

[18] Chladil, H.F.; Clemens, H.; Leitner, H.; Bartels, A.; Gerling, R.; Schimansky, F.-P.; Kremmer, S. Phase Transformations in High Niobium and Carbon Containing γ-TiAl Based Alloys. *Intermetallics* **2006**, *14*, 1194–1198. https://doi.org/10.1016/j.intermet.2005.11.016

[19] Yeoh, L.A.; Liss, K.-D.; Bartels, A.; Chladil, H.; Avdeev, M.; Clemens, H.; Gerling, R.; Buslaps, T. In Situ High-Energy X-Ray Diffraction Study and Quantitative Phase Analysis in the Alpha plus Gamma Phase Field of Titanium Aluminides. *Scr. Mater.* **2007**, *57*, 1145–1148. https://doi.org/10.1016/j.scriptamat.2007.08.021

[20] Li, X.; Dippenaar, R.; Shiro, A.; Shobu, T.; Higo, Y.; Reid, M.; Suzuki, H.; Akita, K.; Funakoshi, K.-I.; Liss, K.-D. Lattice Parameter Evolution during Heating of Ti-45Al-7.5Nb-0.25/0.5C Alloys under Atmospheric and High Pressures. *Intermetallics* **2018**, *102*, 120–131. https://doi.org/10.1016/j.intermet.2018.08.011

[21] Liss, K.-D.; Funakoshi, K.-I.; Dippenaar, R.J.; Higo, Y.; Shiro, A.; Reid, M.; Suzuki, H.; Shobu, T.; Akita, K. Hydrostatic Compression Behavior and High-Pressure Stabilized β-Phase in γ-Based Titanium Aluminide Intermetallics. *Metals* **2016**, *6*, 165. https://doi.org/10.3390/met6070165

[22] Zhao, W.; Han, J.-K.; Kuzminova, Y.O.; Evlashin, S.A.; Zhilyaev, A.P.; Pesin, A.M.; Jang, J.; Liss, K.-D.; Kawasaki, M. Significance of Grain Refinement on Micro-Mechanical Properties and Structures of Additively-Manufactured CoCrFeNi High-Entropy Alloy. *Mater. Sci. Eng. A* **2021**, *807*, 140898. https://doi.org/10.1016/j.msea.2021.140898

[23] Paul, A.; Liu, X.; Kawasaki, M.; Liss, K.-D. Inverted Magnetic Response in Severe Plastically Deformed Nanostructured High-Entropy Alloy. *Appl. Phys. Lett.* **2023**, *122*, 052402. https://doi.org/10.1063/5.0138040

[24] Kawasaki, M.; Han, J.-K.; Liu, X.; Moon, S.-C.; Liss, K.-D. Synchrotron High-Energy X-Ray & Neutron Diffraction, and Laser-Scanning Confocal Microscopy: In-Situ Characterization Techniques for Bulk Nanocrystalline Metals. *Mater. Trans.* **2023**, *advpub*, MT-MF2022022. https://doi.org/10.2320/matertrans.MT-MF2022022

[25] Yan, K.; Bhattacharyya, D.; Lian, Q.; Kabra, S.; Kawasaki, M.; Carr, D.G.; Callaghan, M.D.; Avdeev, M.; Li, H.; Wang, Y.; et al. Martensitic Phase Transformation and Deformation Behavior of Fe–Mn–C–Al Twinning-Induced Plasticity Steel during High-Pressure Torsion. *Adv. Eng. Mater.* **2014**, *16*, 927–932. https://doi.org/10.1002/adem.201300488

[26] Li, X.; Dippenaar, R.J.; Han, J.-K.; Kawasaki, M.; Liss, K.-D. Phase Transformation and Structure Evolution of a Ti-45Al-7.5Nb Alloy Processed by High-Pressure Torsion. *J. Alloys Compd.* **2019**, *787*, 1149–1157. https://doi.org/10.1016/j.jallcom.2019.02.174

[27] Asta, M.; de Fontaine, D.; van Schilfgaarde, M. First-Principles Study of Phase Stability of Ti–Al Intermetallic Compounds. *J. Mater. Res.* **1993**, *8*, 2554–2568. https://doi.org/10.1557/JMR.1993.2554

[28] Liss, K.-D.; Liu, X.; Li, X.; Han, J.-K.; Dippenaar, R.J.; Kawasaki, M. On the Thermal Evolution of High-Pressure Torsion Processed Titanium Aluminide. *Mater. Lett.* **2021**, *304*, 130650. https://doi.org/10.1016/j.matlet.2021.130650

[29] Kawasaki, M.; Han, H.-K.; Moon, S.-C.; Liss, K.-D. In-Situ Heating Observations on Microstructure Relaxation of Ultrafine-Grained High-Entropy Alloys Using Neutron Diffraction and Laser-Scanning Confocal Microscopy. *Mater. Res. Forum LLC* **2023**.

[30] Liu, X.; Han, J.-K.; Onuki, Y.; Kuzminova, Y.O.; Evlashin, S.A.; Kawasaki, M.; Liss, K.-D. In Situ Neutron Diffraction Investigating Microstructure and Texture Evolution upon Heating of

Nanostructured CoCrFeNi High-Entropy Alloy. *Adv. Eng. Mater.* **2023**, *25*, 2201256. https://doi.org/10.1002/adem.202201256

[31] Nakajima, K.; Kawakita, Y.; Itoh, S.; Abe, J.; Aizawa, K.; Aoki, H.; Endo, H.; Fujita, M.; Funakoshi, K.; Gong, W.; et al. Materials and Life Science Experimental Facility (MLF) at the Japan Proton Accelerator Research Complex II: Neutron Scattering Instruments. *Quantum Beam Sci.* **2017**, *1*, 9. https://doi.org/10.3390/qubs1030009

[32] Liss, K.-D. Materials and Life Science with Quantum Beams at the Japan Proton Accelerator Research Complex. *Quantum Beam Sci.* **2018**, *2*, 10. https://doi.org/10.3390/qubs2020010

[33] Takada, H.; Haga, K.; Teshigawara, M.; Aso, T.; Meigo, S.-I.; Kogawa, H.; Naoe, T.; Wakui, T.; Ooi, M.; Harada, M.; et al. Materials and Life Science Experimental Facility at the Japan Proton Accelerator Research Complex I: Pulsed Spallation Neutron Source. *Quantum Beam Sci.* **2017**, *1*, 8. https://doi.org/10.3390/qubs1020008

[34] Kawasaki, M.; Han, J.-K.; Liu, X.; Onuki, Y.; Kuzminova, Y.O.; Evlashin, S.A.; Pesin, A.M.; Zhilyaev, A.P.; Liss, K.-D. In Situ Heating Neutron and X-Ray Diffraction Analyses for Revealing Structural Evolution during Postprinting Treatments of Additive-Manufactured 316L Stainless Steel. *Adv. Eng. Mater.* **2022**, *24*, 2100968. https://doi.org/10.1002/adem.202100968

[35] Liss, K.-D.; Bartels, A.; Clemens, H.; Bystrzanowski, S.; Stark, A.; Buslaps, T.; Schimansky, F.-P.; Gerling, R.; Scheu, C.; Schreyer, A. Recrystallization and Phase Transitions in a Gamma-TiAl-Based Alloy as Observed by Ex Situ and in Situ High-Energy X-Ray Diffraction. *Acta Mater.* **2006**, *54*, 3721–3735. https://doi.org/10.1016/j.actamat.2006.04.004

[36] Yan, K.; Liss, K.-D.; Garbe, U.; Daniels, J.; Kirstein, O.; Li, H.; Dippenaar, R. From Single Grains to Texture. *Adv. Eng. Mater.* **2009**, *11*, 771–773. https://doi.org/10.1002/adem.200900163

[37] Liss, K.-D.; Garbe, U.; Li, H.; Schambron, T.; Almer, J.D.; Yan, K. In Situ Observation of Dynamic Recrystallization in the Bulk of Zirconium Alloy. *Adv. Eng. Mater.* **2009**, *11*, 637–640. https://doi.org/10.1002/adem.200900094

[38] Liss, K.-D.; Yan, K. Thermo-Mechanical Processing in a Synchrotron Beam. *Mater. Sci. Eng. - Struct. Mater. Prop. Microstruct. Process.* **2010**, *528*, 11–27. https://doi.org/10.1016/j.msea.2010.06.017

[39] Liss, K.-D.; Schmoelzer, T.; Yan, K.; Reid, M.; Peel, M.; Dippenaar, R.; Clemens, H. In Situ Study of Dynamic Recrystallization and Hot Deformation Behavior of a Multiphase Titanium Aluminide Alloy. *J. Appl. Phys.* **2009**, *106*, 113526-113526–6. https://doi.org/doi:10.1063/1.3266177

[40] Liss, K.-D. Thermo-Mechanical Processing in a Synchrotron Beam - from Simple Metals to Multiphase Alloys and Intermetallics. *World J. Eng.* **2010**, *7*, P438,1-4.

[41] Liss, K.D. Thermo-Mechanical Processing in a Synchrotron Beam. *Mater. Sci. Forum* **2012**, *715–716*, 102–102. https://doi.org/10.4028/www.scientific.net/MSF.715-716.102

Materials Research Proceedings **32** (2023) 41-52 | https://doi.org/10.21741/9781644902615-4

Ultra-severe plastic deformation for room-temperature superplasticity and superfunctionality

Kaveh Edalati[1,a*]

[1]WPI International Institute for carbon-Neutral Energy Research (WPI-I2CNER), Kyushu University, Fukuoka, Japan

[a]kaveh.edalati@kyudai.jp

Keywords: Severe Plastic Deformation (SPD), High-Pressure Torsions (HPT), Superplasticity, Strain Rate Sensitivity, Functional Properties

Abstract. Ultra-severe plastic deformation (ulta-SPD) is a terminology used for the introduction of extremely large shear strains (over 1000) to material so that the thickness of sheared phases geometrically reaches the subnanometer level. Under such extreme shearing conditions, new nanostructured phases with unique properties are formed even from the immiscible systems. Various metallic alloys and ceramics were developed by this concept for different applications such as room-temperature superplasticity, room-temperature hydrogen storage, photocatalytic hydrogen production, photocatalytic carbon dioxide conversion, etc. This article reviews recent advances regarding ultra-SPD with a focus on low-temperature superplasticity, which was reported for the first time at room temperature in aluminum and magnesium alloys.

Introduction

Severe plastic deformation (SPD) is a popular technology to introduce large plastic strain in a workpiece [1,2]. There are various SPD methods such as high-pressure torsion (HPT), equal-channel angular pressing (ECAP) and accumulative roll-bonding (ARB) [3,4]. The main target to employ SPD is the generation of ultrafine grains (UFG) [5,6]. Formation of such UFG structures leads to strengthening and enhanced functional properties [7,8]. A review paper written by 49 experts in SPD referred to these SPD-processed UFG materials as superfunctional materials due to their excellent functional properties [9]. SPD has been used since ancient times, but its scientific principles were documented by Bridgman in the 1930s [10,11], Segal et al. in the 1970s [12], and Valiev et al. in the 1980s [13]. For the latest progress in the SPD field, the readers are referred to review papers published in a special issue of Materials Transactions [14].

Ultra-SPD is one of the new directions which is getting attention to synthesize new materials with promising properties [15]. This manuscript reviews recent advances in the application of ultra-SPD with a focus on room-temperature superplasticity, which could be realized for the first time in aluminum and magnesium alloys by ultra-SPD [16,17].

Ultra-Severe Plastic Deformation

The concept of ultra-SPD was suggested in 2017 as an attempt to find a strategy to synthesize new materials [18]. Ultra-SPD is defined as SPD processing in which the shear strain is drastically increased so that the thickness of sheared phases reaches the subnanometer level, as schematically shown in Fig. 1a [15]. The necessary shear strain depends on the size of phases and their co-deformation, but the applied shear strain is at least 1000 [15]. Besides the objective of material synthesis, ultra-SPD has also raised questions regarding the behavior of metallic materials in the steady state, where the strength is supposed to remain constant with increasing shear strain [19].

High-Pressure Torsion for Ultra-Severe Plastic Deformation

Ultra-SPD studies use HPT, a process schematically shown in Fig. 1b. The method, in which a disc [20] or ring [21] is strained under pressure by torsion, has some merits for ultra-SPD. First, shear strain in HPT ($\gamma = 2\pi rN/h$; γ: shear strain, r: radial distance from the rotation center, N: rotation numbers, h: thickness of disc- or ring-shaped specimen) can be increased with almost no limits by increasing the rotation numbers [9-11]. Second, pressure in HPT is high, and thus it is applicable to hard metals (Hf [22], Mo [23], W [24]), glasses [25-27], semiconductors (Si [28,29], Ge [30,31]), oxides (TiO_2 [32], ZrO_2 [33], $BiTaO_3$ [34]), and multi-component alloys (intermetallics [35-37], single-phase high-entropy alloys [38-40], dual-phase high-entropy alloys [41-43]). Third, it is applicable for cold consolidation of powders [44-46], because of high pressure/strain and minor temperature rise [47,48]. Fourth, the method uses the advantages of high pressure/strain to control phase transformations [49-51]. Fifth, the method continuously applies strain, resulting in a high concentration of defects [52-54] and ultra-fast dynamic interdiffusion [55-57].

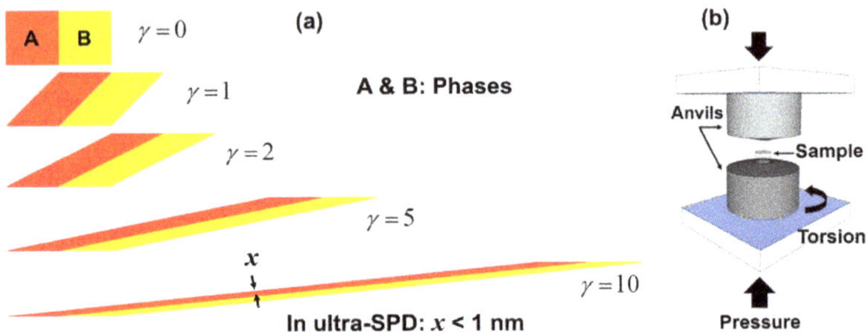

Fig. 1. (a) Schematic illustration of (a) ultra-SPD, and (b) HPT.

Main Findings by Ultra-Severe Plastic Deformation

Ultra-SPD is promising to synthesize a variety of materials including conventional binary/ternary alloys [15], intermetallic compounds [18], high-entropy alloys [18], hydrides [58], and high-entropy ceramics [59]. Such materials show promising properties like high strength, room-temperature superplasticity, hydrogen storage, biocompatibility, superconductivity, thermal stability, and photocatalysis, as reviewed in [60]. A summary of materials processed by ultra-SPD and their characteristics are given in Table 1 [61-88]. These properties are briefly summarized below, but room-temperature superplasticity is discussed in the next section.

SPD-processed materials usually show low thermal stability due to a large fraction of defects [89-93], but the materials processed by ultra-SPD can have high thermal stability. For example, the application of ultra-SPD to produce supersaturated Al-Ca [69], Al-Fe [70], Al-Zr [75], and Al-La-Ce [76] alloys, leads not only to high thermal stability but also to age hardening.

Ultra-SPD is effective in the stabilization of phases that do not exist in the phase diagram such as BCC, HCP and FCC phases in the immiscible Mg-Ti [62], Mg-Zr [63] and Mg-Hf [64] systems.

Table 1. Summary of the application of ultra-SPD to various materials.

Alloy	Shear Strain	Characteristics	Ref.
Mg–Li	7800	Superplasticity at room temperature	[16]
Mg_2X (X: 21 elements)	5500	Storing hydrogen	[61]
Mg–Ti	5500	Storing hydrogen	[62]
Mg–Zr	55000	Storing hydrogen in new phases	[63]
Mg–Hf	3900	Biocompatible new phases	[64]
Mg–V–Cr	50000	Storing hydrogen	[65]
Mg_4NiPd	59000	Storing hydrogen at room temperature	[66]
MgTiVCrFe	12000	Storing hydrogen	[67]
$MgTiH_4$	17000	Storing hydrogen	[68]

Al–Ca	39000	Thermal stability	[69]
Al–Fe	39000	Thermal stability	[70]
AlNi	4700	Enhanced microhardness	[71]
Al_3Ni	4700	Enhanced microhardness	[72,73]
Al–Cu	3900	Quick interdiffusion	[74]
Al–Zn	7800	Superplasticity at room temperature	[17]
Al–Zr	39000	Thermal stability and age hardening	[75]
Al–La–Ce	39000	Thermal stability and age hardening	[76]
TiAl	2000	Enhanced strength/plasticity	[77]
TiV	5500	Storing hydrogen with no activation treatment	[78]
Ti-Nb	5900	Biocompatibility with good strength/elasticity	[79]
TiZrHfNbTa	2000	Biocompatibility with good strength/elasticity	[80]
$TiZHfNbTaO_{11}$	7800	Photocatalytic H_2 evolution and CO_2 conversion	[81,82]
$TiZrHfNbTaO_6N_3$	3900	Photocatalytic H_2 evolution and CO_2 conversion	[83,84]
$TiZrNbTaWO_{12}$	3900	Photocatalytic O_2 evolution	[85]
FeNi	3900	Quick phase transition	[86]
Ni_2AlTi	4700	Enhanced microhardness	[87]
Nb-Ti	3900	Superconductivity	[88]

Intermetallic compounds with nanograin sizes are of interest, but these materials should be produced by two-step procedures like high-temperature sintering of nanipowders [93-96]. However, ultra-SPD is a single-step process to produce bulk nano-intermetallics like Mg_2X (X: other elements) for hydrogen storage [61], AlNi [71] and Al_3Ni [72,73] with high hardness, TiAl with high strength and plasticity [77], $L1_0$-FeNi [86] and Ni_2AlTi [87].

Orthopedic materials should have biocompatibility, high strength and low elastic modulus. Ti-based biomaterials suffer from low strength and high modulus [97], but SPD can improve their strength [98-101]. Ultra-SPD was successfully used to synthesize alloys with a good combination of high strength and low elastic modulus from the Ti-Zr-Hf-Nb-Ta biocompatible system [79,80].

Nanostructured Nb-Ti superconductors are commercially fabricated by several repetitions of wire drawing and long-time annealing [102-105]. Ultra-SPD followed by short annealing is a fast method to produce such superconductors with properties comparable with commercial ones [88].

The design and synthesis of magnesium alloys that can reversibly store hydrogen at room temperature is a long-term challenge [106,107]. Ultra-SPD not only contributed to the synthesis of various hydrogen-storing compounds such as Mg_2X intermetallics (X: other elements) [61], immiscible Mg–Ti [62,68], Mg–Zr [63] and Mg–Hf [66] alloys, ternary Mg–V–Cr alloy [65] and high-entropy MgVTiCrFe alloy [67], but also it introduced Mg4Ni1pd as the first Mg-based material with reversible hydrogen storage performance at room temperature [66].

Ultra-SPD introduced the first high-entropy photocatalysts for hydrogen and oxygen production from water splitting as well as for CO_2 conversion. The high-entropy oxides $TiZrHfNbTaO_{11}$ [81,82] and $TiZrNbTaWO_{12}$ [85] and the high-entropy oxynitride $TiZrHfNbTaO_6N_3$ [83,84] are the first high-entropy photocatalysts with high activity.

Room-Temperature Superplasticity

Superplasticity is defined as the capability of a material for large tensile deformations over 400%. Such elongations can be achieved at homologous temperatures over 0.5, where grain-boundary sliding is dominant. Superplasticity is quantified by the creep equation with a strain rate sensitivity of ~0.5 [108]. The equation suggests two solutions to achieve low-temperature superplasticity: (i) grain size reduction which was used to attain low-temperature superplasticity by SPD processing [109-111]; and (ii) acceleration of grain boundary diffusion by grain boundary engineering, which was used to attain room-temperature superplasticity in Mg-Li and Al-Zn by ultra-SPD [16,17].

The application of ultra-SPD (200 HPT turns) to a two-phase Mg-8wt.%Li alloy (Mg-rich α phase + Li-rich β) resulted in 440% elongation at room temperature (Fig. 2a) with a strain-rate sensitivity of 0.37 (Fig. 2b) [16]. However, the extrusion and SPD process for 5 HPT turns did not lead to such an elongation in agreement with earlier studies [112]. Superplastic deformation after ultra-SPD is promising because room temperature corresponds to a homologous temperature of 0.35 for the alloy. Room-temperature superplasticity occurs due to fast grain-boundary diffusion by increasing the fraction of α/β interphase boundaries and the segregation of lithium along the α/α grain boundaries, as shown in Fig. 2d using high-angle annular dark-field (HAADF) image and in Figs. 2d and 2e using energy-filtered transmission electron microscopy (EFTEM).

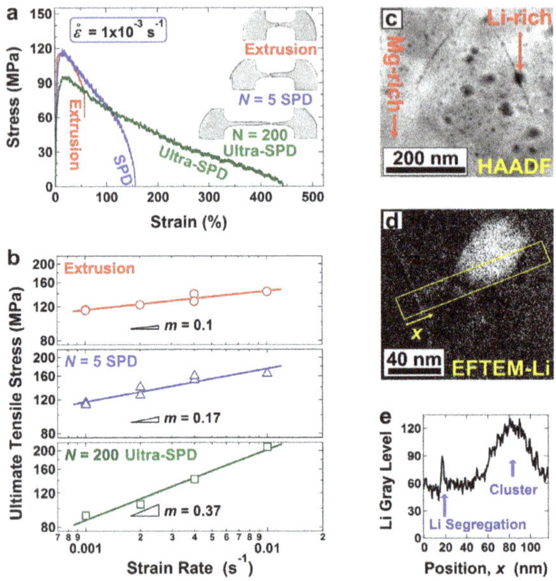

Fig. 2. (a) Stress-strain curves, (b) ultimate tensile strength versus strain rate for Mg-Li alloy processed by extrusion, SPD with 5 HPT turns and ultra-SPD with 200 HPT turns (m: strain-rate sensitivity). (c) HAADF image, (d) EFTEM-Li image and (e) gray level versus position shown in (b) for Mg-Li after ultra-SPD [16].

Ultra-SPD (200 HPT turns) was applied to an Al-30at%Zn alloy with two phases of Al-rich α and Zn-rich η [17]. The tensile behavior of the material was examined after ultra-SPD and after 100 days of aging at room temperature. As shown in Fig. 3a, the homogenized sample shows limited plasticity, but elongation is 480% after ultra-SPD and 330% after aging. It should be noted that this alloy did not show room-temperature superplasticity after SPD using 20 HPT turns [113]. The strain-rate sensitivity for ultra-SPD-processed samples reached a high value of 0.41, as shown in Fig. 3b. Superplasticity after ultra-SPD at room temperature (i.e. the homologous temperature of 0.36) is due to fast grain-boundary diffusion by increasing the fraction of α/η. interphase boundaries and the segregation of zinc along the α/α grain boundaries, as shown in Fig. 3c and 3d using HAADF image and in Figs. 3e and 3f using energy-dispersive X-ray spectroscopy (EDS).

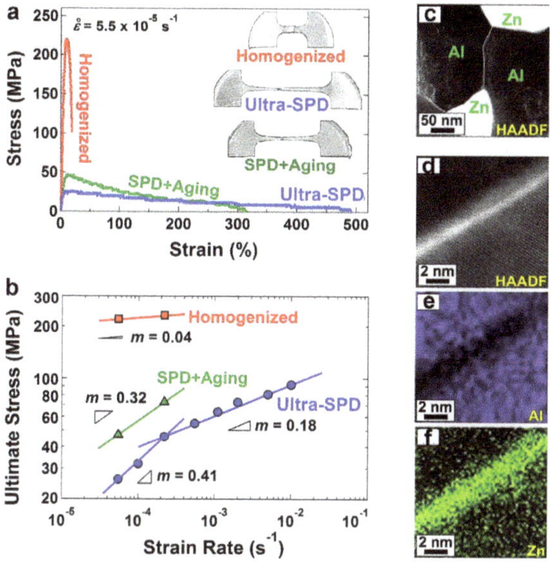

Fig. 3. (a) Stress-strain curves, (b) ultimate tensile strength versus strain rate for Mg-Li alloy processed by homogenization, ultra-SPD with 200 HPT turns and ultra-SPD followed by 100 day aging (m: strain-rate sensitivity). (c, d) HAADF images, and (e, f) EDS elemental mappings of region in

Summary
The concept of ultra-SPD is effective to develop various superfunctional materials for energy, electrical and biomedical applications. One of the interesting results achieved by ultra-SPD is the development of first room-temperature-superplastic aluminum and magnesium alloys.

References

[1] R.Z. Valiev, R.K. Islamgaliev, I.V. Alexandrov, Bulk nanostructured materials from severe plastic deformation, Prog. Mater. Sci. 45 (2000) 103-189. https://doi.org/10.1016/S0079-6425(99)00007-9

[2] R.Z. Valiev, Y. Estrin, Z. Horita, T.G. Langdon, M.J. Zehetbauer, Y.T. Zhu, Producing bulk ultrafine-grained materials by severe plastic deformation, JOM 58(4) (2006) 33-39. https://doi.org/10.1007/s11837-006-0213-7

[3] A. Azushima, R. Kopp, A. Korhonen, D.Y. Yang, F. Micari, G.D. Lahoti, P. Groche, J. Yanagimoto, N. Tsuji, A. Rosochowski, A. Yanagida, Severe plastic deformation (SPD) processes for metals, CRIP Ann. Mauuf. Technol. 57 (2008) 716-735. https://doi.org/10.1016/j.cirp.2008.09.005

[4] V. Segal, Review: modes and processes of severe plastic deformation (SPD), Materials 11 (2018) 1175. https://doi.org/10.3390/ma11071175

[5] R. Pippan, S. Scheriau, A. Taylor, M. Hafok, A. Hohenwarter, A. Bachmaier, Saturation of fragmentation during severe plastic deformation, Annu. Rev. Mater. Res. 40 (2010) 319-343. https://doi.org/10.1146/annurev-matsci-070909-104445

[6] M.J. Starink, X.C. Cheng, S. Yang, Hardening of pure metals by high-pressure torsion: A physically based model employing volume-averaged defect evolutions, Acta Mater. 61 (2013) 183-192. https://doi.org/10.1016/j.actamat.2012.09.048

[7] M. Zehetbauer, R. Grossinger, H. Krenn, M. Krystian, R. Pippan, P. Rogl, T. Waitz, R. Wurschum, Bulk nanostructured functional materials by severe plastic deformation, Adv. Eng. Mater. 12 (2010) 692-700. https://doi.org/10.1002/adem.201000119

[8] Y. Estrin, A. Vinogradov, Extreme grain refinement by severe plastic deformation: a wealth of challenging science, Acta Mater. 61 (2013) 782-817. https://doi.org/10.1016/j.actamat.2012.10.038
[9] K. Edalati, A. Bachmaier, V.A. Beloshenko, Y. Beygelzimer, V.D. Blank, W.J. Botta, K. Bryła, J. Čížek, S. Divinski, N.A. Enikeev, Y. Estrin, G. Faraji, R.B. Figueiredo, M. Fuji, T. Furuta, T. Grosdidier, J. Gubicza, A. Hohenwarter, Z. Horita, J. Huot, Y. Ikoma, M. Janeček, M. Kawasaki, P. Krăl, S. Kuramoto, T.G. Langdon, D.R. Leiva, V.I. Levitas, A. Mazilkin, M. Mito, H. Miyamoto, T. Nishizaki, R. Pippan, V.V. Popov, E.N. Popova, G. Purcek, O. Renk, Á. Révész, X. Sauvage, V. Sklenicka, W. Skrotzki, B.B. Straumal, S. Suwas, L.S. Toth, N. Tsuji, R.Z. Valiev, G. Wilde, M.J. Zehetbauer, X. Zhu, Nanomaterials by severe plastic deformation: review of historical developments and recent advances, Mater. Res. Lett. 10 (2022) 163-256. https://doi.org/10.1080/21663831.2022.2029779
[10] P.W. Bridgman, Effects of high shearing stress combined with high hydrostatic pressure, Phys. Rev. 48 (1935) 825-847. https://doi.org/10.1103/PhysRev.48.825
[11] K. Edalati, Z. Horita, A review on high-pressure torsion (HPT) from 1935 to 1988, Mater. Sci. Eng. A 652 (2016) 325-352. https://doi.org/10.1016/j.msea.2015.11.074
[12] V.M. Segal, V.I. Reznikov, A.E. Drobyshevskiy, V.I. Kopylov, Plastic working of metals by simple shear, Russ. Metall. 1 (1981) 99-105.
[13] R.Z. Valiev, O.A. Kaibyshev, R.I. Kuznetsov, R.S. Musalimov, N.K. Tsenev, Low-temperature superplasticity of metallic materials, Dokl. Akad. Nauk. SSSR 301 (1988) 864-866.
[14] K. Edalati, Z. Horita, Special issue on severe plastic deformation for nanomaterials with advanced functionality, Mater. Trans. 60 (2019) 1103. https://doi.org/10.2320/matertrans.MPR2019904
[15] K. Edalati, T. Masuda, M. Arita, M. Furui, X. Sauvage, Z. Horita, R.Z. Valiev, Room-temperature superplasticity in an ultrafine-grained magnesium alloy, Sci. Rep. 7 (2017) 2662. https://doi.org/10.1038/s41598-017-02846-2
[16] K. Edalati, Z. Horita, R.Z. Valiev, Transition from poor ductility to room-temperature superplasticity in a nanostructured aluminum alloy, Sci. Rep. 8 (2018) 6740. https://doi.org/10.1038/s41598-018-25140-1
[17] K. Edalati, Metallurgical alchemy by ultra-severe plastic deformation via high-pressure torsion process, Mater. Trans. 60 (2019) 1221-1229. https://doi.org/10.2320/matertrans.MF201914
[18] K. Edalati, R. Uehiro, K. Fujiwara, Y. Ikeda, H.W. Li, X. Sauvage, R.Z. Valiev, E. Akiba, I. Tanaka, Z. Horita, Ultra-severe plastic deformation: evolution of microstructure, phase transformation and hardness in immiscible magnesium-based systems, Mater. Sci. Eng. A 701 (2017) 158-166. https://doi.org/10.1016/j.msea.2017.06.076
[19] J.G. Sevillano, Dynamic steady state by unlimited unidirectional plastic deformation of crystalline materials deforming by dislocation glide at low to moderate temperatures, Metals 10 (2020) 66. https://doi.org/10.3390/met10010066
[20] A.P. Zhilyae, T.G. Langdon, Using high-pressure torsion for metal processing: fundamentals and applications, Prog. Mater. Sci. 53 (2008) 893-979. https://doi.org/10.1016/j.pmatsci.2008.03.002
[21] K. Edalati, Z. Horita, Scaling-up of high pressure torsion using ring shape, Mater. Trans. 50 (2009) 92-95. https://doi.org/10.2320/matertrans.MD200822
[22] K. Edalati, Z. Horita, Y. Mine, High-pressure torsion of hafnium, Mater. Sci. Eng. A 527 (2010) 2136-2141. https://doi.org/10.1016/j.msea.2009.11.060

[23] S. Lee, K. Edalati, Z. Horita, Microstructures and mechanical properties of pure V and Mo processed by high-pressure torsion, Mater. Trans. 51 (2010) 1072-1079. https://doi.org/10.2320/matertrans.M2009375
[24] L. Chen, L. Ping, T. Ye, L. Lingfeng, X. Kemin, Z. Meng, Observations on the ductility and thermostability of tungsten processed from micropowder by improved high-pressure torsion, Rare Met. Mater. Eng. 45 (2016) 3089-3094. https://doi.org/10.1016/S1875-5372(17)30059-0
[25] K. Edalati, Y. Yokoyama, Z. Horita, High-pressure torsion of machining chips and bulk discs of amorphous Zr50Cu30Al10Ni10, Mater. Trans. 51 (2010) 23-26. https://doi.org/10.2320/matertrans.MB200914
[26] Y.B. Wang, D.D. Qu, X.H. Wang, Y. Cao, X.Z. Liao, M. Kawasaki, S.P. Ringer, Z.W. Shan, T.G. Langdon, J. Shen, Introducing a strain-hardening capability to improve the ductility of bulk metallic glasses via severe plastic deformation, Acta Mater. 60 (2012) 253-260. https://doi.org/10.1016/j.actamat.2011.09.026
[27] Á. Révész, Z. Kovács, Severe plastic deformation of amorphous alloys, Mater. Trans. 60 (2019) 1283-1293. https://doi.org/10.2320/matertrans.MF201917
[28] V.D. Blank, M.Y. Popov, B.A. Kulnitskiy, The effect of severe plastic deformations on phase transitions and structure of solids, Mater. Trans. 60 (2019) 1500-1505. https://doi.org/10.2320/matertrans.MF201942
[29] Y. Ikoma, K. Hayano, K. Edalati, K. Saito, Q. Guo, Z. Horita, Phase transformation and nanograin refinement of silicon by processing through high-pressure torsion, Appl. Phys. Lett. 101 (2012) 121908. https://doi.org/10.1063/1.4754574
[30] K. Edalati, Z. Horita, Correlations between hardness and atomic bond parameters of pure metals and semi-metals after processing by high-pressure torsion, Scr. Mater. 64 (2011) 161-164. https://doi.org/10.1016/j.scriptamat.2010.09.034
[31] Y. Ikoma, Severe plastic deformation of semiconductor materials using high-pressure torsion, Mater. Trans. 60 (2019) 1168-1176. https://doi.org/10.2320/matertrans.MF201907
[32] H. Razavi-Khosroshahi, K. Edalati, M. Arita, Z. Horita, M. Fuji, Plastic strain and grain size effect on high-pressure phase transformations in nanostructured TiO2 ceramics, Scr. Mater. 124 (2016) 59-62. https://doi.org/10.1016/j.scriptamat.2016.06.022
[33] Q. Wang, K. Edalati, Y. Koganemaru, S. Nakamura, M. Watanabe, T. Ishihara, Z. Horita, Photocatalytic hydrogen generation on low-bandgap black zirconia (ZrO2) produced by high-pressure torsion, J. Mater. Chem. A 8 (2020) 3643-3650. https://doi.org/10.1039/C9TA11839J
[34] K. Edalati, M. Arimura, Y. Ikoma, T. Daio, M. Miyata, D.J. Smith, Z. Horita, Plastic deformation of BaTiO3 ceramics by high-pressure torsion and changes in phase transformations, optical and dielectric properties, Mater. Res. Lett. 3 (2015) 216-221. https://doi.org/10.1080/21663831.2015.1065454
[35] J.Y. Huang, Y.T. Zhu, X.Z. Liao, R.Z. Valiev, Amorphization of TiNi induced by high-pressure torsion, Phil. Mag. Lett. 84 (2004) 183-190. https://doi.org/10.1080/09500830310001657353
[36] C. Gammer, C. Mangler, H.P. Karnthaler, C. Rentenberger, Growth of nanosized chemically ordered domains in intermetallic FeAl made nanocrystalline by severe plastic deformation, Scr. Mater. 65 (2011) 57-60. https://doi.org/10.1016/j.scriptamat.2011.03.002
[37] D.G. Morris, M.A. Muñoz-Morris, Microstructural refinement in alloys and intermetallics by severe plastic deformation, J. Alloys. Compd. 536 (2012) S180-S185. https://doi.org/10.1016/j.jallcom.2011.10.069

[38] A. Kilmametov, R. Kulagin, A. Mazilkin, S. Seils, T. Boll, M. Heilmaier, H. Hahn, High-pressure torsion driven mechanical alloying of CoCrFeMnNi high entropy alloy, Scr. Mater. 158 (2019) 29-33. https://doi.org/10.1016/j.scriptamat.2018.08.031
[39] J. Gubicza, A. Heczel, M. Kawasaki, J.K. Han, Y. Zhao, Y. Xue, S. Huang, J.L. Lábár, Evolution of microstructure and hardness in Hf25Nb25Ti25Zr25 high-entropy alloy during high-pressure torsion, J. Alloys Compd. 788 (2019) 318-328. https://doi.org/10.1016/j.jallcom.2019.02.220
[40] W. Skrotzki, A. Pukenas, E. Odor, B. Joni, T. Ungar, B. Völker, A. Hohenwarter, R. Pippan, E.P. George, Microstructure, texture, and strength development during high-pressure torsion of CrMnFeCoNi high-entropy alloy, Crystals 10 (2020) 336. https://doi.org/10.3390/cryst10040336
[41] P. Edalati, R. Floriano, Y. Tang, A. Mohammadi, K. Danielle Pereira, A. Ducati Luchessi, K. Edalati, Ultrahigh hardness and biocompatibility of high-entropy alloy TiAlFeCoNi processed by high-pressure torsion, Mater. Sci. Eng. C 112 (2020) 110908. https://doi.org/10.1016/j.msec.2020.110908
[42] P. Edalati, A. Mohammadi, M. Ketabchi, K. Edalati, Ultrahigh hardness in nanostructured dual-phase high-entropy alloy AlCrFeCoNiNb developed by high-pressure torsion, J. Alloys Compd. 884 (2021) 161101. https://doi.org/10.1016/j.jallcom.2021.161101
[43] P. Edalati, A. Mohammadi, M. Ketabchi, K. Edalati, Microstructure and microhardness of dual-phase high-entropy alloy by high-pressure torsion: twins and stacking faults in FCC and dislocations in BCC, J. Alloys Compd. 894 (2022) 162413. https://doi.org/10.1016/j.jallcom.2021.162413
[44] A.P. Zhilyaev, G. Ringot, Y. Huang, J.M. Cabrera, T.G. Langdon, Mechanical behavior and microstructure properties of titanium powder consolidated by high-pressure torsion, Mater. Sci. Eng. A 688 (2017) 498-504. https://doi.org/10.1016/j.msea.2017.02.032
[45] K. Edalati, Z. Horita, H. Fujiwara, K. Ameyama, Cold consolidation of ball-milled titanium powders using high-pressure torsion, Metall. Mater. Trans. A 41 (2010) 3308-3317. https://doi.org/10.1007/s11661-010-0400-6
[46] S. Panda, J.J. Fundenberger, Y. Zhao, J. Zou, L.S. Toth, T. Grosdidier, Effect of initial powder type on the hydrogen storage properties of high-pressure torsion consolidated Mg, Int. J. Hydrog. Energy 42 (2017) 22438-22448. https://doi.org/10.1016/j.ijhydene.2017.05.097
[47] R.B. Figueiredo, P.H.R. Pereira, M.T.P. Aguilar, P.R. Cetlin, T.G. Langdon, Using finite element modeling to examine the temperature distribution in quasi-constrained high-pressure torsion. Acta Mater. 60 (2012) 3190-3198. https://doi.org/10.1016/j.actamat.2012.02.027
[48] K. Edalati, Y. Hashiguchi, P.H.R. Pereira, Z. Horita, T.G. Langdon, Effect of temperature rise on microstructural evolution during high-pressure torsion, Mater. Sci. Eng. A 714 (2018) 167-171. https://doi.org/10.1016/j.msea.2017.12.095
[49] K. Edalati, S. Toh, M. Arita, M. Watanabe, Z. Horita, High-pressure torsion of pure cobalt: hcp-fcc phase transformations and twinning during severe plastic deformation, Appl. Phys. Lett. 102 (2013) 181902. https://doi.org/10.1063/1.4804273
[50] V.I. Levita,: High-pressure phase transformations under severe plastic deformation by torsion in rotational anvils, Mater. Trans. 60 (2019) 1294-1301. https://doi.org/10.2320/matertrans.MF201923
[51] A. Mazilkin, B. Straumal, A. Kilmametov, P. Straumal, B. Baretzky, Phase transformations induced by severe plastic deformation, Mater. Trans. 60 (2019) 1489-1499. https://doi.org/10.2320/matertrans.MF201938
[52] X. Sauvage, A. Duchaussoy, G. Zaher, Strain induced segregations in severely deformed materials, Mater. Trans. 60 (2019) 1151-1158. https://doi.org/10.2320/matertrans.MF201919

[53] J. Gubicza, Lattice defects and their influence on the mechanical properties of bulk materials processed by severe plastic deformation, Mater. Trans. 60 (2019) 1230-1242. https://doi.org/10.2320/matertrans.MF201909
[54] J. Čížek, M. Janeček, T. Vlasák, B. Smola, O. Melikhova, R.K. Islamgaliev, S.V. Dobatkin, The development of vacancies during severe plastic deformation, Mater. Trans. 60 (2019) 1533-1542. https://doi.org/10.2320/matertrans.MF201937
[55] G. Wilde, S. Divinski, Grain boundaries and diffusion phenomena in severely deformed materials, Mater. Trans. 60 (2019) 1302-1315. https://doi.org/10.2320/matertrans.MF201934
[56] A. Bachmaier, R. Pippan, High-pressure torsion deformation induced phase transformations and formations: new material combinations and advanced properties, Mater. Trans. 60 (2019) 1256-1269. https://doi.org/10.2320/matertrans.MF201930
[57] J.K. Han, J.I. Jang, T.G. Langdon, M. Kawasaki, Bulk-state reactions and improving the mechanical properties of metals through high-pressure torsion, Mater. Trans. 60 (2019) 1131-1138. https://doi.org/10.2320/matertrans.MF201908
[58] K. Edalati, E. Akiba, Z. Horita, High-pressure torsion for new hydrogen storage materials, Sci. Technol. Adv. Mater. 19 (2018) 185-193. https://doi.org/10.1080/14686996.2018.1435131
[59] S. Akrami, P. Edalati, M. Fuji, K. Edalati, High-entropy ceramics: review of principles, production and applications, Mater. Sci. Eng. R 146 (2021) 100644. https://doi.org/10.1016/j.mser.2021.100644
[60] K. Edalati, Superfunctional materials by ultra-severe plastic deformation, Materials 16 (2023) 587. https://doi.org/10.3390/ma16020587
[61] H. Emami, K. Edalati, A. Staykov, T. Hongo, H. Iwaoka, Z. Horita, E. Akiba, Solid-state reactions and hydrogen storage in magnesium mixed with various elements by high-pressure torsion: experiments and first-principles calculations, RCS Adv. 6 (2016) 11665-11674. https://doi.org/10.1039/C5RA23728A
[62] K. Edalati, H. Emami, A. Staykov, D.J. Smith, E. Akiba, Z. Horita, Formation of metastable phases in magnesium-titanium system by high-pressure torsion and their hydrogen storage performance, Acta Mater. 50 (2015) 150-156. https://doi.org/10.1016/j.actamat.2015.07.060
[63] K. Edalati, H. Emami, Y. Ikeda, H. Iwaoka, I. Tanaka, E. Akiba, Z. Horita, New nanostructured phases with reversible hydrogen storage capability in immiscible magnesium-zirconium system produced by high-pressure torsion, Acta Mater. 108 (2016) 293-303. https://doi.org/10.1016/j.actamat.2016.02.026
[64] E.I. Lopez-Gomez, K. Edalati, D.D. Coimbroo, F.J. Antiqueira, G. Zepon, J.M. Cubero-Sesin, W.J. Botta, FCC phase formation in immiscible Mg-Hf (magnesium-hafnium) system by high-pressure torsion, AIP Adv. 10 (2020) 055222. https://doi.org/10.1063/5.0009456
[65] K. Fujiwara, R. Uehiro, K. Edalati, H.W. Li, R. Floriano, E. Akiba, Z. Horita, New Mg-V-Cr BCC alloys synthesized by high-pressure torsion and ball milling, Mater. Trans. 59 (2018) 741-746. https://doi.org/10.2320/matertrans.M2018001
[66] K. Edalati, R. Uehiro, Y. Ikeda, H.W. Li, H. Emami, Y. Filinchuk, M. Arita, X. Sauvage, I. Tanaka, E. Akiba, Z. Horita, Design and synthesis of a magnesium alloy for room temperature hydrogen storage, Acta Mater. 149 (2018) 88-96. https://doi.org/10.1016/j.actamat.2018.02.033
[67] M.O. de Marco, Y. Li, H.W. Li, K. Edalati, R. Floriano, Mechanical synthesis and hydrogen storage characterization of MgVCr and MgVTiCrFe high-entropy alloy, Adv. Eng. Mater. 22 (2020) 1901079. https://doi.org/10.1002/adem.201901079
[68] K. Kitabayashi, K. Edalati, H.W. Li, E. Akiba, Z. Horita, Phase transformations in MgH2-TiH2 hydrogen storage system by high-pressure torsion process, Adv. Eng. Mater. 22 (2020) 1900027. https://doi.org/10.1002/adem.201900027

[69] X. Sauvage, F. Cuvilly, A. Russell, K. Edalati, Understanding the role of Ca segregation on thermal stability, electrical resistivity and mechanical strength of nanostructured aluminum, Mater. Sci. Eng. A 798 (2020) 140108. https://doi.org/10.1016/j.msea.2020.140108
[70] A. Duchaussoy, X. Sauvage, K. Edalati, Z. Horita, G. Renou, A. Deschamps, F. De Geuser, Structure and mechanical behavior of ultrafine-grained aluminum-iron alloy stabilized by nanoscaled intermetallic particles, Acta Mater. 167 (2019) 89-102. https://doi.org/10.1016/j.actamat.2019.01.027
[71] K. Edalati, S. Toh, M. Watanabe, Z. Horita, In-situ production of bulk intermetallic-based nanocompsites and nanostructured intermetallics by high-pressure torsion, Scr. Mater. 66 (2012) 386-389. https://doi.org/10.1016/j.scriptamat.2011.11.039
[72] A. Alhamidi, K. Edalati, Z. Horita, Production of nanograined intermetallics using high-pressure torsion, Mater. Res. 16 (2013) 672-678. https://doi.org/10.1590/S1516-14392013005000057
[73] A. Alhamidi, K. Edalati, Z. Horita, Effect of temperature on solid-state formation of bulk nanograined intermetallics during high-pressure torsion, Philos. Mag. 94 (2014) 876-887. https://doi.org/10.1080/14786435.2013.868945
[74] K. Oh-ishi, K. Edalati, H.S. Kim, K. Hono, Z. Horita, High-pressure torsion for enhanced atomic diffusion and promoting solid-state reactions in aluminum-copper system, Acta Mater. 61 (2013) 3482-3489. https://doi.org/10.1016/j.actamat.2013.02.042
[75] A. Mohammadi, N.A. Enikeev, M.Y. Murashkin, M. Arita, K. Edalati, Developing age-hardenable Al-Zr alloy by ultra-severe plastic deformation: significance of supersaturation, segregation and precipitation on hardening and electrical conductivity, Acta Mater. 203 (2021) 116503. https://doi.org/10.1016/j.actamat.2020.116503
[76] A. Mohammadi, N.A. Enikeev, M.Y. Murashkin, M. Arita, K. Edalati, Examination of inverse Hall-Petch relation in nanostructured aluminum alloys by ultra-severe plastic deformation, J. Mater. Sci. Technol. 91 (2021) 78-89. https://doi.org/10.1016/j.jmst.2021.01.096
[77] K. Edalati, S. Toh, H. Iwaoka, M. Watanabe, Z. Horita, D. Kashioka, K. Kishida, H. Inui, Ultrahigh strength and high plasticity in TiAl intermetallics with bimodal grain structure and nanotwins, Scr. Mater. 67 (2012) 814-817. https://doi.org/10.1016/j.scriptamat.2012.07.030
[78] K. Edalati, H. Shao, H. Emami, H. Iwaoka, Z. Horita, E. Akiba, Activation of titanium-vanadium alloy for hydrogen storage by introduction of nanograins and edge dislocations using high-pressure torsion, Int. J. Hydrog. Energy 41 (2016) 8917-8924. https://doi.org/10.1016/j.ijhydene.2016.03.146
[79] A. Campos-Quirós, J.M. Cubero-Sesín, K. Edalati, Synthesis of nanostructured biomaterials by high-pressure torsion: effect of niobium content on microstructure and mechanical properties of Ti-Nb alloys, Mater. Sci. Eng. A 795 (2020) 139972. https://doi.org/10.1016/j.msea.2020.139972
[80] J. González-Masís, J.M. Cubero-Sesin, A. Campos-Quirós, K. Edalati, Synthesis of biocompatible high-entropy alloy TiNbZrTaHf by high-pressure torsion, Mater. Sci. Eng. A 825 (2021) 141869. https://doi.org/10.1016/j.msea.2021.141869
[81] P. Edalati, Q. Wang, H. Razavi-Khosroshahi, M. Fuji, T. Ishihara, K. Edalati, Photocatalytic hydrogen evolution on a high-entropy oxide, J. Mater. Chem. A 8 (2020) 3814-3821. https://doi.org/10.1039/C9TA12846H
[82] S. Akrami, Y. Murakami, M. Watanabe, T. Ishihara, M. Arita, M. Fuji, K. Edalati, Defective high-entropy oxide photocatalyst with high activity for CO2 conversion, Appl. Catal. B 303 (2022) 120896. https://doi.org/10.1016/j.apcatb.2021.120896

[83] P. Edalati, X.F. Shen, M. Watanabe, T. Ishihara, M. Arita, M. Fuji, K. Edalati, High-entropy oxynitride as low-bandgap and stable photocatalyst for hydrogen production, J. Mater. Chem. A 9 (2021) 15076-15086. https://doi.org/10.1039/D1TA03861C
[84] S. Akrami, P. Edalati, Y. Shundo, M. Watanabe, T. Ishihara, M. Fuji, K. Edalati, Significant CO2 photoreduction on a high-entropy oxynitride, Chem. Eng. J. 449 (2022) 137800. https://doi.org/10.1016/j.cej.2022.137800
[85] P. Edalati, Y. Itagoe, H. Ishihara, T. Ishihara, H. Emami, M. Arita, M. Fuji, K. Edalati, Visible-light photocatalytic oxygen production on a high-entropy oxide with multiple-heterojunction introduction, J. Photochem. Photobiol. A 433 (2022) 114167. https://doi.org/10.1016/j.jphotochem.2022.114167
[86] S. Lee, K. Edalati, H. Iwaoka, Z. Horita, T. Ohtsuki, T. Ohkochi, M. Kotsugi, T. Kojima, M. Mizuguchi, K. Takanashi, Formation of FeNi with L10-ordered structure using high-pressure torsion, Philos. Mag. Lett. 94 (2014) 639-646. https://doi.org/10.1080/09500839.2014.955546
[87] K. Edalati, T. Daio, Z. Horita, K. Kishida, H. Inui, Evolution of lattice defects, disordered/ordered phase transformations and mechanical properties in Ni-Al-Ti intermetallics by high-pressure torsion, J. Alloys Compd. 563 (2013) 221-228. https://doi.org/10.1016/j.jallcom.2013.02.128
[88] K. Edalati, T. Daio, S. Lee, Z. Horita, T. Nishizaki, T. Akune, T. Nojima, T. Sasaki, High strength and superconductivity in nanostructured niobium-titanium alloy by high-pressure torsion and annealing: significance of elemental decomposition and supersaturation, Acta Mater. 80 (2014) 149-158. https://doi.org/10.1016/j.actamat.2014.07.065
[89] J. Cubicza, N.Q. Chinh, J.L. Labar, Z. Hegedus, T.G. Langdon, Principles of self-annealing in silver processed by equal-channel angular pressing: the significance of a very low stacking fault energy, Mater. Sci. Eng. A 527 (2010) 752-760. https://doi.org/10.1016/j.msea.2009.08.071
[90] V.V. Popov, E.N. Popova, D.D. Kuznetsov, A.V. Stolbovskii, V.P. Pilyugin, Thermal stability of nickel structure obtained by high-pressure torsion in liquid nitrogen, Phys. Met. Metallogr. 115 (2014) 682-691. https://doi.org/10.1134/S0031918X14070060
[91] Y. Huang, S. Sabbaghianrad, A.I. Almazrouee, K.J. Al-Fadhalah, S.N. Alhajeri, T.G. Langdon, The significance of self-annealing at room temperature in high purity copper processed by high-pressure torsion, Mater. Sci. Eng. A 656 (2016) 55-66. https://doi.org/10.1016/j.msea.2016.01.027
[92] J.A. Muñoz, O.F. Higuera, A.H. Expósito, A. Boulaajaj, R.E. Bolmaro, F.D. Dumitru, P.R. Calvillo, A.M. Jorge Jr, J.M. Cabrera, Thermal stability of ARMCO iron processed by ECAP, Int. J. Adv. Manufact. Technol. 98 (2018) 2917-2932. https://doi.org/10.1007/s00170-018-2353-7
[93] H. Gleiter, Nanocrystalline materials, Prog. Mater. Sci. 33 (1989) 223-315. https://doi.org/10.1016/0079-6425(89)90001-7
[94] J.A. Haber, J.L. Crane, W.E. Buhro, C.A. Frey, S.M.L. Sastry, J.J. Balbach, M.S. Conradi, Chemical synthesis of nanocrystalline titanium and nickel aluminides from the metal chlorides and lithium aluminum hydride, Adv. Mater. 8 (1996) 163-166. https://doi.org/10.1002/adma.19960080215
[95] Y.B. Pithawalla, M.S. El-Shall, S.C. Deevi, Synthesis and characterization of nanocrystalline iron aluminide particles, Intermetallics 8 (2000) 1225-1231. https://doi.org/10.1016/S0966-9795(00)00076-5
[96] S. Suryanarayana, Mechanical alloying and milling, Prog. Mater. Sci. 46 (2001) 1-184. https://doi.org/10.1016/S0079-6425(99)00010-9
[97] M. Long, H.J. Rack, Titanium alloys in total joint replacement - a materials science perspective, Biomaterials. 19 (1998) 1621-1639. https://doi.org/10.1016/S0142-9612(97)00146-4

[98] R. Valiev, I.P. Semenova, E. Jakushina, V.V. Latysh, H.J. Rack, T.C. Lowe, J. Petruželka, L. Dluhoš, D. Hrušák, J. Sochová, Nanostructured SPD processed titanium for medical implants, Mater. Sci. Forum 584-586 (2008) 49-54. https://doi.org/10.4028/www.scientific.net/MSF.584-586.49
[99] Y. Estrin, H.E. Kim, R. Lapovok, H.P. Ng, J.H. Jo, Mechanical strength and biocompatibility of ultrafine-grained commercial purity titanium, BioMed Res. Int. 2013 (2013) 914764. https://doi.org/10.1155/2013/914764
[100] A.P. Zhilyaev, Y. Huang, J.M. Cabrera, T.G. Langdon, Influence of inhomogeneity on mechanical properties of commercially pure titanium processed by HPT, Defect Diffus. Forum 385 (2018) 284-289. https://doi.org/10.4028/www.scientific.net/DDF.385.284
[101] T.C. Lowe, R.Z. Valiev, X. Li, B.R. Ewing, Commercialization of bulk nanostructured metals and alloys, MRS Bull. 46 (2021) 265-272. https://doi.org/10.1557/s43577-021-00060-0
[102] C. Meingast, D.C. Larbalestier, Quantitative description of a very high critical current density Nb-Ti superconductor during its final optimization strain. II. Flux pinning mechanisms, J. Appl. Phys. 66 (1989) 5971-5983. https://doi.org/10.1063/1.343625
[103] L.D. Cooley, P.D. Jablonski, P.J. Lee, D.C. Larbalestier, Strongly enhanced critical current density in Nb 47 wt.% Ti having a highly aligned microstructure, Appl. Phys. Lett. 58 (1991) 2984-2986. https://doi.org/10.1063/1.104689
[104] R.W. Heussner, P.D. Jablonski, P.J. Lee, D.C. Larbalestier, Properties of rod-based artificial pinning center Nb-Ti superconductors, IEEE Trans. Appl. Supercon. 5 (1995) 1705-1708. https://doi.org/10.1109/77.402905
[105] V.A. Beloshenko, V.V. Chishko, Deformation-heat treatment of Nb-Ti superconductors using severe plastic deformation methods, Phys. Met. Metallogr. 114 (2013) 992-1002. https://doi.org/10.1134/S0031918X13090032
[106] I.P. Jain, C. Lal, A. Jain, Hydrogen storage in Mg: a most promising material, Int. J. Hydrogen Energy 35 (2010) 5133-5144. https://doi.org/10.1016/j.ijhydene.2009.08.088
[107] Á. Révész, M. Gajdics, High-pressure torsion of non-equilibrium hydrogen storage materials: a review, Energies 14 (2021) 819. https://doi.org/10.3390/en14040819
[108] T.G. Langdon, Seventy-five years of superplasticity: historic developments and new opportunities, J. Mater. Sci. 44 (2009) 5998-6010. https://doi.org/10.1007/s10853-009-3780-5
[109] R.Z. Valiev, M.Y. Murashkin, A. Kilmametov, B. Straumal, N.Q. Chinh, T.G. Langdon, Unusual super-ductility at room temperature in an ultrafine-grained aluminum alloy, J. Mater. Sci. 45 (2010) 4718-4724. https://doi.org/10.1007/s10853-010-4588-z
[110] M. Kawasaki, T.G. Langdon, The contribution of severe plastic deformation to research on superplasticity, Mater. Trans. 60 (2019) 1123-1130. https://doi.org/10.2320/matertrans.MF201915
[111] M. Demirtas, G. Purcek, Room temperature superplaticity in fine/ultrafine grained materials subjected to severe plastic deformation, Mater. Trans. 60 (2019) 1159-1167. https://doi.org/10.2320/matertrans.MF201922
[112] R. Wu, Y. Yan, G. Wang, L.E. Murr, W. Han, Z. Zhang, M. Zhang, Recent progress in magnesium-lithium alloys, Int. Mater. Rev. 60 (2015) 65-100. https://doi.org/10.1179/1743280414Y.0000000044
[113] A. Alhamidi, K. Edalati, Z. Horita, S. Hirosawa, K. Matsuda, D. Terada, Softening by severe plastic deformation and hardening by annealing of aluminum-zinc alloy: significance of elemental and spinodal decompositions, Mater. Sci. Eng. A 610 (2014) 17-27. https://doi.org/10.1016/j.msea.2014.05.026

Regular Talks

An investigation of forming of Ti-6Al-4V at lower temperatures

Hosam Elrakayby[1,a,*], Diego Gonzalez[1,b], Paranjayee Mandal[1,2,c] and Paul Blackwell[1,d]

[1]Advanced Forming Research Centre, University of Strathclyde, 85 Inchinnan Drive, Inchinnan, Renfrewshire PA4 9LJ, Scotland

[2]Now at BorgWarner Inc., Courteney Road, Gillingham ME8 0RU, England

[a]hosam.elrakayby@strath.ac.uk, [b]diego.gonzalez@strath.ac.uk, [c]pmandal@borgwarner.com, [d]paul.blackwell@strath.ac.uk

Keywords: Superplastic Forming, Titanium, Ti-6Al-4V, Alpha Case

Abstract. Forming components of titanium alloys via superplastic forming at elevated temperatures while being exposed to oxygen leads to surface oxidation. The hard oxide layer formed on the surface of components is referred to as the alpha case. This alpha case layer requires post-form processing to remove it, thus increasing overall manufacturing costs and times. Superplastic forming at lower temperatures can significantly reduce the formation of the alpha case and has other benefits, such as life extension of tooling and less energy consumption. This paper shows the work done in terms of forming non-commercial components at temperatures significantly lower than the traditional ones, proving that forming at those temperatures is readily achievable. Forming pressures and tonnage need to be readjusted because of the increase in the flow stresses of the material. This paper also illustrates the implementation of a microstructural-based model to predict the hot forming behaviour of commercial titanium alloy Ti-6Al-4V (Ti64) during forming at 800°C. The material model is implemented into a commercial finite element software, Abaqus, and PAM-STAMP to obtain the optimal pressure cycle to form a non-commercial research component at 800 °C with a view to minimizing alpha case formation.

1. Introduction

Titanium alloys in general, and commercial Ti64 in particular, are optimal candidates for aerospace sector applications due their high strength-to-weight ratio, corrosion resistance and fatigue strength [1]. For a long time, the aerospace sector has been forming Ti64 through SPF to obtain near-net complex shapes without springback and residual stresses. Typically, Ti64 is superplastically formed at optimum SPF conditions, with temperatures close to 900 °C [2–4]. The exposure of titanium alloys to air/Argon at a high temperature (>650°C) during the SPF process frequently results in surface oxidation [5]. This oxygen-enriched layer, referred as alpha case, is hard, brittle and very abrasive in nature, thus producing excessive tool wear. Alpha case can be removed by machining or chemical etching, which increases the costs and the times of manufacturing (and etching makes use of environmentally harmful chemicals).

Superplastic forming (SPF) manufacturers in the Aerospace industry pursue lower forming temperatures due to the associated benefits in terms of alpha case thickness reduction, tooling and heating element life extension, less energy consumption together with the possibility of using cheaper platen and die materials [6]. Previous work [7,8] was done to obtain ultra fine grained Ti64 through severe plastic deformation to improve the superplasticity of Ti64 material at lower temperatures.

This work investigates the possibility of forming Ti64 superplastically at 800 °C, well below the optimum conditions. Tensile testing has been performed at 800 °C (ASTM E2448) and subsequent microstructural analysis of the samples has been fed into a microstructural-based material model. This material model is then implemented into a commercial finite element software Abaqus and PAM-STAMP to obtain the optimal pressure cycle to form a non-commercial component.

2. Experimental Methodology

TIMETAL® 6-4 sheet (Ti64) material (produced by TIMET) with a bimodal microstructure was used in this work. The average grain size and primary α-phase volume fraction were ~2.3 μm and 90%, respectively.

Mechanical Characterization. A series of strain rate jump and interrupted tensile tests were conducted to identify the optimal forming conditions for Ti64 material within the temperature range under consideration. Strain rate jump tensile tests were conducted at temperatures of 750, 775, 800 and 825 °C and strain rates of 10^{-4} s^{-1}, $2x10^{-4}$ s^{-1} and $5x10^{-4}$ s^{-1}. Once the optimum forming window was defined, a small number of interrupted tests at a predefined strain (0.7) and temperature were carried out. One set of the interrupted tests were followed by air-cooling outside the furnace mainly for studying the alpha case formed and the other set was water-quenched to study the grain size after testing. To mimic the forming conditions during forming components in a press, all tensile tests samples were coated with boron nitride (BN) and tested in an air environment. All tensile samples were cut by wire electro-discharge machining (EDM) with geometries following the ASTM E2448 standard.

Microstructural Analysis. The as-received and tested interrupted tensile samples were characterised for microstructure using a FEI 650 Quanta SEM to assess the presence of voids and/or cavities, alpha-beta volume fraction and grain size analysis. A Leica DM1200M optical microscope was used to image alpha case formed on interrupted tensile test samples after etching with 100 mL water, 25 mL ethanol and 2 g ammonium bifluoride. Table 1 summarises the testing conditions of these investigated samples.

Table 1 Conditions of samples for grain size measurements

Sample	Measurement	Temperature (°C)	Interrupted True Strain	Strain rate (s^{-1})	Soak Time	Coating	Cooling
C1	After Deformation (tensile test)	800	0.3	$1x10^{-4}$	50 min before starting tensile test (dwell time)	Boron Nitride	Water-Quenched
C2			0.7				
C3			0.3	$5x10^{-4}$			
C4			0.7				
C5	Before Deformation (heat treatment only)	800	N/A	N/A	50	Boron Nitride	Water-Quenched
C6					100		
C7					165		
C8					50		Air-Cooling
C9					100		
C10					165		

3. Finite Element Analysis

Constitutive Equations of the Material Model. For numerical modelling of the SPF process, a microstructural-based model was used to represent the material behaviour. Eq. 1 was used to model the grain size evolution during the forming process and the visco-plastic response of the material was expressed by a power-law based model in Eq. 2. [3,9–11]

$$\dot{d} = \dot{d}_{static} + \dot{d}_{dynamic} = \frac{\rho}{d^{\mu_1}} + \frac{\phi}{d^{\mu_2}}\dot{\varepsilon} \qquad (1)$$

Where ρ, μ_1, φ and μ_2 are temperature-dependent material parameters. $\dot{\varepsilon}$ is the strain rate.

$$\dot{\varepsilon} = A\frac{\sigma^n}{d^p} \qquad (2)$$

Where A, n and p are temperature-dependent material parameters.

Modelling of Microstructural and Mechanical Behaviour. The modelling of the Ti64 response during the high temperature tensile tests has been carried out by capturing numerically the grain size evolution at different testing conditions by obtaining the parameters of Eq. 1. Also, the flow curves of the tensile tests were fitted numerically by using Eq. 2 to optimise the parameters.

Modelling the SPF Process for a Research Demonstrator. The material model developed was used to obtain the pressure cycles to form a non-commercial research demonstrator. Numerical simulations were carried out by implementing the constitutive equations into Abaqus software via the CREEP subroutine. Also, numerical simulations were carried out in PAM-STAMP software using the built-in material data base. The purpose of having simulations in both software platforms was to compare the pressure cycles obtained, as each software platform is using different algorithms in controlling the strain rate in the formed material. Modelling was completed at different target strain rates of $1x10^{-3}$ and $5x10^{-4}$ s^{-1} to obtain the pressure cycles with limiting maximum pressure to 30 bar and 60 bar.

4. Forming Trials of Research Demonstrator

An ACB 200T press was used to complete the forming trials. Ti64 sheet material was cut into size of (500 x 500 x 1.2) mm then was sprayed by Boron Nitride. The press was heated to the target temperature then the sheet is loaded into the press and held for 15 min to stabilise the temperature in the press and confirm that the sheet reached the target temperature. Once the sheet reached the target temperature, the gas pressure was supplied based on the pressure curve obtained from the simulation.

After the forming cycle was completed, the pressure was released from the press over 5~10 min before unloading the part from the press.

5.Results and Discussion

Fig. 1 summarizes the flow curves obtained for the different test conditions and Fig. 2 summarises the calculated m-values for each test condition. It can be observed from the flow curves, that flow stress decreases with increasing the temperature and decreasing strain rates. Also, hardening behaviour was noted at the higher temperature (800 °C) and slower strain rates of $1x10^{-4}$ and $2x10^{-4}$ s^{-1}. while softening behaviour was observed at lower temperatures (750 and 775 °C) and highest strain rate; $5x10^{-4}$ s^{-1}. The m-values for the lower temperatures (750 and 775°C) are around 0.45 at different strain rates and true strains conditions. By increasing the temperature to 800 °C, the m-values increased to be around 0.5 at different strain rates and true strains. Even at this temperature and at lowest strain rate, the m-value was around 0.6. At 825 °C, m-values decreased to be around 0.47. Based on these results, 800 °C was selected to be the optimal condition for investigating the formability of Ti64 further through interrupted tensile tests.

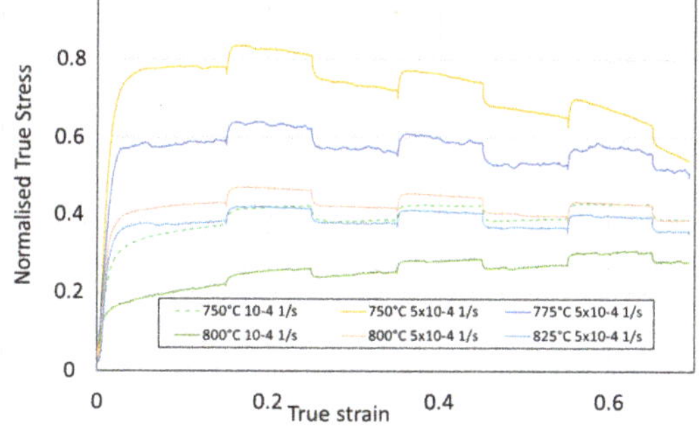

Fig. 1 Flow curves from jump tests

Materials Research Forum LLC
https://doi.org/10.21741/9781644902615-5

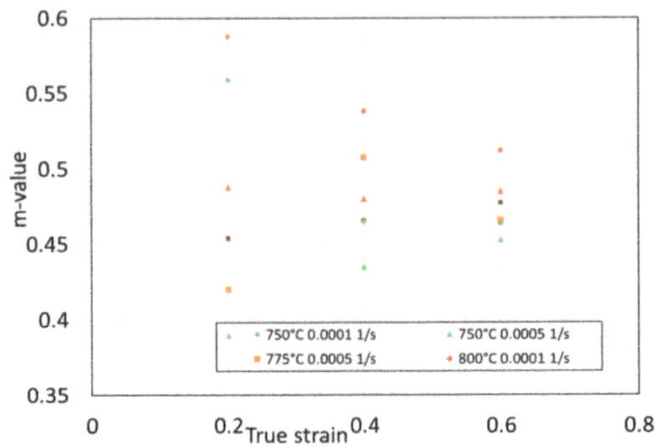

Fig. 2 m-values calculated from jump tests

Fig. 3 shows the flow curves obtained from the interrupted tensile tests. For the air-cooling condition a test was conducted at 900 °C to compare the alpha case generated at different temperatures. The flow curves at 800 °C agrees well with those obtained from the jump tests. Hardening behaviour can be noticed at lower strain rates, while curves are flattened at higher strain rates.

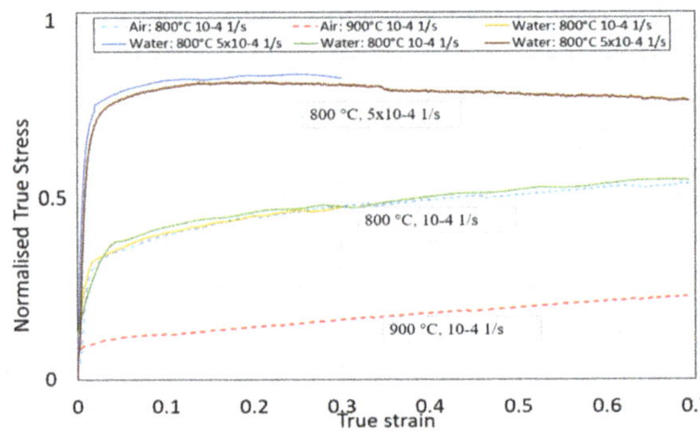

Fig. 3 Flow curves from interrupted tensile tests

Fig. 4 shows SEM images of different samples and Fig. 5 shows the average grain size and aspect ratio measured before and after the deformation at 800 °C. The as-received material has an average grain size for the alpha phase around 2.32 μm. The aspect ratio was measured at 1.87 and almost 95% alpha phase was observed as globular in shape. No grains were observed larger than 10 μm. Heat-treated samples without any deformation at 800°C show an average grain size of ~2.21 – 2.61 μm after water-quenching and ~2.09 – 2.57 μm after air-cooling. The aspect ratio was found to be in the range of ~1.71 – 1.79. Overall, no noticeable change is observed in the grain size after heat-treatment although the grains tended to globularise in all cases.

Deformed tensile samples tested at 800 °C followed by water-quenching showed an average grain size of ~2.33 – 3.05 μm, with an aspect ratio of 1.42 – 1.65 depending on the strain level and strain rate used. The typical microstructure contained fewer alpha clusters as compared to the as-received and heat-treated samples, rather, individual grain growth was observed, which led to a higher average grain size particularly for the slower strain rates. The average grain size was ~3 μm for the lower strain rate, which then decreased to ~2.33 – 2.58 μm for the faster strain rate. With slower strain rates, the grains receive sufficient time and energy to grow.

Fig. 6 shows optical images of alpha case formed on interrupted tensile test sample (strain rate of 1×10^{-4} s^{-1} and 0.7 true strain) at 800 °C and 900 °C. Sample tested at 800 °C and 900 °C revealed alpha case formation with thickness ranges 4 - 18 μm and 7 - 20 μm, respectively.

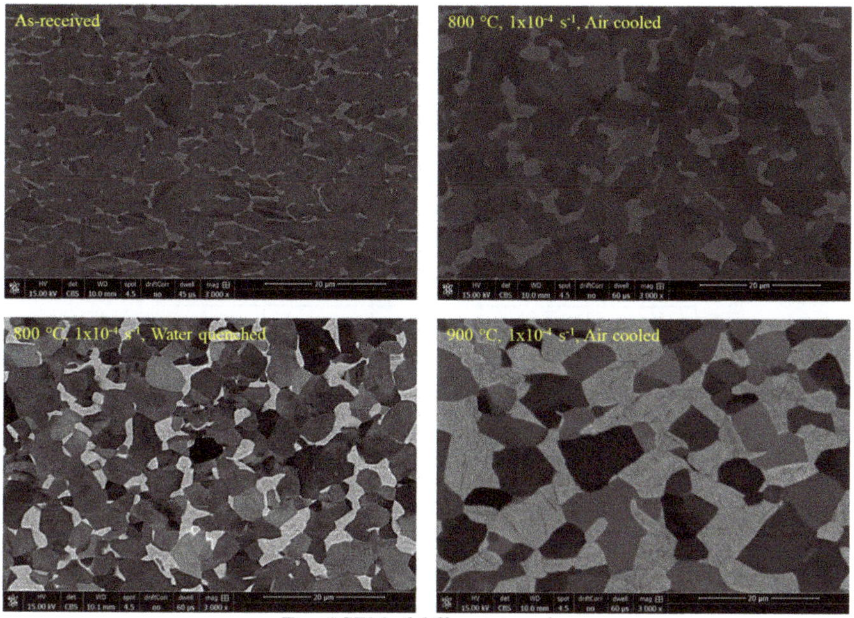

Fig. 4 SEM of different samples

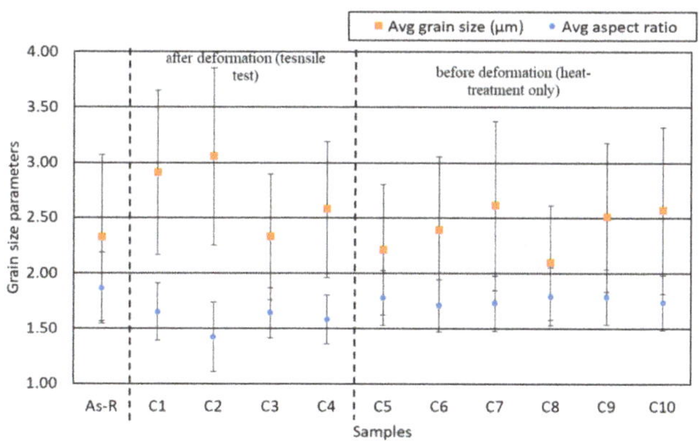

Fig. 5 Average grain size and aspect ratio measured before and after the deformation at 800 °C

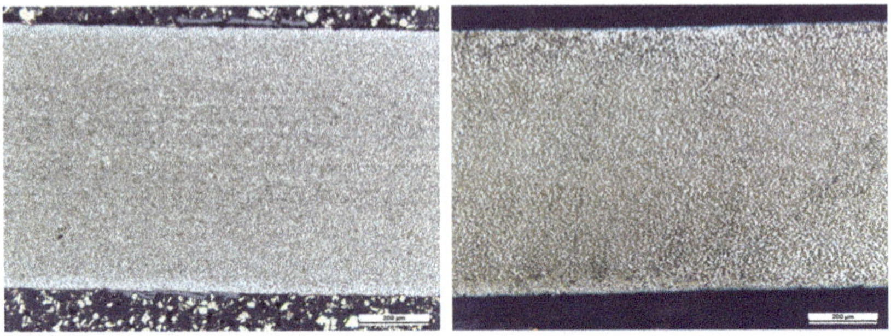

(a) 800 °C, $1x10^{-4}$ s^{-1}, Air cooled
Alpha case thickness: 4 – 18 µm

(b) 900 °C, $1x10^{-4}$ s^{-1}, Air cooled
Alpha case thickness: 7 – 20 µm

Fig. 6 Optical microscopy for alpha case measurement at (a) 800 and (b) 900 °C

Table 2 Material model parameters

Parameter	μ_1	μ_2	n	p
Value	0.86	2.13	3	1.5

Fig. 7 shows the fitting for the grain size data with time at 800 °C and strain rates $1x10^{-4}$, $5x10^{-5}$ s^{-1} and also in the static condition with no deformation. The model parameters obtained are summarised in Table 2. Fig. 8 shows the fitting for the flow data at 800 °C and strain rates $1x10^{-4}$ and $5x10^{-5}$ s^{-1}. The model parameters obtained are summarised in table 2. The fitting of the curves for the grain size and flow behaviour is based on a least-squares approach.

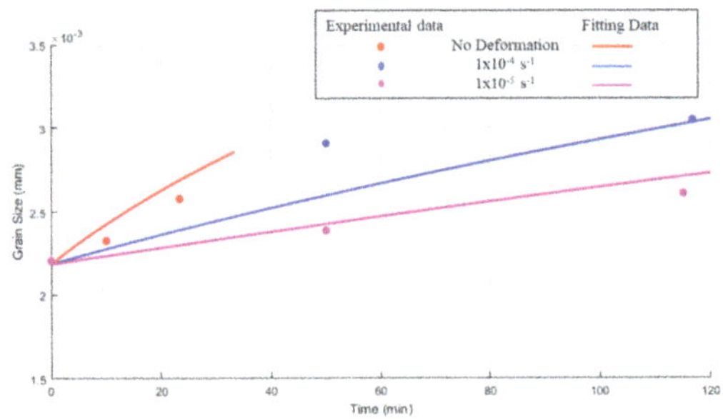

Fig. 7 Fitting of grain size data against time

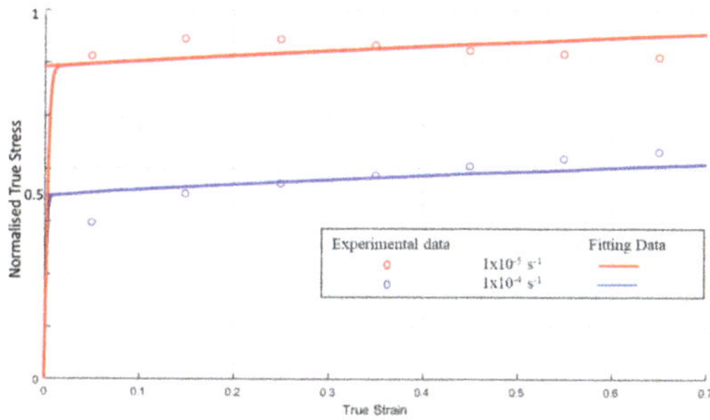

Fig. 8 Fitting of flow stress data against true strain

Table 3 summarises the different conditions used for completing the numerical simulation to obtain the pressure curves for forming trials. Fig. 9 shows the pressure curves from the numerical simulation. Increasing the maximum pressure or decreasing the controlled strain rate leads to longer cycles. At 800 °C, the forming cycle is 61 min at 60 bar and a controlled strain rate of $1x10^{-3}$ s^{-1}. Also, at 800 °C, the forming cycle is 146 min at 30 bar and controlled strain rate of $5x10^{-4}$ s^{-1}. Thus, a reduction of 58% in cycle time can be achieved. Fig. 10 shows the finite element calculations for the grain size and thickness for trial 1. Ti64 sheet materials were cut into blanks of 500 x 500 mm and sprayed with boron nitride for the forming trials. Fig. 11 shows the formed parts. It was noticed that all parts were fully formed.

In terms of pressure cycle prediction, PAMSTAMP software has a more robust methodology in comparison to Abaqus, as a statistical approach is used in order to define the maximum strain-rate over all the integration points of the component at any instant of the forming process. This results in more precise and, as a consequence, shorter pressure cycles. Indeed, very similar pressure cycles have been predicted by both software platforms despite the fact that the targeted strain-rate imposed in Abaqus was 0.001 s^{-1}, twice the targeted strain-rate imposed in PAMSTAMP.

Table 3 Modelling conditions for SPF process

	Condition			
	Maximum Pressure (bar)	Temperature (°C)	Controlled Strain rate (s^{-1})	other
Trial 1	60	800	0.001	Abaqus
Trial 2	30	800	0.001	Abaqus, Ironing
Trial 3	30	800	0.001	Abaqus, No ironing
Trial 4	30	800	0.0005	Abaqus
Trial 5	60	800	0.0005	PAMSTAMP

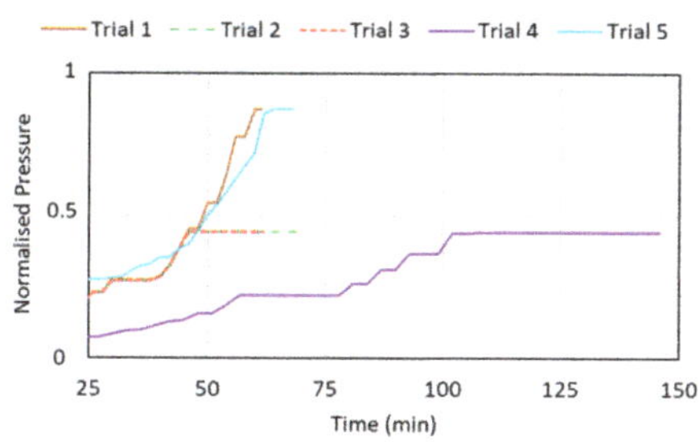

Fig. 9 Pressure curves obtained from numerical simulations

Materials Research Proceedings **32** (2023) 55-63
https://doi.org/10.21741/9781644902615-5

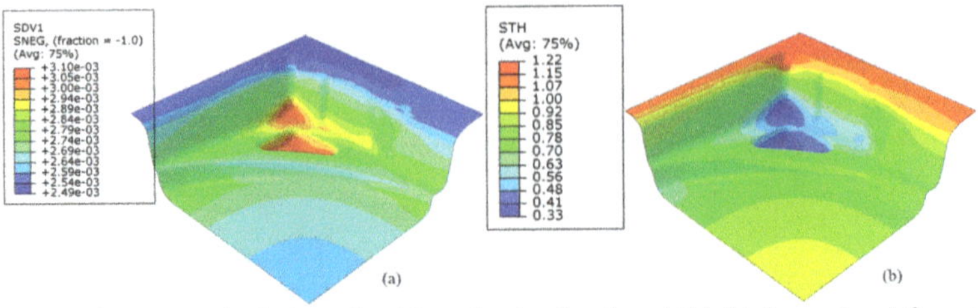

Fig. 10 Finite elements calculations for (a) grain size (mm) and (b) thickness (mm) from trial 1

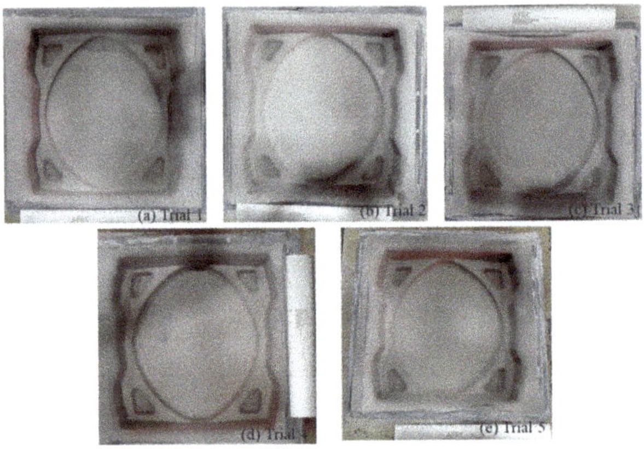

Fig. 11 SPF formed parts (a) trial 1 (b) trial 2 (c) trial 3 (d) trial 4 (e) trial 5

Conclusions

Based on its multiple benefits, potential reductions in the temperature of SPF processes have been analysed in the present work using Ti64 material from TIMET. Experimental and modelling work have shown the possibility of a temperature reduction of at least 100 °C for the component here analysed with respect to traditional SPF forming temperatures. The Ti64 material has been analysed from a mechanical testing and a microstructural point of view. Results from the tensile tests at high temperature show the outstanding behaviour of this material from an SPF point of view, with high values of the strain rate sensitivity (close to 0.6 in many cases), as well as absence of failure for the temperature ranges here analysed (from to 750 °C to 800 °C). The microstructure of the as-received material has proven to be a key aspect to enable the temperature reduction, with an as-received average grain size of 2.3 μm. Absence of cavitation has been verified for the multiple process conditions here analysed. Modelling work using a microstructural model has proven the possibility of forming a complex component shape at the temperatures of interest.

Different components have been formed based on the pressure cycles predicted by the simulation work, validating the modelling as well as the general approach for the temperature reduction. In addition, a time reduction in the pressure cycle has been achieved by the usage of advanced software for SPF simulation.

As it has been shown in this work, the pressure values for any proposed temperature reduction will increase in value, in the present work pressures up to 60 bar would be required (also the tonnage associated to the pressure cycle), as the material develops higher internal stresses in comparison to SPF at conventional temperatures. Any temperature reduction attempt would need to consider if the pressure and the hydraulic systems of the press will be able to deliver the required pressure and tonnage increments. As the reader can guess, the answer to this question is dependent on the amount of temperature reduction and the component characteristics (overall size, radii of the features, etc.). There could be cases where more powerful systems would be required to be either installed or designed (if the press is in the design stage). A cost analysis would solve for each case if the associated capital expenditure costs were compensated by all the benefits of a lower temperature process: reduction on the energy costs, cheaper material for dies and platens (also an increase in their life), less maintenance costs, longer production shifts and less consumables.

References

[1] R.R. Boyer, An overview on the use of titanium in the aerospace industry, Mater. Sci. Eng. A. 213 (1996) 103-114. https://doi.org/10.1016/0921-5093(96)10233-1

[2] E. Alabort, P. Kontis, D. Barba, K. Dragnevski, R.C. Reed, On the mechanisms of superplasticity in Ti-6Al-4V, Acta Mater. 105 (2016) 449-463. https://doi.org/10.1016/j.actamat.2015.12.003

[3] E. Alabort, D. Putman, R.C. Reed, Superplasticity in Ti-6Al-4V: Characterisation, modelling and applications, Acta Mater. 95 (2015) 428-442. https://doi.org/10.1016/j.actamat.2015.04.056

[4] P. Mandal, A. Gomez-Gallegos, D. Gonzalez, H. Elrakayby, P. Blackwell, Superplastic Behaviour of Ti54M and Ti64, MATEC Web Conf. 321 (2020) 04028. https://doi.org/10.1051/matecconf/202032104028

[5] F. Pitt, M. Ramulu, Influence of grain size and microstructure on oxidation rates in titanium alloy Ti-6Al-4V under superplastic forming conditions, J. Mater. Eng. Perform. 13 (2004) 727-734. https://doi.org/10.1361/10599490421394

[6] G.A. Salishchev, R.M. Galeyev, O.R. Valiakhmetov, R. V. Safiullin, R.Y. Lutfullin, O.N. Senkov, F.H. Froes, O.A. Kaibyshev, Development of Ti-6Al-4V sheet with low temperature superplastic properties, J. Mater. Process. Technol. 116 (2001) 265-268. https://doi.org/10.1016/S0924-0136(01)01037-8

[7] R.S. Mishra, V. V. Stolyarov, C. Echer, R.Z. Valiev, A.K. Mukherjee, Mechanical behavior and superplasticity of a severe plastic deformation processed nanocrystalline Ti-6Al-4V alloy, Mater. Sci. Eng. A. 298 (2001) 44-50. https://doi.org/10.1016/S0921-5093(00)01338-1

[8] H. Matsumoto, K. Yoshida, S.H. Lee, Y. Ono, A. Chiba, Ti-6Al-4V alloy with an ultrafine-grained microstructure exhibiting low-temperature-high-strain-rate superplasticity, Mater. Lett. 98 (2013) 209-212. https://doi.org/10.1016/j.matlet.2013.02.033

[9] J. Lin, T. Zhu, L. Zhan, Constitutive equations for modelling superplastic forming of metals, in: Superplast. Form. Adv. Met. Mater., Elsevier, 2011: pp. 154-183. https://doi.org/10.1533/9780857092779.2.154

[10] T.G. Langdon, Grain boundary sliding revisited: Developments in sliding over four decades, J. Mater. Sci. 41 (2006) 597-609. https://doi.org/10.1007/s10853-006-6476-0

[11] H. Elrakayby, D. Gonzalez, A. Gomez-Gallegos, P. Mandal, N. Zuelli, A Study on Modelling the Superplastic Behaviour of Ti54M Alloy, Solid State Phenom. 306 (2020) 9-14. https://doi.org/10.4028/www.scientific.net/SSP.306.9

Fabrication and superplastic forming property research of TiB whisker reinforced titanium alloy matrix composite sheet

Mingjie Fu[1,a*], Fuxin Wang[1,b], Zhen Li[1,c], Yingying Liu[1,d]

[1]No.1 Chaoyang street, Chaoyang District, Beijing, China

[a]mingjie.fu@outlook.com, [b]wangfx1216@126.com, [c]lizhen90420@163.com, [d]m18335100146@163.com

Keywords: Titanium Matrix Composite, Sheet, Superplastic Forming, Diffusion Bonding, Microstructure

Abstract. Applied at elevated temperature compared with in situ-whisker reinforced titanium matrix composite(TMC) higher than matrix titanium alloy with 50~150℃. The advantage of TMC compared with Ti-Al intermetallic compound is lower cost, better weldable property, simple fabrication process; lower density compared with super alloy. One of important demand of TMC sheet for high temperature application is used for higher speed craft with lighter part fabricated by SPF/DB technology. This paper illustrated TiBw/TA15+ TMC sheet fabrication and SPF/DB processing property, and analyzed effects of rolling processing on the microstructure and tensile properties. Wide sheet has been fabricated successfully. Forming mechanism of TMC is revealed, which is different from conventional titanium alloy, fined microstructure is unnecessary for TMC, four layers SPF/DB component also can be formed for TMC sheet.

1. Introduction

More than 40 years passed from the first 600°C high temperature titanium alloy(IMI834) successfully developed, because decrease capability of the anti-oxidation and stability of the microstructure, then application at elevate temperature no beyond 600°C applied in the aero engine. The focus research on high temperature low density alloy turn to titanium-aluminium intermetallic compound. However, the extreme brittleness and lower fracture toughness still not very satisfied the requirement of engine part, including manufacture suitability, just a few parts was successful applied. Traditionally, strengthen way for high temperature titanium alloy at high temperature are solution, α2 phase and silicides. the in situ titanium matrix composite could keep increasing the strength at higher temperature than matrix through add another one or two second phase as TiB or TiC, usually 50~150°C higher, and temperature and mechanical properties could be optimized of in situ TMC, which is a very prospective high temperature structural material.

In this paper, Si and B are added into Ti-6.5Al-2Zr-1Mo-1V alloy to fabricate the in situ TMC sheet, and further investigate the SPF/DB combined processing experiments; the four layers structure part is successfully developed.

2. Experimental

The in situ TMC reinforced with 3.5 vol.% TiB/ Ti-6.5Al-2Zr-1Mo-1V cast ingot was prepared by vaccum arc remleting(VAR), and totally remelted three times. Several billet sections were cross worked by multiple upset and redraw operations at β single field and α+β phase field with a total 9 times reheating and working, to create a more homogeneous billet structure and higher ratio of TiB whiskers broken as much as possible, then billet is rolled on a 1200mm wide roller machine at preheating temperature of β_T-20°C, β_T+50°C, β_T+100°C respectively, and finally the thickness of 1.2mm and 1000mm wideness sheet is fabricated with the same surface quality as matrix alloy, no defect was found by ultrasonic inspection. Annealing treatment applied at 860~940°C/1h AC.

Superplasticity and diffusion bonding processes is examined with 900~960°C with uniaxial superplastic tensile and gas pressure diffusion bonding respectively. The optimized parameters are used to form a four layers SPF/DB combined structure.

Metallographic preparation through grinding and polishing, followed by immersion etching in Kroll's reagent. Microstructure was studied using ZEISS optical micrograph and SEM.

3. Results and discussion

3.1 Rolling temperature effects on microstructure and tensile property

Hot rolling operation is done at 20°C below β transus, the microstructure is combined with primary α and β transformed with a-laths, and primary α distributed align to the rolling direction, the formed micro voids due to the long TiB whiskers fracture without filling with matrix. With rolling temperature increasing, microstructure turn to be complete β transformed with α-laths. Distribution of TiB whiskers is related homogeneity, but few micro voids still can be found. Furthermore, the micro voids decrease obviously with increasing rolling temperature, as shown in Fig. 1.

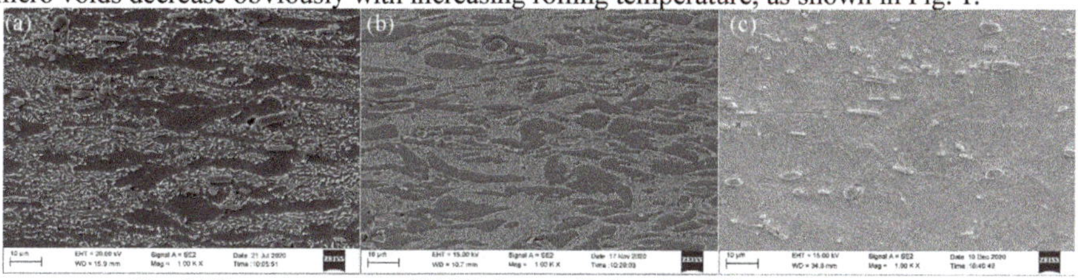

Fig.1 Microstructures of TMC sheet rolled under different temperatures (a) β_T-20 ℃, (b) β_T+50 ℃, (c) β_T+100 ℃

Fig.2 shows tensile properties results at RM and 650°C, better RT ductility and middle high temperature ultra tensile strength are obtained which rolled at 20°C below β transus, the average is 7.8% and 457MPa respectively. However, the highest average of UTS is 523MPa at 650°C which is rolled at 100°C above β transus. For the whisker reinforced TMC, the ductility at room temperature is one of key technical difficulties. Above results show micro or nano dimension of TiB whiskers could improve the ductility at RT.

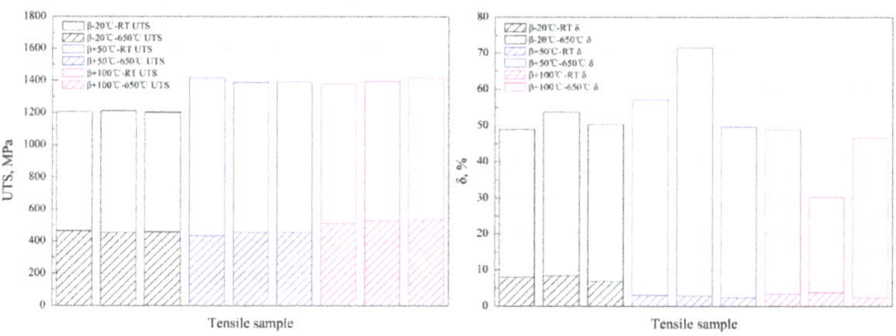

Fig.2 Tensile properties at RT and 650℃ of rolled condition (a) UTS (b) δ

2.2 Effects of annealing temperature on microstructure and tensile properties

Based on the sheet rolled at β_T-20°C, compared with microstructure of rolled sheet, after 860°C, 900°C and 940°C annealing for 1h, the obviously changes happened in second precipitation α-laths, it becomes wider and even being to equiaxed, and contents of α-laths decrease with increasing anneal temperature. However, TiB whisker is very stable, as shown in Fig. 3.

Materials Research Forum LLC
https://doi.org/10.21741/9781644902615-6

Fig. 4 shows tensile results at RT and 650°C, 900°C/1h is the optimized anneal parameters according to balance of 650°C strength and RT ductility. Tensile ductility at RT and ultra strength at 650°C are 6.9% and 580MPa respectively.

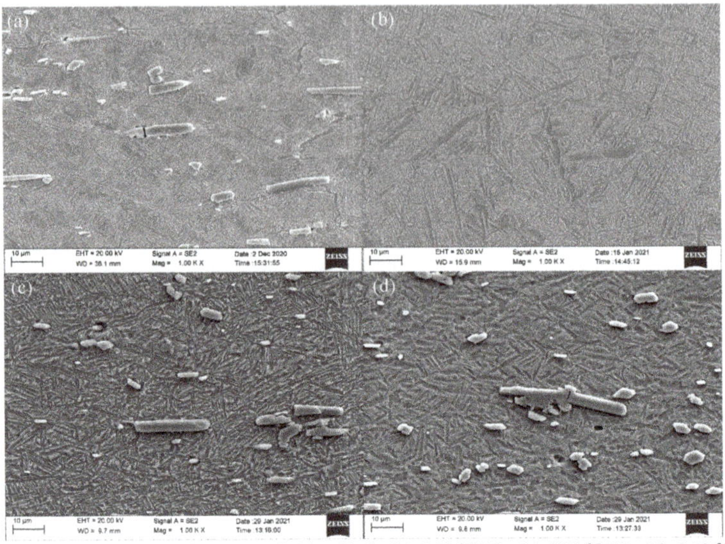

Fig.3 Microstructures after anneal (a)rolled; (b) 860 ℃/1h, (c) 900 ℃/1h, (d) 940 ℃/1h

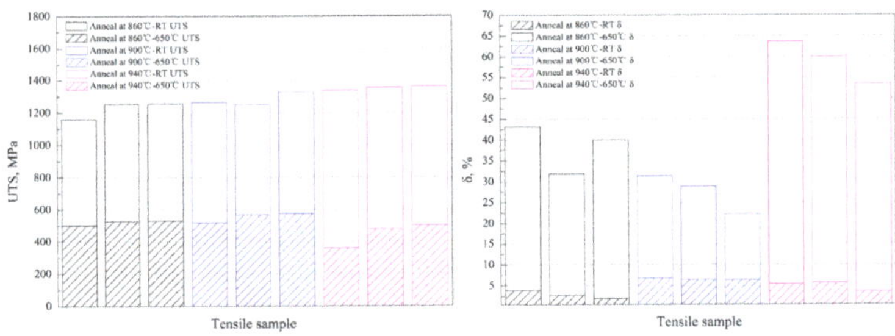

Fig.4 Tensile properties at RT and 650 ℃ after annealing at different temperature (a) UTS (b) δ

2.3 Superplasticity evaluation

According our previous and present researches, this article combined the results of uniaxial superplastic tensiles. Compared superplastic tensile results with 5.0 vol.% and 3.5 vol.% TiB whisker content, the best ductility of 569% obtained at 960°C、 $1\times10^{-3}s^{-1}$ with 3.5 vol.% TiB whisker, and 439% obtained at 940°C、 $5\times10^{-3}s^{-1}$ with 5 vol.% TiB whisker. The other results are shown in Fig. 5. The highest ductility obtained in 3.5vol.% TiB sheet, but average of ductility of 5.0 vol.% is higher than 3.5 vol.% at the same tensile conditions.

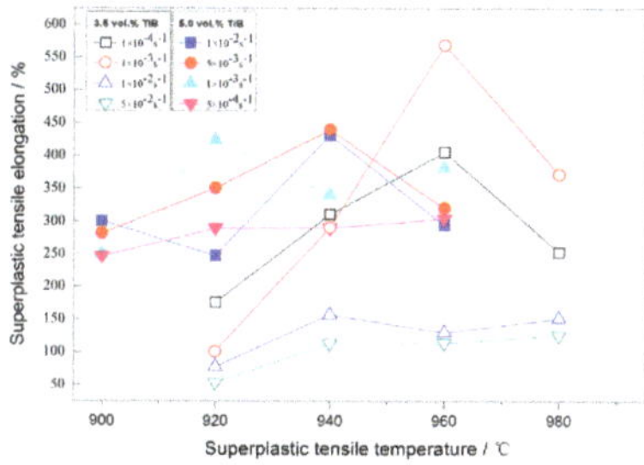

Fig.5 Effect of temperature and strain rate on ductility of composites sheet with 3.5 vol% and 5.0 vol.% TiB whisker

Fig.6 shows comparison of 3.5 vol% and 5.0% TiB whisker strain-stress curves as the same superplastic tensile parameters. Obviously, the flow stress of 3.5 vol% TiB whisker is lower than 5.0 vol % TiB whisker. However, the elongations with strain rate $1\times10^{-2}s^{-1}$ of 5.0 vol % TiB whisker have better elongations.

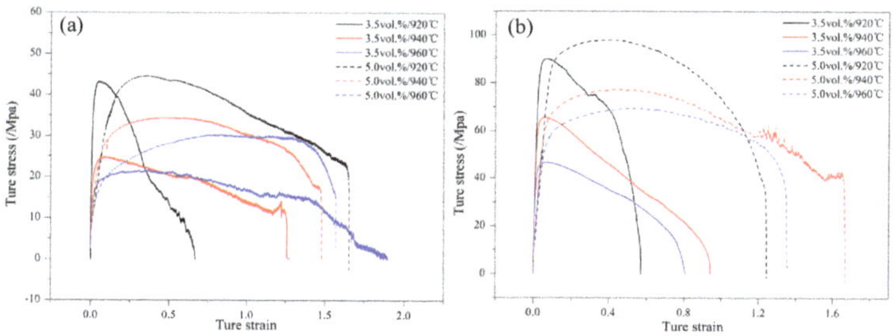

Fig.6 True stress-strain curves of the TiB_w/TA15 sheet at different temperature (a)$1\times10^{-3}s^{-1}$, (b)$1\times10^{-2}s^{-1}$

During superplastic tensile deformation, microstructure evolution of matrix alloy is deomained by α→β phase transformation and dynamic recrystallization, signigicant dynamic recrystallization occurred at second α lath phase, and sphered feature, which is almost the same as matrix alloy. Superplasticity of matrix will effect on the changes of TiB whiskers, the higher superplasticity of matrix, the better consistent deformation between whiskers and matrix, TiB whiskers are difficult to fracture during deformation. Much more debonding defects will occur at interficail between whiskers and matrix when decreasing compatibility deformation.

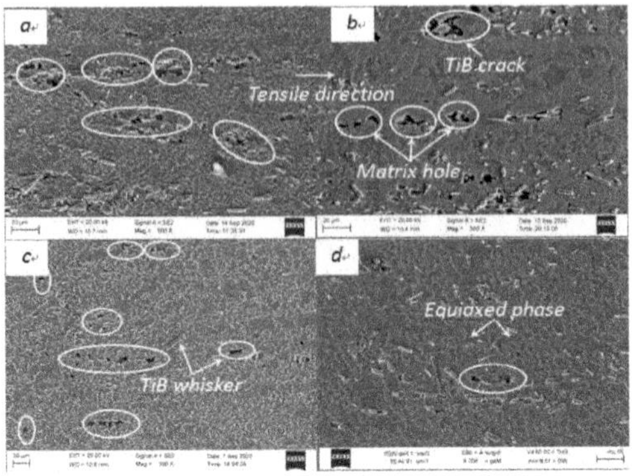

Fig.7 Superplastical microstructures deformed at different conditions
(a)900 ℃/5×$10^{-3}s^{-1}$, (b)960 ℃/5×$10^{-3}s^{-1}$, (c) 920 ℃/1×$10^{-2}s^{-1}$, (d) 920 ℃/5×$10^{-4}s^{-1}$

Significant effect of deformation strain rate on micro holes at higher strain rates, due to the worse consistent deformation at higher strain rate, dislocations will be stopped at whiskers, micro voids are easy to occur at here. Dynamic recreystillation not only could release the stress concentration, but also could close-up micro voids.

2.4 Diffusion bonding property and mechanism

TiB could increase the β transus with 30~40°C(1040°C), so diffusion bonding temperature of TMC also will increase accordingly. The optimized diffusion bonding temperature of matrix alloy is 920°C, and 1.0MPa gas pressure with 2 hours. Result of diffusion bonding rate examined by OM as shown in table1, the 95% diffusion bonding rate could be obtain by 950°C+2.5MPa and 960°C+1.6MPa with 3h.

Table 1 Diffusion bonding rate result with different parameters

No.	Temprature	Gas pressure	Time	Suface condition	Diffusion bonding rate
0	920°C	1.0MPa	2h	normal	>95%
1	950℃	1.6MPa	3h	normal	unbonded
2	950℃	1.6MPa	3h	grinding	30%
3	960℃	1.6MPa	3h	normal	60%
4	950℃	2.5MPa	3h	grinding	>95%
5	960℃	1.6MPa	3h	grinding	>95%

Fig. 8 shows the typical diffusion bonding microstructures with different rate. In fig.1(a), fine equiaxed grain with 100% diffusion bonding rate could be observed without TiB phase. Compared with as-received microstructure of TMC sheet, grains of matrix all turn to be more equiaxed with increasing of gas pressure and temperature, because of furnace cooling. Fairly uniform distribution of TiB whiskers with length less than 10μm. Interface with or without TiB whisker, the diffusion bonding mechanism the same as matrix if without TiB whisker, and the complete interface formed with TiB whisker due to the TiB whisker could continue react with matrix at phase boundary above 900°C. As result, we could expect all type of TiB whisker reinforced α or α+β type titanium matrix composites could be diffusion bonded. Surface roughness is a very critical effecting factor from experimental results; lower roughness could decrease parameters for diffusion bonding, because of TiB whisker existing like as a "needle", which will block the first stage of physic contact at high temperature.

As we know, fine equiaxed grain is the best microstructure for SPF and DB process, because of the theory of grain boundary sliding for superplasticity and density of grain boundary for diffusion bonding. However, present results revealed another microstructure also could have a good quality of bonding interface, the mechanism maybe more about phase boundary for diffusion of atom. Further research will prove the actual microstructure at diffusion bonding temperature through fast cooling to keep microstructure at high temperature. However, we have already investigated the extent of equiaxed grains depend on the pressure, which means that the outer pressure is a driven factor for transforming from α lath to equiaxed grains.

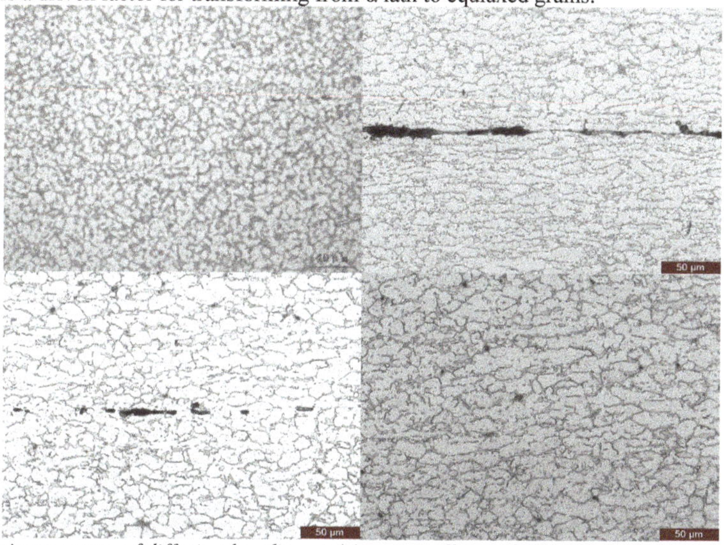

Fig. 8 Mcirostructure of diffusion bonding at (a) matrix alloy with 920°C+1.0MPa+2h; (b) TMC with 950°C+1.6MPa+3h; (c) TMC with normal surface at 960°C+1.6MPa+3h (d) TMC with grinding surface at 950°C+2.5MPa+3h

2.5 Four layers structure formed with SPF/DB

According the superplastic and diffusion bonding test results, parameters with 950~970°C, gas pressure with 1.6~2.0MPa, forming or diffusion bonding time with 2~3h. All other processing is same as traditional titanium alloys like as Ti64. The component with multiple ribs was successfully formed as showed in Fig. 9. Diffusion bonding interface between inner and outer sheets and ribs shows in Fig.3. Very good diffusion bonding quality could be observed, also finite decrease of thickness of inner sheet and without any maro holes formed in biggest deformation area of ribs.

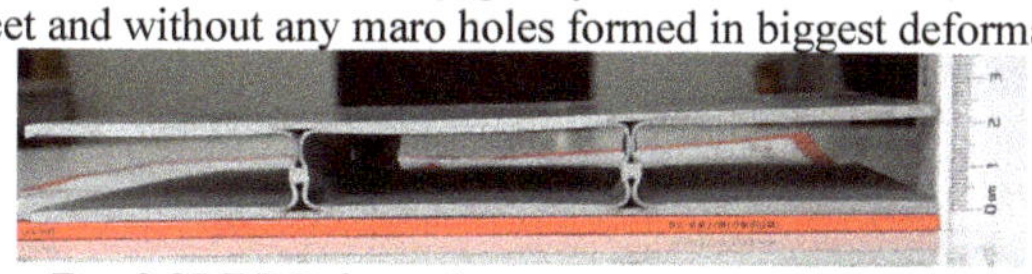

Fig. 9 SPF/DB formed component of in situ TMC

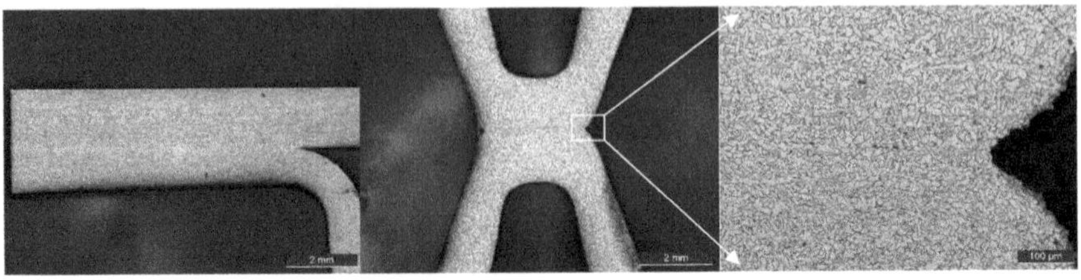

Fig.10 Macrostructure of diffusion bonding at surface and rib

4. Summary

(1) TiB whisker reinforced TMC is successfully fabricated with 1000mm wideness and 1.2mm thickness. Rolling processing is the critical factor to effect on mechnical properties. Rolling at above 100°C from beta transformation temperature have a good tensile strength at RT and 650°C, but the ductility is related lower at RT.

(2) Obtained the best superplastic elongation with 569% and optimized diffusion bonding parameters with 95% rate, the combined process is used to successfully form multiple ribs component.

(3) Mechanism of diffusion bonding of in situ TMC depends on the phase boundary transformation and pressure.

(4) Evolution of matrix microstructure same as without TiB whisker, however, due to deformation compatibility between TiB whisker and matrix, after an extent deformation, the micro voids will be formed at the interface of matrix and whisker.

References

[1] L.J. Huang, L. Geng, X.H. Peng, Strengthening and toughening mechanisms of the second phase in titanium alloys and titanium matrix composites, J. Materials China. 2019 38(03) 214-222.

[2] G.H. Wu, Z.Y. Kuang, Opportunities and Challenges for Metal Matrix Composites in the Context of Equipment Upgrading, J. China Engineering Science. 2020 22(2) 79-90. https://doi.org/10.15302/J-SSCAE-2020.02.012

[3] J. Ni, H. Chai, K. Shi, Research progress of titanium matrix composites reinforced by particle , J. Materials reports. 2019 22(Z2) 369-373.

[4] F.X. Wang, M.J. Fu, J.H. Qian, et al, Microstructure and Mechanical Properties of TiBw/TA15 Composite Sheet , J. Aeronautical Manufacturing Technology. 2021 64(10) 95-101.

[5] Y.Y. Liu, M.J. Fu, F.X. Wang, Research on Heat Treatment Process of TiBw /TA15 Composite, J. Aeronautical Manufacturing Technology. 2021 64(14) 40-48.

[6] Y.Y. Liu, J.H. Qian, M.J. Fu, et al, Superplastic Deformation Behavior of TiBw/TA15 Composite Sheet, J. Aeronautical Manufacturing Technology. 2022 65(4) 80-86.

[7] Y.Y. Liu, M.J. Fu, F.X. Wang, et al, Structure evolution of laser welding of TiBw/TA15 composites, J. Materials Science and Technology. 2022 30(5) 60-68.

[8] H. Qiao, Y.S. Zeng, F.X. Wang, et al, Research on Thermal Deformation Behavior of As-cast and Forged TiBw/TA15 Composite, J. Titanium Industry Progress. 1(2023) 1-9.

[9] L. Huang, M. Qian, Z. Liu, et al, In situ preparation of TiB nanowires for high-performance Ti metal matrix nanocomposites, J. Journal of Alloys and Compounds. 735 (2018) 2640-2645. https://doi.org/10.1016/j.jallcom.2017.11.238

[10] K.R. Ravi Chandran, K.B. Panda, S. Sahay. TiBw-reinforced Ti composites: Processing, properties, application prospects, and research needs, J. JOM, 2004 56(5) 42-48. https://doi.org/10.1007/s11837-004-0127-1

[11] W.J. Lu, L. Xiao, K. Geng, et al, Growth mechanism of in situ synthesized TiBw in titanium matrix composites prepared by common casting technique, J. Materials Characterization. 2008 59(7) 912-919. https://doi.org/10.1016/j.matchar.2007.07.016

[12] R. Srinivasan, D. Miracle, S. Tamirisakandala, Direct rolling of as-cast Ti-6Al-4V modified with trace additions of boron, J. Materials Science and Engineering: A. 2008. 487(1-2) 541-551. https://doi.org/10.1016/j.msea.2007.10.053

[13] Z.D. Lu, C.J. Zhang, H. Feng, et al, Effect of heat treatment on microstructure and tensile properties of 2 vol.% TiCp/near-β Ti composite processed by isothermal multidirectional forging, J. Materials Science and Engineering: A. 22 (2019) 138064.1-138064.7. https://doi.org/10.1016/j.msea.2019.138064

[14] H.A. Rastegari, S. Asgari, S.M. Abbasi, Producing Ti6Al4V/TiC composite with good ductility by vacuum induction melting furnace and hot rolling process, J. Materials & Design. 2011 32(10) 5010-5014. https://doi.org/10.1016/j.matdes.2011.06.009

[15] J. Qu, C. Zhang, S. Zhang, et al, Relationships among reinforcement volume fraction, microstructure and tensile properties of (TiBw+TiCp)/Ti composites after (α + β) forging, J. Materials Science and Engineering: A. 701 (2017) 16-23. https://doi.org/10.1016/j.msea.2017.06.061

[16] L.J. Huang, L. Geng, B. Wang, et al, Effects of extrusion and heat treatment on the microstructure and tensile properties of in situ TiBw/Ti6Al4V composite with a network architecture, J. Composites Part A: Applied Science and Manufacturing. 2012 43(3) 486-491. https://doi.org/10.1016/j.compositesa.2011.11.014

[17] C. Zhang, F. Kong, S. Xiao, et al, Evolution of microstructural characteristic and tensile properties during preparation of TiB/Ti composite sheet, J. Materials & Design (1980-2015). 36 (2012) 505-510. https://doi.org/10.1016/j.matdes.2011.11.060

[18] C. Zhang, C. Guo, S. Zhang, et al, The effect of rolling temperature on the microstructure and mechanical properties of 5 vol.% (TiBw+TiCp)/Ti composites, J. JOM. 2020 72 (3) 1376-1383. https://doi.org/10.1007/s11837-019-03972-0

[19] R. Zhang, D. Wang, L. Huang, et al, Effects of heat treatment on microstructure and high temperature tensile properties of TiBw/TA15 composite billet with network architecture, J. Materials Science and Engineering A. 679 (2017) 314-322. https://doi.org/10.1016/j.msea.2016.10.041

[20] C.J. Zhang, High-temperature deformation behavior and microstructure and mechanical properties of (TiB+TiC)/Ti composites. Ph.D. Thesis, Harbin Institute of Technology, China, 2013

[21] C.J. Lu, Research on superplastic deformation behaviors and mechanisms of as-extruded TiBw/TC4 composite, Master's thesis, Harbin Institute of Technology, China, 2014.

[22] A. Yousefiani, J.C. Earthman, F.A. Mohamed, Formation of cavity stringers during superplastic deformation, J. Acta Materialia. 1998 46(10) 3557-3570. https://doi.org/10.1016/S1359-6454(98)00030-5

Superplasticity of Ti-6Al-7Nb alloys with different initial microstructures processed by high-pressure torsion

Kathy González Jiménez[1,a], Joaquín E. González Hernández[1,b], Jorge M. Cubero-Sesin[1,c*] Jeremy Barrantes Rodríguez[1,d] and Zenji Horita[2-4,e]

[1]Centro de Investigación y Extensión en Materiales (CIEMTEC), Escuela de Ciencia e Ingeniería de los Materiales, Instituto Tecnológico de Costa Rica, Cartago 159-7050, Costa Rica

[2]Graduate School of Engineering, Kyushu Institute of Technology, Kitakyushu 804-8550, Japan

[3]Magnesium Research Center, Kumamoto University, Kumamoto 860-8555 Japan

[4]Synchrotron Light Application Center, Saga University, Saga 840-8502, Japan

[a]kathygon18@estudiantec.cr, [b]jegonzalez@itcr.ac.cr, [c]jcubero@itcr.ac.cr, [d]jerbara1@estudiantec.cr and [e]horita.zenji.688@m.kyushu-u.ac.jp

Keywords: High-Pressure Torsion, Ultrafine Grained, Nanostructure, Superplasticity, Strain Rate Sensitivity, Activation Energy

Abstract. Ti-6Al-7Nb alloy was quenched below and above the β-phase field and processed by high-pressure torsion (HPT) to produce different grain sizes in the ultra-fine range. The influence of temperature and strain rate over the superplastic properties, activation energy and strain rate sensitivity were evaluated by high temperature tensile tests. The mechanical properties were complemented with Vickers microhardness measurement of heat treated and HPT-processed samples. The crystallographic analysis was performed through XRD and EBSD studies in the grip and tip of the high temperature tensile specimens. Microhardness measurements show significant hardening with respect to the imposed strain after HPT. Meanwhile, high superplastic elongation of 1200% was accomplished at 850 °C with a strain rate of $2x10^{-3}$ s^{-1}. EBSD studies of the failed specimens suggest that the tip part exhibited dynamic recrystallization during tensile test, which produced a larger grain size of ~5.1 μm and higher fraction of β phase of 3.7%, with respect to the grip part under static recrystallization, which attained ~3.4 μm grain size and 3.3% β phase fraction.

Introduction

Titanium alloys are widely used in aerospace industry and medical implants due to their workability [1–3], low density (60% of steel), thermal stability, high mechanical and corrosion resistance [4–7]. Titanium alloys are classified according to their crystal structure with respect to the presence of phases: α (HCP), β (BCC) and duplex α+β [4–7]. The α+β alloys have enough content of β-stabilizer to allow the nucleation of β-phase particles at room temperature, these duplex microstructures provide balanced properties and can be modified through heat treatments (increases strength 30% to 50% in comparison with annealed condition or overaged) [7]. Ti-6Al-7Nb is an α+β alloy with enough Nb content to produce a microstructure that can serve as replacement to the commonly used Ti-6Al-4V alloy, and avoid the toxicity associated with V^+ ion release [1–7]. On the other hand, superplastic properties of metals and alloys provide an excellent option for fabrication of parts with complex shape at lower temperatures and higher strain rates than conventional hot working, which results in lower energy consumption [8]. Superplasticity is characterized by high elongation to failure (above 400%) during a high temperature tensile test [9]. Superplastic behavior requires three main conditions: (1) high deformation temperature $T > 0.5T_m$ (T_m: melting temperature), (2) grain size < 10 μm and (3) low strain rate, in the order of

Materials Research Proceedings **32** (2023) 72-80
https://doi.org/10.21741/9781644902615-7

10^{-3} to 10^{-5} s^{-1} [10–11]. The strain rate sensitivity index, *m*, describes the capability of plastic deformation. It has been observed that superplastic flow requires m ~ 0.5 [12–14]

Eq. 1 shows the constitutive equation for superplastic behavior of microcrystalline materials, equation used to find the activation energy of the process [13–15].

$$\dot{\varepsilon} = A\left(\frac{b}{d}\right)^P \left(\frac{\sigma}{G}\right)^n e^{\frac{-Q}{RT}} \quad (1)$$

Where $\dot{\varepsilon}$ is the strain rate, A is a dimensional constant, b is the Burger vector, d is the grain size, P is the inverse exponent of the grain size, σ is the yield stress, G is the shear modulus of elasticity, n = 1/m, and $e^{-Q/RT}$ denotes the grain boundary diffusion coefficient, where Q is the activation energy, R is the ideal gas constant and T is the temperature. Eq.1 can be rewritten in such a way that it becomes a linear equation, as shown by Eq. 2 [13–15].:

$$\ln \dot{\varepsilon} = \ln\left[A\left(\frac{b}{d}\right)^P \left(\frac{\sigma}{G}\right)^n\right] - \frac{Q}{R}T^{-1} \quad (2)$$

Severe Plastic Deformation (SPD) methods are an excellent option for producing nanostructured alloys with enhanced properties. SPD processing can be described as metal forming techniques that induce large shear strains under high hydrostatic pressure at low temperatures (commonly less than 0.4T melting temperature), capable of producing ultrafine grained (UFG) structures [16–21]. The most effective methods to produce grain refinement is High-Pressure Torsion (HPT) [17–20], in which a disc is subjected to torsional straining under a high hydrostatic pressure. Several studies reported the superplastic elongation in titanium alloys: 987% in Ti55 alloy [22], a range between 768% and 2100% for Ti-6Al-4V [23–25] and 820% for UFG Ti-6Al-4V[26]. UFG Ti-6Al-7Nb produced by HPT attained 930% [27] and 400% [28] of superplastic elongation. This work describes the influence of strain rate and temperature over the superplastic properties of nanostructured Ti-6Al-7Nb produced by quenching and HPT, through the calculation of strain rate sensitivity (m) and activation energy (Q).

Experimental procedures

Discs of 10 mm in diameter and 1.5 mm in thickness were sliced from commercial Ti-6Al-7Nb bar by electrical discharge machining (EDM). Then the specimens were subjected to heat treatment at 700°C for 2 h in air atmosphere and cooled in cold water. The cooling in cold water from 700 °C did not have a major effect in the microstructure of a slowly cooled Ti-6Al-7Nb, so it is referred to as annealed condition (A). Following, samples were heat treated at 1060°C (which is above β-transus) above martensite start (Ms) for 1 h and quenched in cold water. This condition is referred to as quenched (Q). HPT was conducted on 10 mm diameter discs from A and Q conditions at room temperature, under 6 GPa for N = 10 revolutions at ω = 1 rpm. Vickers microhardness measurements were performed with 500 μm spacing in radial directions from the center of the disc for: annealed (A), annealed+quenched (A+Q) and annealed, quenched and HPT-processed samples (A+Q+HPT). Tensile test specimens were also extracted by EDM from a location 2 mm from the center of the 10 mm disc. The dimensions of the gage section were 1.5 mm length, 0.7 mm width and 0.5-0.7 mm thickness, as shown schematically in the inset of Fig. 4. High temperature tensile tests were conducted in air atmosphere combining temperatures of 700 °C, 750 °C, 800 °C, 850 °C and 900 °C with initial strain rates of 1×10^{-3} s^{-1}, 2×10^{-3} s^{-1}, 5×10^{-3} s^{-1}, 1×10^{-2} s^{-1} and 2×10^{-2} s^{-1}. These tests allowed the calculation of strain rate sensitivity (m) and activation energy of Ti-6Al-7Nb for comparison with reported literature and stablish the deformation mechanism during tensile test.

Metallographic analysis was performed in optical microscope for annealed and quenched conditions, the discs were polished to a mirror-like finish and etched with Kroll's reagent. X-ray diffraction (XRD) was carried out to identify the main phases in heat treated, HPT-processed and failed tensile specimen samples, using Cu-Kα radiation (λ = 1.54016 Å), scanning step of 0.02° and speed of 1°/min. Transmission electron microscopy (TEM) observations were carried out at 200 kV from 3 mm discs extracted from the periphery of the HPT-processed disks, thinned by twin-jet electropolishing in a solution of 90% acetic acid and 10% perchloric acid under 20 V. After high temperature tensile test, the failed specimens were selected for crystal structure and phase fraction analysis using electron backscatter diffraction (EBSD) in grip and tip sections. EBSD samples were mechanically polished to a mirror-like finish followed by electropolishing using the same electrolyte and conditions as for the TEM specimens.

Results and discussion

Fig. 1a shows the initial microstructure after annealing at 700 °C, which is a typical α+β microstructure, consisting of equiaxed grains of α-phase with β-phase particles at grain boundaries. Fig. 1b shows the as-quenched condition from 1060 °C, which consists of a combination of α grains with regions of basket-weave structure of acicular martensite α′ and β platelets. β-transus temperature for Ti-6Al-7Nb is 1010 °C, thus quenching from 1060 °C allowed the formation of the martensitic structures [29].

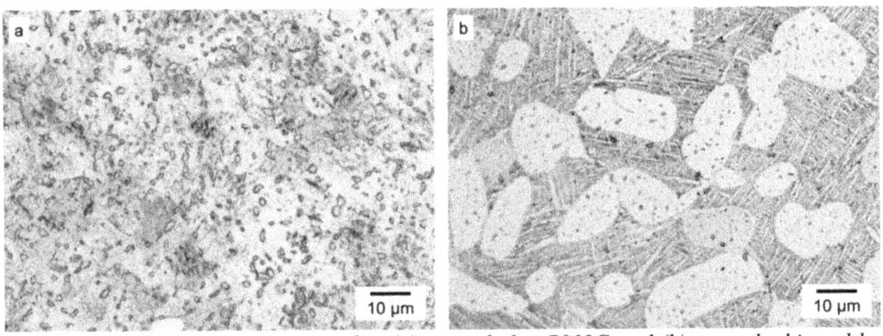

Fig. 1. *Ti-6Al-7Nb microstructures after (a) annealed at 700°C and (b) quenched in cold water from 1060°C.*

Fig. 2 shows a comparison of XRD profiles of A, A+HPT, A+Q, A+Q+HPT conditions, as well as the tip and grip section of A+Q+HPT specimens tested at 850°C with strain rate of $2x10^{-3}$ s^{-1}. Peaks of α and β are observed in A sample. The intensity of the β-phase peaks is decreased beyond the resolution of the XRD for A+HPT sample, which is associated with the refinement of the β-phase particles by HPT. The A+Q and A+Q+HPT samples, show similar behavior, in this case also associated with the fine β platelets that are part of the martensitic structure formed by quenching. The presence of α′ is overlapped with the α peaks in these cases as well. Increased peak broadening can be observed in the cases after HPT processing due to grain refinement and high density of lattice defects [16-18]. The A+Q+HPT tip and grip section profiles show the presence of both α and β peaks with less peak broadening due to the recrystallization and grain growth that occurs during high temperature tensile test.

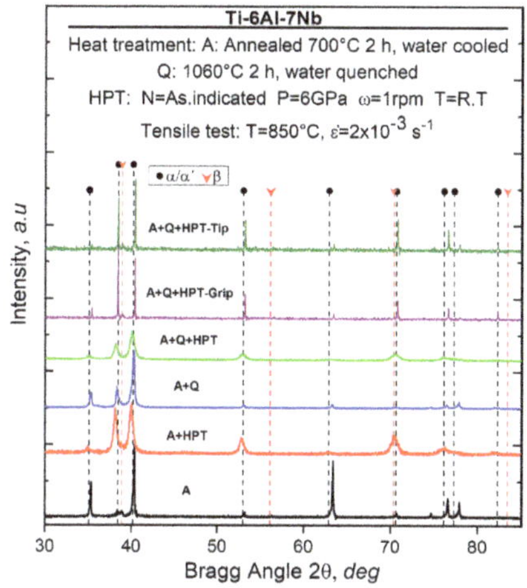

Fig. 2. *XRD profiles before and after HPT and after tensile testing*

Fig. 3 shows Vickers microhardness profiles of A, A+Q, A+HPT and A+Q+HPT samples with respect to the distance from the center of the discs. Dotted lines correspond to the hardness levels of the microstructures in the initial states (after heat treatment: A and A+Q). The A+Q sample attained higher hardness than A, since the resulting microstructure is finer [24]. After HPT, both conditions A+HPT and A+Q+HPT exhibited an increase in microhardness up to ~350 HV and ~400 HV in the edge of the disc respectively, strongly associated with grain refinement provided by HPT. Examination of TEM images also shown in Fig. 3, reveal that the A+Q+HPT had a smaller grain size with respect to A+HPT, it was estimated ~50-70 nm from dark-field images.
High temperature tensile test results are shown in Fig. 4a for the case of A+Q+HPT samples tested at an initial strain rate of $2x10^{-3}$ s^{-1}. This strain rate attained the highest elongations at any given temperature, but in general all samples processed by HPT attained superplastic elongation > 400%, when tested at 800 °C or higher, as shown by selected samples in Fig. 4b. Superplastic elongation increased with testing temperature at the expense of yield stress. Superplastic elongation of A+Q+HPT samples between 800 °C and 900 °C increased from 700% up to 1200% with maximum magnitude reached at 850 °C with a strain rate of $2x10^{-3}s^{-1}$. Meanwhile, A+HPT samples tested at 850°C with $2x10^{-3}$ s^{-1} strain rate reached 900% elongation. This suggests there are differences in microstructural evolution and grain refinement from the initial microstructures. Tensile test at 700°C with a strain rate of $1x10^{-2}$ s^{-1} exhibited the lowest elongation, with not much difference with respect to the quenching.

Materials Research Proceedings **32** (2023) 72-80 https://doi.org/10.21741/9781644902615-7

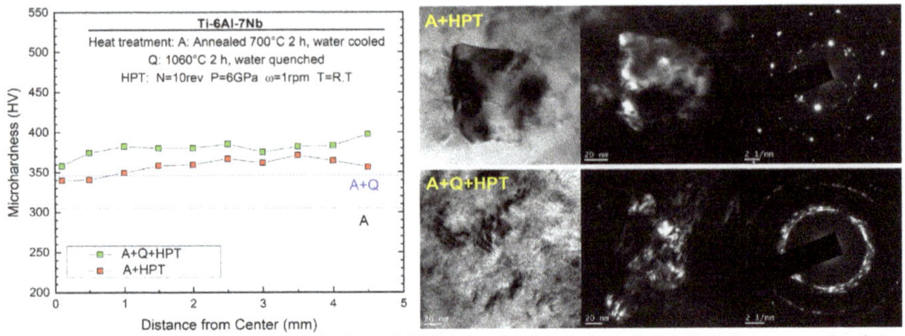

Fig. 3. *Vickers microhardness (left) and TEM bright-field, dark-field, SAED images (right) of heat-treated samples after HPT processing.*

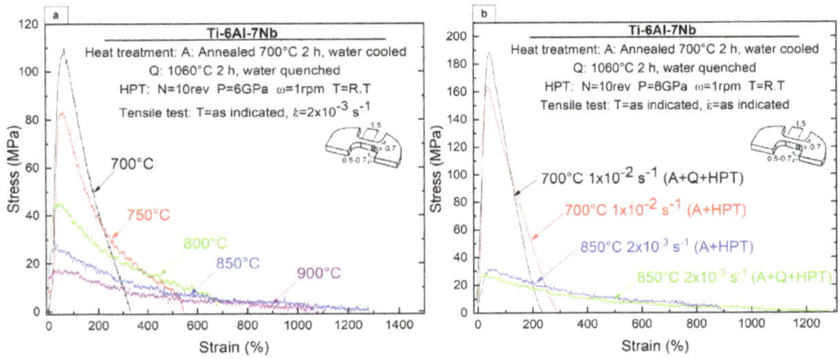

Fig. 4. *Stress-strain curves at different temperatures and strain rates: a) A+Q+HPT at a strain rate of $2x10^{-3}$ s^{-1} and b) A+HPT and A+Q+HPT at 700°C and 850°C at a strain rate of $1x10^{-2}$ s^{-1} and $2x10^{-3}$ s^{-1} respectively*

Fig. 5a shows the plot for maximum stress versus initial strain rate of each of the testing temperatures using logarithmic scale. The best fit lines were estimated for each temperature and consequently the strain rate sensitivity parameter was estimated as the slope of each line $m = (\partial ln\sigma/(\partial ln\dot{\varepsilon})_{\varepsilon,T}$. The values obtained were: 0.275, 0.342, 0.390, 0.388 and 0.462 for 700°C, 750°C, 800°C, 850°C and 900°C respectively. These values are very close to usual magnitudes observed in superplastic region of titanium alloys with m ≥ 0.3 and the grain boundary sliding deformation mechanism [13,25].

Fig. 5b. shows the plot of initial strain rate in logarithmic scale against the inverse of the temperature. Following Eq. 2, the data were adjusted to a linear plot in terms of $\ln(\dot{\varepsilon})$ vs T^{-1}, with σ constant so that the activation energy associated with the process can be obtained from the slope of the fitted line. For the A+Q+HPT samples, the activation energy achieved 277.89 ± 24.15 kJ/mol. This result is slightly higher than 174 kJ/mol which is ordinarily obtained for Ti-6Al-4V α+β alloys [24,30], above self-diffusion of α and β Ti (169 and 153 kJ/mol respectively [30]) and still below the activation energy for dislocation displacement deformation of α and β Ti (306 and 317 kJ/mol respectively [30]). All this suggests that the deformation process present is driven mainly by grain boundary sliding, with a higher activation energy than reported, probably due to other processes that are occurring simultaneously to grain boundary sliding, such as dynamic recrystallization and dislocation accommodation.

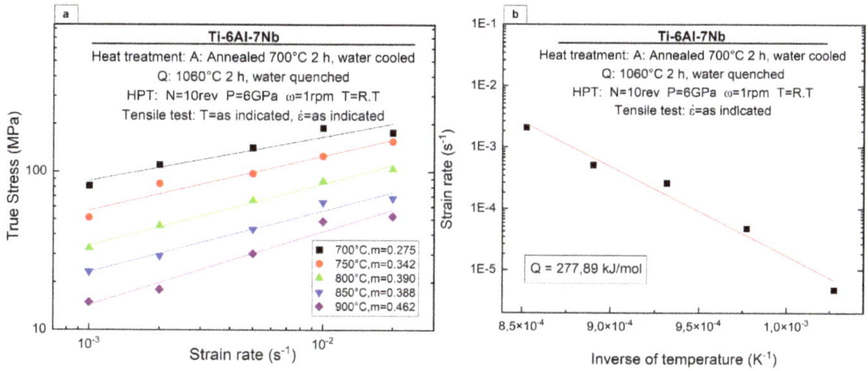

Fig. 5. *Linear fit of (a) maximum stress versus initial strain rate in tensile tests at different temperatures and of (b) initial strain rate versus inverse of temperature*

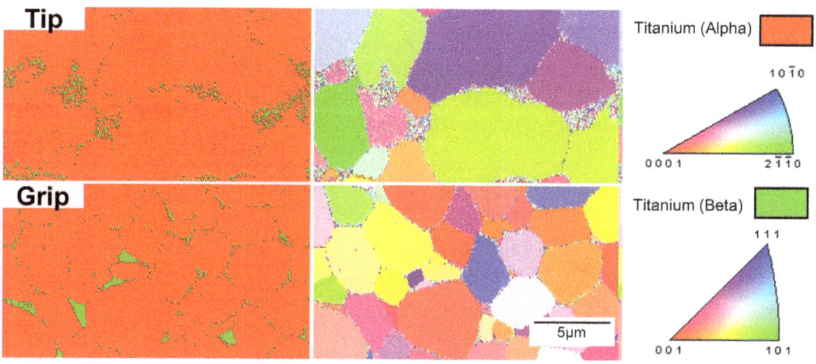

Fig. 6. *EBSD for sample A+Q+HPT after tensile stress at 850°C with strain rate of 2x10^{-3} s^{-1}.*

EBSD analysis was performed in the grip and tip sections of the A+Q+HPT specimen after the high-temperature tensile test at 850°C and strain rate of $2\text{x}10^{-3}$ s^{-1}. Fig. 6 shows the corresponding phase and orientation mapping results. The tip region exhibited a larger grain size ~5.1 μm when compared to the grip region of ~3.4 μm, suggesting that the tip part is subjected to dynamic recrystallization during tensile test, and thus higher energy is generated in the fault zone to allow the grain growth compared to the non-deformed zone, as well as recrystallization of a higher fraction of β phase of ~3.7%, with respect to the grip part of ~3.3%. Therefore, the accommodation mechanisms generated when the A+Q+HPT nanostructure recrystallizes to a typical α+β structure during tensile test can explain the higher activation energy and increased superplastic elongation with respect to the A+HPT condition.

Summary

In this work, Ti-6Al-7Nb alloy was heat treated in two conditions and processed through HPT to produce ultrafine-grained structures. Duplex α+β microstructure was observed in annealed condition and combination of α grains with regions of acicular grains of martensite α′+ β in quenched condition. After HPT processing, both conditions A+HPT and A+Q+HPT exhibited an increase of microhardness measurements up to ~350 HV and ~400 HV. High temperature tensile tests were conducted at different temperatures and strain rates to calculate activation energy and strain rate sensitivity. Superplastic elongation of nanostructured Ti-6Al-7Nb reached 900% and 1200% for A+HPT and A+Q+HPT samples, respectively with a strain rate of $2x10^{-3}$ s^{-1} at 850°C. Activation energy of A+Q+HPT was calculated (277.89 ± 24.15) kJ/mol which suggests grain boundary sliding is not the single deformation mechanism assisting the superplastic elongation. EBSD analysis suggest the occurrence of dynamic recrystallization during tensile test in the tip of failed tensile test specimen, which resulted in larger grain size with respect to the grip part. Thus, accommodation of deformation is enhanced by the finer initial microstructure by quenching which is refined to smaller grains by HPT processing when compared to the annealed initial condition.

References

[1] M. Niinomi, Mechanical biocompatibilities of titanium alloys for biomedical applications, J. Mech. Behav. Biomed. Mater. 1 (2008) 30-42. https://doi.org/10.1016/j.jmbbm.2007.07.001

[2] M. Balazic, J. Kopac, M.J. Jackson, W. Ahmed, Review: titanium and titanium alloy applications in medicine, Int. J. Nano and Biomaterials. 1 (2007) 33-34. https://doi.org/10.1504/IJNBM.2007.016517

[3] A.S. Hoffman, Introduction, The Diversity and Versatility of Biomaterials, in: B.D. Ratner, A.S. Hoffman, F.J. Schoen, J.E Lemons (Eds.), Biomaterials Science, An Introduction to Materials: Third Edition, Elsevier, Oxford, 2013, pp. 63-64. https://doi.org/10.1016/B978-0-08-087780-8.00007-3

[4] F.H. Froes, Titanium: Physical Metallurgy, Processing, and Applications, ASM International, 2015. https://doi.org/10.31399/asm.tb.tpmpa.9781627083188

[5] M.J. Donachie, Titanium: A Technical Guide, second ed., ASM International, 2000. https://doi.org/10.31399/asm.tb.ttg2.9781627082693

[6] I. Polmear, D. StJohn, J.F. Nie, M. Qian, Titanium Alloys, in: I. Polmear, D. StJohn, J.F. Nie, M. Qian (Eds.), Light Alloys, Metallurgy of the Light Metals, Elsevier, 2017, pp. 369-460. https://doi.org/10.1016/B978-0-08-099431-4.00007-5

[7] S. Lampman, Titanium and Its Alloys for Biomedical Implants, in: R.J Narayan (Eds.), Materials for Medical Devices: Volume 23, ASM International, 2012, pp.223-236. https://doi.org/10.31399/asm.hb.v23.a0005674

[8] M. Naito, T.Yokoyama, M.Hosokawa, K. Nogi, Nanoparticle Technology Handbook, third ed., Elsevier, 2008

[9] T.G. Langdon, Seventy-five years of superplasticity: Historic developments and new opportunities, J. Mater. Sci. 44 (2009) 5998-6010. https://doi.org/10.1007/s10853-009-3780-5

[10] T.G. Langdon, Superplasticity of bulk nanostructured materials, in: M.J. Zehetbauer, Y.T. Zhu (Eds.), Bulk Nanostructured Materials, Wiley-VCH, Germany. 2009 pp.455-468. https://doi.org/10.1002/9783527626892.ch20

[11] Y.H.Z. Zhao, Advanced Mechanical Properties and Deformation Mechanisms of Bulk Nanostructured Materials, Trans. Tech. Publications, 2011.

[12] M.T. Pérez-Prado, M.E. Kassner, Superplasticity, in: M.E. Kassner (Eds), Fundamentals of Creep in Metals and Alloys, Elsevier, 2008: pp. 135-152. https://doi.org/10.1016/B978-0-08-047561-5.00006-3

[13] F.A. Mohamed, Micrograin superplasticity: Characteristics and utilization, Materials. 4 (2011) 1194-1223. https://doi.org/10.3390/ma4061194

[14] M. Kawasaki, T.G. Langdon, Principles of superplasticity in ultrafine-grained materials, J. Mater. Sci. 42 (2007) 1782-1796. https://doi.org/10.1007/s10853-006-0954-2

[15] A. Alhamidi, Z. Horita, Grain refinement and high strain rate superplasticity in alumunium 2024 alloy processed by high-pressure torsion, Mater. Sci. Eng. A. 622 (2015) 139-145. https://doi.org/10.1016/j.msea.2014.11.009

[16] R.Z. Valiev, R.K. Islamgaliev, I. V Alexandrov, Bulk nanostructured materials from severe plastic deformation, Prog. Mater. Sci. 45 (2000) 103-189. https://doi.org/10.1016/S0079-6425(99)00007-9

[17] R.Z. Valiev, Y. Estrin, Z. Horita, T.G. Langdon, M.J. Zehetbauer, Y.T. Zhu, Producing Bulk Ultrafine-Grained Materials by Severe Plastic Deformation, JOM. 58 (2006) 33-39. https://doi.org/10.1007/s11837-006-0213-7

[18] R.Z. Valiev, Y. Estrin, Z. Horita, T.G. Langdon, M.J. Zehetbauer, Y.T Zhu, Producing Bulk Ultrafine-Grained Materials by Severe Plastic Deformation: Ten Years Later, JOM. 68 (2016) 1216-1226. https://doi.org/10.1007/s11837-016-1820-6

[19] A. Azushima, R. Kopp, A. Korhonen, D.Y. Yang, F. Micari, G.D. Lahoti, P. Groche, J. Yanagimoto, N. Tsuji, A. Rosochowski, A. Yanagida, Severe plastic deformation (SPD) processes for metals, CIRP annals. 57 (2008) 716-735. https://doi.org/10.1016/j.cirp.2008.09.005

[20] A.P. Zhilyaev, T.G. Langdon, Using high-pressure torsion for metal processing: Fundamentals and applications, Prog Mater Sci. 53 (2008) 893-979. https://doi.org/10.1016/j.pmatsci.2008.03.002

[21] G. Faraji, H.S. Kim, H.T. Kashi, Severe Plastic Deformation: Methods, Processing and Properties, first ed., Elsevier, 2018. https://doi.org/10.1016/B978-0-12-813518-1.00003-5

[22] Z. Liu, P. Li, L. Xiong, T. Liu, L. He, High-temperature tensile deformation behavior and microstructure evolution of Ti55 titanium alloy, Mater. Sci. Eng. A. 680 (2017) 259-269. https://doi.org/10.1016/j.msea.2016.10.095

[23] G. Zhou, L. Chen, L. Liu, H. Liu, H. Peng, Y. Zhong, Low-temperature superplasticity and deformation mechanism of Ti-6Al-4V alloy, Materials. 11 (2018) 1212. https://doi.org/10.3390/ma11071212

[24] E. Alabort, P. Kontis, D. Barba, K. Dragnevski, R.C. Reed, On the mechanisms of superplasticity in Ti-6Al-4V, Acta Mater. 105 (2016) 449-463. https://doi.org/10.1016/j.actamat.2015.12.003

[25] J. Sieniawski, M. Motyka, Superplasticity in titanium alloys, JAMME. 24 (2007)123-130.

[26] W. Zhang, H. Ding, P.H.R. Pereira, Y. Huang, T.G. Langdon, Grain refinement and superplastic flow in a fully lamellar Ti-6Al-4V alloy processed by high-pressure torsion, Mater. Sci. Eng. A. 732 (2018) 398-405. https://doi.org/10.1016/j.msea.2018.07.010

[27] M. Ashida, P. Chen, H. Doi, Y. Tsutsumi, T. Hanawa, Z. Horita, Superplasticity in the Ti-6Al-7Nb alloy processed by high-pressure torsion, Mater. Sci. Eng. A. 640 (2015) 449-453. https://doi.org/10.1016/j.msea.2015.06.020

[28] J.M. Cubero-Sesin, J.E. Gonzalez-Hernandez, E. Ulate-Kolitsky, K. Edalati, Z. Horita, Superplasticity of nanostructured Ti-6Al-7Nb alloy with equiaxed and lamellar initial microstructures processed by High-Pressure Torsion, IOP Conf. Ser.: Mater. Sci. Eng. IOP Publishing, 194 (2017) 1-6. https://doi.org/10.1088/1757-899X/194/1/012041

[29] J. Sieniawski, W. Ziaja, K. Kubiak, M. Motyk, Microstructure and Mechanical Properties of High Strength Two-Phase Titanium Alloys, in: J. Sieniawski, W. Ziaja, K. Kubiak, M. Motyk (Eds), Titanium Alloys -Advances in Properties Control, InTech, 2013, pp.69-80 https://doi.org/10.5772/56197

[30] J. Shen, Y. Sun, Y. Ning, H. Yu, Z. Yao, L. Hu, Superplasticity induced by the competitive DRX between BCC beta and HCP alpha in Ti-4Al-3V-2Mo-2Fe alloy, Mater. Charact. 153 (2019) 304-317. https://doi.org/10.1016/j.matchar.2019.05.014

Materials Research Proceedings **32** (2023) 81-86 https://doi.org/10.21741/9781644902615-8

Precise forming of Ti60 alloy by superplastic forming /diffusion bonding

Jiang Shaosong[1,a,*], Kang Liangwei[1,b], Li Yang[1,c]

[1]State Key Laboratory of Metal Precision Hot Processing, Harbin Institute of Technology, Harbin, China

[a]jiangshaosong@hit.edu.cn, [b]15513547207@163.com, [c]2468462178@qq.com

Keywords: Ti60, Diffusion Bonging, Superplastic Forming, Four-Storey Structure, High-Temperature Titanium Alloy

Abstract. The study investigated the high-temperature tensile behavior of Ti60 alloy at different temperatures (940°C to 1040°C) and different strain rates ($0.1s^{-1}$ to $0.0005s^{-1}$). Ti60 exhibited superplasticity with the best deformation temperature ranging from 940°C to 960°C. The elongation of the samples decreased significantly at the temperatures over 980°C due to the growth of grain size. In addition, the study investigated the influence of diffusion bonding temperature and pressure on the microstructure and mechanical properties of the joint. The results showed that there was no microscopic gaps or cracks at the interface of the diffusion bonding joints performed at 2.5MPa, 920°C and 2 hours. Thereafter, a four-layer structure of Ti60 alloy was fabricated through the SPF/DB. The component with uniform thickness distribution, high surface quality, high diffusion bonding quality was achieved.

Introduction

Titanium and its alloys have wide applications in various fields such as medicine, power generation, automotive manufacturing, and aviation due to their special properties such as good corrosion resistance, low density, and excellent mechanical properties at high temperatures [1]. Compared with steel and other hard alloys, titanium alloy is widely used in aerospace, chemical industry and automobile fields because of its high specific strength, corrosion resistance and excellent fracture toughness. With the requirements of high temperature resistant materials, a lot of new materials such as IMI834, Ti1100, BT18Y, Ti60 and other near-α high temperature titanium alloys serving at temperatures up to 600°C have been developed. As a potential candidate material for airfoil blade and wheel, Ti60 alloy has been widely concerned by scholars. Compared with IMI834, more Si elements were added to Ti60 alloy, which improved the creep property. A small amount of Ta and C elements were added to improve the heat resistance and widen the α+β phase field.

Ti60 alloy is a hige temperature alloy with high deformation resistance, making it difficult to use traditional processing methods. However, it has been reported that Ti60 has superplasticity at high temperatures [6]. Therefore, SPF and DB can be used for processing. SPF involves forming metal sheets with fine grain or two-phase microstructure at high temperatures, using gas pressure with tools or molds at lower strain rates [7]. Compared with traditional processing techniques such as forging, casting, and milling, SPF/DB posseses unique advantages, including greater design freedom, lower mold costs, elimination of rebound, and production of complex geometries in a single manufacturing step [8].

Wanjara [9] studied the isothermal hot working flow stress behavior of IMI834 at temperatures of 950°C, and the results showed that the phase exhibited greater flow stress softening at higher strain rates. Biallas and Gerhard [10] investigated the effect of environment on the deformation behavior of high temperature titanium alloy IMI834, cracks were found to initiate at slip bands.

Xiufeng Gao, Shaoping Chen et al.[11]used external current to increase atomic activity and performed diffusion bonding on Ti/Ni, and the results showed that the growth of all phases was enhanced by the application of a direct current. However, there are few investigations on the deformation behavior at high temperatures and diffusion bonding parameters of Ti60 alloy. Thus, the purpose of this paper is to study the high temperature deformation behavior of Ti60 and influence of pressure and temperature on the quality of diffusion bonding joints. Besides, a four-layer hollow structural component without obvious defects was manufactured.

Organization of the Text

Material

The Ti60 plates with the thickness of 2mm are provided by Litai Metal Co. Ltd. Their chemical composition is shown in Table 1. The sufficient α stable element and a certain amount of β stable element make it fully solid solution strengthened, and the isomorphous β element further improves its thermal stability.

Table1 Chemical composion of Ti60 titanium alloy (wt%)

Element	Al	Sn	Zr	Mo	Si	Nb	Ta	C	Ti
wt%	5.7	4	3.5	0.4	0.4	0.4	1.0	0.05	allowance

High temperature tensile test

The high-temperature tensile tests were conducted to investigate the high-temperature deformation behavior of Ti60 alloy. The temperature range was from 940°C to 1040°C, and the strain rates range was from $10^{-1}s^{-1}$ to $10^{-4}s^{-1}$.

Figure 1 shows the true stress-strain curves of Ti60 alloy at different temperatures and strain rates. The flow stress decreased with the increase of temperature, and the material exhibited flow softening behavior after the peak of flow stress. The flow stress increased with the increase of strain rate (from $10^{-1}s^{-1}$ to $10^{-4}s^{-1}$). All curves in Figure 1 showed similar fluctuation trends which could be roughly divided into three stages. The first stage was the initial deformation stage before the peak stress; the second stage started with the appearance of the peak stress to the occurrence of necking ; the last stage started from the occurrence of necking until fracture, which was more obvious in Figure 1 (a), directly reflecting the fracture of Ti60. When the strain rate was $0.005s^{-1}$, the flow stress approached to a steady state indicating that there were sufficient softening mechanisms to balance the strain hardening, such as the dynamic recovery and dynamic recrystallization.

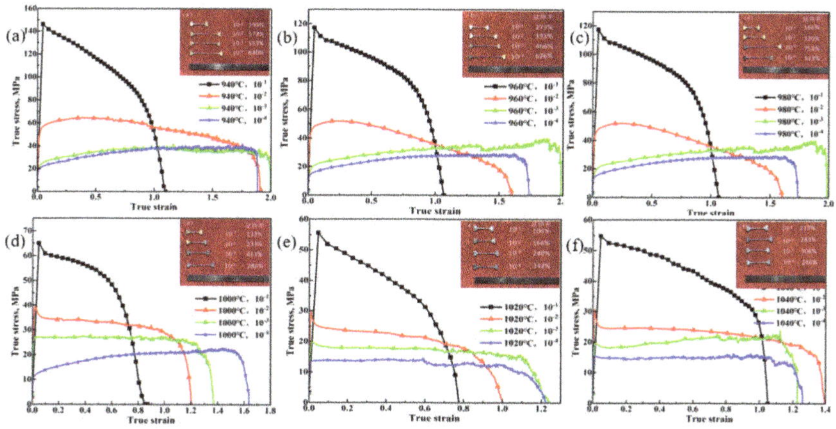

Fig. 1 Real stress-strain curves at different strain rates (a)940℃ (b) 960℃ (c) 980℃ (d) 1000℃ (e) 1020℃ (f) 1040℃.

The dynamic recovery and dynamic recrystallization are the main softening mechanisms for Ti60 alloy at high temperature deformation[14,15]. The degree of softening depends on the the variation of temperaturses which changes the microstructure, such as grain sizes or dislocation densit. Figures 2 shows the trends of elongation and tensile strength at different temperatures. When the temperature is 980°C and the strain rate is $10^{-4}s^{-1}$, the elongation of the material is the highest (780%). The elongation of the materials generally decreases with the decrease of temperature. When the temperature is 980°C and 1000°C, the elongation of the material increases when the strain rate decreases.

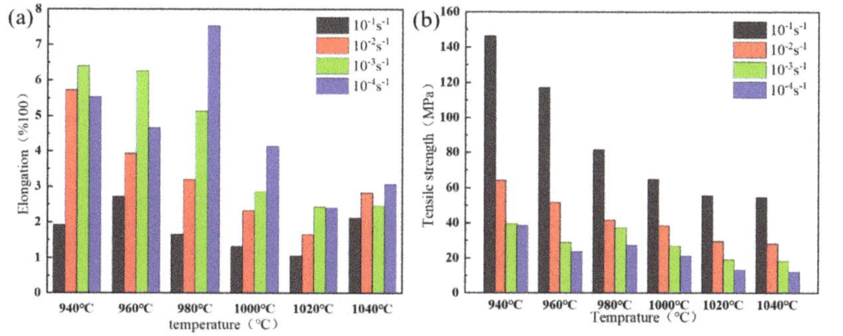

Fig. 2 (a) The variation trend of elongation at different temperatures after fracture (b) The trend of tensile strength with temperature.

Fig. 2 (b) shows the tensile strength of samples at different temperatures. It can be seen that the tensile strength decreases with the increase of temperature at the same strain rate, and the tensile strength increases with increasing of strain rates at the same temperature. Therefore, for the deformation behavior of Ti60, the temperature and strain rate are important parameters. The maximum tensile strength (148.4Mpa) is achieved at 940°C and $0.001s^{-1}$.

Materials Research Proceedings **32** (2023) 81-86
https://doi.org/10.21741/9781644902615-8

Diffusion bonding

This experiment aimed to investigate the influence of several experimental parameters on the joint performance by using different temperatures and pressures to obtain optimal joint quality. Prior to the experiment, the contact surfaces of the Ti60 alloy were polished by 180# sandpaper, the plate surface was also cleaned by ultrasonic to remove oil stains.

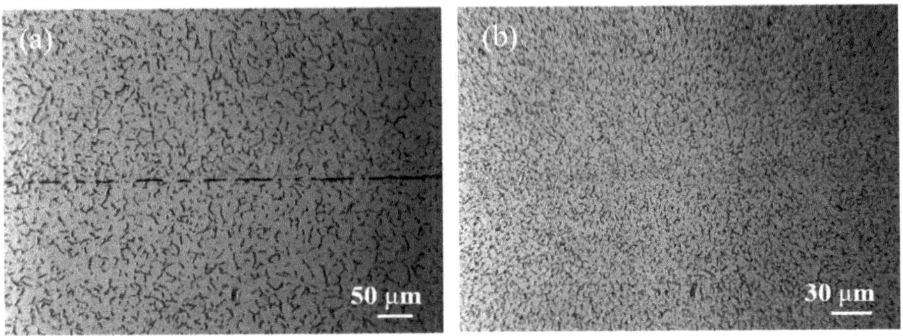

Fig. 3 microstructure of the diffusion-bonded joint at (a) 900℃, 2.5Mpa,1.5h. (b) 920℃, 2.5Mpa,2h.

In Figure 3 (a), due to the insufficient diffusion temperature and holding time, there were several unconnected areas at the diffusion-bonded joints. With the increase of temperature and the extension of holding time, the atomic diffusion of the sample became more sufficient and the joint ratio could reach over 90%. However, if the temperature is too high, the grains size of the Ti60 will increase, resulting in embrittlement of the material.

At 920°C, 2.5 MPa for 2 hours, the surface of the joint was smooth. Due to the high pressure of the diffusion bonding, the material underwent large plastic deformation at the interface, which ensured the tight connection of the joint. The interface of the joints almost disappeared, and there were no pores or cracks in the connection interface.

Four-storey structure

The lightweight four-layer structure with high toughness saves materials and maks it highly valuable in the aerospace industry. Based on the above experimental results, the optimal temperature for DB was 920°C. Therefore, Ti60 sheets were diffusion-bonded at 920°C and 2.5 MPa for about 2 hours. Subsequently, SPF was performed at 980°C with a strain rate of $10^{-4}s^{-1}$.

According to the high-temperature tensile and diffusion bonding experimental parameters described above, Ti60 four-layer structure components were processed, as shown in Figure 4. The surface of the component was flat and smooth, with no obvious defects such as burrs, and the dimensional accuracy was high. The vertical ribs structure of the component could be observed at the wire-electrode cutting position, as shown in Figure 4. The quality of the vertical ribs was well, and the joint in the diffusion bonding area of the core sheet was tight.

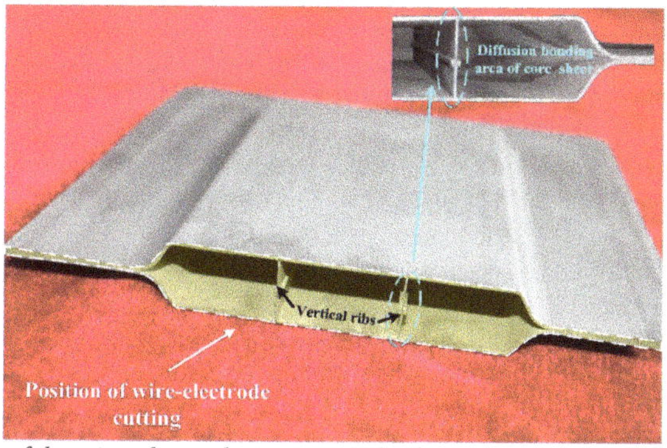

Fig. 4 the position of the wire-electrode cutting on the four-layer structure and the vertical ribs.

Figure 5 shows the defects that occurred during the SPF/DB process. It shows the failure of diffusion bonding due to insufficient holding time of the core sheet, resulting in the disconnection of the core sheet and the Face sheet. And it shows the rupture of the core sheet at the location of excessive thinning during the SPF process, causing insufficient air pressure and failure of the experiment.

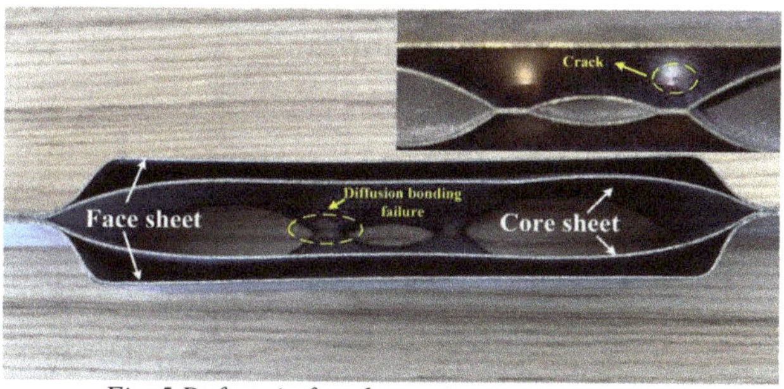

Fig. 5 Defects in four-layer structure components

The outer surface of the four-layer structure was smooth and flat, with no obvious collapse, and the cross-sectional shape was regular without any significant defects. There were no obvious unconnected parts in the diffusion bonding area. The Ti60 four-layer structure produced in this paper ensures a lightweight structure with excellent temperature resistance performance. It ensures the continuous improvement of spacecraft speed and performance.

Summary

The superplasticity of Ti60 alloy was studied in the temperature range from 940°C to 1040°C and the strain rate range from $10^{-1}s^{-1}$ to$10^{-4}s^{-1}$. The diffusion bonding of Ti60 alloy was also studied, and a four-layer structure was fabricated. The following conclusions can be drawn: At 980°C and a strain rate of $1\times10^{-4}s^{-1}$, Ti60 can achieve the maximum elongation rate (780%). The materials tensile strength is inversely proportional to the temperature and directly proportional to the strain rate. A stable joint surface can be obtained by insulation at 920°C and 2.5Mpa for 2 hours. Based on the above experimental parameters, the Ti60 four-layer structure was produced, and with a high surface quality and no significant defects.

References

[1] HUANG L. J., YANG F. Y., HU H. T., et al. TiB whiskers reinforced high temperature titanium Ti60 alloy composites with novel network microstructure, J. Mater Design. 51 (2013) 421-426. https://doi.org/10.1016/j.matdes.2013.04.048

[2] XU Z. Y., CHEN X. Z., ZHOU Z. S., et al. Electrochemical machining of high-temperature titanium alloy Ti60, J. Proc Cirp. 42 (2016) 125-130. https://doi.org/10.1016/j.procir.2016.02.206

[3] SHAO X., GUO X. L., HAN Y. F., et al. Characterization of the diffusion bonding behavior of pure Ti and Ni with different surface roughness during hot pressing, J. Mater Design. 65 (2015) 1001-1010. https://doi.org/10.1016/j.matdes.2014.09.071

[4] JIA W J., ZENG W. D., LIU J. R., et al. Influence of thermal exposure on the tensile properties and microstructures of Ti60 titanium alloy, J. Mat Sci Eng a-Struct. 530 (2011) 511-518. https://doi.org/10.1016/j.msea.2011.10.011

[5] LIU J., TAN M. J., AUE-U-LAN Y., et al. Superplastic-like forming of Ti-6Al-4V alloy, J. Int J Adv Manuf Tech. 69 (2013) 1097-1104. https://doi.org/10.1007/s00170-013-5101-z

[6] Ding Yuan, Shuaiqi Shao, Chunhuan Guo, et al. Grain Refining Of Ti-6al-4v Alloy Fabricated By Laser And Wire Additive Manufacturing Assisted With Ultrasonic Vibration, J. Ultrasonics Sonochemistry. 73 (2021) 105472 https://doi.org/10.1016/j.ultsonch.2021.105472

[7] PENG P., JIANG S., QIN Z., et al. Superplastic Forming and Reaction Diffusion Bonding Process of Hollow Structural Component for Mg-Gd-Y-Zn-Zr Rare Earth Magnesium Alloy, J. Metals. 12 (2022) 1. https://doi.org/10.3390/met12010152

[8] YANG J. L., WANG G. F., JIAO X. Y., et al. High-temperature deformation behavior of the extruded Ti-22Al-25Nb alloy fabricated by powder metallurgy, J. Mater Charact. 137 (2018) 170-179. https://doi.org/10.1016/j.matchar.2018.01.019

[9] WANJARA P., JAHAZI M., MONAJATI H., et al. Hot working behavior of near-α alloy IMI834, J. Materials Science & Engineering A. 396 (2005) 50-60. https://doi.org/10.1016/j.msea.2004.12.005

[10] BIALLAS G., ESSERT M., MAIER H. J., et al. Influence of environment on fatigue mechanisms in high-temperature titanium alloy IMI834, J. Int J Fatigue. 27 (2005) 1485-1493. https://doi.org/10.1016/j.ijfatigue.2005.06.009

[11] GAO X. F., CHEN S. P., DONG F., et al. Diffusion bonding of Ti/Ni under the influence of an electric current: mechanism and bond structure, J. J Mater Sci. 52 (2017) 3535-3544. https://doi.org/10.1007/s10853-016-0648-3

[12] EROGLU M., KHAN T. I., ORHAN N.. Diffusion bonding between Ti-6Al-4V alloy and microduplex stainless steel with copper interlayer, J. Mater Sci Tech-Lond. 18 (2002) 68-72. https://doi.org/10.1179/026708301125000230

[13] JIA W. J., ZENG W. D., ZHOU Y. G., et al. High-temperature deformation behavior of Ti60 titanium alloy, J. Mat Sci Eng a-Struct. 528 (2011) 4068-4074. https://doi.org/10.1016/j.msea.2011.01.113

[14] DENG J., LIN Y. C., LI S. S., et al. Hot tensile deformation and fracture behaviors of AZ31 magnesium alloy, J. Mater Design. 49 (2013) 209-219. https://doi.org/10.1016/j.matdes.2013.01.023

[15] LIN Y. C., DENG J., JIANG Y. Q., et al. Effects of initial delta phase on hot tensile deformation behaviors and fracture characteristics of a typical Ni-based superalloy, J. Mat Sci Eng a-Struct. 598 (2014) 251-262. https://doi.org/10.1016/j.msea.2014.01.029

Materials Research Proceedings 32 (2023) 87-94 https://doi.org/10.21741/9781644902615-9

Numerical simulation and experimental trials on superplastic forming for thin-walled structure of Ti2AlNb alloy

NING ZHANG[1,2,3,a*], YAOQI WANG[1,2,3,b], YANHONG MU[1,2,3,c] and ZHEN LI[1,2,3,d]

[1] AVIC Manufacturing Technology Institute, Beijing 100024, China

[2] Aeronautical Key Laboratory for Plastic Forming Technology, Beijing 100024, China

[3] Beijing Key Laboratory of Digital Forming Technology and Equipment, Beijing 100024, China

[a*]zn64112@163.com, [b]54805819@qq.com, [c]muyh001@avic.com, [d]lizhen90420@163.com

Keywords: Ti2AlNb Alloy, Superplastic Forming, Constitutive Equation, Numerical Simulation

Abstract. The difficult formability of Ti2AlNb alloy leads to the difficulty of traditional manufacturing methods to form the complex thin-walled components, limiting the wide application of Ti2AlNb alloy in aerospace. In this paper the superplastic forming process of Ti2AlNb alloy with thin-walled structure was studied. The superplastic tensile tests of Ti2AlNb alloy were carried out at varying temperature ranging from 920°C to 980°C and strain rate ranging from $0.001s^{-1}$ to $0.03s^{-1}$. The effect of the deformation-temperature and strain rate on the superplastic deformation behaviour was analysed, showing that the flow stress and elongation increase with the increasing temperature and the flow stress decreases, the elongation increases with the decrease of the strain rate. The elongation reaches the maximum value of 525% at the temperature of 940°C and the strain rate of $0.001s^{-1}$. The constitutive equation taking into account the effect of strain was developed, which was implemented into the FE-based software Abaqus to simulate the superplastic forming of the thin-walled structure. The optimal values of the slab thickness and loading pressure-time curve were obtained by simulation and the superplastic forming tests of thin-walled structure were performed. The comparison between the numerical and experimental data with regard to thickness variation verified the accuracy of the finite element model. The research results provide a reference basis for the preparation of complex thin-walled structures of Ti2AlNb alloy in aerospace.

Introduction

When hypersonic aircraft flies at high Mach speed, severe aerodynamic heating phenomenon may occur, resulting in a sharp rise in the surface temperature of the aircraft, which puts forward more stringent requirements for the material properties of components, especially for high-temperature structural materials with low density, high specific strength and good high-temperature performance[1-3]. Ti2AlNb alloy is a new Ti-Al intermetallic compound material type with an ordered orthogonal structure. It has excellent high-temperature properties (specific strength, specific stiffness, creep resistance at high temperatures, oxidation resistance, heat resistance and flame retardant properties). It can be used for a long time in the range of 600 ~ 750°C, and its density is lower than that of nickel-based superalloys. It has become an aerospace lightweight high-temperature structural material with the most potential to replace nickel-based superalloy [4-7].

However, Ti2AlNb alloy is a refractory material with poor plasticity at room temperature, high deformation resistance at high temperature and demanding deformation conditions, which seriously restrict its development and industrial application. Superplastic forming technology is a process that takes advantage of the excellent formability of materials in the superplastic state to extend the materials under appropriate stress and strain rate by applying air pressure. This process

Materials Research Proceedings **32** (2023) 87-94 https://doi.org/10.21741/9781644902615-9

is suitable for the forming of hard-to-deform materials, especially thin-walled and complex surface components [8-10].

In this paper, the superplastic deformation behavior of Ti2AlNb alloy was studied by high temperature superplastic tensile test, and the superplastic forming constitutive equation of the material was established. The numerical simulation of the superplastic forming process of the thin-walled structure of Ti2AlNb alloy was carried out by using the finite element software Abaqus to study the forming condition and the change of wall thickness to optimize the forming load. The pressure-time curve is obtained to provide theoretical guidance for the forming test of structural parts.

Establishment of constitutive equation

Superplastic deformation behavior

The experimental material is a 1.5mm thick Ti2AlNb alloy plate with a grain size of 11-12 grades and an average size of<8 μm. The high temperature superplastic tensile test was carried out on LETRYDL-20T electronic universal tensile test machine. The experimental deformation temperature was 920 °C, 940 °C, 960 °C, 980 °C and the strain rate was 0.001 s^{-1}, 0.003 s^{-1}, 0.01 s^{-1}, 0.03 s^{-1}. The speed of the collet was kept constant during the stretching process. The obtained true stress-strain curves are shown in Fig. 1. At the same strain rate, the elongation and peak stress decrease with the increase of deformation temperature. The maximum elongation of Ti2AlNb alloy is 430 % at 920 °C and 0.003 s^{-1}. When the deformation temperature increases from 920 °C to 940 °C, the superplastic elongation of Ti2AlNb alloy decreases from 430 % to 428 %, and the peak stress decreases from 69.3 MPa to 56.3 MPa. The elongation at 940°C is not much different from that at 920 °C, but the peak stress is reduced by 23 %.

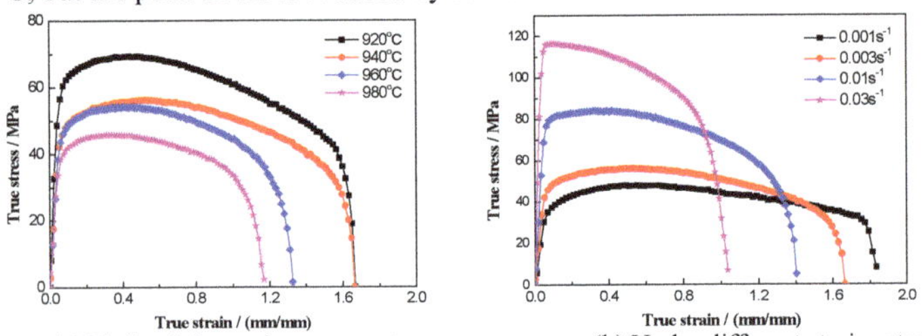

(a) Under deformation temperatures (b) Under different strain rates

Fig. 1 True stress-stain curves of Ti2AlNb alloy

At the same deformation temperature, the elongation decreases and the peak stress increases with the increase in strain rate. The maximum elongation of Ti2AlNb alloy is 525 % at 940 °C and 0.001 s^{-1}. When the strain rate increases from 0.001 s^{-1} to 0.03 s^{-1}, the superplastic elongation of Ti2AlNb alloy decreases from 525 % to 182 %. In contrast, the peak stress increases from 48 MPa to 116 MPa, indicating that the superplasticity of the material is sensitive to the change of strain rate.

Constitutive equation

The temperature during superplastic deformation is constant. Therefore, this paper temporarily ignores the influence of temperature and adopts the following Eq. (1) to describe the constitutive equation of materials:

$$\sigma = K\varepsilon^{n}\dot{\varepsilon}^{m} \qquad (1)$$

Where: K is the strength coefficient; n is strain hardening index; m is the strain rate sensitivity index.

Logarithm of both sides of the above equation, we get:

$$\ln\sigma = \ln K + n\ln\varepsilon + m\ln\dot{\varepsilon} \tag{2}$$

Under specific strain, the m value is the slope of the lnσ-ln$\dot{\varepsilon}$ curve. True stress-strain curves of material at 940°C are selected to calculate the m values for different strains. The relationship between m values and strains is shown in Fig. 2. The m value decreases gradually with the increase of the strain, showing a linear relationship. Therefore, the expression of m value can be obtained by fitting it:

$$m = 0.334 - 0.155\varepsilon \tag{3}$$

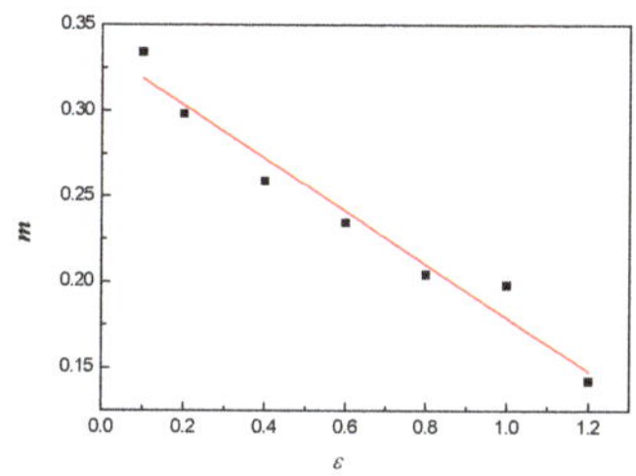

Fig. 2 m-ε curve

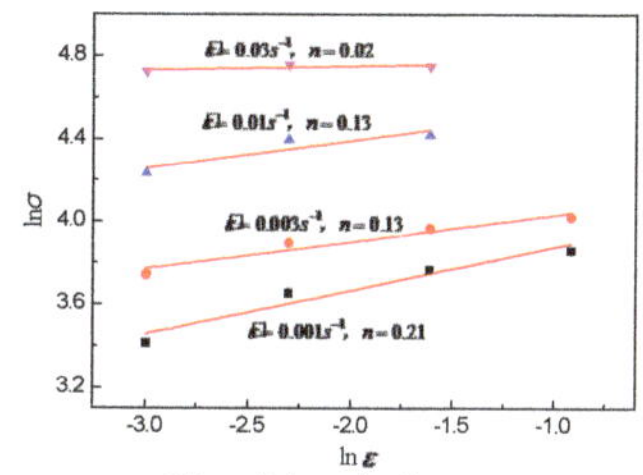

Fig. 3 lnσ-lnε curves

When the strain rate is constant, n is the slope of the lnσ-lnε curve from Eq. (2). The true stress-strain curves under different strain rates at 940°C show that when the strain is greater than 0.4, the stress gradually decreases with the increase of strain, and the dynamic softening plays a dominant role. Thus, n value can be calculated only from the strain value between 0.05 and 0.4, as shown in Fig. 3. Taking the average value of n calculated by different strain values, $n = 0.1225$ is obtained.

Substitute the calculated values of m and n into Eq. (2), and the following can be obtained:

$$\ln\sigma = \ln K + 0.1225\ln\varepsilon + (0.334 - 0.155\varepsilon)\ln\dot{\varepsilon} \tag{4}$$

The K values corresponding to different strains under four strain rates are calculated respectively. The results show that the K values under the same strain have little difference, that is, the K value is not sensitive to strain rate. Therefore, the K value at each strain is the average value of K calculated at the same strain at different strain rates. Draw K-ε curve, as shown in Fig. 4, and the expression of K is obtained by fitting the least square method:

$$K = 498.71 - 482\varepsilon + 123.70\varepsilon^2 \tag{5}$$

Materials Research Forum LLC
https://doi.org/10.21741/9781644902615-9

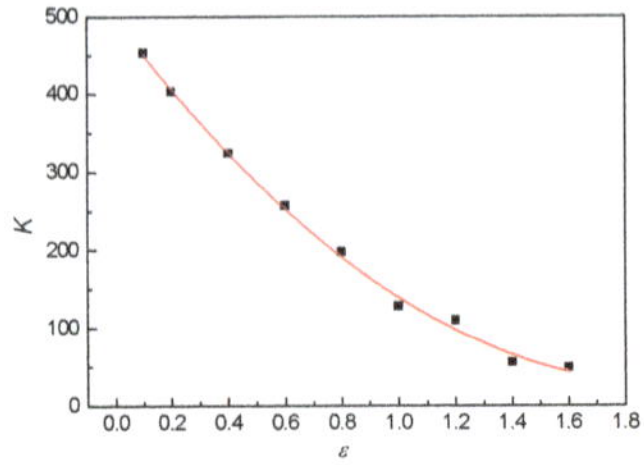

Fig. 4 K-ε curve

Therefore, the constitutive equation of superplastic formation of Ti2AlNb alloy at 940°C is obtained as follows:

$$\begin{cases} \sigma = K\varepsilon^n \dot{\varepsilon}^m \\ m = 0.334 - 0.155\varepsilon \\ K = 498.71 - 482\varepsilon + 123.70\varepsilon^2 \end{cases} \tag{6}$$

Where $n = 0.1225$. The experimental and computational results of flow stress is shown in Fig. 5 which indicates that the constitutive equation considering the effect of strain agrees well with the experimental data.

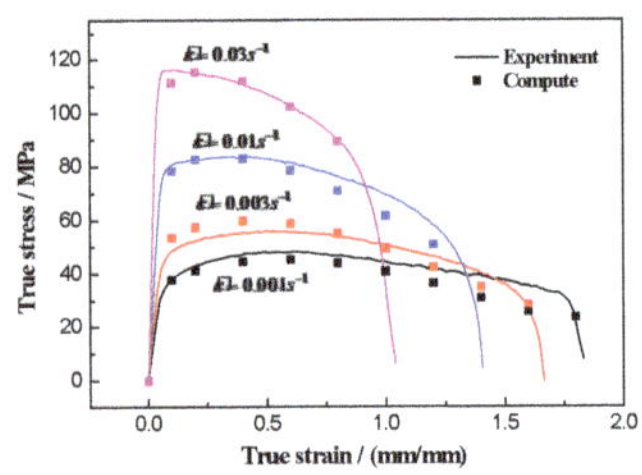

Fig. 5 Comparison of computational data (symbols) with experimental results (solid curves)

Establishment of finite element model of superplastic forming

FEM model

The finite element software Abaqus was used to simulate the superplastic bulging processes of of Ti2AlNb alloy thin-walled structural part. The dimension of the part is 300mm×300mm×15mm, and it is symmetrical, as shown in the Fig. 6. In order to simplify the calculation, only 1/4 of the member was taken for simulation, the plate size is 150mm×150mm, and the plate thickness is 1.5mm. The FEM model was established with Catia, and then imported into Abaqus software as shown in Fig.6. The plate was defined as shell elements and the die as discrete solid elements.

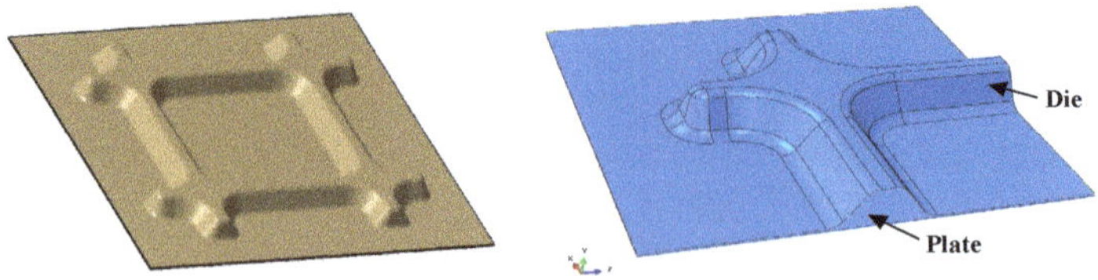

Fig. 6 Thin-walled structural part and FEM model

Material parameter setting

In the superplastic forming simulation of Abaqus software, the material properties can be defined through the Creep model, but the constitutive equation established in this paper is difficult to be converted into the creep constitutive model inherent in the software. Therefore, the Abaqus software needs to be redeveloped, and the superplastic constitutive equation established in this paper is written into the Creep user subroutine through Fortran programming language, so as to achieve accurate simulation of the superplastic forming process.

Pressure-time optimization control

During the superplastic forming, properly controlling the pressure-time curve will make the maximum strain rate always near the target strain rate, which can not only shorten the forming time, but also reduce the uneven wall thickness. In Abaqus software, through load control algorithm, the solver will automatically adjust the loading of pressure in real time to ensure that the maximum strain rate is maintained near the target strain rate [11]. Abaqus software can automatically adjust the pressure during superplastic forming based on the maximum load and target strain rate set during the simulation.

In this paper, the maximum loads were set as 1.5 MPa, 2.0 MPa and 2.5 MPa respectively. According to the superplastic tensile test results of Ti2AlNb alloy, the target strain rate was set as 0.001 s^{-1}. The superplastic forming effects of Ti2AlNb alloy thin-walled structural parts under different loads were analyzed, and the optimal forming load parameters were obtained.

Simulation results and experimental verification

Simulation results of superplastic forming process

Fig. 7 shows the wall thickness distribution of Ti2AlNb alloy thin-walled structural part at different moments of superplastic forming under the pressure of 2.5 MPa. The superplastic forming of thin-walled structural parts can be divided into three parts, namely, plane zone I , bar zone II and fillet zone III. When the forming time is 300 s (Fig. 7a), the plate is in the free expanding state, and have not been in contact with the die, so the forming speed is faster, and the thinning of the plates is less. With the progress of deformation, the plate gradually contacts with the die. When the forming time is 600 s, the plane zone I of plate is completely contact with the die, and the thickness is distributed between 1.33-1.35 mm, as shown in Fig. 7b.

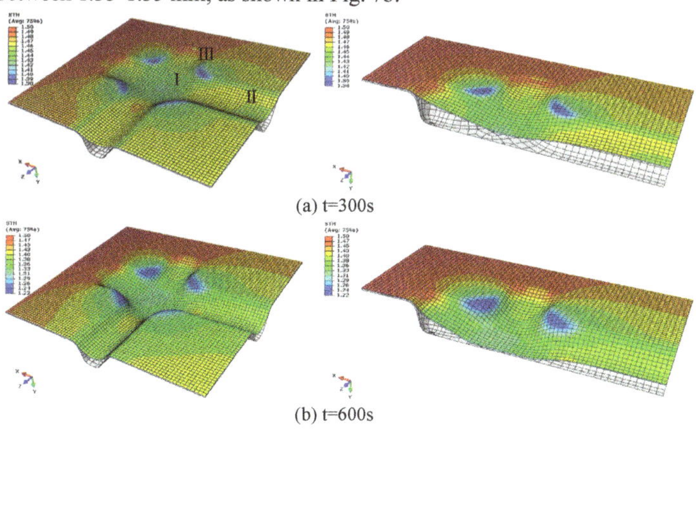

(a) t=300s

(b) t=600s

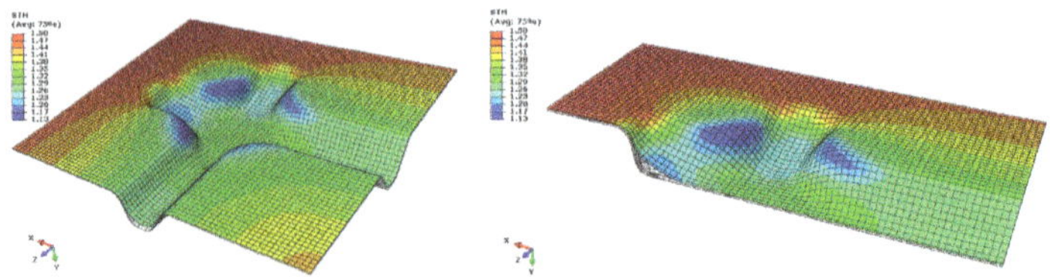

(c) t=900s

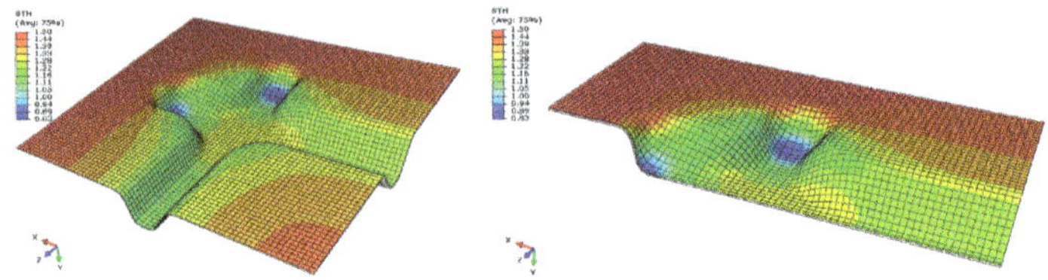

(d) t=1980s

Fig. 7 Superplastic forming process and wall thickness distribution of thin-walled structural part

When the forming time is 900 s, the bar zone II of plate is also fully attached to the die, and the thickness distribution is between 1.26-1.28 mm. The thickness of plane zone I is between 1.29-1.32mm. But the fillet zone III is still not attached, indicating that the round corner is not easy to form. At the time of 1980s, superplastic forming ended and the fillet zone III is fully pasted.

Due to the latest pasting, the plate has a large thinning, with a minimum wall thickness of about 0.83mm and a thinning rate of 44.7%. It can be predicted that when the thickness of the plate is small, cracking defects are most likely to occur here.

The pressure-time curve calculated automatically by Abaqus software according to the constant strain rate is shown in FIig. 8. This curve can provide a reference for the pressure loading of the subsequent thin-walled parts forming test, and realize the precise control of the superplastic forming process, so as to obtain the structural parts with uniform wall thickness distribution.

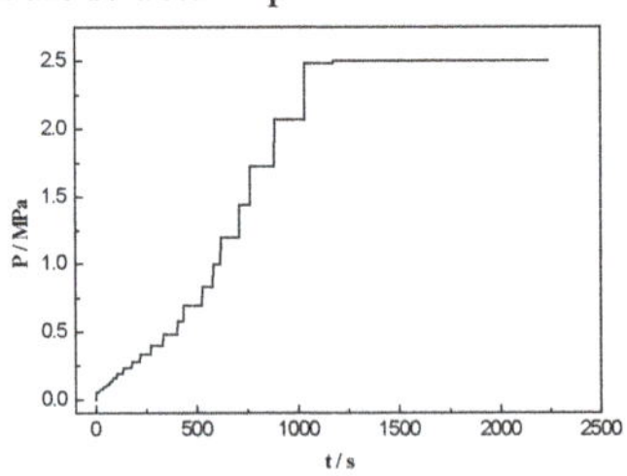

Fig. 8 Pressure-time curve

Influence of forming pressure on superplastic forming

Fig. 9 shows the wall thickness distribution nephogram of Ti2AlNb alloy thin-walled structural part under different pressures. When the pressure is less than 1.5 MPa, it can be seen from the profile that at the end of the simulation, the plate is not fully attached in the fillet zone III. While when the pressure is greater than 2.0 MPa, the plate is fully attached, and the superplastic forming is completed. When the pressure is 2.0 MPa and 2.5 MPa, the wall thickness distribution and thinning are basically the same.

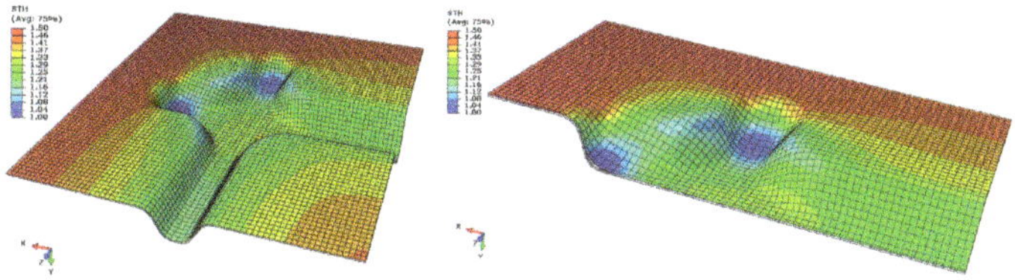

(a) 1.5MPa

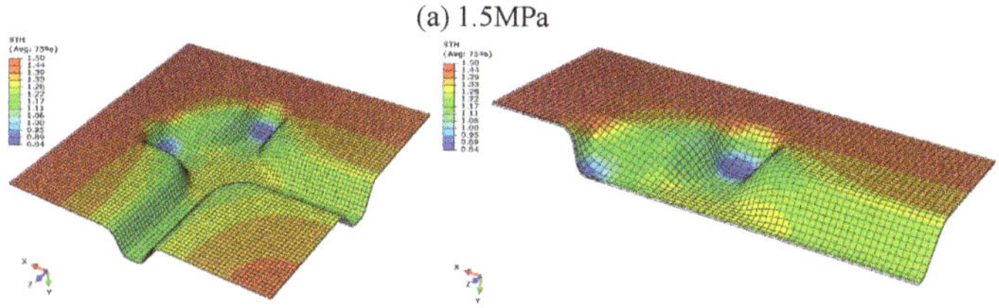

(b) 2.0MPa

Fig. 9 Wall thickness distribution of thin-walled structural part under different pressures

When the pressure is 2.0 MPa, the forming time of thin-walled structural part is 2520 s, which is 540 s longer than that when the pressure is 2.5 MPa. Therefore, with the increase of pressure, the superplastic forming time decreases. In order to shorten the forming time of parts, under the premise of not affecting the wall thickness distribution, 2.5 MPa is preferred to be formed.

Experimental study

The superplastic forming experiments of thin-walled structure was carried out at the temperature of 940 °C, strain rate of 0.001 s^{-1} and forming pressure of 2.5 MPa. The pressure-time curve obtained by finite element simulation was used for pressure loading. Fig. 10 shows the superplastic formed thin-walled structure of Ti2AlNb alloy. The part is well formed, and the rounded corners is completely pasted without cracking phenomenon.

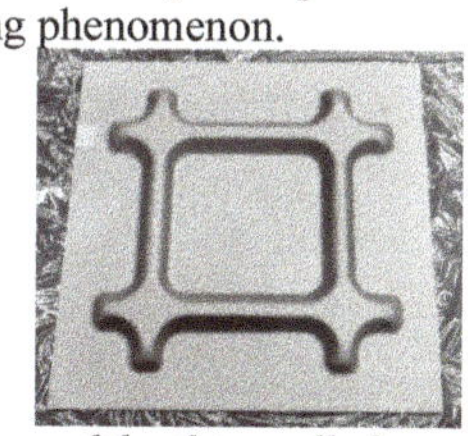

Fig. 10 Picture of the thin-walled structural part

The thickness of the bar area of thin-walled structural part ranges from 1.27 mm to 1.30 mm, and the wall thickness at the rounded corner is the thinnest, about 0.85 mm, which is basically consistent with the simulation results.

Summary

(1) At the same strain rate, the elongation and peak stress of Ti2AlNb alloy decrease with the increase of deformation temperature. With the increase of strain rate, the elongation decreases and the peak stress increases. When the deformation temperature is 940°C and strain rate is $0.001s^{-1}$, the maximum elongation of Ti2AlNb alloy is 525%. The constitutive equation considering the strain factor was established.

(2) Using Abaqus software, the finite element numerical simulation of superplastic forming of Ti2AlNb alloy thin-walled structural part was carried out. As a result, the pressure-time curve was

obtained, and the pressure parameters were optimized. The wall thickness distribution was uniform and the forming time was short at 2.5MPa, which provided a theoretical basis for the selection of actual superplastic forming parameters.

(3) The forming expriment of Ti2AlNb alloy thin-walled structural part was carried out at the superplastic forming temperature of 940 °C and the maximum atmospheric pressure of 2.5 MPa. The wall thickness of the component was evenly distributed and no cracking phenomenon was found, which was basically consistent with the simulation results.

References

[1] T.A. Barber, J. Majdalani, B.A. Maicke. Review of gaps, obstacles, and technological challenges in hypersonic applications, International Journal of Energetic Materials and Chemical Propulsion, 2018, 17(1): pp. 13-55. https://doi.org/10.1615/IntJEnergeticMaterialsChemProp.2018022006

[2] H.Y. Huang, L.J. Su, C.S. Lei, LI Jian, E.S. Zhang, W.J. Li, J.Y. Yang, Y.M. Zhao, Y.C. Pei, H. Zhang. Reusable thermal protective materials : application and research progress, Acta Aeronauticaet Astronautica Sinica, 2020, 41(12): pp. 023716-1- 023716-35.

[3] X. Ma, J.W. Zhang, X.B. Liang, Z.W. Zeng. Microstructure and mechanical properties of Ti-22Al-25Nb alloy during hot mechanical processing, Rare Metal Materials and Engineering, 2017, 46(5): pp. 1375-1379.

[4] J. Kumpfert. Intermetallic alloys based on orthorhombic titanium aluminide, Advanced Engineering Materials, 2001, 3(11): pp. 851-864. https://doi.org/10.1002/1527-2648(200111)3:11<851::AID-ADEM851>3.0.CO;2-G

[5] Y.P. Zheng, W.D. Zeng, D. Li, X.B. Liang, J.W. Zhang, X. Ma. Effect of orthorhombic case on the creep rupture of Ti-22Al-25Nb (at.%) orthorhombic alloy, Materials Science and Engineering A, 2017, 696: pp. 529-535. https://doi.org/10.1016/j.msea.2017.05.003

[6] S.H. Huang, B. Shao, W.C. Xu, D.B. Shan, B. Guo, Y.Y. Zong. Deformation behavior and dynamic recrystallization of Ti-22Al-25Nb alloy at 750~990°C, Advanced Engineering Materials , 2020, 22(4): pp. 1-8. https://doi.org/10.1002/adem.201901231

[7] L. Carrino, V. Paradiso, S. Franchitti, A. Squillace, S. Russo. Superplastic forming/diffusion bonding of a titanium alloy for the realization of an aircraft structural component in multi-sheets configuration, Key Engineering Materials, 2012, 504-506: pp. 717-722. https://doi.org/10.4028/www.scientific.net/KEM.504-506.717

[8] G. Liu, Y. Wu, J. Zhao, K. Wang, S.J. Yuan. Formability determination of titanium alloy tube for high pressure pneumatic forming at elevated temperature, Procedia Engineering, 2014, 81: pp. 2243-2248. https://doi.org/10.1016/j.proeng.2014.10.315

[9] S.M.R. Tabatabaei, M. Safari, R.S. Esfahani, A.A. Sichani, M.N.P. Moghadam. Experimental and numerical investigation of cold bulge forming of titanium alloy Ti55, World Journal of Mechanics, 2013(9): pp. 323-327. https://doi.org/10.4236/wjm.2013.39035

[10] J.L. Ma, Y.S. Liu, P. Li, K.M. Xue. The influence of temperature for thickness distribution of Ti2AlNb alloy sheet formed by superplastic free bulging, Journal of Harbin Institute of Technology, 2016, 48(5): pp. 172-176.

[11] Z.F. Zhou, M.H. Chen, D.G. Wang, N. Wang. Superplastic forming and diffusion bonding numerical simulation for core sheets of TC4 titanium alloy rudder, Hot Working Technology, 2014, 43(3): pp. 190-195.

Superplasticity in Advanced Materials - ICSAM 2023 Materials Research Forum LLC
Materials Research Proceedings **32** (2023) 95-102 https://doi.org/10.21741/9781644902615-10

Effect of friction stir forming parameters on mechanical properties and plastic flow of material in the mechanical interlocking of optical fiber and SP-700 superplastic titanium alloy

Hamed MOFIDI TABATABAEI[1,a*], Takahiro OHASHI[1,b] and Tadashi NISHIHARA[1,c]

[1] School of Science and Engineering, Kokushikan University, Setagaya 4-28-1, Setagaya Ward, 154-8515, Japan

[a]mofidih@kokushikan.ac.jp, [b]tohashi@kokushikan.ac.jp, [c]nisihara@kokushikan.ac.jp

Keywords: Friction Stir Welding, Friction Stir Forming, Titanium Alloy, Superplasticity, SP-700, Optical Fiber, Mechanical Interlock, Dissimilar Material Joining

Abstract. The present study discusses the fluidity of titanium alloy during friction stir forming (FSF) and investigates the effect of process parameters on the mechanical properties of titanium alloy in the mechanical interlocking of optical fiber and SP-700 superplastic titanium alloy. In this study, a guide slit is provided in a titanium alloy plate, an optical fiber is placed inside the slit, and FSF is performed on the surface of the titanium alloy. By performing FSF, titanium softens and material plastically deforms inside the slit, which results in mechanically joining optical fiber and titanium alloy. Experimental results of evaluating the mechanical properties after FSF revealed that the hardness value of the material that flowed into the slit was significantly higher than that of the base material, which confirms that the stirring zone has flowed into the slit around the fiber. It was also confirmed that the temperature becomes unstable during the process for a travel speed of 50 mm/min or less, also above 800 mm/min. This resulted in insufficient material flow. As a result, inhomogeneous structures were confirmed. Tensile test results after FSF showed that the strength was lower than that of the base metal (67% to 55% of the base metal) for most process parameters. Although no significant changes in strength were observed by changing the travel speed, it was confirmed that the strength has an increasing tendency as the rotation speed increases, which is considered to be related to the grain refinement of the material. It was also concluded that the limitations in the present study such as insufficient material flow and reduced strength and also damage to the embedded optical fiber can be improved by pre-heating the workpiece and controlling the FSF process parameters which require further experiments and temperature measurements during FSF. In this study, the developed composite can give us hope for embedding FBG sensors inside high melting point alloys to simultaneously measure deformation, strain, pressure, and temperature and create new functional smart materials.

Introduction

Recently, a new metal forming method using the principles of Friction Stir Welding (FSW) [1] called "Friction Stir Forming (FSF)" [2] has gained attention in the field of dissimilar material joining [3-7]. Prof. Tadashi Nishihara of Kokushikan University, the developer of this method, proposed placing the workpiece on the mold and applying friction stir on the backside of the workpiece to transfer the mold shape onto the workpiece, as shown in Fig. 1.

The development of Fiber Bragg Grating (FBG) sensors is essential for achieving lightweight smart parts in aircraft and intelligent aviation control. The development of FBG sensors requires the embedding of fiber optic sensors (FOS) into the substrate of components. However, embedding fibers flawlessly into high-melting-point alloys like titanium-based materials is challenging. As aircraft systems become more complex and high-performance, the need for aircraft lightweighting and fuel efficiency continues to rise. Consequently, the development of innovative technologies to improve the quality, reliability, and health monitoring capability of aircraft systems has become crucial. However, research on embedding FOS into metals is very limited. High-melting-point metals like titanium can protect sensors in harsh environments, but the embedding process into metals involves stringent conditions, and methods like casting or hot pressing cannot be easily employed [3]. Our previous study has proposed a novel method for joining dissimilar materials, titanium alloys, and optical fibers, using friction stir forming and aimed to experimentally investigate the development of novel composite material with embedded optical fibers inside titanium alloy using FSF without compromising the fiber's characteristics which opens possibilities for the development of new functional materials [3]. By introducing this novel composite material as the base material for smart structures, it is expected to function as a sensor for detecting the internal state (temperature and strain) of the host titanium. In previous research, the authors demonstrated the potential of embedding optical fibers inside titanium alloys and discussed the fluidity of titanium during friction stir forming.

In the present study, further investigation is conducted on the material's fluidity during the process, aiming to examine the influence of process parameters in friction stir forming on the mechanical properties of titanium alloys.

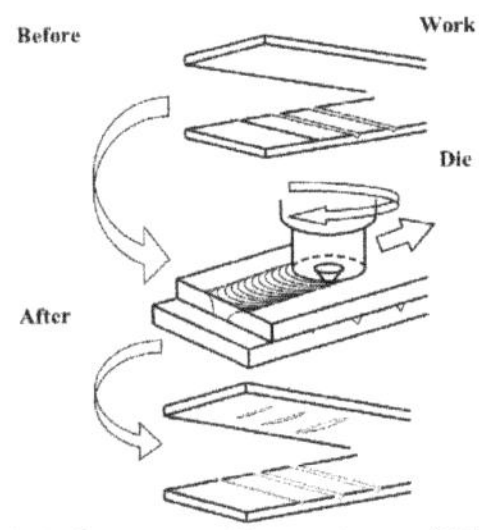

Fig. 1 Schematic Drawing of FSF [2]

Experimental Procedure

The methodology employed in this study is similar to previous research [3]. As shown in Fig. 2, a guide slit is created in an SP-700 titanium alloy plate, and optical fibers are placed within it. Ti-4.5Al-3V-2Mo-2Fe (commercially known as SP-700) is a beta-rich superplastic titanium alloy with fine structures developed for the aerospace industry [9]. Using the tool depicted in Fig. 1 (c) which was made of pure tungsten, friction stir forming (FSF) is performed on the titanium alloy's surface. FSF induces a plastic flow of the material within the slit, mechanically joining the optical fibers inside the titanium alloy.

Previous research results have revealed that during FSF, the material flows more on the retreating side compared to the advancing side. Therefore, in this experiment, the probe center was offset by 2 mm from the center of the slit and offset towards the advancing side while performing FSF. When the probe center was placed directly above the slit and FSF was performed, the optical fiber became brittle and tended to break due to the effect of heat input [1].

Materials Research Forum LLC
https://doi.org/10.21741/9781644902615-10

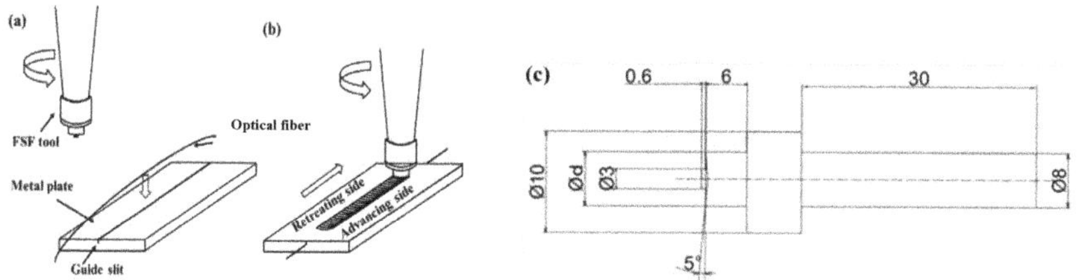

Fig. 2 Schematic of the experimental procedure (a) rotating tool is plunged into the workpiece, (b) tool is traversed along the guide slit to mechanically interlock titanium and optical fiber [1], (c) Dimension of FSF tool (d= 8, 10, 12 mm)

Experimental Results

Material flow and hardness distribution around the optical fiber. It was confirmed in previous studies that as the feed rate increases, the heat input decreases, and especially at speeds of 200 and 400 mm/min, it was confirmed that the frictional heat does not reach the back of the test specimen. However, it was also confirmed that the material was sufficiently filled within the slit even at high feed rate parameters. It is believed that the material flow during FSF is influenced not only by heat input but also by pressure. With higher feed rates, the pressure increases, which is expected to affect the material flow. Additionally, it was observed that burrs are more likely to occur and the surface becomes rough when the feed rate is low at 50 mm/min or high at 800 mm/min [1].

To investigate the change in material flow, FSF was conducted by changing the tool shoulder diameter. Due to the instability of temperature during FSF caused by process parameters, the specimens were cut into four sections after FSF, and cross-sectional microstructure observations were conducted using a metallographic microscope.

As observed in Figure 3 (a), the optical fiber remained intact without damage, but voids were observed, indicating that the material did not flow completely around the optical fiber. Additionally, at the beginning of the process, insufficient heating resulted in insufficient material flow, leaving slits.

Furthermore, hardness tests were conducted, revealing that the hardness value near the stirred region adjacent to the optical fiber was significantly higher than that of the base material. The hardness value of the base material near the optical fiber was found to be equivalent to the original base material hardness. This result suggests that the stirred region flows within the slit, while regions other than the stirred region are not affected by the heat during FSF. In Figures 3 (b)-(d), a uniform material flow can be observed, and there is no significant change in hardness values. This is attributed to the stable heating during the process.

However, it was observed that the temperature becomes unstable during the process at feed rates below 50 mm/min and above 800 mm/min, resulting in heterogeneous material flow. Figure 4 shows the hardness values around the fiber when using a tool shoulder diameter of 8 mm in the FSF experiment, and compared to a tool shoulder diameter of 10 mm, no significant changes were observed. However, it was confirmed that reducing the shoulder diameter reduces the size of the thermomechanically affected zone (TMAZ) and heat-affected zone (HAZ), leading to minimal changes in the structure below the shoulder.

Materials Research Proceedings **32** (2023) 95-102 https://doi.org/10.21741/9781644902615-10

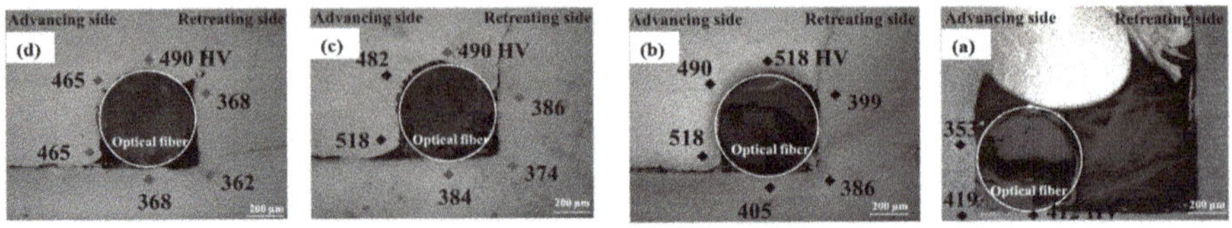

Fig. 3 Cross section of embedded fiber for (a) starting region, (b) and (c) middle region, (d) finishing region (Tool shoulder diameter: 10 mm)

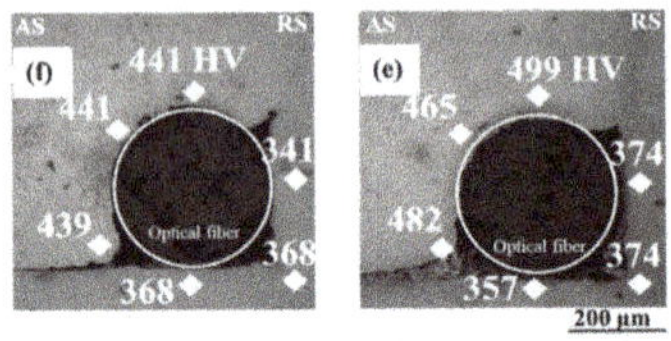

Fig. 4 Cross-section observation of embedded fiber within the middle zone (Tool shoulder diameter: 8 mm)

Evaluation of tensile strength after FSF. In experiments where FSF was performed with an optical fiber installed inside a slit with a width of 1.0 mm and a depth of 1.7 mm, the tensile strength perpendicular to the feed direction may decrease due to insufficient material flow. Therefore, rectangular tensile test specimens perpendicular to the feed direction were created after FSF, by varying the process parameters of the tool and subjected to tensile testing. The creation of the tensile test specimens referred to ASTM E 646-98. The results are shown in Figure 11. In all parameters, lower strengths (ranging from 55% to 67% of the base material strength) were obtained compared to the base material strength, and as confirmed in Figure 5, fracture occurred near the guide slit. It is believed that this can be improved by performing multipass, but further investigation is necessary.

Furthermore, as shown in Figure 5 (right), it was observed that the FSF tool was damaged and the probe section wore out under conditions of high heat input.

(a)

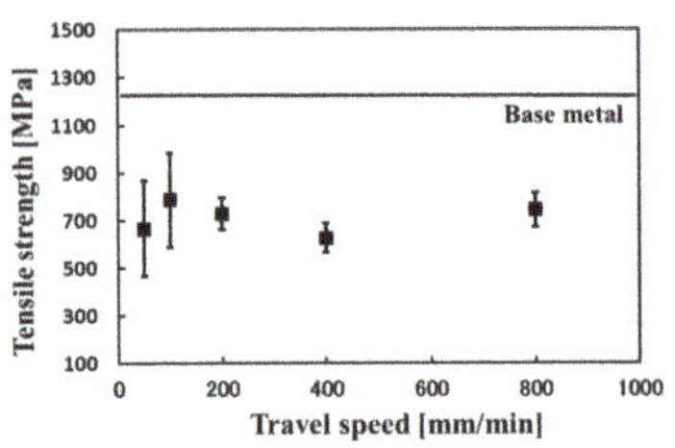

(b)

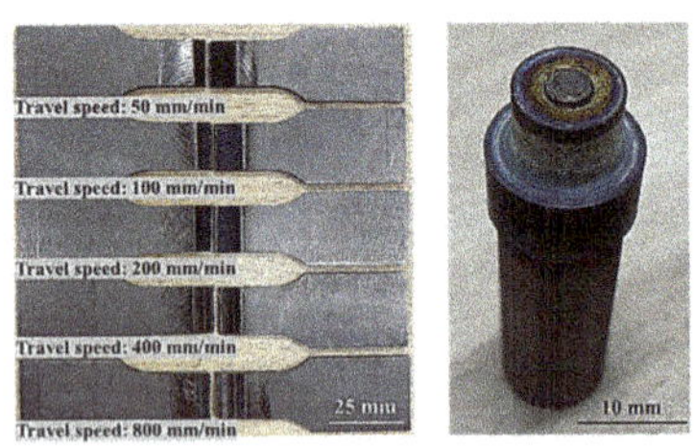

Fig. 5 (a) Relation between tensile strength and FSF tool travel speed (Perpendicular to the direction of processing) (b) Fig. 12 Tensile tested specimens (left) and tungsten tool after FSF (right)

Fig. 6 (a) shows the relationship between tensile strength and tool rotation speed (tensile test results perpendicular to the FSF process direction). As can be seen from the graph, an increasing trend in tensile strength was observed with an increase in rotation speed. This is believed to be related to the heat input during FSF. When the rotation speed is low, the material flow is insufficient, resulting in voids remaining in the slit, which is thought to lead to a decrease in tensile strength.

Fig. 6 (b) shows the relationship between tensile strength and tool travel speed (tensile test results parallel to the FSF process direction). Comparing it with the tensile test results perpendicular to the process direction shown in Figure 11, a significant increase in tensile strength is observed. The strength variation due to different feed rates is believed to be related to the plastic flow of the material inside the slit and the refinement of the grain structure in the stirred region due to the stirring of the probe during FSF. Additionally, in the tensile tests conducted after FSF of Ti-6Al-4V titanium alloy, relatively lower tensile strength was obtained compared to the SP-700 alloy. This is considered to be due to insufficient plastic flow of the material and inadequate material flow into the slit compared to the SP-700 titanium alloy.

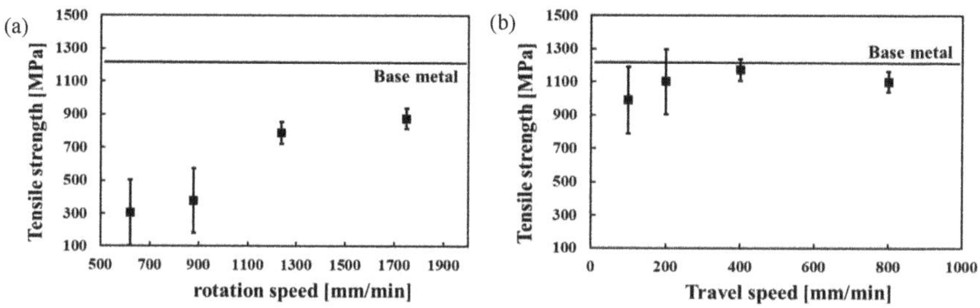

Fig. 6 (a) Relation between tensile strength and FSF tool rotation speed (Perpendicular to the direction of processing), (b) Relation between tensile strength and FSF tool travel speed (Parallel to the direction of processing)

Fig. 7 (a) shows the hardness distribution across the cross-section for different feed rates after FSF. The variation in feed rate causes changes in heat input, which affects the size of the surface grain structure, and this is believed to manifest as changes in hardness values (it is known that hardness increases as grain size decrease due to the Hall-Petch relationship [10-11]).

Furthermore, Fig. 7 (b) presents the hardness distribution across the cross-section for different rotation speeds. As can be seen from the graph, there was no significant change in hardness values due to insufficient stirring at low rotation speeds. At a rotation speed of 1240 rpm, the hardness increased as a result of grain refinement, while at a rotation speed of 1750 rpm, it is believed that the heat input caused coarsening of the grain structure, leading to a decrease in hardness compared to the 1240 rpm rotation speed.

Fig. 8 shows hardness values according to the distance from the surface. As you approach the surface, there is a tendency for the hardness value to increase, which is believed to be related to the differences in crystal grain refinement observed in the micrographs due to stirring.

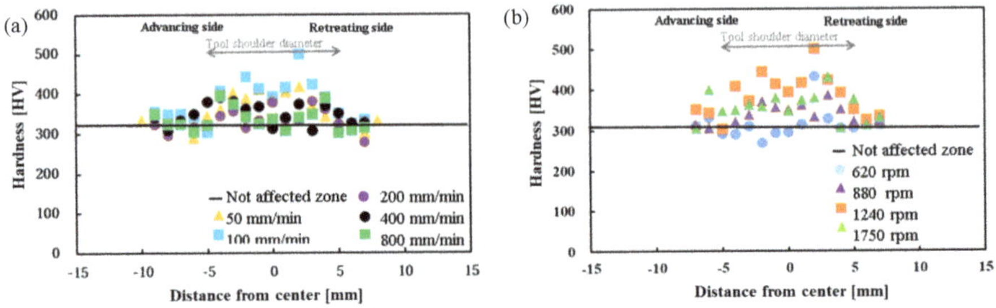

Fig.7 (a) Hardness distribution of cross-section after FSF for different travel speeds (0.5 mm under surface), (b) Hardness distribution of cross-section after FSF for different rotation speeds (0.5 mm under surface)

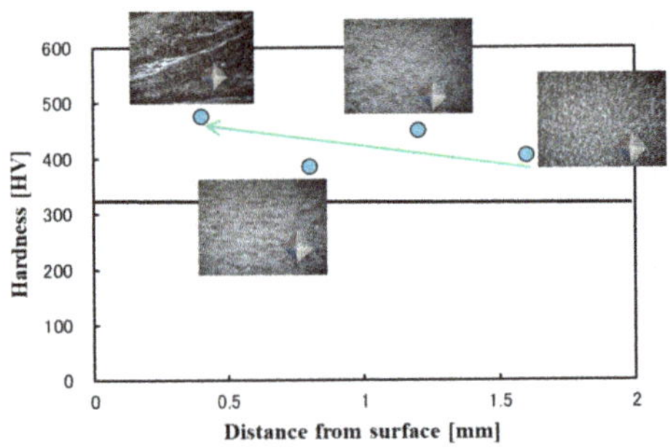

Fig.8 Hardness distribution for different distances from the surface (1240 rpm 100 mm/min)

Temperature measurement during the FSF process. In this experiment, a type K thermocouple was directly installed in the slit (a section with a depth of 1.7 mm) of the test material to create a system that enables temperature measurement inside the material during the FSF process. However, there was a significant variation in the results, and it was not possible to accurately measure the temperature changes with parameter adjustments. The temperature at which the superplasticity of SP-700 titanium alloy occurs is around 1050 K. When the rotational speed was 1240 rpm and the feed rate was 50 mm/min, the maximum temperature measured was 512°C, but it did not meet the temperature indicating the superplastic phenomenon of titanium alloy. By changing the process parameters during the FSF process, it is possible to vary the heat input and control the temperature during the process, which can potentially control material flow and structure. However, further investigation regarding additional temperature measurements is necessary.

Summary

In this study, a mechanical bonding method for optical fibers and titanium alloys using FSF was proposed, and the possibility of forming new functional composite alloys was experimentally investigated. The challenges encountered in this study were insufficient material flow due to the process parameters of FSF and cases where excessive heating resulted in fiber damage. In the future, the development of a system that accurately measures the temperature distribution of the material during FSF will be necessary. However, the potential for smart material development through the mechanical bonding of titanium alloys and optical fibers using FSF has been demonstrated. Furthermore, the relationship between FSF process parameters and the mechanical properties of titanium alloys was examined, and the following findings were obtained:

1. It was revealed that the surface condition, microstructure, and mechanical properties of titanium alloys depend on the process parameters. It was confirmed that feed rates below 50 mm/min and above 800 mm/min lead to temperature instability during the process, resulting in uneven material flow.
2. Hardness tests showed that the hardness value near the stirred region was significantly higher than that of the base material, while the hardness value of the base material near the fiber was comparable to the original base material. This result indicates that the stirred region is undergoing flow within the slit, and areas other than the stirred region are not affected by the heat during FSF.
3. Tensile test results in the feed direction of the tool showed that lower strength values (ranging from 55% to 67% of the base material strength) were obtained in all parameters.
4. Significant changes in strength due to variations in feed rate were not observed, but a decreasing trend in strength was observed as the rotation speed decreased. This is believed to be related to the grain refinement of the material.
5. It is believed that the issue of strength reduction due to insufficient material flow can be improved by performing multipass, but further investigation is needed.
6. The results of material flow at high feed rates suggest that in addition to the effect of heating, pressure and the velocity dependence of superplastic titanium alloys during FSF may also influence material flow, indicating the occurrence of superplasticity during the process.
7. The FSF tool made of pure tungsten was damaged under high heating conditions, particularly because the probe part experienced wear. Therefore, the selection of tool materials for FSF of titanium alloys needs to be considered.

Acknowledgments
Part of his study was conducted with the support of the AMADA Foundation's "Encouragement Research Grant" (Young Researchers Category) (AF-2019044-C2), and we would like to express our gratitude for their support.

References

[1] W.M. Thomas et al. "Friction Stir Welding", International Patent Application No. PCT/GB92/ 02203.

[2] T. Nishihara, "Development of Friction Stir Forming", Mater. Sci. Forum, 426-432 (2003) 2971-2978. https://doi.org/10.4028/www.scientific.net/MSF.426-432.2971

[3] Hamed Mofidi Tabatabaei, Takahiro Ohashi, Tadashi Nishihara, "Innovative Method of Embedding Optical Fiber inside Titanium Alloy", Utilizing Friction Stir Forming, Key Engineering Materials, 934, pp. 67-73. https://doi.org/10.4028/p-683ks0

[4] H. M. Tabatabaei et al., "Friction Stir Forming for Mechanical Interlocking of Ultra-Thin Stainless Steel Strands and Aluminum Alloys", Defect and Diffusion Forum, 382, pp.114-119, 2018. https://doi.org/10.4028/www.scientific.net/DDF.382.114

[5] Hamed Mofidi Tabatabaei, Tadashi Nishihara, Friction stir forming for mechanical interlocking of insulated copper wire and Zn-22Al superplastic alloy, Weld World, 61:47-55, December 2016. https://doi.org/10.1007/s40194-016-0406-9

[6] Hamed Mofidi Tabatabaei, Takahiro Hara, Tadashi Nishihara, Production of a Superplastic Vibration-Damping Steel Sheet Composite Using Friction Stir Forming, Materials Science Forum 838-839:574-580, January 2016. https://doi.org/10.4028/www.scientific.net/MSF.838-839.574

[7] Takahiro Ohashi, Tadashi Nishihara, Hamed Mofidi Tabatabaei, Mechanical Joining Utilizing Friction Stir Forming, Materials Science Forum, 1016:1058-1064, January 2021. https://doi.org/10.4028/www.scientific.net/MSF.1016.1058

[8] Takahiro Ohashi, Hamed Mofidi Tabatabaei, Tadashi Nishihara, Cylindrical extrusions on A5083 aluminum alloy plate fabricated by friction stir forming, AIP Conference Proceedings, 1896(1):080002, October 2017. https://doi.org/10.1063/1.5008082

[9] A. Ogawa, M. Niikura, C. Ouchi, K. Minikawa, and M. Yamada, "Development and Applications of Titanium Alloy SP-700 with High Formability," Journal of Testing and Evaluation 24, no. 2 (1996): 100-109.

[10] Hall E.O., "The deformation and ageing of mild steel: III Discussion of results," Proceedings of the Physical Society 64 (1951) 747-753. https://doi.org/10.1088/0370-1301/64/9/303

[11] Petch N.J., "The cleavage strength of polycrystals," J. Iron and Steel Inst. 174 (1953) 25-28.

Mechanical and microstructural analysis of Ti-6Al-4V material in a wide range of superplastic forming conditions

Alexander Sancho[1,a,*], Paranjayee Mandal[1,2,b], Craig Knowles[1,c], Diego Gonzalez[1,d]

[1]Advanced Forming Research Centre, University of Strathclyde, 85 Inchinnan Drive, Inchinnan, Renfrewshire PA4 9LJ, Scotland

[2]Now at BorgWarner Inc., Courteney Road, Gillingham ME8 0RU, England

[a]alexander.sancho@strath.ac.uk, [b]paranjayee.mandal@strath.ac.uk, [c]craig.knowles@strath.ac.uk, [d]diego.gonzalez@strath.ac.uk

Keywords: Superplasticity, Forming, Titanium, Ti-6Al-4V, Microstructure, Tensile, Strain-Rate Sensitivity

Abstract. In order to accurately define the superplastic forming (SPF) conditions of Ti-6Al-4V material, an understanding of the stress-strain behaviour, the initial microstructure, and their evolution during superplastic deformation are required. Ti-6Al-4V material with microstructure beneficial for SPF was superplastically tested according to the ASTM E2448 standard considering a wide range of forming conditions in terms of temperatures (750°C – 830°C) and strain-rates (seven strain-rates ranging from $5 \cdot 10^{-5}\ s^{-1}$ to $5 \cdot 10^{-2}\ s^{-1}$) – some of the tests of the 3x7 matrix are considered "extreme" conditions from an SPF point of view. The material showed improved superplastic behaviour, which was evident from the stress levels and strain-rate sensitivity values as estimated from the flow curves obtained for the different conditions. In comparison with other commercial alloys or results from similar analyses published in the last decades, low stress values and high strain-rate sensitivity (m) values were obtained despite the low temperatures and high strain-rates used in this analysis. The tests were interrupted when 0.5 true strain (65% engineering strain) was achieved followed by quenching, as this was the maximum local strain achieved when forming the component of interest. Samples did not show any sign of premature necking or failure, with the exception of the two most "extreme" cases. Particularly for the lower strain-rates (below $10^{-4}\ s^{-1}$), some level of material hardening associated with a minimum grain growth was observed in the flow curves. In contrast, a noticeable material softening was observed for the higher strain-rate conditions (above $5 \cdot 10^{-3}\ s^{-1}$), associated with the microstructural changes occurring due to dynamic recrystallization. These higher strain-rates led to formation of submicron-sized grains, which could have helped in the superplastic response of the material under these strain-rate conditions. Intermediate strain-rates ($5 \cdot 10^{-4}\ s^{-1}$ and $10^{-3}\ s^{-1}$) showed a different type of response in terms of microstructural behaviour (and flow curve outline) depending on the testing temperature. A negligible amount of cavitation was observed in the samples tested under extreme conditions.

Introduction

Ti-6Al-4V is a titanium alloy that has traditionally been used in the hot forming of aeronautic components. Examples of these hot manufacturing processes include hot creep forming and superplastic forming (SPF) [1, 2]. The industry is currently exploring and implementing methodologies for temperature and forming time reductions of the SPF process, as this has a real impact in production. Typical SPF temperatures in industry have been traditionally found in the range 850°C to 900°C. There are multiple advantages in that temperature reduction. Among others: energy savings (and the associated carbon footprint reduction), the possibility of using cheaper alloys for the tooling (generally very expensive Nickel alloys due to the Cr, Co, and Ni contents), increase in the tooling life (with less re-machining operations during its lifetime), the possibility to increase batch sizes before tooling needs to be replaced for cleaning (i.e., increase of productivity and reduction of maintenance) and cost reductions in terms of consumables (for example, the life of the cartridges increases at lower temperatures).

In order to achieve temperature and forming time reductions, a key parameter is the grain size of the material. Fine grain Ti-6Al-4V alloys present better superplastic behaviour (higher strain-rate sensitivities) than alloys with coarser grains, even at lower temperatures and faster strain-rates [3 – 6]. The refined grain structure tends to enhance grain boundary sliding allowing for larger plastic deformations [5, 7].

The aim of this work was to test a fine-grain Ti-6Al-4V alloy in a wide range of strain-rates and temperatures in order to understand the new set of forming conditions for faster and colder SPF forming. The targeted temperatures are relatively low compared to traditional SPF temperatures for titanium (750°C to 830°C). This research work was funded by The Boeing Company.

Results

This section will show the obtained results in terms of mechanical testing and microstructural analysis of the Ti-6Al-4V material supplied by The Boeing Company as per AMS4911 and used in this work. This information will help to understand the material behaviour in terms of stress levels, strain-rate sensitivity and formability for each temperature and strain-rate condition, also the microstructure behaviour when the material is deformed at temperature, as well as the formability limits of the material. In addition, mechanical and microstructural results will be later used for the definition of SPF forming conditions.

- **Tensile Testing:**

Tensile testing at elevated temperature was conducted using a Zwick 250 Universal testing machine and a three zones clamshell furnace attached to the machine. The three zones furnace is controlled based on the temperature readings of three thermocouples attached to the sample. The gauge length of the samples was coated with glass lubricant (Prince Minerals, pink glass lubricant) in order to minimize as much as possible the appearance of alpha-case during the test. The initial thickness of test specimens was 2.03mm (0.080"). Table 1 collects the testing conditions in terms of temperature and strain-rate.

Table 1. Test matrix with three temperatures and seven strain-rates. Tests interrupted at 0.5 true strain.

Temperature (T)	Strain-rate (s^{-1})	True strain (ε)
Low temp. - T = 750°C Mid. temp. (T between 750°C - 830°C) High temp. - T = 830°C	$\dot{\varepsilon} = 5 \cdot 10^{-5}$	$\varepsilon = 0.5$
	$\dot{\varepsilon} = 10^{-4}$	
	$\dot{\varepsilon} = 5 \cdot 10^{-4}$	
	$\dot{\varepsilon} = 10^{-3}$	
	$\dot{\varepsilon} = 5 \cdot 10^{-3}$	
	$\dot{\varepsilon} = 10^{-2}$	
	$\dot{\varepsilon} = 5 \cdot 10^{-2}$	

Three different temperatures were used in the tests: 750°C (1382°F), 830°C (1526°F) and a middle temperature in the range between those two values. Seven strain-rate values have been defined for the material testing, covering a wide range of values, from $\dot{\varepsilon} = 5 \cdot 10^{-5} s^{-1}$ to $\dot{\varepsilon} = 5 \cdot 10^{-2} s^{-1}$. The idea for this wide range was to explore all the posibilities in terms of the strain-rate of the deformation undergone by the material (although the SPF forming cycle will be defined to run as fast as possible, there will be material points in the component that will see very low strain-rates at some point of the forming process).

The tests have been conducted as per the ASTM E2448 standard, increasing the strain-rate value of the test by 20% during specific ranges of strain. These so-called jump strain-rate tests allow to estimate the strain-rate sensitivity of the material for each test condition. Tests were interrupted at 0.5 true strain (65% engineering), as the maximum deformation achieved in the component of analysis was below this value.

Since microstructural analysis of specific tested samples was subsequently conducted, the samples were water quenched once the 0.5 true strain was achieved. This water quenching, which took no more than 6 seconds, "freezes" the existing microstructure at high temperature to be later analysed via SEM microscopy.

Figure 1 collects the tensile test results for 750°C. As the reader can see, there is material hardening for the slowest strain-rates associated to grain growth. For the highest strain-rates, material softening takes place due to grain refinement (linked to dynamic recrystallization phenomena). For the fastest strain-rate, $\dot{\varepsilon} = 5 \cdot 10^{-2} s^{-1}$, fracture occured in one of the two tests.

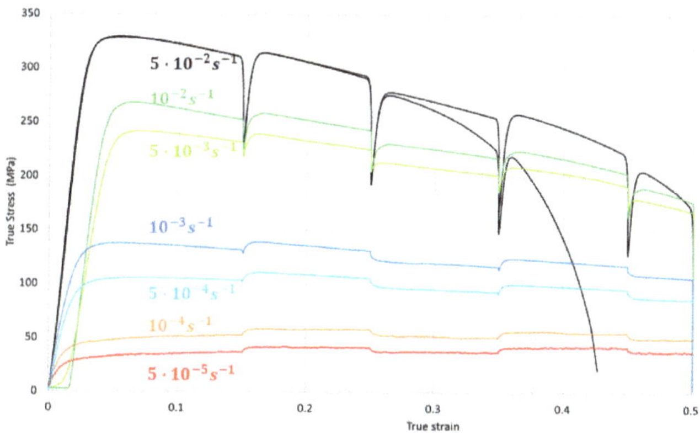

Figure 1. Flow curves obtained from the jump strain-rate tests at 750 °C.

Figure 2 collects the tensile test results for the temperature within the range between 750°C and 830°C. Similarly to the previous temperature, material hardening occurs for the slowest strain-rates whereas softening happens for the highest strain-rates. As the temperature has been increased, lower stress values than in the 750°C case have been achieved. No failure happened for any of the tests at this temperature (and samples did not show any sign of premature necking for the fastest strain-rates either).

 Materials Research Forum LLC
https://doi.org/10.21741/9781644902615-11

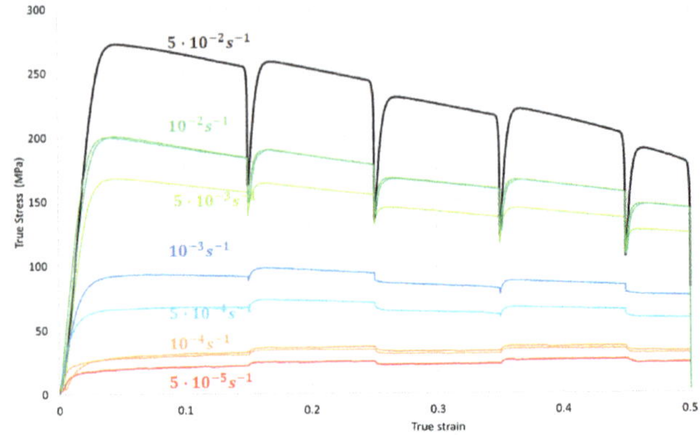

Figure 2. Flow curves obtained from the jump strain-rate tests at the temperature within the range between 750°C and 830°C.

Figure 3 collects the tensile test results for 830°C. As expected, lower stress values than in the previous temperatures were obtained.

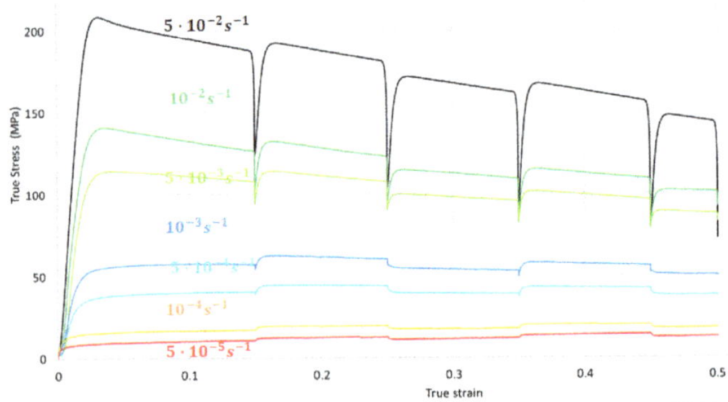

Figure 3. Flow curves obtained from the jump strain-rate tests at 830 °C.

From the jump strain-rate curves, strain-rate sensitivity values have been extracted for each strain-rate and for the three temperatures. Figure 4 collects the strain-rate sensitivity versus strain-rate (in logarithmic scale). As the reader can see, the strain-rate sensitivity increases as the strain-rate decreases (with the exception of the case with the lowest temperature and the lowest strain-rate). It is also evident that for the temperatures used in the present analysis, the strain-rate sensitivity improves as the temperature increases (reaching values close to 0.7 at 830°C for the slowest strain-rate).

Materials Research Forum LLC
https://doi.org/10.21741/9781644902615-11

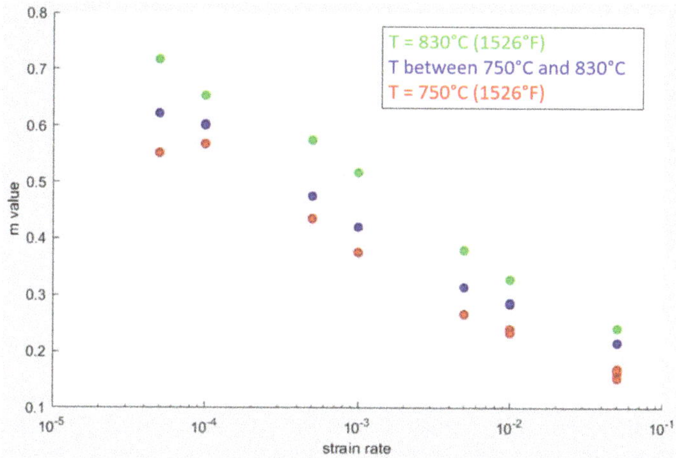

Figure 4. Comparison of strain-rate sensitivity values obtained for the three temperatures.

- **Microstructure Analysis:**

Analysis has been done on Ti-6Al-4V as-received material as well as on tested samples on both longitudinal (L) and transverse (T) directions of the samples via SEM microscopy. Several SEM micrographs were randomly obtained from different areas of each sample in order to have a representative analysis.

The analysis of the as-received material showed an average grain size of 2.4μm and a volumetric fraction of alpha and beta phases of 86% and 14%, respectively. The reader can see in Figure 5 an example of the as-received material where the globularity and the small grain size of the microstructure can be appreciated.

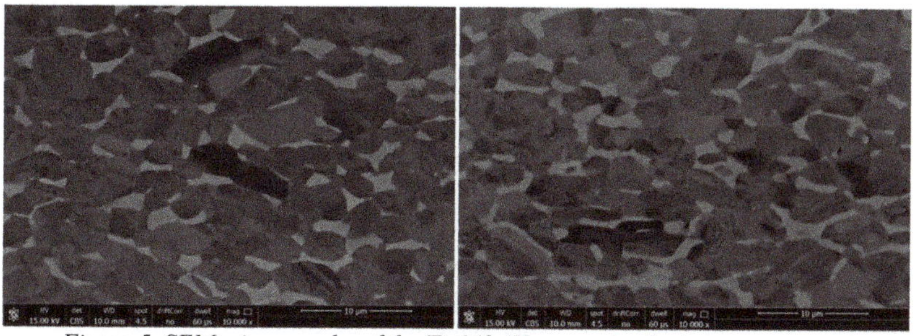

Figure 5. SEM micrographs of the Ti-6Al-4V as-received microstructure.

The SEM analysis of the tested samples confirmed the previously inferred concussions coming from the tensile tests. The average grain size increased for the slow strain-rates due the microstructure being at temperature during periods of time (the higher the temperature, the bigger the final grain size). On the contrary, for the fast strain-rates, dynamic recrystallization phenomena induced a decrease in the grain size. Figure 6 supports those conclusions for the strain-rates of $\dot{\varepsilon} = 5 \cdot 10^{-5} s^{-1}$ and $\dot{\varepsilon} = 10^{-2} s^{-1}$.

Materials Research Forum LLC
https://doi.org/10.21741/9781644902615-11

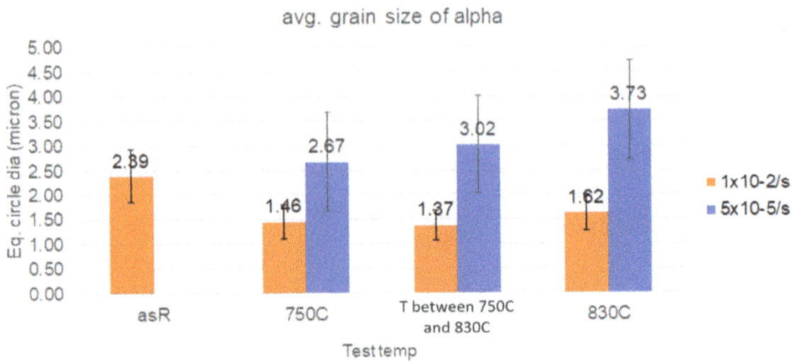

Figure 6. Average grain size of the tested samples at $5 \cdot 10^{-5} s^{-1}$ *and* $\dot{\varepsilon} = 10^{-2} s^{-1}$.

In regard with the volumetric fraction of alpha and beta, as expected, part of the alpha phase transforms into beta. The higher the temperature and the time of exposure, the higher the increase of the volumetric fraction of beta phase (see Figure 7). As temperatures are still far away from the beta-transus temperature, this transformation does not equalise the amount of both phases (this generally happening at temperatures close to 900°C).

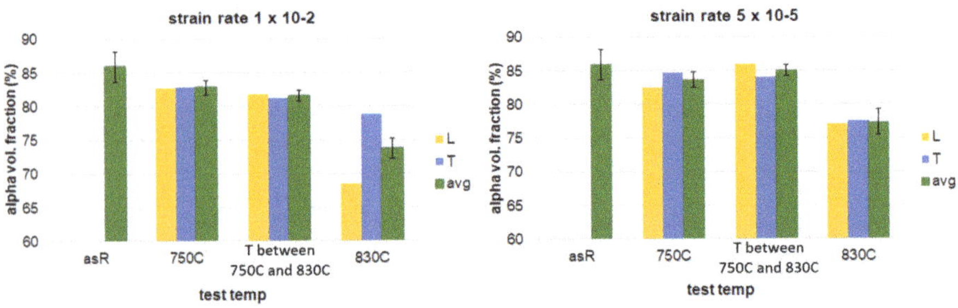

Figure 7. Alpha and beta volumetric fractions in the tested samples at $5 \cdot 10^{-5} s^{-1}$ *and* $\dot{\varepsilon} = 10^{-2} s^{-1}$.

Regarding cavitation, it can be stated that only extreme testing conditions (fast strain-rates above $\dot{\varepsilon} = 5 \cdot 10^{-3} \mathrm{s}^{-1}$) have developed some level of cavitation, always below the 0.3% volumetric fraction. Figure 8 shows two examples (at 750°C and the temperature in the 750°C - 830°C range) of SEM micrographs with some level of cavitation for the fastest strain-rate $\dot{\varepsilon} = 5 \cdot 10^{-2} \mathrm{s}^{-1}$.

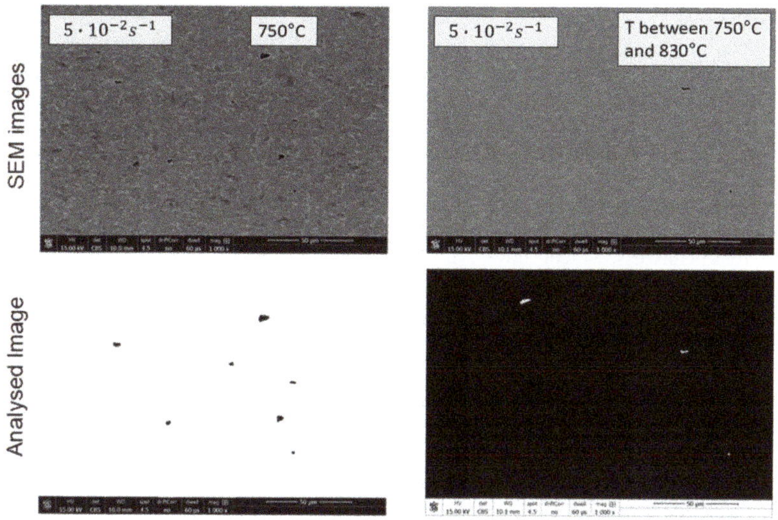

Figure 8. SEM micrographs showing negligible levels of cavitation at 750°C and at the temperature in the 750°C - 830°C range (fastest strain-rate $5 \cdot 10^{-2}s^{-1}$*).*

A number of sub-micron grains are observed within the α colonies in the samples tested at fast strain-rates, regardless of the test temperature (see Figure 9). This phenomenon, which was not observed for the slow strain-rate tests, could have helped in the superplastic response of the material under the fast strain-rate conditions.

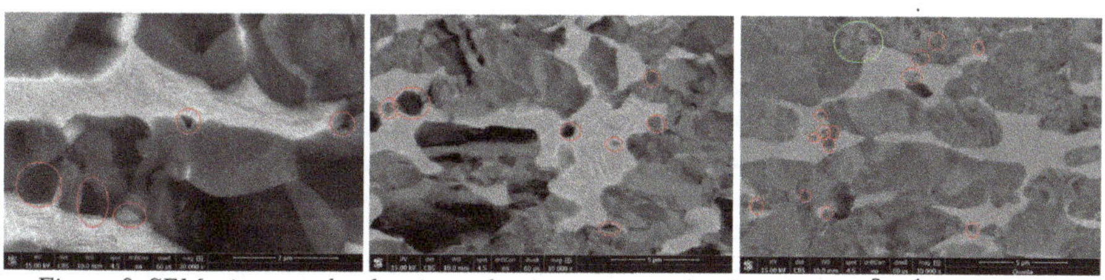

Figure 9. SEM micrographs showing sub-micron grains developed at $10^{-2}s^{-1}$ *for the three temperatures.*

- **Forming Parameters Definition:**

The tensile testing at temperature has allowed to define a material model for the Ti-6Al-4V material of analysis to be used in FE models of SPF. This modelling work will allow to define SPF forming conditions working at lower temperatures and shorter cycles. The limits in the temperature and time reduction would be established by the presence of cavitation in the material, which could compromise the final component in-service performance. Conditions for the presence of cavitation have been defined by the microstructural analysis of this work.

Summary

Tensile tests have been performed on Ti-6Al-4V material with an as-received average grain size of 2.4 μm. The tests were run at three different temperatures (750°C, 830°C and a third temperature in between those two) and seven strain-rates (ranging from $\dot{\varepsilon} = 5 \cdot 10^{-5}\ s^{-1}$ to $\dot{\varepsilon} = 5 \cdot 10^{-2}\ s^{-1}$). The tests were interrupted and specimens quenched when 0.5 true strain was achieved in order to subsequently analyse the microstructure.

For the lowest temperature tested (750°C), strain-rate sensitivities between 0.16 for the quickest test ($\dot{\varepsilon} = 5 \cdot 10^{-2}\ s^{-1}$) and 0.57 for the slowest ($\dot{\varepsilon} = 5 \cdot 10^{-5}\ s^{-1}$) were obtained. Those results were only slightly lower than the ones obtained for the highest temperature tested (830°C), which showed sensitivities between 0.23 for $\dot{\varepsilon} = 5 \cdot 10^{-2}\ s^{-1}$ and 0.71 for $\dot{\varepsilon} = 5 \cdot 10^{-5}\ s^{-1}$. The values of strain-rate sensitivity obtained and the absence of significant necking in almost all the samples denote an outstanding SPF behaviour of the material at such low temperatures and fast strain-rates.

The microstructure assessment showed that apart from the overall grain distribution, during the deformation process clusters of sub-micron size grains appeared in some of the grain boundaries. This remarkable behaviour is thought to have contributed to the grain boundary sliding and the enhancement of superplastic properties. Additionally, presence of cavitation has been analysed in order to define the forming conditions to be avoided. This material has shown cavitation presence only under the extreme testing conditions, and its volumetric fraction can be considered as negligible.

Acknowledgements
We would like to thank The Boeing Company for funding this research as part of project AFRC-DIRF-1172. In particular, we thank Larry Hefti, David Milliken and Nicola Zuelli for their contribution. We also thank the laboratory technicians at the Advanced Forming Research Centre (AFRC) for their contribution running the experimental tests presented in this study.

References

[1] J. Beal, R. Boyer, and D. Sanders, Forming of Titanium and Titanium Alloys. ASM Handbook, Volume 14B: Metalworking: Sheet Forming, S.L. Semiatin, pp. 656-669. https://doi.org/10.31399/asm.hb.v14b.a0005146

[2] L.D. Hefti, Innovations in the Superplastic Forming and Diffusion Bonded Process, J. Mater. Eng. and Perf. 17 (2008) 178-182. https://doi.org/10.1007/s11665-007-9178-0

[3] Y.G. Ko et al., Low-Temperature Superplasticity of Ultra-Fine-Grained Ti-6Al-4V Processed by Equal-Channel Angular Pressing, Metall. and Mater. Trans. A. 37A (2006) 381-391. https://doi.org/10.1007/s11661-006-0008-z

[4] H. Matsumoto et al., Ti-6Al-4V alloy with an ultrafine-grained microstructure exhibiting low-temperature-high-strain-rate superplasticity, Mater. Letters 98 (2013) 209-212. https://doi.org/10.1016/j.matlet.2013.02.033

[5] E. Alabort et al., Alloys-by-design: Application to titanium alloys for optimal superplasticity, Acta Mater. 178 (2019) 275-287. https://doi.org/10.1016/j.actamat.2019.07.026

[6] L. Despax et al., Influence of strain rate and temperature on the deformation mechanisms of a fine-grained Ti-6Al-4V alloy, Mater. Sci. and Eng. A. 790 (2020) 139718. https://doi.org/10.1016/j.msea.2020.139718

[7] J.S. Kim et al., Microstructural analysis on boundary sliding and its accommodation mode during superplastic deformation of Ti-6Al-4V alloy, Mater. Sci. and Eng. A. 263 (1999) 272-280. https://doi.org/10.1016/S0921-5093(98)01157-5

Effects of deformation conditions on the superplastic deformation behavior of LZ91 Mg-Li alloy under electric field

Bao Meng [a*], Feng Pan [b], Jianqiang Yang [c], Dongsheng Li [d], Min Wan [e]

School of Mechanical Engineering and Automation, Beihang University, Beijing, 100191, PR China

[a]mengbao@buaa.edu.cn, [b]buaapanfeng@gmail.com, [c]jianqiangyang@buaa.edu.cn, [d]lidongs@buaa.edu.cn, [e]mwan@buaa.edu.cn

Keywords: Superplastic Deformation Behavior, LZ91 Mg-Li Alloy, Electric Field Assisted

Abstract. The electro-superplastic effect (ESP effect) can enhance the superplastic deformation ability of alloy, and can also make alloy with poor plasticity have superplastic properties. Diverse grain sizes of LZ91 Mg-Li alloy were successfully prepared through equal channel angular pressing (ECAP) process. In order to explore the superplastic deformation behavior of LZ91 Mg-Li alloy under the electric current, an electric field assisted superplastic uniaxial tensile test platform was designed and fabricated. A decreasing constant voltage electrification scheme was proposed, and the experiments under different current densities, initial strain rates and grain sizes were carried out. The results indicate that the true strain-stress curve of LZ91 Mg-Li alloy gradually comes to steady with the increase of current density, presenting a steady-state rheological characteristic. The initial strain rate has a significant effect on the superplastic deformation behavior of LZ91 Mg-Li alloy under high voltage condition. For the fine-grained LZ91 Mg-Li alloy, the electric field can effectively reduce the superplastic deformation temperature and considerably enhance the elongation. This paper enriches the understanding of the superplastic deformation behavior of LZ91 Mg-Li alloy under the action of electric field.

Introduction

Mg-Li alloy is the lightest structural metal, when utilized as a structural material, it can significantly lower the mass of structural parts. The Mg-Li alloy also has the advantages of high specific strength, specific stiffness, excellent damping shock absorption, and good thermal conductivity [1]. The pulsed electric field can decrease the forming temperature of Mg-Li alloy, and reduce its flow stress. Applying an electric field during the superplastic deformation can improve Mg-Li alloy's superplasticity. Therefore, Applying an electric field into the superplastic deformation can enhance the material's superplastic deformation capacity, which is called the ESP effect. The ESP effect can decrease the superplastic deformation temperature and increase its deformation rate, which has attracted considerable scholarly interest domestically and internationally. Li Yao [2] discovered that the elongation and strain rate sensitivity coefficient of Zn-22A1 alloy were significantly enhanced by studying the superplastic properties of Zn-22A1 alloy under the influence of pulse current. Yoshida et al. [3] achieve low-temperature and high-velocity superplastic flow in structural ceramics by employing a reinforced electric field during deformation. They found that the elongation is increased from 5% to 135% after applying 120 V/cm^2 DC to ceramic materials. Wang et al. [4] who successfully welded copper and steel alloys under an external electric field, demonstrated that electric fields can enhance superplasticity.

In this paper, Mg-Li alloy bar with different grain sizes were prepared by equal channel angular pressing (ECAP). In order to investigate the effect of various current densities, initial strain rates, and grain sizes on the mechanical properties of LZ91 Mg-Li alloy, a series of electric field assisted superplastic tensile tests are conducted.

Experimental setup and methodology

Experimental specimen. LZ91 Mg-Li alloy was selected as the experimental material, and the chemical composition is shown in Table 1. The shape and dimensions of the tensile specimen are shown in Fig. 1.

Table 1 Chemical composition of LZ91 Mg-Li alloy (wt.%)

Li	Zn	Mg	Mn	Fe	Cu	Ni	Si
9.21	0.97	base	0.018	0.003	0.0002	0.0003	0.007

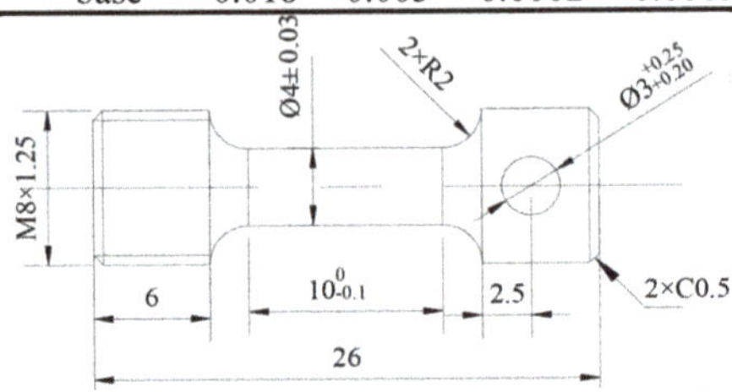

Fig. 1 Tensile specimen dimensions (unit: mm)

Experimental setup. As shown in Fig. 2, The experimental platform consists of an electric field assisted superplastic tensile die, a video extensometer, an infrared thermal imager (FLIR A615), a power supply (SOYI-242000ML), and a fan. The whole set of dies and the cupping machine are connected by a screw thread that holds them together. The power supply is connected to the H59 brass die which has excellent electrical conductivity. The die attached to the cupped machine is made of 45 steel. The insulated material between H59 brass and steel 45 is bake lite.

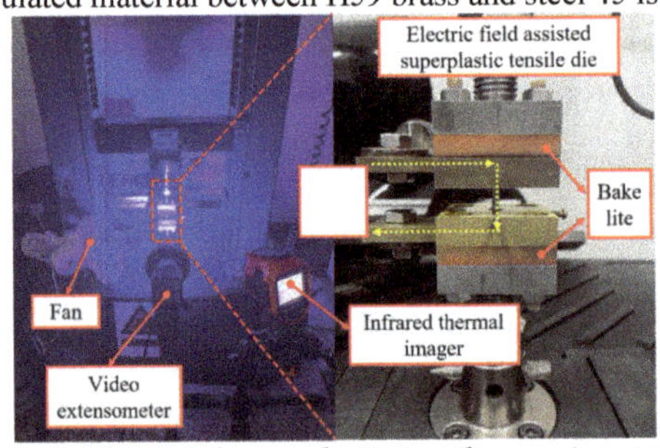

Fig. 2 Electric field assisted superplastic tensile experimental platform

The LZ91 Mg-Li alloy bar was extruded in 0, 4, and 8 passes using ECAP technology to produce specimens with varying grain sizes (β-Li phase) of 18.34, 1.27 and 0.88 μm The B_C path is used to extrude the LZ91 Mg-Li alloy bar in this experiment.

Experimental methodology. During the superplastic deformation process, the relative stability of the specimen temperature is essential. When a constant current is applied to the spacemen, the temperature of the specimen rises rapidly, and it is impossible to maintain a constant temperature. When a constant voltage is applied to the specimen, the rate of temperature rising is reduced compared with the constant current, but the temperature is still not constant. Therefore, a decreasing constant voltage power-on scheme is proposed. The scheme is that power on for a period of time, lower the voltage value and then energized for the same time repeatedly until the specimen is broken.

Materials Research Forum LLC
https://doi.org/10.21741/9781644902615-12

Results and discussions

The effect of current density on the superplastic deformation behavior of the Mg-Li alloy. The unextruded specimen was installed between the electric field assisted superplastic tensile dies, and five tensile tests were performed at the initial strain rate of 8×10^{-4} s^{-1}. Power-on time was set to 330 s→300 s→300 s→300 s→500 s, and the relevant parameters of the experiments are shown in Table 2. In various experimental situations, the elongation of the specimens is diverse. Thus, the time when the specimen fractures is also different. Table 2 displays the fracture time of each experimental specimen. The current density is constantly changing, so the average value of the current density during the stretching process is used as the reference value for the current density.

Table 2. Experimental parameters of the original specimens

Average current density [A·mm^2]	Power-on settings [V]	Temperatur e range [°C]	σ_b [MPa]	δ [%]
17.82	2→1.7→1.5→1.3(break)→1.1	165~272	22.35	89
18.53	2.1→1.8→1.6→1.4(break)→1.2	185~285	16.05	102
16.69	2.2→1.9→1.7→1.5→1.3(break)	174~321	19.31	95
20.62	2.3→2.0→1.8→1.6→1.4(break)	190~332	13.11	105
22.39	2.4→2.1→1.9→1.7→1.5(break)	224~377	9.31	105

Note 1: "Break" in the table indicates that the specimen has fractured under this voltage value
Note 2: Tensile strength σ_b and elongation δ

Fig. 3 displays the true stress-strain curves of the specimens under various current densities. In the initial tensile stage, the flow stress of the specimen rises rapidly with increasing strain. Meanwhile, the internal dislocations of the specimen multiply and interact with each other, and dynamic recrystallization within the specimen is substantially weaker than the hardening effect. Thus, the specimen exhibits a more obvious hardening phenomenon [5,6]. After the stress peaks, the impact of dynamic recrystallization and recovery within the specimen is enhanced. Dynamic recrystallization can effectively reduce the dislocation density, and lower the effect of work hardening on the specimen, which gradually decreases the deformation resistance of the specimen and the slope of the true stress-strain curve [7-9].

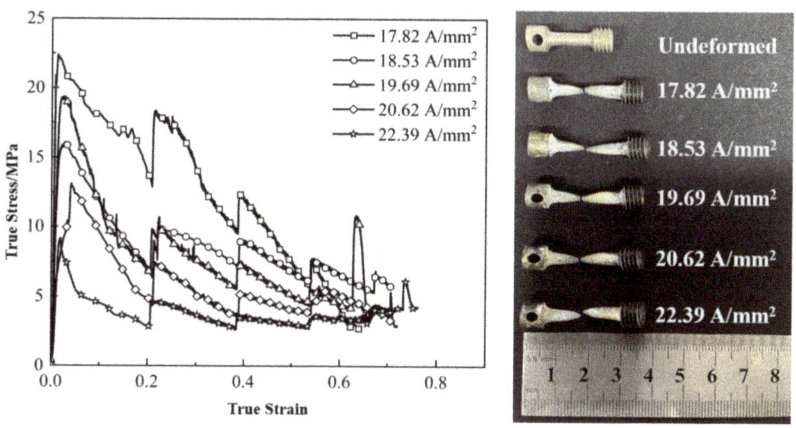

(a) True stress-strain curve (b) Tensile samples

Fig. 3 Experimental results of the initial specimens under different current densities

Materials Research Forum LLC
https://doi.org/10.21741/9781644902615-12

In Fig. 3, it is evident that all true stress-strain curves are "zigzags". This is because the voltage decreases during the stretching process, and the temperature of specimen will drop abruptly at the moment of voltage conversion, causing the true stress-strain curve to rise at the moment of voltage conversion. Thus, each stress surge represents a change in the voltage value. As the current density goes up, the temperature of the specimen grows as well. When the temperature goes up, the atomic kinetic energy and the activity of dislocations increase. Therefore, the LZ91 Mg-Li is more likely to deform under the action of external forces.

Table 2 displays the tensile strength σ_b and elongation of the original specimen under various energized conditions. Table 2 reveals that, the tensile strength of specimen decreases and the elongation rises as the current density increases. Maximum superplastic elongation is 105% under the current density of 20.62 A/mm^2. As the strain increases, the true stress-strain curve of 20.62 A/mm^2 gradually flattens and exhibits steady-state rheological characteristics. The true stress-strain curve of 22.39 A/mm^2 also showed steady-state rheological characteristics. However, compared to the current density of 20.62 A/mm^2, the elongation of Mg-Li alloy didn't get any better, which meant that the deformation temperature of Mg-Li alloy at a current density of 20.62 A/mm^2 had exceeded the optimal superplastic deformation temperature. The difference of elongation at electric field is not very obvious, and the maximum difference is only 16%, which indicates that the ESP effect has little effect on the elongation. The Joule heating effect generated by the current is still more influential on the elongation of the coarse grain LZ91 Mg-Li alloy.

The effect of initial strain rate on the superplastic deformation behavior of the Mg-Li alloy. The relevant experiment parameters are presented in Table 3. In different strain rate experiments, the magnitude of change in current density is not uniform, so voltage is used as the electric field parameter.

Table 3 Experimental parameters of the original specimens with various initial strain rates

Power-on settings [V]	strain rate [s^{-1}]	Temperature range [°C]	σ_b [MPa]	δ [%]
2.1→1.8→1.6 →1.4→1.2	4×10^{-3}	179~260	29.16	64
	8×10^{-4}	185~285	15.90	105
	4×10^{-4}	167~320	19.71	97
2.2→1.9→1.7 →1.5→1.3	4×10^{-3}	170~258	33.30	70
	8×10^{-4}	174~321	19.22	95
	4×10^{-4}	188~302	10.33	110

Fig. 4 shows the true stress-strain curves under different initial strain rates. As Fig. 4 (a) shows, the curve of 8×10^{-4} s^{-1}and 4×10^{-4} s^{-1} are relatively close and show steady-state rheological characteristics at low voltage. This phenomenon indicates that the superplastic deformation behavior of LZ91 Mg-Li alloy is less affected by the initial strain rate at low voltage. However, at high voltage, as shown in Fig. 5, the flow stress of the material decreases significantly as the initial strain rate decreases. This indicates that under high voltage, the initial strain rate has a great influence on the superplastic deformation behavior of the material. Fig. 5 (a) shows the true stress-strain curve with an initial strain rate of 4×10^{-4} s^{-1} "upturns". This is because the deformation temperature is higher, and the grain of Mg-Li alloy continues to get coarser at a low initial strain rate. This slows down the process of grain boundary slippage. At the same time, the internal recovery and dynamic recrystallization of material alternate, significantly reducing the density of dislocations, leading to dislocation plugging and entanglement, which makes the material less likely to deform.

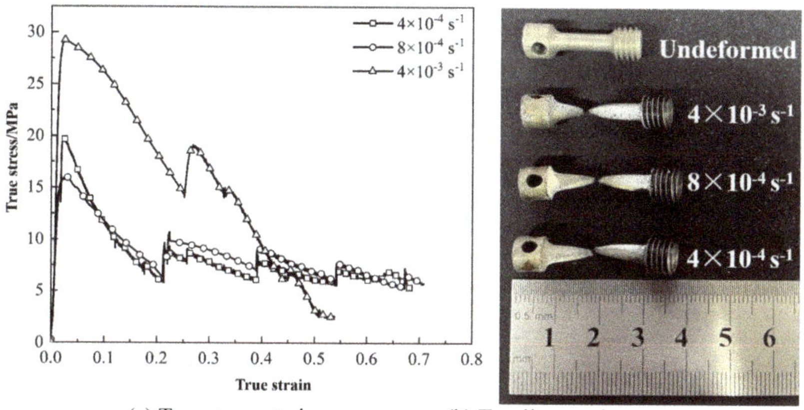

(a) True stress-strain curve (b) Tensile samples

Fig. 4 Experimental results of the initial specimens with different initial strain rates at low voltage

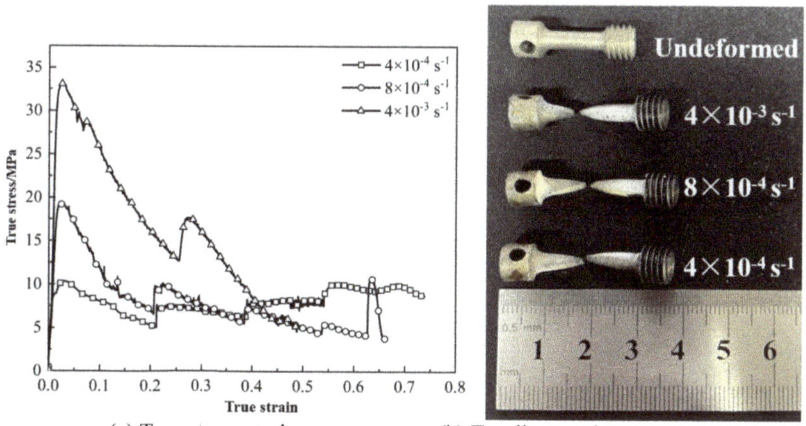

(a) True stress-strain curve (b) Tensile samples

Fig. 5 Experimental results of the initial specimens with different initial strain rates at high voltage

Table 3 lists the tensile strength and elongation of the LZ91 Mg-Li alloy at different initial strain rates. It can be noticed that the tensile strength of Mg-Li alloy declines with the decreasing initial strain rate, which is more obvious at high voltage. When the initial strain rate drops at low voltage, the elongation of Mg-Li alloy increases first and then decreases. The decrease in elongation can be attributed to the fact that a low voltage makes the electro-plasticity effect less apparent. Besides, long-term energization causes high-temperature oxidation of LZ91 Mg-Li alloy. Thus, the material goes into the necking stage early which causes high temperature and cracks in some places. At high voltage, the elongation of Mg-Li alloy increases with the decrease of the initial strain rate, and the maximum elongation of 110% is reached at the initial strain rate of 4×10^{-4} s^{-1}, which also proves that higher current density is more effective for improving the superplasticity of materials.

The effect of initial grain size on the superplastic deformation behavior of the Mg-Li alloy. Mg-Li alloy with various grain sizes (β-Li phase) of 18.34, 1.27 and 0.88 μm were utilized to conduct the experiments. The average grain size of β-Li phase is the reference value. The relevant parameters of the experiment are shown in Table 4.

Table 4 Experimental parameters of specimens with different grain sizes

Power-on settings [V]	Grain size [μm]	Temperature range [°C]	σ_b [MPa]	δ [%]
2.0→1.7→1.5 →1.3→1.1	18.34	179~260	22.20	89
	1.27	185~285	41.15	115
	0.88	167~320	19.45	130
2.2→1.9→1.7 →1.5→1.3	18.34	170~258	19.17	95
	1.27	174~321	7.12	148
	0.88	188~302	22.34	133

The true stress-true strain curves of LZ91 Mg-Li alloy with different grain sizes are shown in Fig. 6 and Fig. 7. It can be seen that no matter what energization conditions, the flow stress of fine grain LZ91 Mg-Li alloy decreases more obviously, because the refinement of the grain can reduce the superplastic deformation temperature. Thus, at the superplastic deformation temperature appropriate for the initial LZ91 Mg-Li alloy, the recrystallization of the fine grain Mg-Li alloy is more completely and has exceeded the appropriate superplastic deformation temperature. At the same time, it can be seen from the figure that the effect of fine crystal strengthening is obvious.

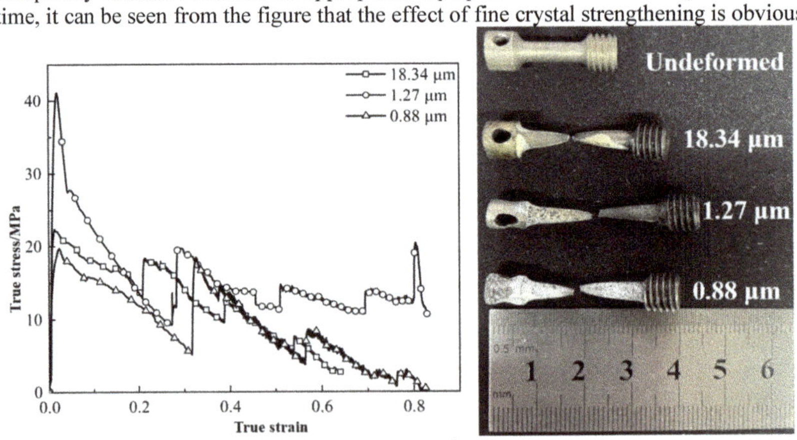

(a) True stress-strain curve (b) Tensile samples

Fig. 6 Experimental results of the specimens with different grain size at low voltage

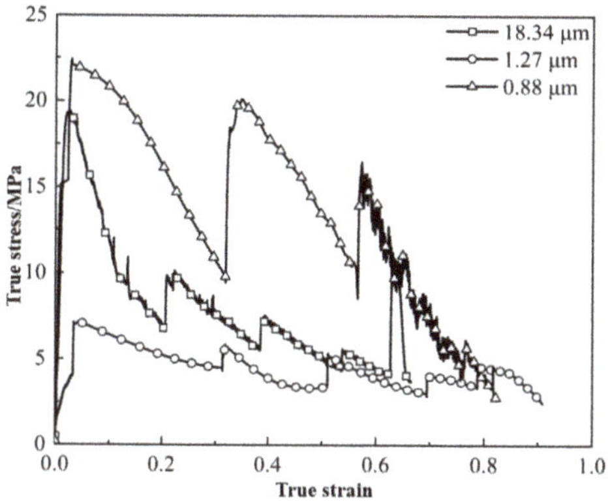

(a) True stress-strain curve

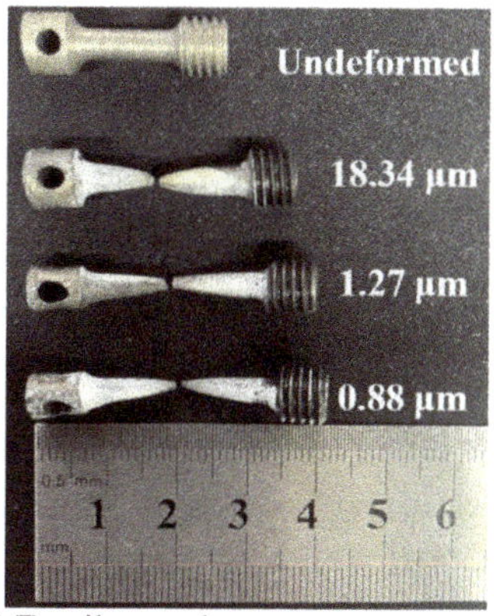

(b) Tensile samples

Fig. 7 Experimental results of the specimens with different grain size at high voltage

As shown in Table 4, the elongation of Mg-Li alloy whose grain size (β-Li phase) is 1.27 μm reaches 148% under the voltage conditions of 2.2 V→1.9 V→1.7 V→1.5 V→1.3 V, and the true stress-strain curve is basically flat. The superplastic temperature of Mg-Li alloy with initial grain size (β-Li phase) of 0.88 μm is low, which indicates that the refinement of the grain can reduce the superplastic temperature of Mg-Li alloy. Therefore, the electric field has a great influence on the elongation on the fine-grained LZ91 Mg-Li alloy.

Conclusions

Through electric field assisted superplastic tensile tests, the effects of current density, initial strain rate, and initial grain size on the superplastic deformation of LZ91 Mg-Li alloy were investigated. The following conclusions can be drawn:

1. The true strain-stress curve of LZ91 Mg-Li alloy gradually comes to steady with the increase of current density, presenting a steady-state rheological characteristic.
2. At low voltage, the initial strain rate has less influence on the superplastic deformation behavior of LZ91 Mg-Li alloy. At high voltage, the elongation of Mg-Li alloy increases as the initial strain rate decreases, reaching a maximum of 110% at an initial strain rate of 4×10^{-4} s^{-1}, and the tendency of true stress-stain curve appears "upturned".
3. For the fine-grained LZ91 Mg-Li alloy, the electric field can effectively reduce the superplastic deformation temperature and considerably enhance the elongation under the same energized conditions.

References

[1] Jia Y. LZ91 Mg-Li alloy superplastic performance and negative angle box parts superplastic forming [D]. Harbin Institute of Technology, 2017.

[2] Li Yao, Dong Xiao Hua. Dynamic analysis of dislocation activation in electro-plastic effect of Zn-22%A1 alloy, Journal of Jiang Han University (Natural Sciences), 2007 (03): 39-41.

[3] Yoshida H. Sasaki Y. Low temperature and high strain rate superplastic flow in structural ceramics induced by strong electric-field [J]. Scripta Material, 2018, 146: 173-177. https://doi.org/10.1016/j.scriptamat.2017.11.042

[4] Wang Yaoli, Wang Guangxin, Zhang keke. Electro-superplastic solid state welding of 40Cr/QCr0.5[J]. Materials (Basel), 2018, 11 (7). https://doi.org/10.3390/ma11071153

[5] Bi D, Zhang J, Cui H X. Calculation of extrusion force of equal channel corner extrusion die[J]. Heavy Machinery, 2001 (6): 34-36.

[6] Li X F, Cao X D, Wang B, et al. Research progress of Ti alloy electro-assisted plastic forming technology[J]. Aerospace Manufacturing Technology, 2021, 64 (17): 22-30.

[7] Ding X F, Shuang Y H, Lin W L, et al. Rheological behavior and constitutive model of extruded Mg alloy[J]. Chinese Journal of Plastic Engineering, 2017, 24 (06): 165-171.

[8] Gao X B, He C M, Yao T, et al. Dynamic softening behavior and microstructure evolution of Al-Mg-Li alloy[J]. Journal of Specimens and Heat Treatment, 2022, 43 (11): 170-179.

[9] G. D Urso, C. Giardini, M. Quarto, Characterization of surfaces obtained by micro-EDM milling on steel and ceramic components, The International Journal of Advanced Manufacturing Technology, 2018 (97): 2077-2085. https://doi.org/10.1007/s00170-018-1962-5

[10] Pu Y, Kou S, Zhang Z, et al. Effects of return son composition, microstructure and mechanical properties of GH4169 super alloy[J], China Foundry, 2017, 14 (4): 244-250. https://doi.org/10.1007/s41230-017-5125-3

[11] Dong D Y, Liu Y, Wang L, et al. Effect of strain rate on dynamic deformation behavior of DP780 steel laser welded joint[J]. Acta Metallurgica Sinica, 2013 ,49 (12): 1493-1500. https://doi.org/10.3724/SP.J.1037.2013.00341

[12] Liu Y, Wang J, He S S, et al. Effect of long-term aging on dynamic tensile deformation behavior of GH4169 alloy[J]. Acta Metallurgica Sinica, 2012, 48 (1): 49-55. https://doi.org/10.3724/SP.J.1037.2011.00435

[13] Liu Y, Li Z Q, Zhao B, et al. Superplastic deformation behavior and constitutive model of TA32 Ti alloy[J]. Rare Metal Specimens and Engineering, 2022, 51 (10): 3752-3761.

[14] Zhang T Y. Study on low-temperature superplastic deformation of fine-grained TC4 alloy[D]. Central South University, 2014.

Integration of HF and SPF technology in an efficient industrial production flow for high rates aeronautical parts

Yves Marcel[1a], Jeremy Pion[1b], Sharan Raj Rangasamy Mahendren[2c]

[1] LAUAK AEROSTRUCTURES FRANCE, 2245 route de Minhotz, 64240, Hasparren, France

[2] LGP, INP/ENIT, University of Toulouse,47 Avenue d'Azereix, F-65016, Tarbes, France

[a]yves.marcel@groupe-lauak.com, [b]jeremy.pion@groupe-lauak.com, [c]sharan-raj.rangasamy-mahendren@enit.fr

Keywords: Titanium, Hot Forming, Superplastic Forming, Ti6Al4V, Ti6242, Airbus, A320NEO, Industrial Challenges, HF/SPF Simulations, HF/SPF Workshop, Logistics, Production Workflow

Abstract. The presentation will illustrate the creation and production ramp-up of the Titanium Hot Forming and Superplastic Forming Technology in a French aeronautical supplier from Industrialization to A320NEO full-rate production. With deformation and elongations well beyond classical « cold » sheet metal forming capabilities (especially with high-strength titanium alloys such as Ti64 and refractory Ti6242), Hot Forming and Superplastic Forming were key assets to be complementary to other standard manufacturing methods already used in the factories of the LAUAK group. Several aspects of the project will be presented, including the installation of HF/SPF workshop and presses, the industrialization steps of Airbus A320NEO pylon parts from tooling design to the qualification, and then integration of the HF & SPF in the full logistics flow of the factory production lines (from the material, blanks, forming, 5-axis laser trimming and machining, control and finally assembly). Some of the industrial challenges faced and overcame will be highlighted and illustrated.

Introduction

Mostly driven by cost and performance improvement objectives, Titanium was introduced in the aeronautical industry and is now used more and more widely. Applications of Titanium alloys increased in the last forty years and can now represent up to 15% of the total weight of a modern airplane.

Uprising in environmental awareness in conjunction with energy prices boosted the need for lighter higher performance materials, especially to accommodate higher temperatures for engines and exhausts. High-temperature resistant titanium alloys such as Ti-6-2-4-2 could be considered as an interesting option to substitute for much heavier nickel-based alloys (i.e. Typ. Inconel) for application of semi-high temperature where Ti-6Al-4V would give up.

Industrialization of parts with such materials could be a real challenge, especially in the context of higher volume productions of bestselling aircraft programs like the A320NEO (Fig.1). This publication will present the development and production activities of LAUAK of refractory Titanium alloy formed parts on engine pylon (Fig 2).

 https://doi.org/10.21741/9781644902615-13

Fig 1:Airbus A320 NEO (2 engine versions available)

Fig 2:A320NEO Pylon with SPF and refractory Titanium applications

LAUAK GROUP

LAUAK Group is highly active in the aeronautical industry, with operation in 10 production sites in 5 Countries with more than 1600 employees.

Specialized in elementary parts and sub-assemblies for major aircraft manufacturers and tier 1 suppliers, the major division (LAUAK AEROSTRUCTURES) is experienced with various activities like machining, sheet metal forming, welding and assembling, both on light alloys and hard metals (Fig3).

Fig 3:LAUAK AEROSTRUCTURES main activities and partners

Titanium Hot Forming (HF) and SuperPlastic Forming (SPF) Workshop

The development of the Titanium applications in the pylons of the Airbus new program A320 NEO (New Engine Option) pushed the need for new production processes and machines in the LAUAK Hasparren Plan (Pays Basque – FRANCE).

A dedicated workshop (Fig 4.) was created in 2018 in order to accommodate two new Hot forming and Superplastic Forming Presses with platen dimensions up to 1.8m x 1.5m and temperatures up to 940°C.

Five years later, a third machine is now installed and more than 10 000 parts are produced per year with either Hot forming or Superplastic Forming technology.

Fig 4:LAUAK HF SPF Workshop with 800T PRESS

Production organization was a key factor to create a functional production unit with high-performance "Means and Tools", but also skilled and trained operators supported by a strong technical team and process experts.

The full integration of the Titanium hot forming workshop (Fig5) in the global production flow of the whole factory is a major element for the best industrial performance achievement.

Good interaction with all "support services" are also a major contributing factor of the global performance: Method, Logistics, Quality, Purchase, IT, Laboratory, Maintenance,... all with daily interaction, especially in the context of recurring high production rates.

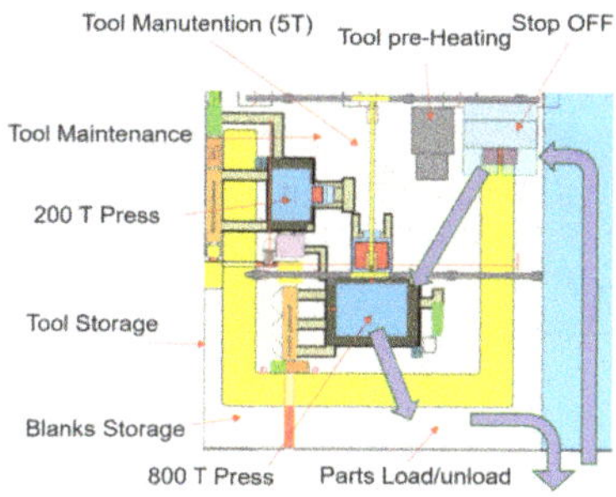

Fig 5:HF/SPF Workshop Flow and environment

Industrialization process

After being awarded new productions and selecting the manufacturing methods, the industrialization process starts with "definition analysis" to design the tools needed for parts production in the most efficient way regarding the global company's main objectives of performance: *Quality, Cost and Delay*.

In Titanium sheet metal applications, Ti6-4 can only be "cold-formed" to 10% elongation. Related to typical part geometry and complexity, Hot Forming is then usually a good option if you need to reach up to 60% elongation. In case of much more complex shapes with higher deformations and elongations (up to 300%), SPF would be the alternative but would require a higher forming temperature and longer processing time.

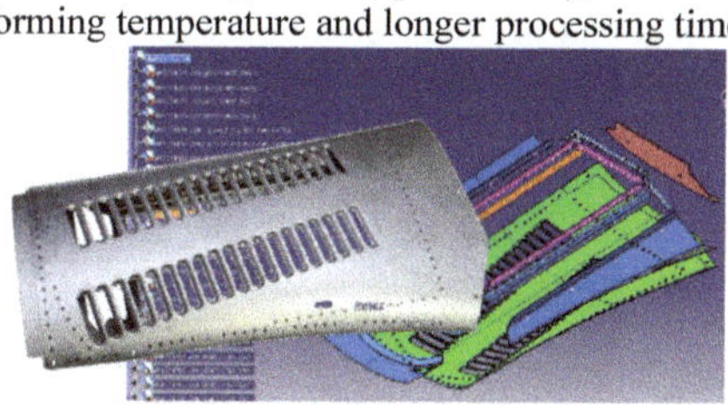

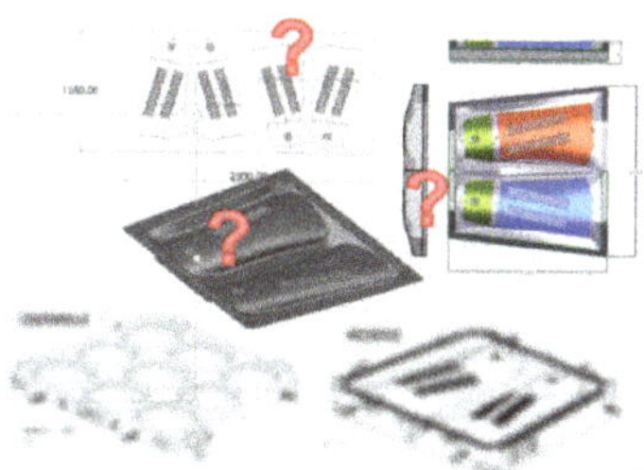

Fig 6:Example of part definition and Tooling configuration analysis

The "manufacturing cost" analysis (Fig 6) is a critical step and is the occasion of many practical choices where you ABSOLUTELY do not want to take wrong decisions or make mistakes, especially for high-rate production situations. Typical choices are related to part-forming technology and configurations, tooling types and materials, calculation of future forming parameters and machine selection.

Mistakes in the definition of hot forming dies worth 100 to 300 k€ could be a nightmare both from an economical and a planning point of view. Leading hypothesis and tool definitions have to be secured and optimized as much as possible during all the industrialization design phase.

As an important part of LAUAK production capabilities "Cold forming" manufacturing alternatives are often evaluated to check parts formability limits in order to get the best understanding of the Hot forming eligibility and industrialization.

Detailed CAD studies are often tested with FEM simulation in order to secure critical hypotheses and decisions regarding tooling designs and choices (Fig 7).

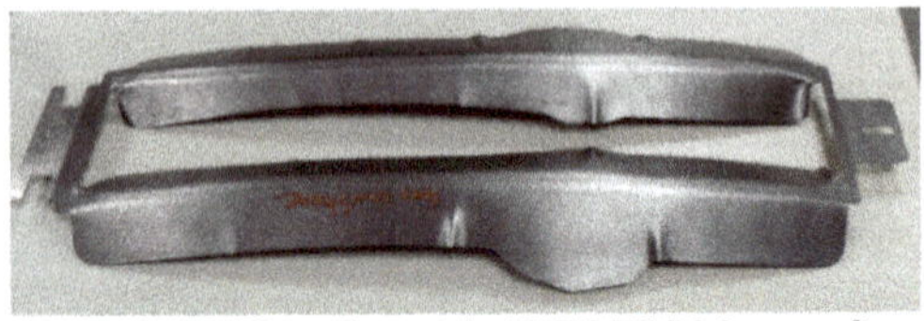

Fig 7: LAUAK studies with AUTOFORM to predict and localize critical zones (left) compared to real part trials (above).

Many specific questions related to HF/SPF tooling studies and detailed design need to be nailed just right in the first attempt in order to carry the project to success (Fig 8).

As an example amongst critical details: thermal differential dilatation of Ni-based (tooling material) and Titanium (part) have to be COMBINED to precisely define a tool machining modifier coefficient (0.9###) in order to target accurate final part dimensions. One of the many challenges to overcome…

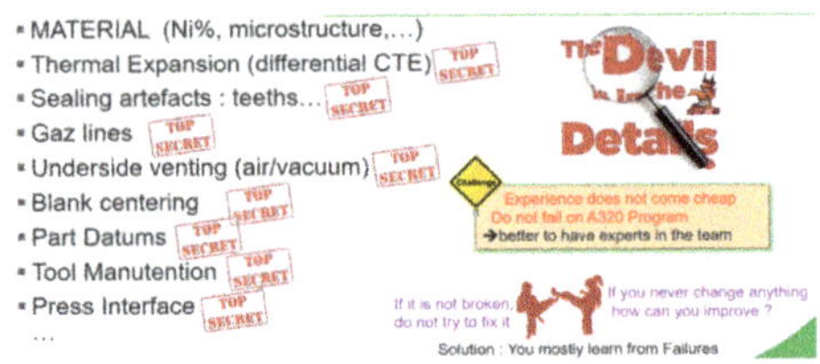

Fig 8:Example of SPF tooling design "hot" topics

Production / Hot Forming & SPF: in the heat and fire thou shall be formed

As an example of Titanium's important components of A320NEO, two main applications of hot forming and superplastic forming in pylons are presented (Fig 9).

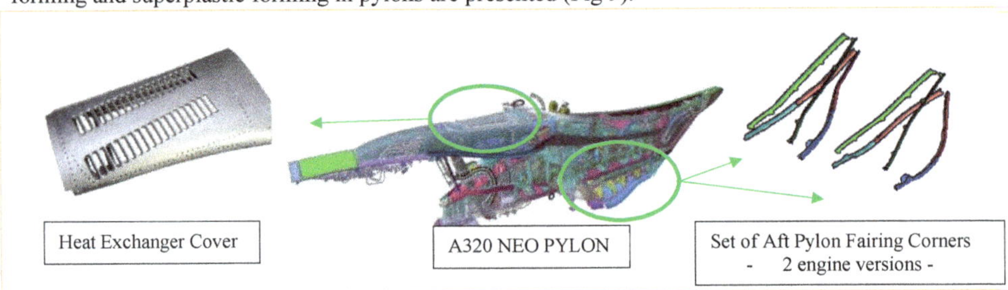

Fig 9 : Examples of applications For LAUAK Titanium SPF and Hot Forming for High Volume A320NEO Productions

The "Heat exchanger cover", with an aerodynamic 3D shape and louvered discharg  s, is produced by superplastic forming of Ti6-4 sheet.

Using the maximum efficiency of our SPF 800T/ 1.8mx1.5m capacity press, each blank is capable of forming several parts per cycle, with more than 10 cycles per day, making the process efficient and capable of high production volumes.

The main challenge was the design of an optimized tool being able to sustain the hard forming conditions with Thermo-mechanical cycling at 920°C + stress caused by 500T + 30 bars applied in the SPF process.

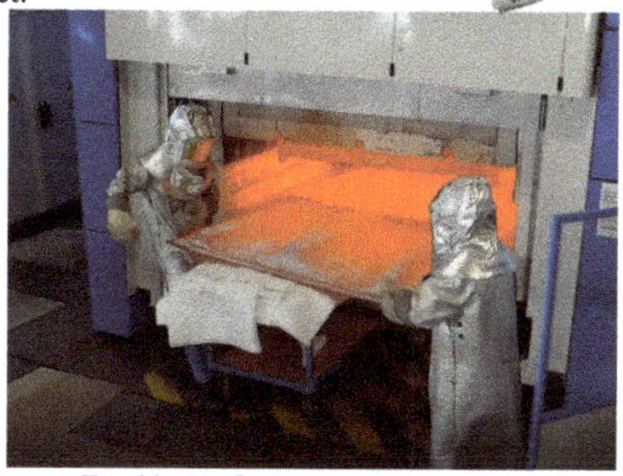

Fig 10: High-temperature SPF blank extraction at 920°C

SPF Production with dimensions more than 1.5m needs special care regarding the hot unloading phase (Fig 10) because forces applied to extract the part from the "sticking hot SPF tool" can be a source of important distortions. *Most of the dimensional defects of large SPF-formed parts are generated between extraction and cooling.* Extra care is needed during the "stripping + handling" phase, including finding the best sequence with special manutention tools, and if needed cooling jigs. Quality of shapes needs to be controlled, usually with templates. With the proper optimization of the process (and it might be tricky to find), the need for final shape adjustments can be completely avoided, and this is what you have to aim for, dealing with high volume A320 parts!

Efficient reliable tool sealing and argon lines are also key for production stability. Adequate tooling care and maintenance ensured our tool's good production capacity for more than 2000 cycles.

Thanks to the quality of the initial industrialization and the commitment of the production staff, The HF SPF LAUAK workshop has been able to steadily form and deliver SPF productions (Fig. 11) towards downstream operations during the last four years.

Fig 11 : Example batch of SPF formed blanks ready to be transferred to next operations (i.e. : Laser 3D Trimming)

The "Aft Pylon Fairing (APF) corners", represents even a greater technical challenge with twisted complex and 3d profiles with deep bending and joggles (Fig 12). The need for a more heat-resistant titanium alloy (Ti6242) due to parts positions close to the engine exhaust makes the forming *even more difficult* with unusually important stress caused on the Ni-based hot forming tools.

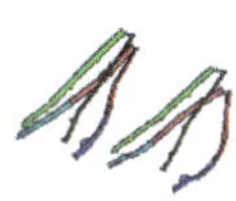

Materials Research Forum LLC
https://doi.org/10.21741/9781644902615-13

With kits of more than twelve different individual parts requiring tools for each operation like forming, laser cutting, fitting, machining, control, … Detailed studies were conducted and more than 70 different toolings were required to design and produce for the full manufacturing process of this work package.

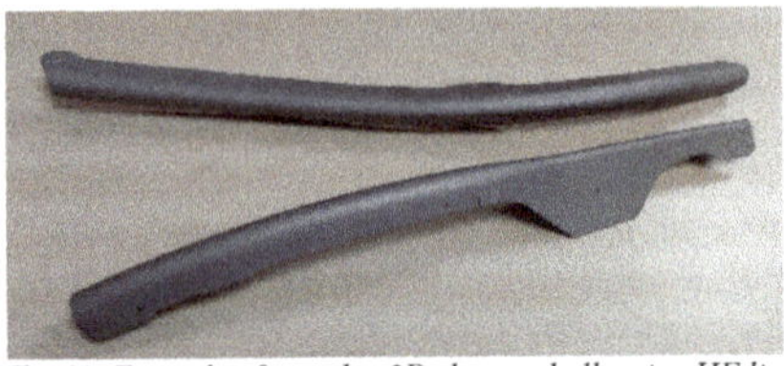
Fig 12: Example of complex 3D shapes challenging HF limits

Hot forming tools were the most challenging to design due to part complexity and unusual extreme HF process parameters of the 6242 material.

All three LAUAK Hot Forming machines (Fig 13;14;15) are successfully contributing to the industrialization with more than a dozen of different HF new tools involved in this project.

Fig13:APF Corners forming on 0.9x0.6m press

Fig14 : Ti6242 HF on 1.2x1.2m Press

Fig15:Ti6242 production on 1.8x1.5 Press

In addition to the 3D shape of most of the parts being really complex, the major challenges for this "APF Corners" application were caused by the refractory nature of the Ti Alloy used. The specific chemical composition (6%Al, 4% Zr, 2% Mo and 2% Sn) being source of significant higher mechanical properties at elevated temperatures than the usual 700-750°C used for HF with Ti Alloys like Ti64.

Some problems were faced and tooling design optimizations were needed to improve deformations trajectories regarding tooling contact causing scratches (Fig.16) and unacceptable tool surface degradation, especially considering this program needed to be conducted for several years with high production volumes.

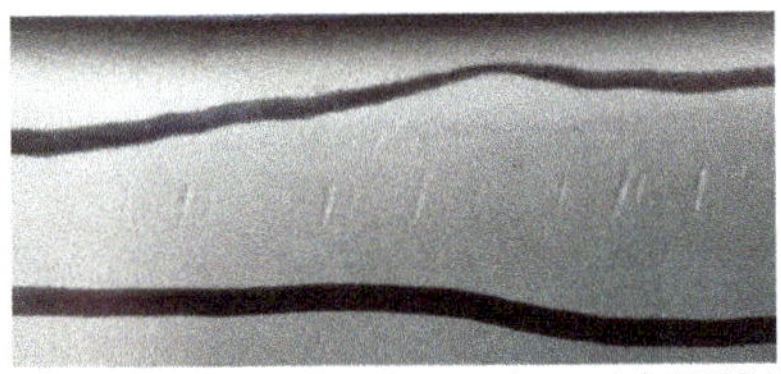
Fig 16: Example of scratches with Ti6242 HF

Optimum forming parameters were explored in order to find the best balance between "*Higher temperature for reducing spring-back and metal toughness*" and "*not too high to maintain minimum part toughness for handling and preserving tool life*"

Serial production started in 2022 in the frame of a careful quality control (Fig17) and process monitoring of the production (Fig 18) to secure the HF/SPF performance and part production flow.

Fig 17: Example of Dim control jigs

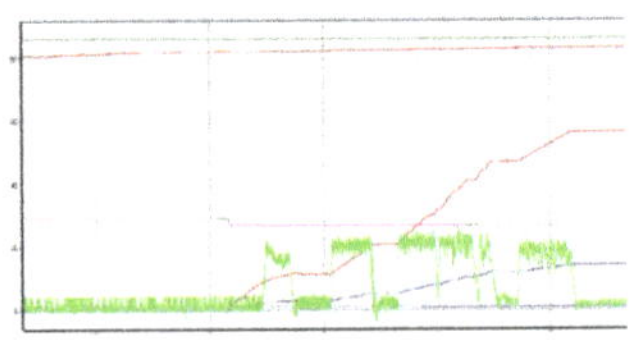
Fig18: Example of process control monitoring

Other contributing factors such as lubrication and tooling maintenance were also identified as key to maintain the best quality of formed production month after month.

Materials Research Proceedings **32** (2023) 119-126 https://doi.org/10.21741/9781644902615-13

There are more things in life than Hot forming: Other Productions steps and Logistics

Of course, the Production of these parts does not start and stop at the hot forming sequence.

From the starting material supply (Titanium alloys sheets) to the finished part controlled, packed and ready to be delivered, many critical steps (Fig 19) need to be conducted in almost every LAUAK workshop, preferably under the management of the Logistics teams to optimize "On Time Delivery" (OTD).

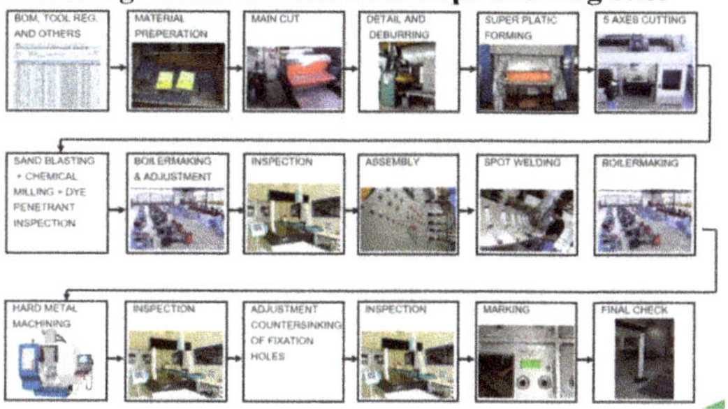

Fig 19 : Full production workflow of Heat Exchanger Cover

Prior hot forming, early operations include Material supply (sometime to be ordered 12 to 18 months in advance for Titanium Material) and laser 2D blank cutting (Fig 20).
Efforts are being made on a daily basis to optimize blank placement on sheets to reduce material loss.

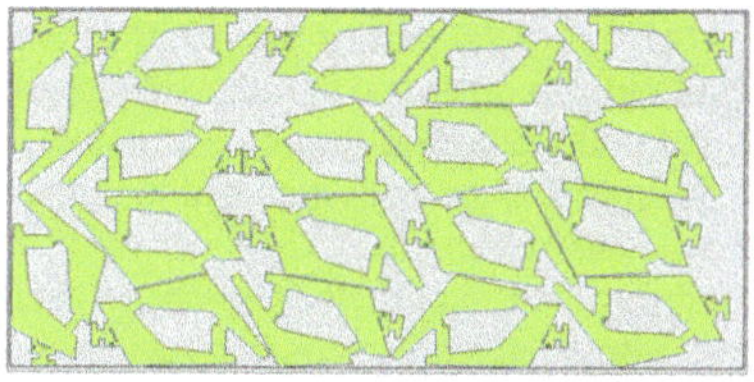

Fig 20: Example of 2D laser cutting blank material optimization

After forming, Main operations are usually 3-axis laser Cutting (Fig 21) and/or machining (Fig 22) (drilling, chamfer, countersink…) to produce the final part contour and interfaces for fasteners ready for assembling (Fig 23).

Fig 21: 5-axis (3D) Laser Cutting

Fig 22: Machining (Contour and Drilling)

Fig 23: Assembling

The efficient use of an "Enterprise Ressource Planer" Software (ERP) is critical not to be lost in the 20 000 "production works orders" (aka job cards, or traveler) currently active at any given time in the plant due to all simultaneous running other productions. Enough anticipation and adhesion to MRP planning are critical for organizing, especially for "high runner" productions.

In addition to our ERP, we launched a project to improve our logistics performance based on geo-localization tags and smart visual planning named "*HORDAGO*" (meaning "*It is there*" – in the Basque language).

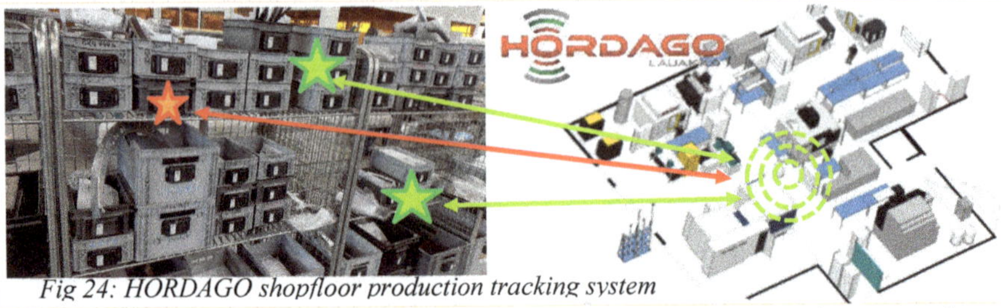

Fig 24: HORDAGO shopfloor production tracking system

Every running production is equipped with wireless tags detected by gates and sensors and can be identified and localized in seconds (Fig 24). Each tag is also linked dynamically to the planification databases and can "call workers for action" via red and green LEDS when needed to help prioritise jobs at the workstation level.

Conclusion

Since 2019, LAUAK has been part of the A320NEO program with several HF and SPF titanium pylon components and demonstrated that these processes can be fully integrated as complementary assets in the global manufacturing frame of a major Tier 1 Aeronautic supplier in addition to all other more classical production activities and technologies such as standard sheet metal forming, welding, machining, assembling,….

Of course, a strong and experienced in-house technical support team is needed to drive all studies and new industrialisation projects taking into account the specifics of Titanium Hot Forming technologies (tooling design, material limits, …)

Finally, to maintain production stability, some specific elements in the Titanium HF and SPF way of life had to be taken into account to steadily sustain the Ramp-UP and the high production volumes associated with the success of the A320. As an example, critical equipment such as SPF Presses need to be kept in the best condition possible by application of consistent Predictive Maintenance to closely monitor machine performance evolutions.

Materials Research Proceedings **32** (2023) 127-140 https://doi.org/10.21741/9781644902615-14

Investigation of the complexities inherent in manufacturing near-unconstrained superplastic parts by experiments and simulation

Bryan Ferguson[1,a], Eric Bol[1,b*], D.G. Sanders[1,c], and M. Ramulu[1,d*]

[1]Department of Mechanical Engineering, Box 352600, University of Washington, Seattle, WA 98195, USA

[a] bjferg@uw.edu, [b] boled84@gmail.com, [c] southernstargold@comcast.net, [d] ramulum@uw.edu

Keywords: Superplastic Forming, Titanium, Finite elements, Simulation

Abstract. Superplastic forming is a sheet metal forming process that has found widespread use in the aerospace industry. It produces parts that are free of residual stresses, dimensionally accurate, and with strains unobtainable using conventional forming methods. When combined with diffusion bonding, a phenomena where under similar processing conditions the material involved will produce a near flawless weld with itself, a variety of reinforcements and internal structures can be produced. In most superplastic parts the material is blow formed up to a die and the material takes on the dimensions of the die with small variations in thicknesses. In this work, however, we investigate a process unique to superplastic forming with diffusion bonding using four sheets. The two outer sheets are formed up to the die while the two inner sheets form a complex sandwich structure. These inner sheets are completely unsupported except at the edges of the part. Superplasticity is stress-history dependent and somewhat chaotic in nature. Therefore, the inner sheets have a large variance in geometry due to the fact that they have only limited constraints and are free to shift and translate as the forming operation progresses. Without rigid constraints on the inner sheets small variations in geometry can be magnified to create large changes in final geometry. The variances in forming are quantified using a variety of techniques to measure the major features of the process including cell wall measurements, gas pathway measurements, and computer vision-based geometry analysis. Two- and three-dimensional finite element simulations of the inner sheet forming process were used to compare the characterization results with idealized geometry. The results of the analysis provide insights into the complexities of manufacturing such internal structures.

Introduction

The subject of this research is a large area, free floating, superplastic structure that needs to form a complex and predictable pattern of interior vertical supports for two exterior sheets of titanium. The variability of this structure is much larger and less predictable than typical superplastic parts due to the stress-history dependence of superplasticity magnifying small variations in geometry to create measurable discrepancies in the finished part. Superplasticity is the ability of a metal, ceramic, or intermetallic to achieve very large strains at elevated temperatures with a controlled strain rate [1,2]. It has become a common production method in the aerospace and automotive industries due to its ability to form complex shapes with no residual stresses. The dominant mechanism for superplasticity is sliding along grain boundaries with some secondary diffusive and dislocation-based accommodation mechanisms [3].

Materials must be fine grained in order to have adequate superplastic properties and titanium is typically fine grained enough without additional processing. The grain structure of titanium, in combination with its expense and manufacturing challenges via other methods, make it an ideal material for superplastic forming. Titanium also has excellent corrosion resistance, a high strength to weight ratio, and large degree of customizability through a variety of alloying elements. There are very few weld nuggets that have superplastic properties, because the weld nugget of resistance welds or other heat-based welds have grains far too elongated to be superplastic. To date the only type of welding that can produce superplastic welds is friction stir welding. Diffusion bonding, often coupled with superplasticity, provides near indistinguishable welds when two surfaces contact each other at the correct processing conditions [4]. Superplastic forming and diffusion bonding allows for the formation of monolithic parts that have the advantages of being a single structure that reduces fasteners and part complexity [5].

Superplastic forming is most commonly used for the production of single sheet structures. Single sheet forming entails clamping the sheet in a die at the edges and blowing the sheet up to the die from one side. Multiple sheet forming also exists and has been used in production parts. Traditionally, the sheets are diffusion bonded before superplastic forming by using a pattern. The pattern is deposited using yttria powder silk screening which prevents diffusion bonding where applied [6]. Gas pathways are included into the pattern with a welded fitting at the side of the sheet. The bonded sheets are then enclosed in a die and gas formed up to it at temperature; however, there are some non-traditional multi-sheet forming patterns that are not diffusion bonded prior to forming. Numerous sheet patterns exists but the subject of this research is the Rockwell four sheet process [7]. This sheet forming pattern has a completely unsupported internal sheet structure that is free to shift and slide.

Superplastic forming is highly sensitive to material properties and processing conditions, specifically the temperature and strain rate applied; however, once the correct conditions have been dialed in, it is a very repeatable process [8]. There can be some variation in thicknesses of formed sheets, but these are mainly dependent on unavoidable variations in material properties. While the geometric constraints on the face sheets are at both the edges and the surface of the die, the core sheet constraints are only at the edges where all four sheets are clamped. These inner sheets are therefore able to move freely with forces being balanced between the cells. With the stress history dependence of superplasticity, slight variations in initial geometry can lead to large changes in formed geometry. The variability of this free-floating core is the subject of this research, both to quantify it and to provide insight into how to cope with the variability in a production setting.

The mechanical behavior of superplasticity is modeled in the form of constitutive equations governing the plastic behavior of the material. Several different forms of constitutive equations have been developed which are typically from empirical, or internal state variable forms. Empirically derived equations map the strain rate to the stress with constants derived from experiments, and have been developed for both single and multi-phase materials. Examples of empirically derived equations use the extended power law creep equation (Eq. 1) [2]:

$$\sigma = K\varepsilon^n\dot{\varepsilon}^m. \tag{1}$$

The constant K, along with the rate constants n and m, are mapped to physical experiments. Grain growth has been included in several models that add increased accuracy to the model [9]. Applications of crystal plasticity implemented in a self-consistent viscoplastic form have also been developed [10]. These equations chose the internal state variables as dislocation density and grain size. With these constitutive equations, finite element simulations of superplasticity have been implemented extensively to model the complex geometry used in the forming operation.

There have been many maps to the experimental using both the hyperbolic sine viscoplastic models [11] and power law creep models, and extensive studies have been conducted on both of these models for titanium. Hyperbolic sine constitutive equations provide a wider range of strain rates for highly variable parts while the power law model does a sufficient job for a smaller range of strain rates [12]. These have had very good correlation between formed parts and the production geometry provided that the material model can accurately simulate the range of processing conditions [13], and the implementation of these models is similar. The stresses are assumed to be mapped to the elastic stiffness tensor to produce the deformation tensor based on the applied pressure and temperature.

Experimental Methods

The Rockwell four sheet SPF/DB process was implemented in this research with the production of rectangular flat panel parts illustrated in Figure 1. Each part has two sheets on the outside that form up to die surfaces and two internal sheets that form a rectangular reinforcement pattern and diffusion bond to themselves and the outer sheets. In this way, an elegant and stiff sandwich structure is fabricated by simply welding a pattern to the interior sheets. The outer sheets, called face sheets, deform very little and take on the shape of the die with minimal thickness variation. The internal sheets that end up forming a rectangular grid of vertical supports are call core sheets, and are created by a rectangular grid of resistance welded lines. These lines form 1" x 2" cells and are made of evenly spaced, but separated, small welds that are not superplastic.

The separations between the core sheet welds allow gas pressure to communicate between the cells from a gas input line at one end of the part and they form an inherent defect call the gas connection path. When the core sheets are pressurized by gas, they balloon out in rectangular sections and eventually deform into each other to form vertical supports (columns) at the weld lines. At the top of each column is a triangular defect called the upper gas path. This pathway is necessary to release the gas that was used to form the face sheets to the die.

a)

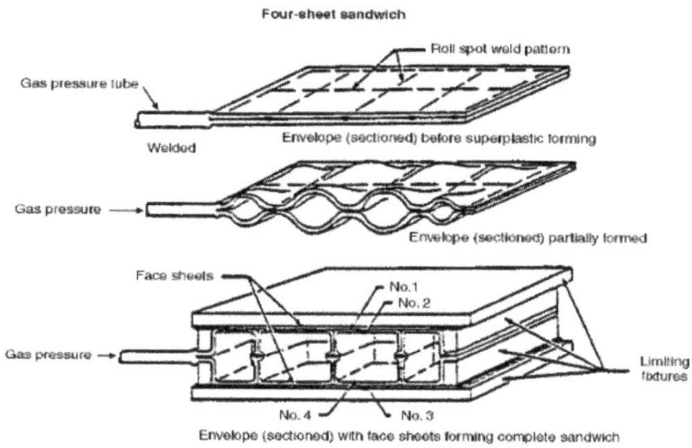

b)

Figure 1: a) Core sheet forming process [7] begins with a resistance weld pattern then gas pressure is applied. The core sheets fold over and eventually contact each other to form a pattern of vertical supports termed columns. The core sheet forming, b) part of the way through the process is shown. The weld lines can be seen as the rectangular pattern of ellipsoidal shapes. The cells will eventually widen to contact each other at the weld nugget lines to form the columns.

To determine how controllable the core sheet formation is, and where and how the variations in geometry of a produced part can occur, both a finite element simulation was performed as well as extensive measurements on the panel geometry. The goal of the simulation was to provide an idealized case where the panel geometry would have minimal variations, and the results compared with the geometry from actual parts. While many experimental panels were produced, only four panels were used in this analysis. Each of these panels, labeled F6-F9, were fabricated with the same Ti-6Al-4V fine grain material from the same manufacturer, and at the same proprietary temperature and pressure schedule. By using the same material and process conditions one can demonstrate the variability inherent in the process. Data was collected on the major examples of variability in the core sheets: column location, column geometry, and geometry of the two different gas pathways. Thickness measurements were taken from mounted micrographs in detail and from cross sections of the cell geometry. Measurements of the two primary defects, the core gas connection holes and the upper gas pathways, were also taken. Variations in overall geometry were additionally measured.

The core sheet formation simulation used the finite element method of the core sheet forming process. An implicit formulation of ABAQUS was employed, in combination with the power law creep model, which only simulates the deformation of the superplastic core material. The deformation of the face sheets is small being at the edges of the panel and is typically not a source of significant thickness variation. Because the deformation of the weld nuggets is extremely small and entirely elastic, only the upper half of a small unit cell of the core sheet superplastic material was simulated. The weld nuggets are modeled as half-ellipsoidal cutouts from the core sheet material. The edge weld nuggets and the crossed weld nugget shown in Figure 2 are held as encastre (i.e. fixed). The interior weld nuggets are not allowed to deform in any way, through the use of a kinematic coupling of the surface nodes to a reference point, but are allowed to translate along their weld line toward or away from the center crossed weld nugget. The face sheet is modeled as a flat, analytically, rigid contact surface with minimal friction. Periodic boundary conditions are applied to the edge nodes in the form of kinematic equations that tie the displacement of each edge node with the node on the opposite side of the mesh.

Materials Research Forum LLC
https://doi.org/10.21741/9781644902615-14

The material model for this simulation uses a strain hardening model based on the power law plasticity model. Material data was used from tensile specimens elongated using the ASTM E2448-06 standard at temperature and a representative strain rate. This material model is used instead of the time hardening model typically used because it has the advantage of being more consistent during large strains. The equation for the strain hardening material model is displayed below (Eq. 2):

$$\dot{\varepsilon}^{cr} = \left(A\, q^{-n}\left[(m+1)\overline{\varepsilon}^{cr}\right]^{m}\right)^{\frac{1}{m+1}}. \tag{2}$$

It relates the equivalent creep strain rate, and the equivalent creep strain, to the deviatoric equivalent stress with some material constants. The goal of the simulation was to use idealized geometry and processing conditions, and therefore several assumptions were made. It was assumed that there is symmetry on the z-axis at the weld nuggets (i.e. the top half of each cell is identical to the bottom half), there is perfect gas distribution and pressure is applied according to a schedule equally everywhere, the material is perfectly isotropic in nature, and the welds are perfectly placed.

a)

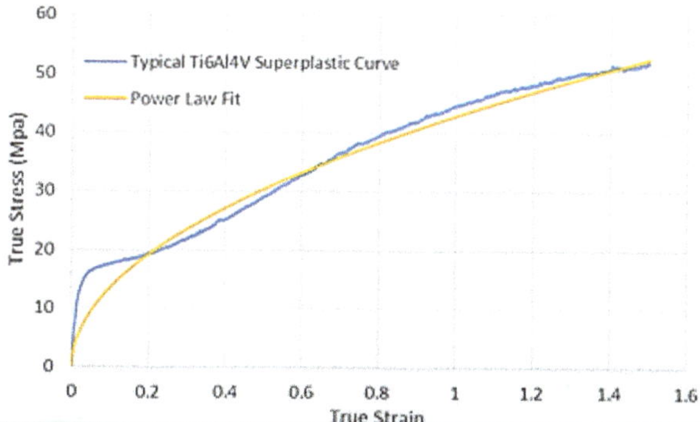

b)

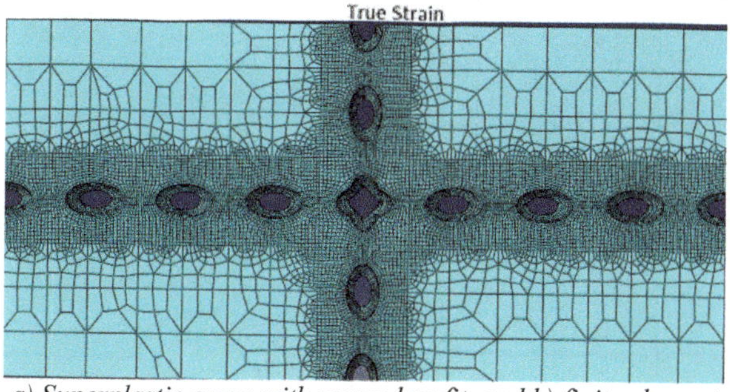

Figure 2: a) Superplastic curve with power law fit, and b) finite element mesh.

To determine the theoretical extent of the effects of different errors on the major dimension of the final product, a series of two-dimensional simulations were conducted with the same material model shown in Figure 2a. However, the mesh and formulation of the model is different from the model previously described where there were no periodic boundary conditions. The boundary conditions are such that the edges form up to vertical rigid walls, and the model is built as a set of three cells each with varying dimensions. The cells were delineated by rigid weld nuggets that were able to translate and rotate but not deform. The cells formed up to a horizontal boundary of a customizable shape which allows for various tests to be conducted. Tests conducted with this setup were to determine the effects of variations in cell dimensions, curvature of the top and bottom of the part, and effects of pressure distributions on the formed geometry.

To determine the geometric trends of the experimental panels several methods were applied and cross sections were taken of the entire sample. The samples were imaged at regular intervals and the thicknesses of each section were measured for each panel examined. Thicknesses and bulk measurements were taken using an X-Y stage microscope (MicroVu). The panels were measured for their cell-to-cell dimensions by recording the dimensions between each weld nugget which is a measure of the cell wandering that can occur in this process. Panels were also measured for the triangular gas pathway at the top of the sheet, as displayed in Figure 3a at the ends of the columns.

Formed samples were machined from each panel using an abrasive waterjet and were sectioned along the column in both the longitudinal direction and the transverse direction. Sections were taken at both the weld nugget and the gas pathway for comparison. Each column section was mounted and polished then the sample sections were imaged using the stage microscope (MicroVu) at regular intervals. Then the images were composited together to provide a macroscopic image of the sample that was thresholded, and processed by the Canny edge detection method to provide lines at the edges of the sample. The vertical location of the digitized edges were subtracted to determine the thickness distribution of the sample. The operations for the threshold and edge detection were developed by OpenCV and the code was implemented in python. Figure 3a shows an original imaged column, the edges produced with Canny edge detection are shown in Figure 3b, and a plot of the edges and resulting thickness distribution is displayed in Figure 3c.

a)
b)
c)

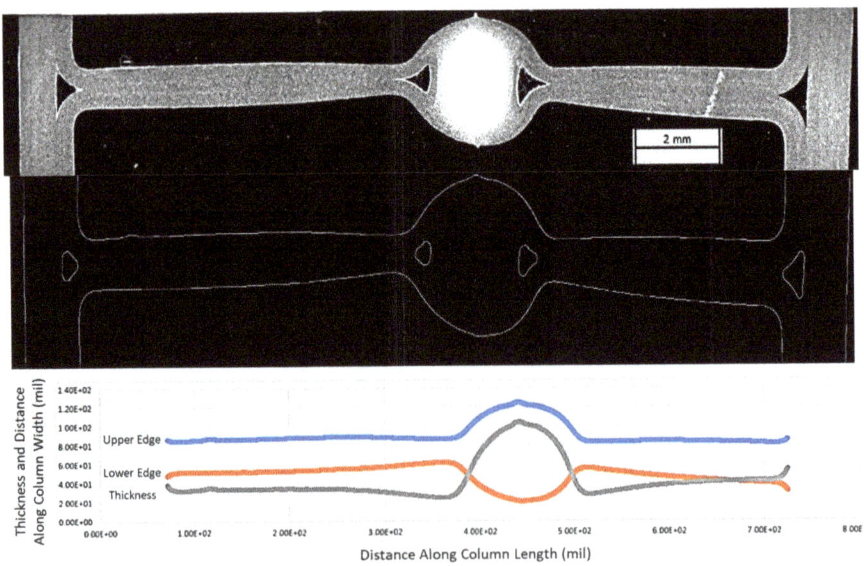

Figure 3: Edge detection of a mounted sample; a) the mounted sample macrograph, b) processed image with canny edge detection, and c) thickness distribution plot after pixel locations are subtracted.

 Materials Research Forum LLC
https://doi.org/10.21741/9781644902615-14

The gas connection pathways in between the different cells have some variation in both size and shape. Samples were measured for the area and aspect ratio of the pathways. The measurements were taken using calibrated images. A chroma key was introduced behind the gas pathways then images were taken. The images were calibrated with a measured length of a single weld nugget. The pathways were digitally isolated from the titanium, separated into a group of pixels representing an individual pathway, then measured using the number of pixels and the weld nugget calibration. Figure 4 has images of the dataset for panels F6 and F9 along with an explanation of the algorithm.

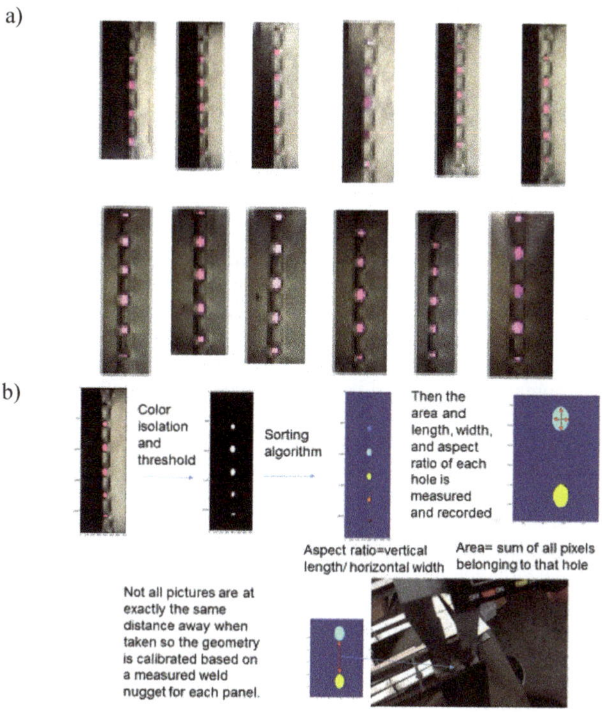

Figure 4: Gas connection pathway measurement process with a) the chroma key dataset for panels F6 and F9, and b) an explanation of the algorithm.

Results

The finite element simulation results were highly symmetric and uniform, which showed very little of the variations that are typical of fabricated panels. In Figure 5, the fully formed simulation is shown, but subsequent figures in this section displaying finite element simulation results were mirrored about the z-axis and replicated along the x and y axes. There are some trends that were exhibited by the experimental results that are detailed in this section, and they are compared with the thickness distributions and the gas pathway statistics.

a) b)

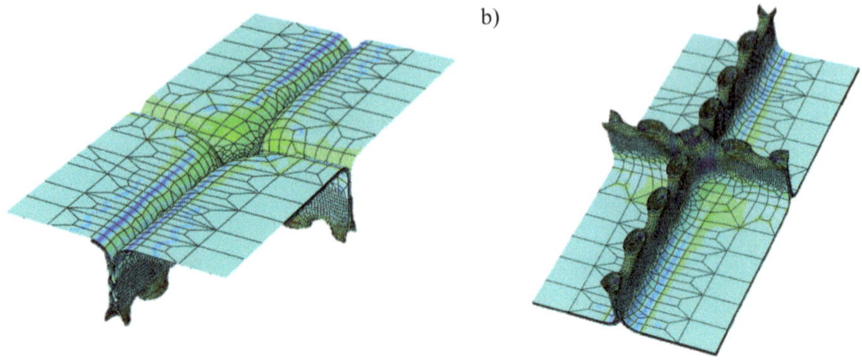

Figure 5: Finite element simulation showing a) the top contacting the face sheets, and b) the internal weld nuggets.

Thinning of the columns is a major problem for this SPF/DB structure because a weakened column can provide a significant failure point. The thickness distributions developed from the edge detection of the mounted samples provide insight into the variance that can develop in finished parts. There is a significant difference between the thickness that occurs at the weld nuggets, and the thicknesses that occurs at the gas connections; therefore, the thickness distributions were consolidated into two groups and analyzed separately. The thickness distributions are best summarized by the plots in Figure 6 which show all the measured thicknesses and the extent of the variability can be observed.

Thickness distributions were also computed for the finite element simulation for comparison for which there was almost no variation in the thickness. The measurements were taken at the midpoint at the center of the weld column and the center of the path column. A standard deviation of 1.2% was observed for the twenty thickness measurements taken at the center column in both the weld nugget area and gas pathway. This shows that with perfect geometry and conditions, there should not be any variation in the column thicknesses. However, as seen in Figure 6, there is a significant amount.

The analysis revealed that there is not always an even stress distribution or perfect weld nugget placement. Length mismatch was also computed between the upper and lower portions of the columns. Ideally the length of both the top and bottom of a column should be equal placing the weld nugget or gas path directly in the center between the two face sheets. This gives the minimum loading to weld nugget area which is beneficial due to the relatively large number of stress concentrations located there. Unequal lengths in the column also leads to unequal thicknesses and can lead to improper processing conditions at thinned sections in the column. The column length mismatch averaged 7.8% of the gas path columns and 9.8% of the weld nugget columns.

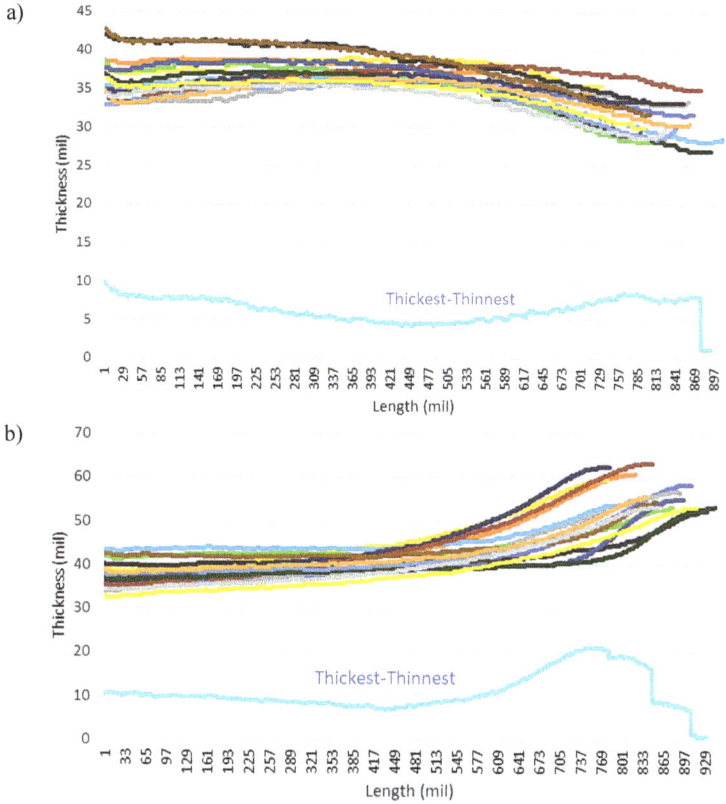

Figure 6: a) Weld nugget column thickness, and b) gas pathway column thickness. There is a large difference between the thickest column versus the thinnest column. These columns support the face sheets and provide the dominant failure points for the finished part. Getting a column that is undersized can be detrimental to the performance of the part.

The gas pathways represent a major defect that is inherent to the core sheet formation. Gas must be able to transfer between the cells in some way but the regular defects can be problematic. They also can be highly inconsistent in size and shape depending on the local loading during forming, which makes them a hard feature to predict. Structurally, the gas pathway and weld nugget locations are not as important as the upper gas path and thickness variations in the columns near the face sheets since the stresses are minimized at the center of the sandwich during bending.

Weld nugget placement is computer controlled and highly accurate in both distance between welds and weld quality. The finite element simulation predicted some key features in the gas pathway geometry, such as the trend of large pathways in the center of the cell, and horizontally shrunken pathways closer to the ends of the cell, as depicted in Figure 7. That is a result of the material transitioning toward the ends to form the perpendicular columns, which in turn causes the weld nuggets to translate, and as a result distorts and sometimes close off the gas pathways at the ends of the cell. From image analysis, the gas pathway dimensions have an average of the 2.055

mm^2 with a deviation of 1 mm^2. The finite element simulation had gas pathways that were significantly larger with an average area of 5.24 mm^2 and a similar relative deviation of 2.33 mm^2.

a) b)

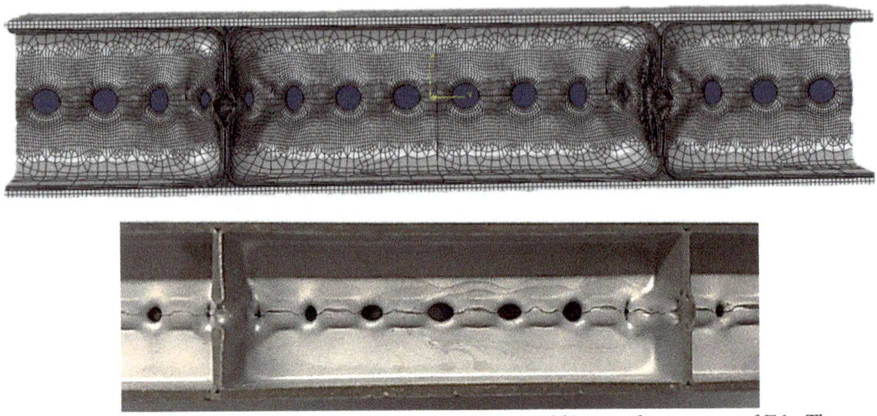

Figure 7: a) FEM results of the gas pathway geometry, and b) actual geometry of F6. The center of the cell has the largest gas pathways which become more distorted as they reach the edges of the cell; perfect gas path geometry is uncommon.

Cell wandering, the process of the column locations not lining up with the initial weld nugget placement, is also an issue in these SPF/DB structures. The variations in the cell size are a natural effect of the stress history dependence of superplasticity. Ideally the cell widths should have a value of 50.8 mm, and in the finite element simulation they do. Despite the precision weld placement, there was still variability in the cell widths. The cells tended to wander with a variation of 1.7% of their average length, as displayed in Figure 8.

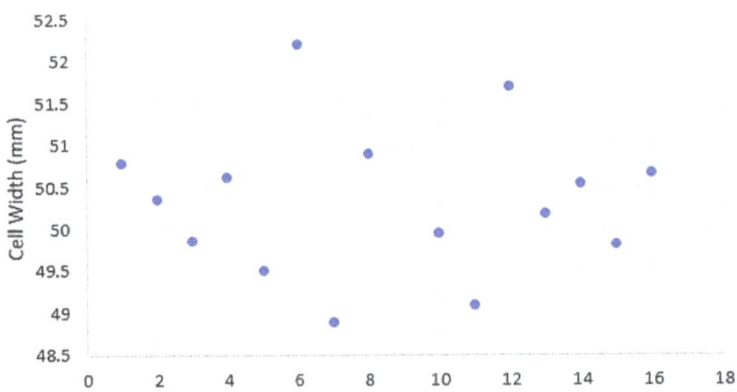

Figure 8: Cell width scatter (50.8 mm ideal).

The effects of upper gas pathway dimensions are arguably the most important characteristic of all the geometric variations in the core sheet. This is because the triangular pathway is directly at diffusion bonds in both the column and the core-face sheet interface. It is also a major stress path in compression, and especially bending where the stresses are maximized at the upper and lower surface of the part, and where failure always occurs at the largest upper gas pathway. The size of the upper gas path in the finite element simulation only varied at the crossed weld nugget area. However, in the experimental panel the trend was to have the smallest pathways closest to the gas inlet, and they get larger as a function of distance from the gas inlet. This is because the material must deform plastically to transmit gasses and the gas inlet is on one side giving the area closest to the gas inlet more time to form than the material far away from the inlet. This has the possible effect of having the failure point be a function of the gas inlet location depending on how the part is loaded. Figure 9 shows the measurements of gas pathway widths along a trimmed panel section.

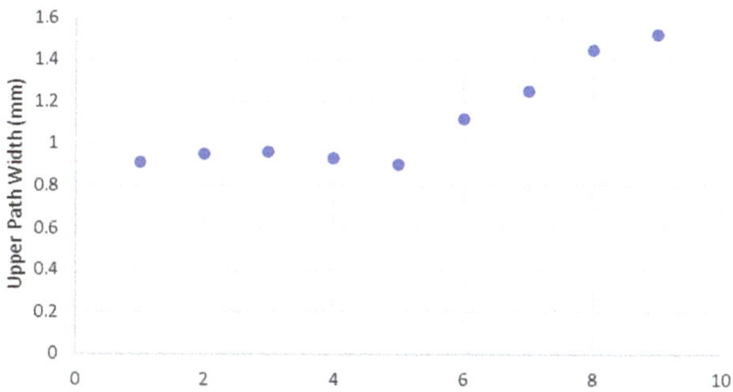

Figure 9: Upper gas pathway width along a trimmed panel section.

Discussion

The purpose of this research was to highlight the unique challenges in producing a consistent part when using an edge supported internal superplastic structure, quantify the extent of the variability in such a structure, and provide insight into possible ways to mitigate problems that could arise. Superplastic forming is a stress history dependent process as is all plasticity. A load causes a change in dimensions of a material which can in turn create larger loads. The difficulties arise in the four-sheet process due to the fact that the internal structures are load balanced and not geometrically constrained. Imperfect geometry causes changes in loading and changes in forming over time.

To illustrate this mechanism, an analysis of cell wander was performed using a two-dimensional simulation and can be viewed in Figure 10. The simulation has four weld nugget areas encompassing three cells along the smaller cross section of the rectangular cell with a dimension of 25.4 mm between cells. The center cell has displaced weld nuggets relative to the outer cells. Both a flat panel and a panel with an expanded (angled) center section were analyzed for normal precise weld nugget location and an error in weld nugget location. The flat panel, with perfect geometry, has perfectly balanced loads on the weld nugget and no increase in error. When the weld nuggets are moved 5% closer together in the center than they should be, the error is increased. This is because the outer cells end up forming quicker since they have a larger surface area exposed to the forming pressure, and therefore a larger stress.

Initially, the tension in the larger cells creates a slight tension in the small center cell that expands it, decreasing the error. However, the outer cells expand outwards first and contact the face sheet which creates a horizontal compression of the center cell since it is smaller and exhibits a smaller horizontal force on the weld nuggets. The weld nuggets relocate to compensate for this differential. By the time the columns are half formed everything has balanced out leaving the center cell even smaller than the weld nugget placement. The 1.27 mm (5%) error in weld nugget placement becomes 2.62 mm (10.3%) after forming. For the angled panel, the center cell is taller than the outer cells and therefore always horizontally expands compressing the outer cells. The offset weld nuggets are 1.27 mm (5%) further apart than desired and this ends up expanding the cell by 4.37 mm (17.2%).

This cell wandering, as well as other geometric defects, are exacerbated in any case where there is uneven loading in the cells. Small defects in material properties, temperature gradients, providing gas pressure from one side of the part, and even gravity can all have effects on the final formed geometry. As shown in Figure 10b and Figure 10c, while flat panels simply increase error, panels with features such as the angle panel are going to have different behavior. Flat panels are not the only shape that the four-sheet process is meant for and it can be molded into a variety of complex contoured shapes for aerospace structures. A complex die surface can also increase the variation in cell geometry by changing the loading conditions in each cell.

Migration in the cells of a production part full of complex contours are going to vary significantly more than a flat panel, and there are techniques that can mitigate some of the variation issues. Weight savings gained from removing fasteners and making a monolithic structure can be utilized to increase the thickness and robustness of the SPF/DB component. This would not decrease the variability of the geometry, but does ensure that there is more than enough material in any location to compensate for complications that may arise due to geometry. Precision placement of the core sheet welds was an additional technique to prevent changes in geometry. Weld placement can be used to compensate for complex curvatures that could cause cell wandering and column thinning.

a)

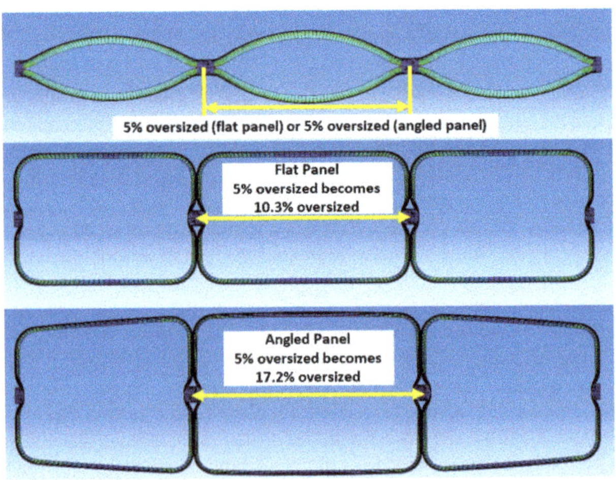

b)
c)

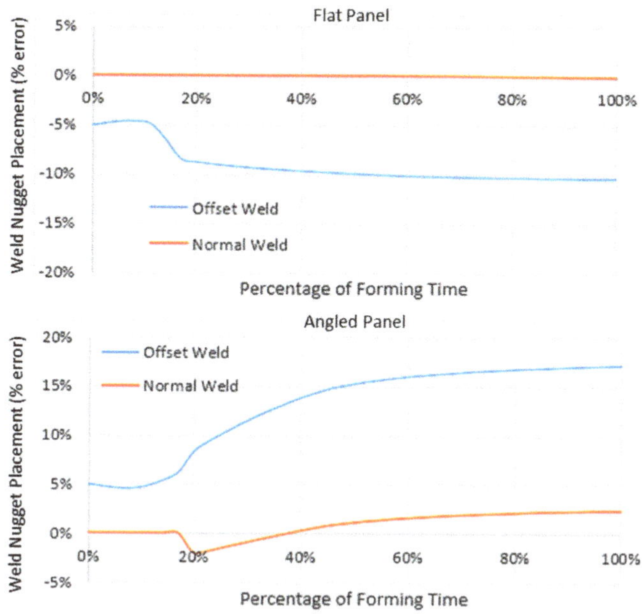

Figure 10: a) Three short (25.4 mm) cell formation in 2D with an undersized/oversized center cell. The flat panel causes an increase in weld error while the angled panel tends to expand the center increasing the error if the welds are oversized but decreasing if undersized. Tracking distance between the center weld nuggets produces the error plots for the b) flat panel, and c) angled panel. These show the predominant source of error occurs near the start of the forming cycle where the partially formed cells are significantly different in size.

Summary

Superplasticity when combined with diffusion bonding is an elegant way to produce high strength metal sandwich parts. The panels can be significantly more cost effective than alternatives; however, there can be challenges in production. This research examined a SPF/DB four sheet design that has a nearly unsupported internal reinforcement structure. The lack of constraints and the extent of the forming strain allow for a large variability in the formed geometry. The internal sheet forming was simulated with the finite element method and compared with the experimental measurements to show that there is significantly more variability in the produced part. This was explained as the effect of large strains combined with superplasticity, a stress history dependent process, which serves to amplify small variations in initial geometry. The effects of this variability can be mitigated by more precise initial geometry, better control over material properties, and recognizing that certain areas of a part are going to have significantly more variation than others. These techniques can be applied to any large plasticity cases where there is loosely constrained forming.

Acknowledgements

The authors sincerely thank The Boeing Company for funding the SPF/DB research at University of Washington. In addition, MR wants to acknowledge the Boeing-Pennell professorship funding and EB would like to acknowledge the Boeing Learning Together Program for educational funding.

References

[1] Hamilton CH, Ghosh AK (1988) Superplastic sheet forming. ASM Handb 14:852-873

[2] Pilling J (1989) Superplasticity in crystalline solids. Inst Met 175-178

[3] Alabort E, Kontis P, Barba D, et al (2016) On the mechanisms of superplasticity in Ti-6Al-4V. Acta Mater 105:449-463. https://doi.org/10.1016/j.actamat.2015.12.003

[4] Pilling J, Livesey DW, Hawkyard JB, Ridley N (1984) Solid state bonding in superplastic Ti-6Al-4V. Met Sci 18:117-122 https://doi.org/10.1179/msc.1984.18.3.117

[5] Edwards P, Ramulu M (2009) Effect of process conditions on superplastic forming behaviour in Ti-6Al-4V friction stir welds. Sci Technol Weld Join 14:669-680. https://doi.org/10.1179/136217109X12464549883330

[6] Hefti LD (2004) Advances in fabricating superplastically formed and diffusion bonded components for aerospace structures. J Mater Eng Perform 13:678-682. https://doi.org/10.1361/10599490421286

[7] Sanders DG, Ramulu M (2004) Examination of superplastic forming combined with diffusion bonding for titanium: Perspective from experience. J Mater Eng Perform 13:744-752. https://doi.org/10.1361/10599490421574

[8] Hales SJ, Wagner JA (1991) Superplastic forming of Al-Li alloys for lightweight, low-cost structures.

[9] Giuliano G (2008) Constitutive equation for superplastic Ti-6Al-4V alloy. Mater Des 29:1330-1333. https://doi.org/10.1016/j.matdes.2007.07.001

[10] Luo J, Li M, Li X, Shi Y (2010) Constitutive model for high temperature deformation of titanium alloys using internal state variables. Mech Mater 42:157-165. https://doi.org/10.1016/j.mechmat.2009.10.004

[11] Khaleel MA, Zbib HM, Nyberg EA (2001) Constitutive modeling of deformation and damage in superplastic materials. Int J Plast 17:277-296. https://doi.org/10.1016/S0749-6419(00)00036-X

[12] Alabort E, Putman D, Reed RC (2015) Superplasticity in Ti-6Al-4V: Characterisation, modelling and applications. Acta Mater 95:428-442. https://doi.org/10.1016/j.actamat.2015.04.056

[13] Carrino L, Giuliano G, Palmieri C (2003) On the optimisation of superplastic forming processes by the finite-element method. J Mater Process Technol 143-144:373-377. https://doi.org/10.1016/S0924-0136(03)00423-0

Superplasticity in Advanced Materials - ICSAM 2023 Materials Research Forum LLC
Materials Research Proceedings **32** (2023) 141-148 https://doi.org/10.21741/9781644902615-15

Formation mechanism and elimination method for surface groove of the hollow lattice structure by superplastic forming

Yanhong Mu[1,2,3,a*], Zhiqiang Li[1,2,3,b], Bing Zhao[1,2,3,c], Feng Yang[1,2,3,d], Ning Zhang[1,2,3,e], Lihua Du[1,2,3,f], Yong Li[4,g]

[1] AVIC Manufacturing Technology Institute, Beijing 100024, China

[2]Aviation Key Laboratory of Science and Technology on Plastic Forming, Beijing 100024, China

[3]Beijing Key Laboratory on Digital Plastic Forming and Equipment, Beijing 100024, China

[4]School of Mechanical Engineering and Automation, Beihang University, Beijing, 100191, China

[a]muyh001@avic.com, [b]zqlee98@126.com, [c]zhao6833@163.com, [d]584551453@qq.com,[e]zn64112@163.com, [f]quietness2007@126.com, [g]liyong19@buaa.edu.cn

Keywords: Lattice Structure, Superplastic Forming, Surface Groove, Double Gas Pressure

Abstract. The titanium alloy hollow lattice structures can be widely applied on the advanced aircrafts due to their excellent mechanical properties and light weight. Superplastic forming and diffusion bonding (SPF/DB) process is an advanced process to manufacture these hollow lattice structures with two outer plates and one inner hollow plate. To further decrease the weight of the structures, thinner outer plate is preferred when compared with the inner plate, however, it would result in the occurrence of surface groove at the intersects between the outer plate and the hollow stiffener during superplastic forming process, leading to the failure of the structures. In this paper, the formation characteristics of the surface groove and the influence of the outer/inner sheet thicknesses on the occurrence of surface grooves was investigated through a well-developed finite element model. Furthermore, an innovative method was presented to eliminate the surface groove during SPF/DB, which makes use of double-sided dynamically adjusted gas pressure on the inner and outer of the plate during superplastic forming process. Based on the developed numerical models, the effects of assistant outer gas pressure parameters on groove formation have been investigated. The effectiveness of the proposed new forming method and the optimized parameters have been validated by performing a SPF/DB forming tests of a typical lattice structures. The method developed in this study provides a new way to improve the forming quality and accuracy of hollow parts by SPF/DB.

Introduction

The titanium alloy hollow lattice structure is a typical ordered topological structure [1-2], with the characteristics of lightweight, high strength and excellent load-bearing performance [3-4]. Meanwhile, the internal connected cavity can also be utilized to integrate multiple functions [5], for example, cooling function with active cooling medium flow through [6], insulation function with filled insulation materials [7], and bulletproof function with filled bulletproof ceramics [8]. Therefore, as a typical load-function integrated structure, titanium alloy hollow lattice structure has broad application prospects in future hypersonic aircraft and helicopters, representing an important direction for the development of aircraft structures.

Superplastic forming/diffusion bonding (SPF/DB) process is an advanced technology to manufacture the titanium alloy hollow lattice structures, which can enable a well metallurgical connection between the face sheet and core sheet through diffusion bonding, and form the hollow structure from bonded plates by superplasticity. Compared with other fabrication methods, it has significant advantages [9]. The use of SPF/DB technology can efficiently manufacture large and complex titanium alloy hollow structures with low cost [10]. One of the most commonly used titanium alloy hollow lattice structure is a three-layer superplastic forming structure, with two outer plates and one inner hollow plate. During SPF/DB, when the layer thickness is small (similar to rib thickness) and the rib angle is large, it is easy to form surface grooves on the formed face sheet [11]. To further decrease the weight of the structures, thinner outer plate is generally preferred when compared with the inner plate, however, it would lead to more significant grooves at the intersects between the outer plate and the hollow stiffener during superplastic forming process. The formation of surface grooves can deteriorate the aerodynamic and load-bearing performance of parts.

In this paper, an SPD/DB finite element model of TC4 alloy hollow lattice structure was established, and the influence of out/inner sheet thickness ratios on the groove was explored through finite element simulation. An innovative method of using double-sided adjusted gas pressure on the panel and assisted sheets was proposed to eliminate the groove during the forming process.

Simulation and experimental method

Numerical modelling

The lattice structure is a complex hollow topological structure, and in order to improve the efficiency of simulation, the cell structure was selected for finite element modelling. Fig. 1(a) shows a schematic diagram of the lattice structure, composed with two panels and truss structure, which is obtained through SPF/DB process. Fig. 1(b) shows a finite element model of superplastic forming with the lattice structure, whose size is 110 * 110mm and the size of connection node is 10 * 10mm. The blanks were set as a deformed body, consisting of upper and lower panels (1mm) and hollow core plates (0.8mm). The tie constraints were applied in the diffusion bonding areas. The upper and lower molds were set as rigid bodies, and the height of two molds was 38mm. The friction coefficient between the molds and blanks was set as 0.1.

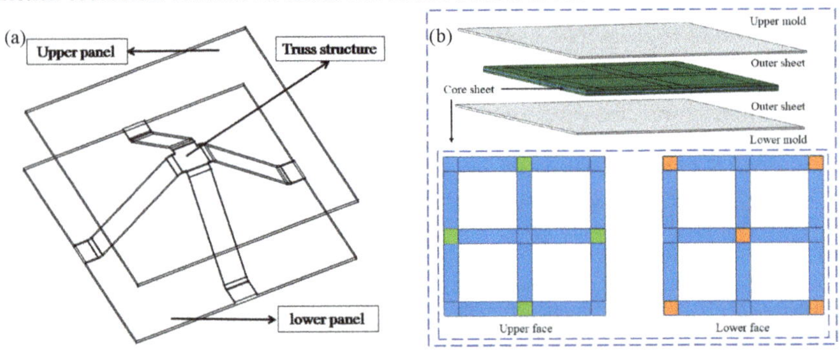

Fig. 1 The schematic diagram of cell lattice (a) and finite element model of the superplastic forming (b)

The material of blanks is Ti6Al4V, whose superplastic deformation behaviour was described using the viscoplastic constitutive model: $\sigma = K\dot{\varepsilon}^m\varepsilon^n$

where K is the material constant, m is the strain rate sensitivity index, and n is the strain hardening index. The values of these constants were calculated according to the true stress-strain curves obtained at different strain rates (5×10^{-4}/s、 1×10^{-3}/s、 3×10^{-3}/s and 1×10^{-2}/s) at 920°C, as: K=591.61, m-0.469 and n=0.177. Fig. 2(a) compares the experimental data with simulation results with the material model, both results are in good agreement with each other. During the SPF/DB process, the gas pressure was set on each surface of the internal cavity, which varied with time, th detailed values are shown in Fig. 2(b).

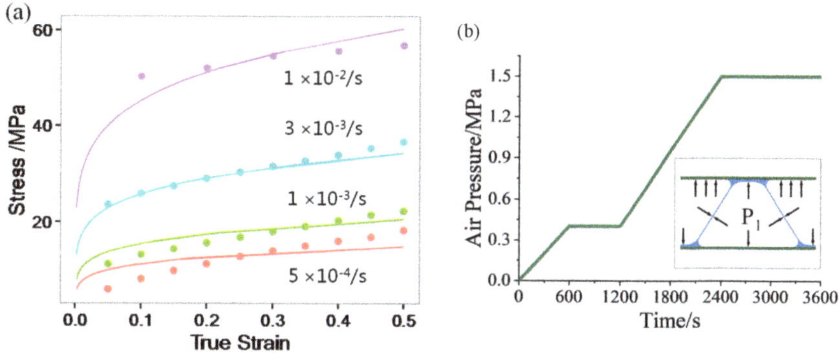

Fig. 2 The true stress-strain curves of Ti6Al4V at 920°C (a) and gas pressure loading curves (b)

SPF/DB tests

A typical lattice structure sample with the size of 420 * 380mm was selected for SPF/DB tests: First of all, the core plate was processed into the hollow shape according to the position of nodes, as shown in Fig. 3(a); Secondly, the stop-off coating was applied at the designated position of the upper and lower panels, so that the corresponding node positions were connected with the core plate by diffusion bonding; Finally, through the superplastic forming process, the upper and lower panels slowly move upwards and downwards to the molds surface, driving the core sheet undergo uniform plastic deformation, until to the formation the three-dimensional lattice structure. The formed lattice structure is shown in the fig. 3(b), and it can be seen that the forming quality of the ribs is intact, and no rupture occurs on the panels, but there are obvious grooves at the nodes.

Fig. 3 The processed core sheet (a) and formed lattice structure (b)

Results and discussion

Formation mechanism and influence factor of the groove

Fig. 4 shows the forming process of the lattice structure by simulation, which can illustrate the formation mechanism of the groove. At the intermediate stage of forming as shown in Fig. 4 (a), the deformation rates in DB zone and non-DB zone are different. In non-DB zone, the inner pressure is directly worked on the outer plate, while in DB zones, multi-layers of plates with truss structures has a much larger stiffness than non-DB zone, and thus would deform slower than the non-DB zones with the same inner pressure. With the progress of the forming process, the layer in non-DB zone would firstly contact with the mold, making it difficult for the layer in DB-zone to fully fit the mold due to the significant changes in stiffness in the connection area and the serious storage of material near the DB zone, as illustrated by the final stage of forming as shown in Fig. 4(b). The grooves then would occur at the connection area between DB and non-DB zones.

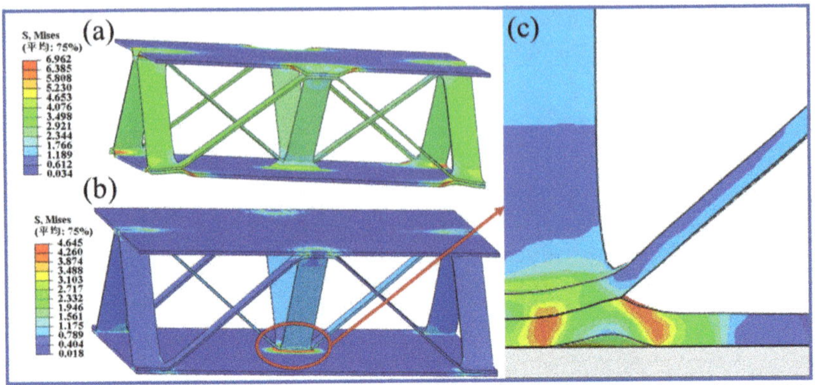

Fig. 4 The simulated forming process of the lattice structure: intermediate stage (a) and final stage (b)

As the occurrence of groove is related to the stiffness changes of the structure, the influence of the outer/inner sheet thickness on the formation of groove was further investigated. The outer/inner sheet thickness ratio (T1/T2) was defined to characterise the thickness effects. Different ratio, 0.75, 1, 1.25, 1.5 and 1.75, have been selected to perform the simulation, based on simulation results, the groove defects was investigated using the maximum groove depth as an evaluation indicator. Fig. 5 shows the maximum groove depths of the panel after forming at the thickness ratios of 0.75, 1, 1.25, 1.5 and 1.75. It can be seen that the groove depth decreases with the increasing outer/inner sheet thickness ratio. When the thickness ratio is 0.75, 1 and 1.25, the groove depth respectively reaches up to 1.566, 1.244 and 0.395mm, which can basically not be eliminated. When the thickness ratio increases to 1.5, the groove depth is 0.126mm, which is possible to be eliminated when continuously increasing the holding pressure and time. At the thickness ratio of 1.75, the groove depth approaches zero. With a thicker outer layer, the difference in stiffness in DB zone and non-DB zone becomes smaller, which could be the reason of the smaller groove being generated.

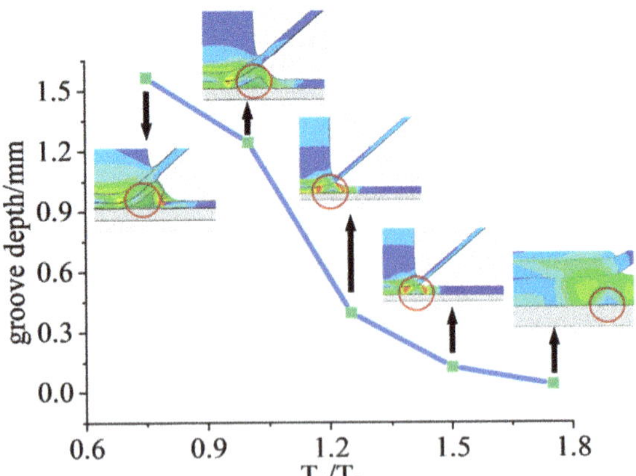

Fig. 5 The maximum groove depths at different outer/inner sheet thickness ratio

Materials Research Proceedings **32** (2023) 141-148
https://doi.org/10.21741/9781644902615-15

It can be concluded that it is a good method to eliminate groove defects by increasing the panels thickness when the thickness of the core sheet is constant. However, in order to achieve weight reduction of the structure, the panel is often designed to be close to the thickness of the core sheet. If thick panels are used for forming, it is still necessary to remove excess materials through chemical milling or mechanical processing, resulting in the increase of manufacturing costs. Therefore, it is still necessary to explore a new forming method for groove suppression when the thickness ratio is not big enough.

Eliminate method on the groove by double gas pressure

Except for changing the thickness of the structures to eliminate groove, another factor that could affect the deformation state in both DB and non-DB zones is the pressure. To further eliminate the strain rate differences in both zones during forming, adjusting the pressure dynamically is another possible way. Hence, an innovative method was presented here to makes use of double-sided dynamically adjusted gas pressure on the panel and assisted sheets to attain the proper thickness ratio during SPF/DB process. The finite element model of the new strategy of SPF/DB with the lattice structure by double gas pressure is shown in Fig. 6 (a), whose difference is that two assisted steel sheets with the thickness of 1.5mm are added contact with panels, and another gas pressure is applied on the surface of the assisted sheets. To verify the effectiveness of the method in this study, a constant out air pressure (1 MPa) was adopted. The outer and inner air pressure loading curves are shown in Fig. 6(b). Corresponding experiments have also been performed based on the method in section 2.2.

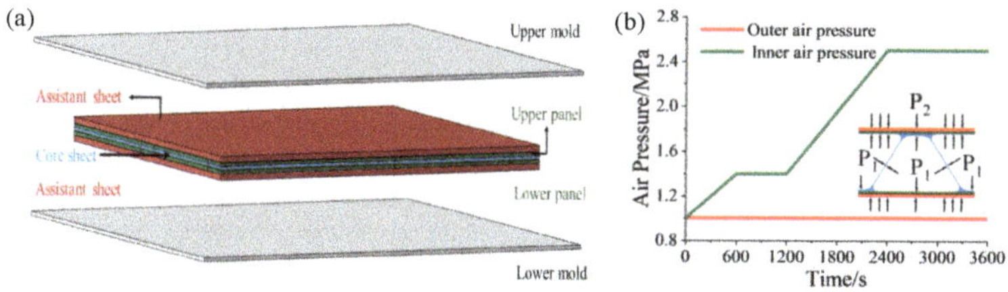

Fig. 6 The finite element model of super-plastic forming with double gas pressure (a) and outer and inner air pressure loading curves (b)

The simulated and experimental results of the lattice structure super-plastic formed using these two methods (single side pressure and double-side pressure) are shown in Fig. 7. At the intermediate stage, the gap between the bonding zone and the mold when formed using double side gas pressure is much smaller than that formed using single side gas pressure. This is because that the assisted sheets are tightly attached to the panels due to the outer gas pressure, which reduces the effect of the "pull" force on the DB zone from the increase of sheet thickness and friction force. At the final stage, the groove does not appear when formed using double side gas pressure, but the groove is obvious when formed using single side gas pressure. The experimental results of typical lattice structure super-plastic forming demonstrate the accuracy of simulation and the effectiveness of groove elimination using double side gas pressure method with assisted sheet.

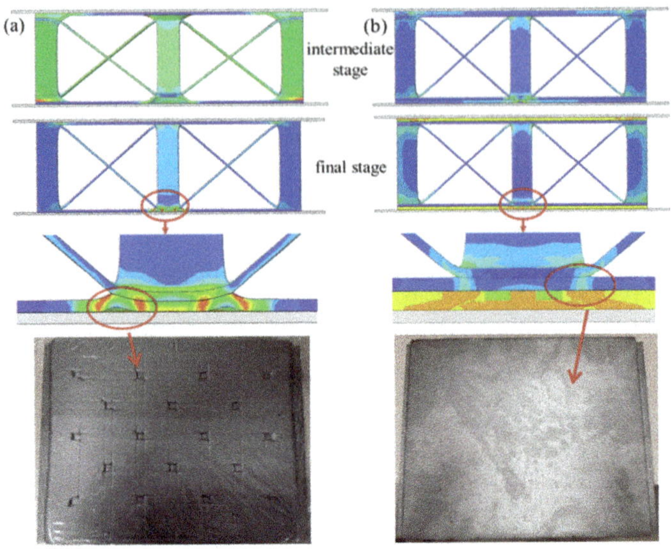

Fig. 7 The simulated and experimental results of the lattice structure super-plastic formed using these two methods: single gas pressure (a) and double gas pressure (b)

Furthermore, the effect of different levels of outer gas pressure on groove formation was investigated by simulations. Fig. 8 shows the maximum groove depths of the panel after forming at the different outer gas pressure of 0, 0.2, 0.5 and 1 MPa. The inner gas pressure was adjusted according to the outer gas pressure in order to maintain the consistency of the pressure differences. It can be seen that the groove depth decreases with the increase of the assistant gas pressure. When the assistant gas pressure is 0, 0.2 and 0.5 MPa, the groove depth is respectively 0.395, 0.144 and 0.034mm. with the assistant gas pressure of 1 MPa, the groove depth was eliminated in a well condition.

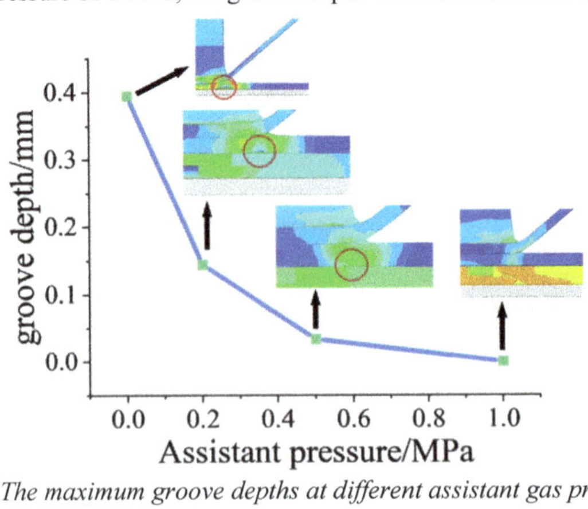

Fig. The maximum groove depths at different assistant gas pressure

Conclusions

(1) The groove would occur during SPF/DB due to the significant stiffness difference between the DB zone and non-DB zone in the hollow structure. The groove depth decreases with the increasing outer/inner sheet thickness ratio, and the groove defects can be eliminated when the thickness ratio reaches to 1.75.

(2) A new method to utilise double-sided adjusted gas pressure (inner gas pressure + assistant outer gas pressure) on the panel and assisted sheets to decrease the stiffness differences has been proposed, both simulation and experiments indicate that it is an effective method to eliminate the surface groove of the lattice structure.

(3) The groove depth decreases with the increase of the assistant outer gas pressure, and the groove depth reaches to zero with the assistant gas pressure of 1 MPa.

References

[1] Zhiqiang Li, Bing Zhao, Jie Shao, Shengjing Liu. Deformation behavior and mechanical properties of periodic topological Ti structures fabricated by superplastic forming/diffusion bonding. International Journal of Lightweight Materials and Manufacture 2 (2019) 1-30. https://doi.org/10.1016/j.ijlmm.2018.11.001

[2] Sahil Dhiman, Malkeet Singh, Sarabjeet Singh Sidhu , Marjan Bahraminasab , Danil Yurievich Pimenov and Tadeusz Mikolajczyk. Cubic Lattice Structures of Ti6Al4V under Compressive Loading: Towards Assessing the Performance for Hard Tissue Implants Alternative. Materials 2021, 14, 3866. https://doi.org/10.3390/ma14143866

[3] Zhang Zhenhua, Niu Chuang, Qian Haifeng, Wang Yuanxin. Impact experiment of six-layer pyramidal lattices sandwich panels subjected to near field underwater explosion. Chinese Journal Of Ship Research, Vol.11, No.4.

[4] Zhihao Du,Kaifeng Zhang.The superplastic forming/diffusion bonding and mechanical property of TA15 alloy for four-layer hollow structure with squad grid.International Journal of Material Forming volume 14, pages1057-1066 (2021). https://doi.org/10.1007/s12289-021-01621-z

[5] P. Bollen,N. Quie'vy, I. Huynen, C. Bailly, C. Detrembleur, J.M. Thomassine and T. Pardoen. Multifunctional architectured materials for electromagnetic absorption. Scripta Materialia 68 (2013) 50-54.[6] Jin Jiang, Ziming Zhang,Haichen Bai,Ruoling Zhang,Guozhu Zhao. Multi-parameter design method for active cooling panel with high thermal load. AIP Conference Proceedings2027, 030173 (2018). https://doi.org/10.1016/j.scriptamat.2012.05.011

[7] Wu Yuan, Jiangtao Wang, Hongwei Song, Te Ma, Wenjun Wu, Junning Li , Chenguang Huang. High-power laser resistance of filled sandwich panel with truss core: An experimental study. Composite Structures, 193 (2018) 53-62. https://doi.org/10.1016/j.compstruct.2018.03.031

[8] Y. Wu, J. Hermes, L. Verbelen, and L. Yang. Impact Energy Absorption Ability Of Thermoplastic Polyurethane (TPU) Cellular Structures Fabricated Via Powder Bed Fusion. Solid Freeform Fabrication 2019: Proceedings of the 30th Annual International Solid Freeform Fabrication Symposium - An Additive Manufacturing Conference.

[9] David J.Sypeck, Haydn, N.G.Wadley. Cellular metal truss core sandwich structures. Advanced engineering materials, 2002, 4 , No.10. https://doi.org/10.1002/1527-2648(20021014)4:10<759::AID-ADEM759>3.0.CO;2-A

[10] Larry D. Hefti. Innovations in the Superplastic Forming and Diffusion Bonded Process. Journal of Materials Engineering and Performance volume 17, pages178-182 (2008). https://doi.org/10.1007/s11665-007-9178-0

[11] Gao Wenjing, Lei Junxiang. Application and Research Progress of Diffusion Bonding Technology in Titanium Alloy Processing[J]. Nonferrous Metal Materials And Engineering, 2017, 38(4): 239-246.

Study on superplastic gas-bulging properties and microstructure evolution of as-rolled 2A97 Al-Li alloy

Zhang Yanling[1,2,3,a*], Liu Jiajia[1,2,3,b], Xu Huiyuan[1,2,3,c], Du Lihua[1,2,3,d]

[1]AVIC Manufacturing Technology Institute, Beijing 100024, China

[2]Aeronautical Key Laboratory for Plastic Forming Technologies, Beijing 100024, China

[3]Beijing Key Laboratory of Digital Plasticity Forming Technology and Equipment, Beijing 100024, China

[a] zhangyanling205@163.com, [b] la.mb@163.com, [c] mail_yefeng@163.com, [d] quietness2007@126.com

Keywords: As-Rolled 2A97 Al-Li Alloy, Superplastic Gas-Bulging, Wall Thickness, Dynamic Recrystallization, Mechanical Properties

Abstract. In this paper, the forming properties of as-rolled 2A97 Al-Li alloy at 370 ~430 ℃ and 0.02~0.08 MPa/min were studied by Superplastic gas-bulging experiment, the microstructure and mechanical properties of 2A97 Al-Li alloy after deformation were evaluated and analysed by means of EBSD, TEM and other testing methods. The results show that the as-rolled 2A97 Al-Li alloy has good superplastic deformation ability, the optimum forming parameters are 390°C, 0.06 MPa/min, and the maximum height of cap cone is 74 mm under these conditions. Dynamic recry -stallization of rolled microstructure occurs during superplastic deformation, the microstructure after superplastic deformation is fine, equiaxed and the average grain size is less than 5 μm. The material keeps good strength and plasticity after superplastic deformation, the tensile fracture shows obvious ductile fracture characteristics, and there are a lot of dimples with regular shape and uniform distribution on the fracture surface. After superplastic forming, it undergoes 520 °C/2h solution treatment, water quenching, and aging at 165 °C/36h, the properties of 2A97 Al- Li were as follows: tensile strength 527MPa, yield strength 441MPa, and elongation 8.3%, respectively.

1. Introduction

Aluminum lithium alloy is a lightweight alloy with low density, high specific strength, and specific stiffness, as well as excellent low-temperature performance, corrosion resistance, and superplasticity. It has excellent comprehensive performance and great development potential, and is considered an ideal structural material for aircraft and ships in the 21st century [1-4]. Although the density of aluminum lithium alloy is about 10% lower than that of conventional aluminum lithium alloy, its elastic modulus has increased by 10%. Replacing conventional aluminum alloys with aluminum lithium alloys can reduce component weight by 10% to 15%, while increasing stiffness by 15% to 20% [5]. Therefore, aluminum lithium alloy, as a lightweight and high-strength structural material, will have broad application prospects in aerospace and other fields in the future. Superplastic forming (SPF) is a method of forming by utilizing the superplasticity of materials under specific temperature and strain rate conditions. It can achieve overall and near net forming of complex structures [6-8]. Research has shown that Al-Li alloy has good superplasticity [9-11], the rolled 2A97 aluminum lithium alloy with high deformation energy obtained through special deformation and heat treatment also has good superplastic deformation ability, with superplastic tensile elongation reaching over 500%. It mainly achieves grain refinement through dynamic recrystallization during superplastic deformation to obtain high elongation [12-15]. There is little research on the superplastic bulging performance of the rolled 2A97 aluminum lithium alloy and the microstructure evolution law of the material during the bulging process.

This article takes the rolled 2A97 as the research object, and studies the superplastic forming performance and wall thickness distribution of the rolled 2A97 aluminum lithium alloy under complex stress states through gas pressure bulging tests of cap shaped and rectangular shaped parts. The optimal bulging process parameters of the material are obtained. At the same time, microstructure evolution laws of materials in different deformation zones are studied using microstructural analysis methods such as OM、EDSB and SEM. And the mechanical properties after superplastic forming are tested.

2. Materials and Experimental Procedure

2.1 Materials

The material in this paper is an Al-Cu-Li alloy supplied by Southwest Aluminum (Group) Co.,Ltd.,China , referred to as 2A97 alloy. The nominal chemical composition of the sheet was Al–3.8Cu–1.4Li–0.5Mg–0.5Zn–0.3Mn–0.2Ti–0.2Si–0.1Zr –0.06Fe–0.006Be (wt %). The material used is 1.5 mm thick warm rolled plates, and was specifically thermomechanical treatment for superplastic forming. After casting and homogenizing, the alloy is hot rolled to 25mm in thickness to break down the as-cast structure. The material is solution treated at 520°C for 2h to take most of the copper into solid solution, and aged at 400°C for 48 h to obtain the uniform distribution and a larger volume fraction of the second phase particles. And then the plate is warm rolled with intermediate annealing to avoid rolling cracking. Finally, retain 20%~30% deformation for cold rolling to as-received state[16]. The microstructure of the as-received 2A97 alloy is shown in Fig.1. It can be seen that there are a certain amount of pancake-shaped grains elongated in the longitudinal rolling direction (RD).

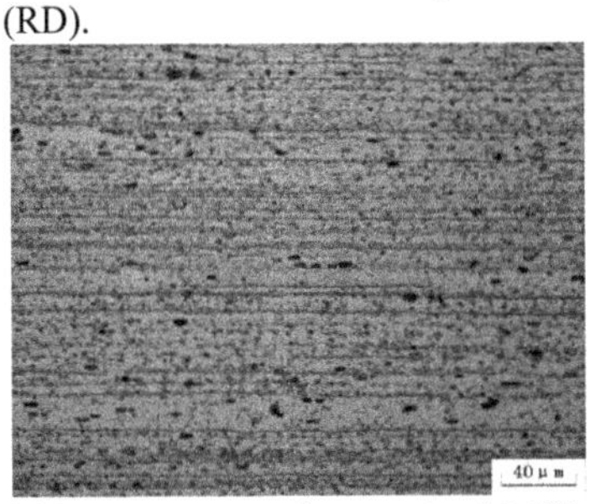

Fig.1 Microstructure of as-received 2A97 alloy

2.2 Experimental Procedure

The superplastic bulging tests of the cap cone and rectangular shaped parts were conducted on a 60t superplastic forming testing machine, which consists of three parts: a hydraulic press, a heating device, and an air source control cabinet. Firstly, the mold was loaded into the forming equipment, and thermocouples were inserted into the upper and lower molds to monitor the mold temperature in real time. When the mold temperature reached the target temperature, the forming sheet was loaded into the forming sheet, which was kept at the forming temperature for 20 minutes before superplastic bulging, After the superplastic bulging test is completed, turn off the heating system and remove the parts. The superplastic bulging process of the cap cone is shown in Fig. 2(a). The superplastic bulging temperature is 390~430 °C, and the gas pressure loading rate is 0.02~0.06 MPa/min. All experiments stopped pressurizing after the cap cone broke. By conducting superplastic bulging tests on cap cones under different temperatures and pressure loading rates, the forming limit, wall thickness distribution characteristics and microstructure evolution of rolled

2A97 aluminum lithium alloy in different deformation conditions were studied. The superplastic forming tests of the rectangular shaped part were conducted under the optimal forming process parameters. The forming height of the rectangular shaped part was 40mm, and the forming process was shown in Fig.2(b). The tensile samples were cut from the bottom of the rectangular shaped part, which are shown in Fig.3 and some samples were subjected to solution treatment at 520 °C/2h. After water quenching, it was artificially aged at 165 °C/36h to test the mechanical properties of the aged material.

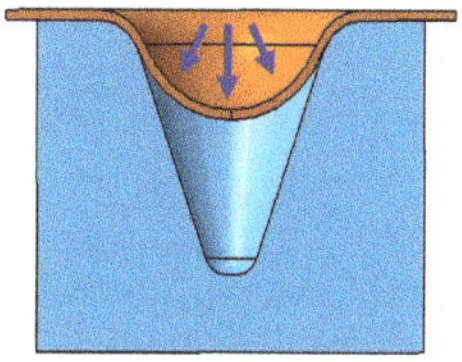

(a) the cap cone

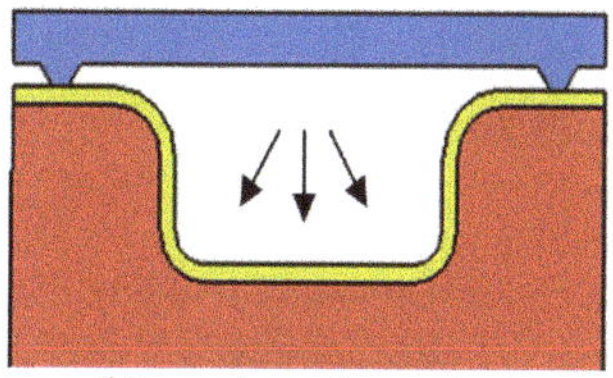

(b) rectangular part

Fig.2 Schematic diagram of the bulging process of the cap cone and rectangular part

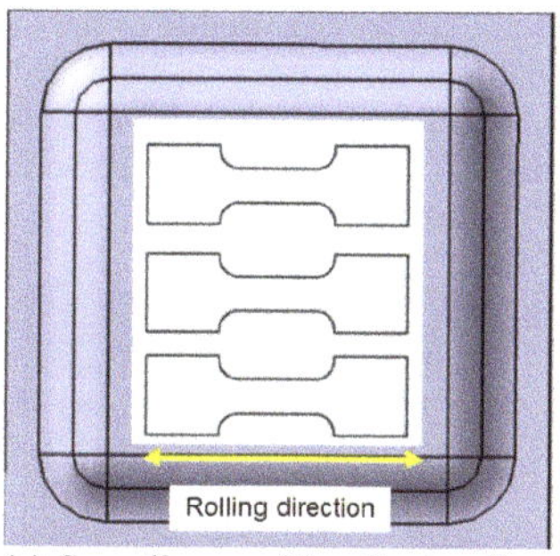

(a) Sampling position for tensile specimens

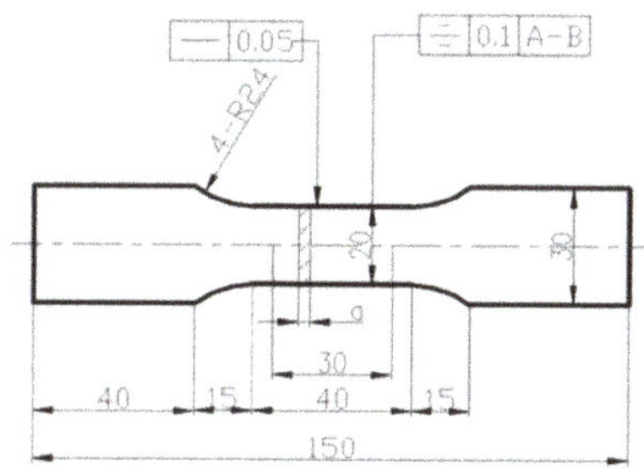

(b) Size of tensile specimens

Fig.3 Dimensions of Room Temperature Tensile Specimen

3. Results and Discussion

3.1 Cap cone forming test results

The test results of free bulging tests of cap cone are shown in Fig.4. From Fig.4 (a), it can be seen that 2A97 Al-Li alloy is relatively sensitive to temperature. When the pressure rate is 0.02 MPa/min, the height of the cap cone decreases with the increase of the temperature. When forming at 390 °C, the forming height is 70.5mm. When the temperature rises to 410 °C and 430 °C, the forming heights decrease to 63.5mm and 57.6mmm respectively. In order to study the effect of pressure rate on the forming performance, bulging tests were conducted at three pressure rates of 0.02MPa/min, 0.04MPa/min, and 0.06MPa/min at 390 °C. The results are shown in Fig. 4 (b). As the increase of pressure rate, the forming height slightly improves. When the pressure rate of superplastic forming increases from 0.02MPa/min to 0.06MPa/min, the forming height increases from 70.5mm to 74mm. The optimal forming process for 2A97 Al-Li alloy was obtained at 390 °C and 0.06MPa/min, and the forming height was 74mm.

Materials Research Forum LLC
https://doi.org/10.21741/9781644902615-16

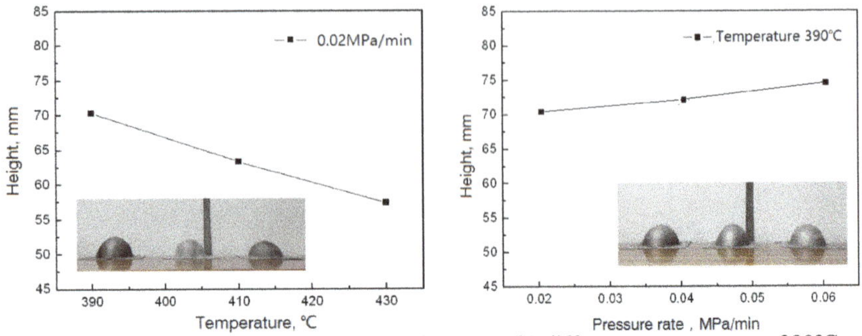

(a) different temperatures at 0.02MPa/min (b) different pressure rate at 390°C

Fig.4 The height of cap cone with deformation conditions

Fig.5 shows the wall thickness distribution of the cap cone after superplastic bulging at 390 °C and 0.06MPa/min along the dome height direction. It can be seen from the figure that the wall thickness decreases uniformly along the height direction, and the surface quality of the formed cap cone is smooth and free of orange peel. This indicates that the deformation is relatively uniform, and the fracture at the fracture area cracks along the rolling direction. The wall thickness of the material in the fracture area is only 0.23mm, it indicates that 2A97 aluminum lithium alloy has good formability and deformation uniformity under this condition.

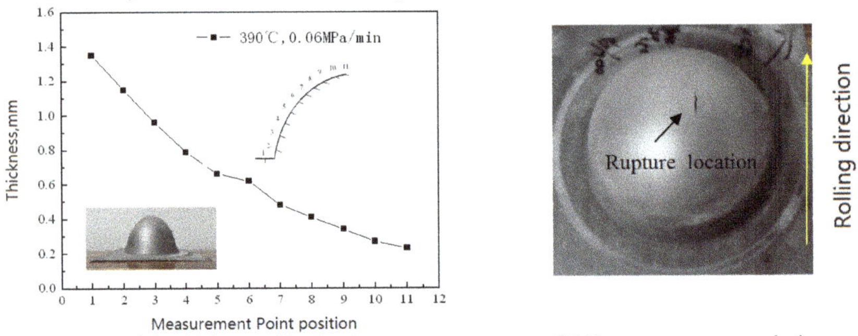

(a) thickness distribution (b) Fracture zone morphology

Fig.5 Experimental results of cap cone forming (390 °C, 0.06MPa/min)

Fig. 6 shows the microstructure corresponding to different strains of the cap cone after superplastic forming at 390 °C and 0.06MPa/min. Figure 16 (a) shows the material structure in the undeformed zone of the cap cone, which has only undergone high-temperature thermal cycling. Static recrystallization has occurred in this region, but the degree of recrystallization is not complete. As the deformation increases, the degree of recrystallization increases. The recrystallization and grain growth exist during the deformation process. When the strain reaches 1.37, the elongated structure almost disappears, the grain morphology tends to be equiaxed, and the structure is more uniform, with an average grain size of about 3.8 μm. Dynamic recrystallization is the main deformation mechanism during the superplastic deformation process of as-rolled 2A97 Al-Li alloy, fine-grained structure and good deformation ability are obtained through it during the deformation.

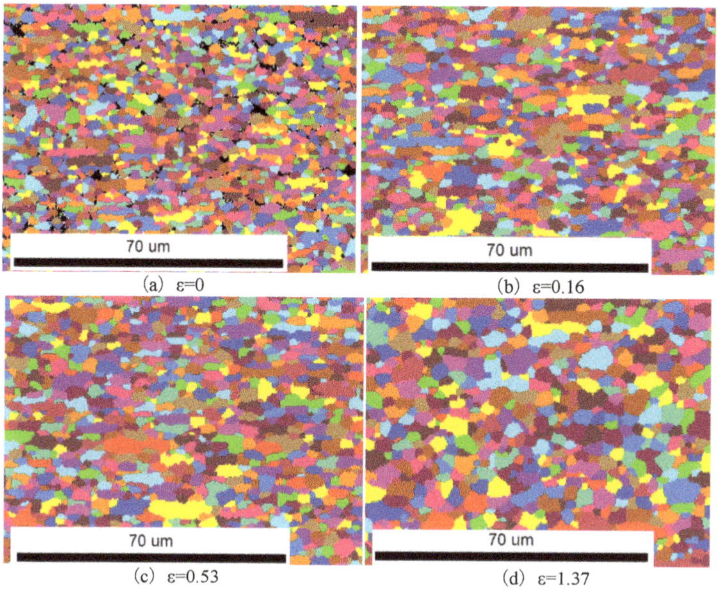

Fig.6 Microstructure of cap cone with different strains (390 °C, 0.06MPa/min))

3.2 Rectangular part forming test results

Rectangular parts of 2A97 Al-Li alloy were formed at temperature 390 °C. The pressure pathway of the rectangular part is first added to 0.6MPa for 10 minutes, and then pressurized to 1.2MPa at a rate of 0.06MPa/min. Under these conditions, the superplastic formed rectangular part and the distribution of wall thickness are shown in Fig.7. It can be seen that the bottom corner position in contact with the concave mold has the largest deformation, with the thinnest wall thickness of 0.78mm and a reduction rate of 42%. The wall thickness at the center of the box bottom position is 1.15mm, because this position first contacts the mold during the forming process, and the friction between the sheet and the mold limits the flow of materials in this area.

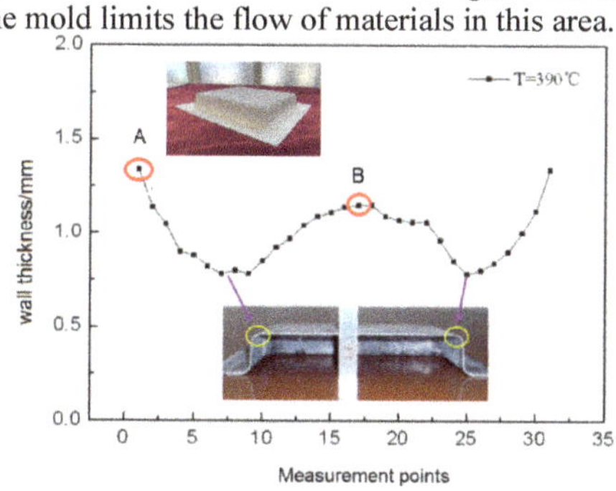

Fig.7 The distribution of wall thickness of rectangular part

Fig.8 shows the microstructure of the undeformed zone (point A in Fig. 7) and the deformed zone (point B in Fig. 7) of the rectangular shaped part. It can be seen that the undeformed zone has

Materials Research Forum LLC
https://doi.org/10.21741/9781644902615-16

undergone significant recrystallization, with very fine recrystallized grains, with an average grain size of only 1.26 μm. Compared with the undeformed zone structure, the deformation zone structure at the bottom of the rectangular part undergoes complete recrystallization and grains grow, but the grains remain fine and equiaxed. The grain boundaries are clearly visible, and the average grain size is about 2.05 μm. This also indicates that during superplastic deformation, not only dynamic recrystallization occurs under the action of deformation force, but also accompanied by the growth of recrystallized grains.

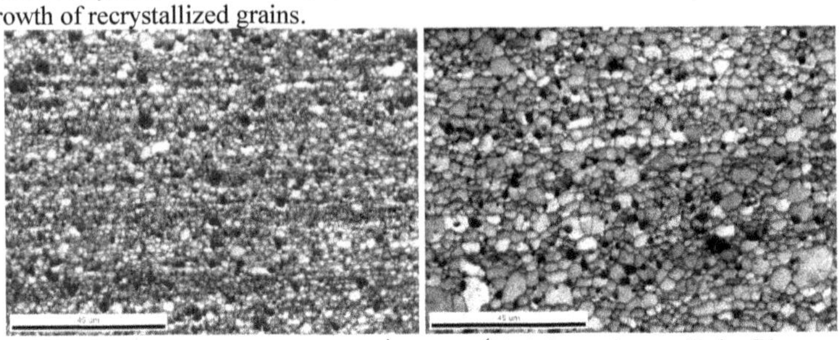

(a) undeformed zone (Point A) (b) deformed zone (Point B)

Fig.8 Organization of rectangular part at different positions

3.3 Performance evaluation after superplastic forming

Take tensile specimen from the rectangular parts is shown in Fig.3. The tensile test results in different states are shown in table 1. It indicates that the original plate is in the rolled state with low performance. The tensile strength in the rolling direction is 339.6 MPa, the yield strength is 316.7 MPa, and the elongation is very low, only 3.8%. After superplastic deformation, the tensile strength remains almost unchanged, and the yield strength slightly decreases. However, the plasticity at room temperature has been greatly improved, the elongation has increased from 3.8% of the original sheet to 14%. After superplastic deformation, the sample was subjected to 520 ℃ /2h solid solution treatment followed by water quenching, then aged by 165 ℃/36h , the properties of 2A97 Al-Li alloy were as follows: R_{m}=527MPa, $R_{p0.2}$=441MPa, A= 8.3%. Aging heat treatment can effectively improve the mechanical properties of 2A97 Al-Li alloy after superplastic deformation.

Table 1 Mechanical properties of 2A97 Al-Li in different states

Material Status	Rm, MPa	Rp0.2, MPa	A, %
Original material	339.6	316.7	3.8
After SPF	345	235.6	14
SPF+520℃/2h+165℃/36h	527	441	8.3

Fig.9 (a) and (b) show the fracture morphology of the specimens after superplastic forming and aging tensile fracture at 520 ℃/2h+165 ℃/36h, respectively. The tensile fracture of 2A97 Al-Li alloy after superplastic forming exhibits typical ductile fracture characteristics, with a large number of deep dimples in the fracture surface, and the size of the dimples is very small, about 1-3 μ m. The shape of the dimples is regular and evenly distributed, mainly due to the uniform and fine microstructure of the material after superplastic forming, which greatly improves its plasticity. After single stage aging at 165 ℃/36h, the fracture surface is convex and uneven, mainly composed of rock sugar like particles. The size of rock sugar particles is relatively small and evenly distributed. The tensile fracture shows obvious intergranular fracture characteristics, belonging to a typical brittle fracture surface. The main reason is that during single stage aging at 165 ℃/36h, a large number of continuously distributed second phases are precipitated at the grain boundaries, and they reduce the strength of grain boundaries.

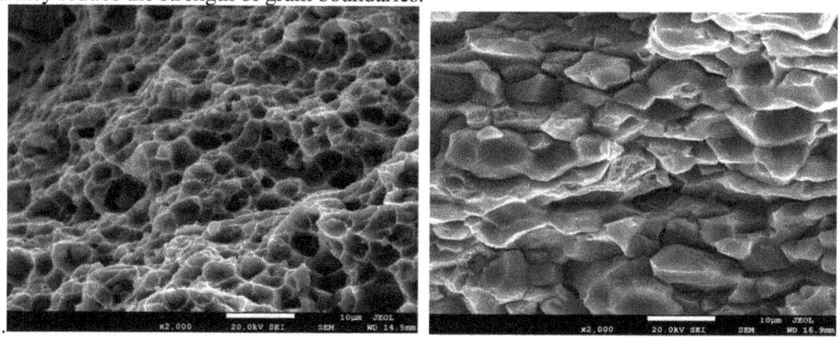

（a） fracture morphology of SPF （b） SPF + 165°C/36h aging

Fig.9 Tensile fracture morphology of 2A97 Al-Li alloy after SPF

4. Summary

(1) The as-rolled 2A97 Al-Li alloy has good superplastic deformation ability, As the temperature decreases and the pressure rate increases, the forming height of the cap cone increases .The optimum forming parameters are 390℃, 0.06 MPa/min, and the maximum height of cap cone is 74 mm under these conditions.

(2) The observation results of different deformation zones of 2A97 Al Li alloy after superplastic deformation indicate that dynamic recrystallization is the main deformation mechanism of as-rolled 2A97 alloy. Fine grain structures are obtained through dynamic recrystallization, with the average grain size of less than 5μm.

(3) Aging heat treatment can effectively improve the mechanical properties of 2A97 Al-Li alloy after superplastic deformation. After being quenched in solution water at 520 °C/2h and aged at 165 °C/36h, the tensile strength, yield strength, and elongation of 2A97 Al-Li alloy respectively are 527MPa, 441MPa and 8.3%。

References

[1] Dejong H F. Properties and Application of Aluminum-Lithium Alloys [J]. Aluminum, 1984,60 (9): 673~679.

[2] N. E. Prasad, A. Gokhale, R. J. H. Wanhill. Aluminum-Lithium Alloys: Processing, Properties, and Applications, 2014, 27.

[3] E. A. Starke, J. T. Staley. Application of modern aluminum alloys to aircraft, Prog. Aerosp. Sci. 32 (1996) 2, 131-172. https://doi.org/10.1016/0376-0421(95)00004-6

[4] I. N. Friedlander, A. G. Bratukhin, V. A. Davydov. Soviet Al-Li alloys of aerospace applications, in: Peters M, Winkler P J,eds. Proceedings of Sixth International Aluminum-Lithium Confere -nce. Garmisclr Partenkirchen: DMG Verlag,1992.

[5] J. T. Staley, J. Liu, W. H. Hunt. Aluminum alloys for aerostructures[J]. Advanced Materials & Processes, 1997, 152(4): 17-20.

[6] S. Hori, M. Tokizane, N. Furushiro. Superplasticity in advanced materials[M]. The Japan Society of Research on Superplasticity. Japan Osaka. 1991.

[7] J. Lin, F.P.E. Dunne. Modelling grain growth evolution and necking in superplastic blow-formi -ng[J]. International Journal of Mechanical Sciences, 2001, 43(3): 595-609. https://doi.org/10.1016/S0020-7403(00)00055-2

[8] T. G. Langdo. The mechanical properties of superplastic materials[J]. Metallurgical Transactio -ns A, 1982, 13(5): 689-701. https://doi.org/10.1007/BF02642383

[9] P. S. Bate, N. Ridley, B. Zhang. Mechanical behaviour and microstructural evolution in superplastic Al-Li-Mg-Cu -Zr AA8090[J]. Acta Materialia, 2007, 55(15): 4995-5006. https://doi.org/10.1016/j.actamat.2007.05.017

[10] W. Fan, B.P. Kashyap, M.C.Chaturvedi. Effects of strain rate and test temperature on flow behaviour and Micro -structural evolution in AA 8090 Al-Li alloy[J]. Metal Science Journal, 2013, 17(4): 431-438. https://doi.org/10.1179/026708301101510005

[11]C. Gao, Y. Luan, J. C. Yu, et al. Effect of thermo-mechanical treatment process on microstructure and mechanical properties of 2A97 Al-Li alloy[J]. Transactions of Nonferrous Metals Society of China, 2014, 24(7): 2196-2202. https://doi.org/10.1016/S1003-6326(14)63332-X

[12] L. Li, L. Y. Ye, Y. L. Deng Y L, et al. Effects of intermediate annealing on grain refinement and superplasticity of 2A97 aluminum-lithium alloy[J]. Chinese Journal of Nonferrous Metals, 2015, 25(1): 36-42. https://doi.org/10.1016/S1003-6326(15)63576-2

[13] L. Jia, X.P, Ren, H. L. Hou, et al. Microstructural evolution and superplastic deformation mechanisms of as-rolled 2A97 alloy at low-temperature[J]. Materials Science and Engineering A, 2019, 759(JUN.24):19-29. https://doi.org/10.1016/j.msea.2019.04.102

[14] L. Jia, X.P, Ren, Y. L. Zhang, H. L. Hou. Forming Limits Analysis of Superplastic 2A97 Al-Li Alloy[J]. Materiali in Tehnologije, 2020, 54(3): 397-405. https://doi.org/10.17222/mit.2019.240

[15] X. Zhang, L. Xie, L. Ye, P. Zhang, Effect of aging treatment on grain refinement and superplasticity of 2A97 aluminum-lithium alloy, Heat Treat. Met. 39 (2014) 2, 88-93.

[16] L. Y. Ye, Y.L.Deng, Y.L.Zhang, L. Liu, L, Xie, X. M. Zhang. China. Patent CN103882351B. (2016).

Superplasticity in Advanced Materials - ICSAM 2023 Materials Research Forum LLC
Materials Research Proceedings **32** (2023) 157-166 https://doi.org/10.21741/9781644902615-17

Superplastic forming process research of front aluminum alloy opening and closing mechanism hatch for multiple units

Ranran Chen[1,a], Guangzhong He[2,b], Kefei Wang[2,c], Guofeng Wang[1,d*]

[1]School of Materials Science and Engineering, Harbin Institute of Technology, Harbin 150001, China

[2]CRRC Changchun Railway Vehicles Co., Ltd., Changchun 130062, China

[a]rrchen1126@163.com,[b]heguangzhong@cccar.com.cn, [c]wangkefei@cccar.com.cn, [d]gfwang@hit.edu.cn

Keywords: Opening and Closing Mechanism Hatch, 5083 Aluminum Alloy, Superplastic Forming, Technological Test

Abstract. In this paper, the forming process research of the front opening and closing mechanism hatch for multiple units was carried out, and the industrial grade 5083 aluminum alloy was selected as the original material. The uniform rapid superplastic forming process combined with hot stamping and direct-reverse superplastic forming was used to form the hatch body skin, the cold stamping and argon arc welding process were used to produce the connection support, and the hatch integral part was obtained by argon arc welding finally, which solved the problem of poor environmental protection and high ratio of traditional FRP hatch and high precision forming of the large complex thin-walled structural part that cannot be achieved by the traditional forming process. After process research and development, finite element analysis, and forming tests, the aluminum alloy hatch with good forming quality was successfully manufactured. The ultimate thinning rate of the part was 23.8%, the overall wall thickness was evenly distributed, the deviation of shape was controlled within 1 mm, and the mechanical properties met the relevant technical requirements.

1. Introduction

High-speed rail has become the main means of transport for people traveling near and far at this stage [1]. With the continuous development in the field of rail transportation, the modern performance requirements for rail vehicles such as safety and comfort, lightweight and high speed, green and environmental protection have been gradually improved, and the engineering and manufacturing technologies of vehicle components have been upgraded and optimized [2].

The opening and closing mechanism hatch, as an important part of the front end of the rolling stock, forms the streamlined shape together with the front hood to ensure aerodynamic performance. At this stage, the hatch is mainly made of FRP, but the difficulties of degradation, environmental pollution, high specific gravity, and insufficient strength make it impossible for the hatch to comply with the sustainable development trend in the field of rail train manufacturing. The aluminum alloy is widely used in the field of forming technology for rail vehicle coverings due to its high strength ratio, environmental friendliness, and machinability [3]. However, the plastic deformation limit of the aluminum alloy at room temperature is low and it is difficult to avoid rebound defects, so cold stamping cannot obtain complex thin-walled structural parts in a single forming [4,5]. Compared with cold stamping, stamping under hot conditions has improved the forming performance, but still cannot meet the quality requirements of complex spatially curved coverings. The superplastic forming process, which takes advantage of the unusually high elongation of aluminum alloys under specific tensile stresses, can solve the problem of large and complex sheet metal parts that cannot be manufactured by conventional stamping [6-8].

However, the traditional superplastic expansion process for forming large and complex three-dimensional components has the disadvantages of long forming cycles, uneven wall thickness distribution and high costs [9,10]. Based on the traditional process, the improved homogeneous rapid superplastic forming technology, which combines hot stamping and superplastic bulging, has shown excellent forming performance [11]. Firstly, hot stamping is used to make the material flow into the cavity of the concave die under the action of the convex die to achieve pre-deformation at high speeds; the reverse superplastic expansion is then used to allow the material to flow into the recesses reserved on the convex surface for further storage and dispersion of the deformation; finally, superplastic direct expansion is used to form localized complex features and difficult to form areas of the component. Therefore, the uniform rapid superplastic forming process can reduce the forming cycle time, improve the wall thickness distribution, reduce processing and manufacturing costs, and reduce energy consumption.

It can be seen that the study of the forming process for the opening and closing mechanism hatch made of aluminum alloy has a positive effect on the healthy and sustainable development of the rail transport industry. In this paper, the complex aluminum alloy opening and closing mechanism hatch of the front end of the intelligent rolling stock will be the object of study, and the forming process of the component will be studied to form a complete process flow and produce the part of qualified quality. It will provide a new manufacturing idea for the opening and closing mechanism hatch, and promote the development of the rail train field in the direction of lightweight and high speed.

2. Experimental

Among the various grades of aluminum alloys, 5083 aluminum alloy shows high specific strength, low density, easy processing and good corrosion resistance, and presents excellent superplastic deformation performance under specific conditions, which has been developed as the preferred material for the superplastic forming process of rail vehicle coverings. Compared with the special aluminum sheet for superplasticity, the industrial state 5083 aluminum alloy has a lower cost and can meet the demand for superplastic deformation capacity of complex thin-walled superplastic components with the composition shown in Tab 1. In addition, the 5083 aluminum alloy is weldable and is mainly joined by means of argon arc welding and laser welding [12]. Liu et al.[13] learned through tests that 5083 aluminum alloy had high strength after argon arc welding, no obvious welding defects, showing good welding quality.

Tab 1 Chemical composition of industrial state 5083 aluminum alloy (wt%)

Elements	**Mg**	**Mn**	**Si**	**Fe**	**Cu**	**Cr**	**Ti**	**Zn**	**Al**
Content	**4.0~4.9**	**0.40~1.0**	**≤ 0.40**	**≤ 0.40**	**≤ 0.10**	**0.05~0.25**	**≤ 0.15**	**≤ 0.25**	**Bal.**

The opening and closing mechanism hatch made of aluminum alloy studied in this project is an important exterior sheet metal part that has a matching connection with the front hood and the movement structure. The hatch includes not only the exterior hatch body skin (Fig.1(a)) but also the connecting support (Fig.1(b)). The uniform rapid superplastic forming technology was used to achieve efficient and precise forming of the large and complex thin-walled hatch body skin. The cold stamping process was used to form the transitional rounded joint profile of the connection support (red area in Fig.1(b)). The side wall plates were welded to the pressed parts using the argon arc welding process to obtain the connection support for the overall structure. Finally, the connection supports were welded to the corresponding reserved positions of the hatch body using argon arc welding to obtain an integral part of the hatch.

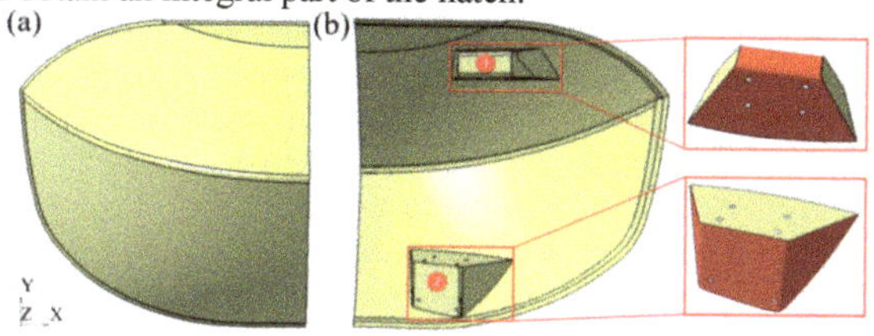

Fig. 1 Three-dimensional model of hatch: (a)the main body skin; (b)the connection support

3. Results and Discussion

3.1 Superplastic forming simulation of the hatch body

Marc software has been widely used in superplastic process simulation due to its powerful non-linear solving calculations and the availability of the dedicated superplastic forming module. The mould surface was extracted, the mould structure was simplified and the finite element geometric model was created by importing the software, as shown in Fig.2. The mould was set up as the rigid body, the sheet was the deformed body, and the coefficient of friction was 0.2. The initial thickness of the sheet was 4.7mm, the cell size was 10 mm×10 mm and the cell type was shell cell. The Backofen equation was used as the material model to describe the relationship between flow stress and strain in metallic materials during the steady-state rheological phase. The industrial state 5083 aluminum alloy exhibited excellent superplastic deformation properties at 480°C and $1\times10^{-3}s^{-1}$ with a material parameter of 155 MPa·s and a strain rate sensitivity factor of 0.35 [14]. The movement of the sheet unit in the sealing area was restricted in the X, Y and Z directions. Superplastic bulging pressure was applied to all unit faces and the contact was controlled using the Coulomb bilinear method.

Fig.2 Finite element geometric model

The wall thickness distribution after hot stamping is shown in Fig.3(a). The minimum thickness on the part profile appeared in the area of maximum stretching depth, with a thickness of 4.293 mm and a thinning rate of 8.7%. The thinning rate was small due to the material flowing into the cavity under the action of the convex die during the stamping process, which carried out a certain degree of replenishment. After hot stamping, superplastic reverse expansion was carried out to a certain extent, with the thickness distribution shown in Fig.3(b). The minimum thickness of the profile was 4.071 mm and the maximum thinning rate was 13.4%, which was an increase in the thinning rate but still within the reasonable value. Finally, superplastic direct expansion was carried out and the material flowed into the concave model cavity until it was fully laminated, with the wall thickness distribution shown in Fig.3(c). The largest reduction in wall thickness of the formed part was at the sharp end of the concave die, with a wall thickness of 3.318mm and a thinning rate of 29.4%, which was the supplementary part of the process, not part of the profile and could be removed by cutting after the actual forming. The thinnest position on the part profile was the deep cavity circular transition area, with a wall thickness value of 3.532mm and a thinning rate of 24.8%, which was still within the thinning rate of ≤25%.

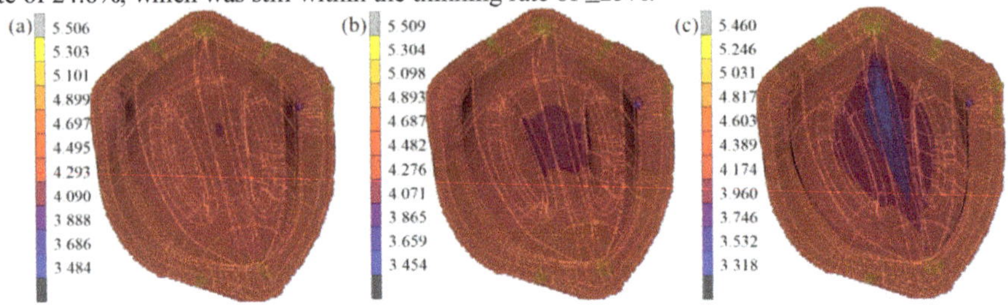

Fig.3 Thickness distribution: (a)hot stamping; (b)superplastic reverse bulging; (c)superplastic direct bulging

3.2 Superplastic forming test of the hatch body

The superplastic mould is shown in Fig.4. To avoid excessive high-temperature flow resistance adversely affecting the final forming quality of the part, the graphite lubricant was applied to the original aluminum alloy sheet. The production trial was carried out at the deformation temperature of 480°C and the deformation rate of $0.001s^{-1}$ using superplastic forming equipment obtained by modification on the basis of the 1000t hydraulic press. The temperature of the sheet was monitored in real-time to reach the target value and then held for a period of time to achieve the uniform temperature distribution in all areas of the sheet. The upper die was then driven down by the hydraulic press at a speed of 5 mm/s to complete the hot stamping. The sealing force was then applied to ensure that the expansion cavity was airtight. The reverse expansion pressure was slowly applied according to Fig.5(a) and the pressure was released when the forming reached a certain level. The direct expansion pressure was then applied according to Fig.5(b) and the strain rate was reasonably controlled so that the material was finally fully fitted to the concave cavity and the formed part was obtained.

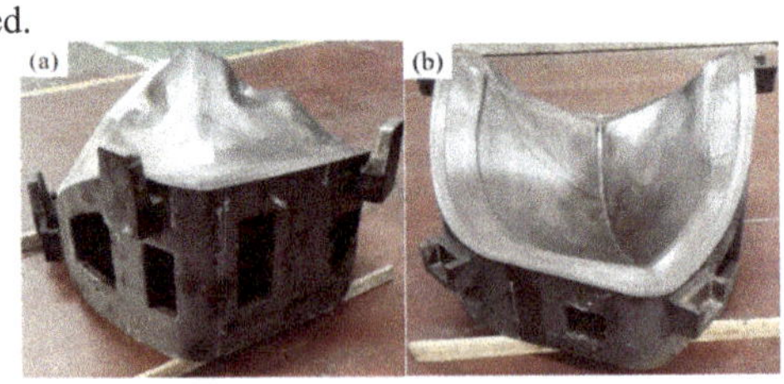

Fig.4 Superplastic mould: (a)punch; (b)die

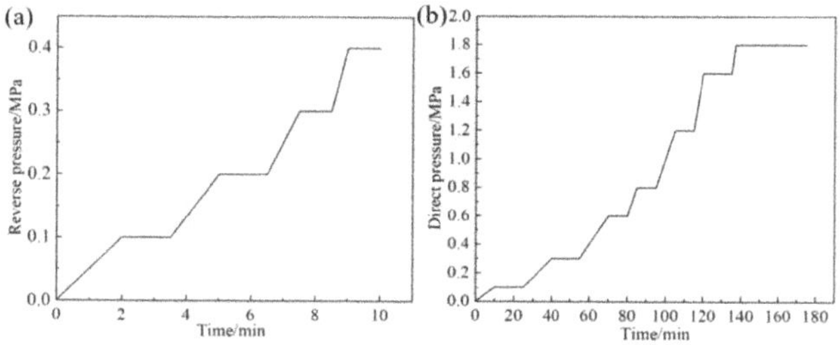

Fig.5 Pressure loading control curve: (a)superplastic reverse bulging; (b)superplastic direct bulging

The superplastic formed part was cut and polished to acquire the hatch body skin as shown in Fig.6. The physical diagram of the test visually showed that the material that flowed into the reverse-expansion cavity in advance had been fully expanded, and the surface quality of the part was good, without obvious scratches, wrinkles, breaks or other defects.

Fig.6 The hatch body skin

3.3 Forming test of the connection support

The cold punching die was mounted on the press for cold stamping. Due to a certain degree of springback in cold punching parts, manual shaping was required to ensure that the shape of the formed part matched the side wall plates to facilitate welding and reduce shape deviations. After straightening the shape, the formed part was trimmed to a certain extent to reduce dimensional deviations, in accordance with the positioning notch reserved on the mould. The cold stamping parts are shown in Fig.7. The side wall panels were finally joined to the formed parts by argon arc welding to obtain the complete joint supports as shown in Fig.8. To avoid defects such as porosity in the weld seam, the surface oxide film was polished and removed before welding to reveal the silvery-white metallic luster [15]. The welding method of ER5356 wire and AC manual tungsten arc welding was used to achieve the butt welding of 5083 aluminum alloys, which was flexible in operation and highly suitable for welding complex structures. By observing the macroscopic shape of the weld seam, it could be seen that the joint was uniform and neat, the weld seam had basically the same width and no significant defects.

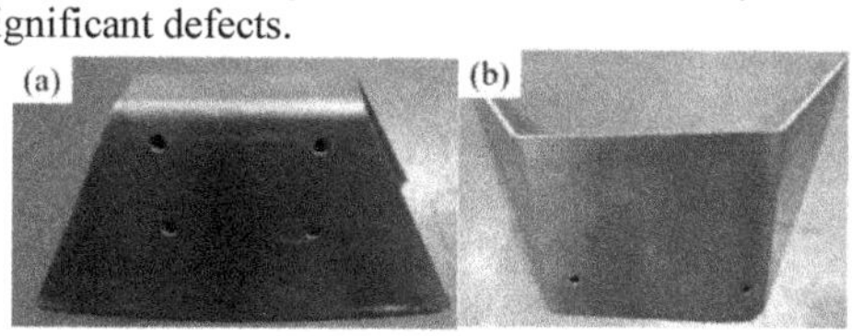

Fig.7 Cold stamping parts: (a)no. 1 connection support; (b)no. 2 connection support

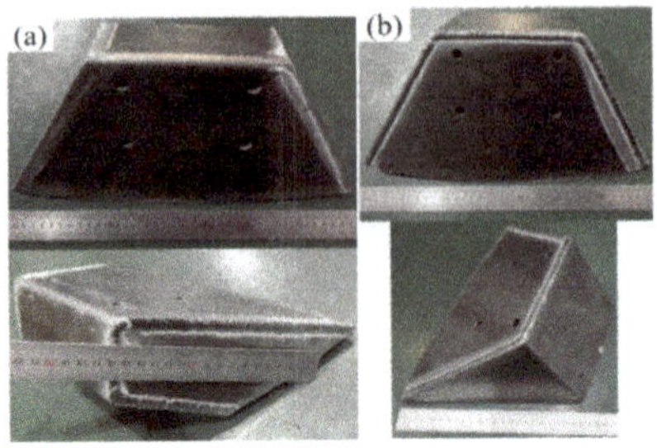

Fig.8 Integral structures of connecting supports: (a)no. 1 connection support; (b)no. 2 connection support

3.4 Welding connection of the integral part for hatch

To achieve the opening and closing action of the hatch, the connecting supports were mechanically connected to the movement mechanism and integrated into the hatch body using the welding process. The hatch body and the connecting supports were made of 5083 aluminum alloy and were connected by argon arc welding. The inspection tool played a positioning role in the welding position, as shown in Fig.9. The welding position of the hatch body and the connecting supports had the same spatial curved surfaces to ensure connection accuracy. Firstly, the connecting supports were placed in the reserved position of the inspection tool, then the hatch body was placed in the inspection tool, covering the connecting supports. The two were fixed and marked for spot welding in position and fully welded to obtain the complete aluminum hatch component as shown in Fig.10. It could be seen that the welding wire was well filled, the welding seam was flat and consistent, no cracks and other defects appeared, and the welding quality was good.

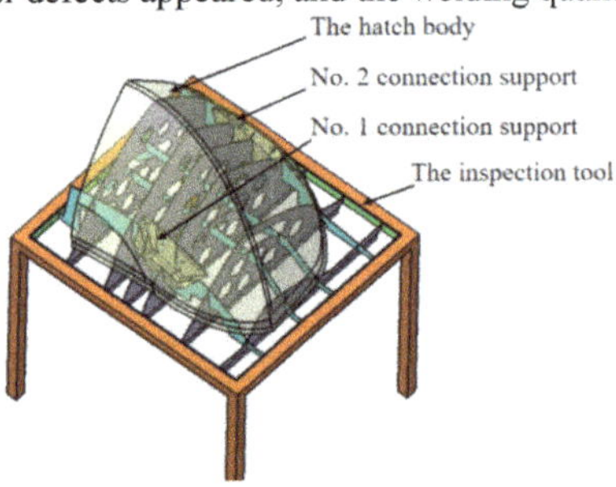

Fig.9 3D model of the inspection tool

 Materials Research Forum LLC
https://doi.org/10.21741/9781644902615-17

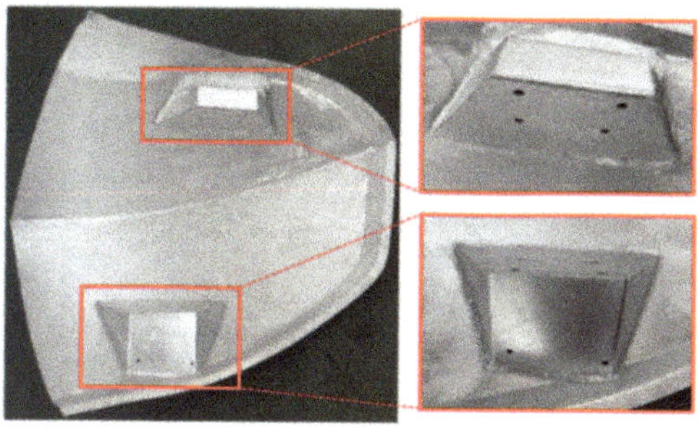

Fig.10 The hatch integral part

3.5 Quality assessment for the part

The hatch was of good quality in appearance with no obvious defects, but further evaluation was required to ensure the strength of the part in use. The wall thickness distribution affects the mechanical properties of the part, and the areas of severe thinning are prone to stress concentration, which can have a detrimental effect on the performance of the component. The ultrasonic thickness gauge was used to test the local feature points of the hatch body and the measurement results are shown in Fig.11. The thinnest thickness was at the bottom deep cavity area, which was consistent with the simulation result, where the deformation was the most intense with the wall thickness value of 3.58mm and the thinning rate of 23.8%. The ultimate thinning rate was controlled within the permitted limits and the overall wall thickness distribution was homogeneous.

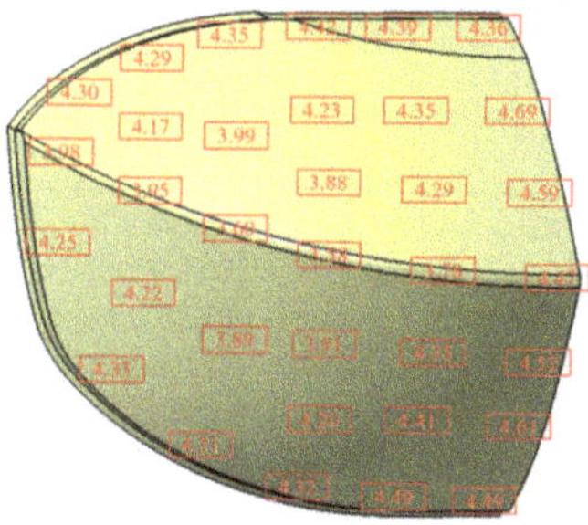

Fig.11 The thickness distribution of superplastic forming part

The final lamination of the part was achieved by using superplastic bulging with excellent deformation properties and good lamination performance. However, further precise measurements with the aid of gauges and moulds were required to ensure shape and dimensional accuracy, the assemblability and coordination with the front hood. The whole hatch was placed on the inspection tool as shown in Fig.12. After measuring, the shape and dimensional deviations were controlled within 1mm, showing good shape and dimensional accuracy.

Fig.12 The hatch placed on the inspection tool

After the superplastic deformation of the aluminum alloy, the microstructure will change and the mechanical properties will change to some extent. To ensure that the formed part has sufficient strength for service at room temperature, the samples were taken from the formed part and tested for tensile mechanical properties. The results of the mechanical properties tests are shown in Tab 2. It could be seen that mechanical property parameters had been reduced, but none of them had been reduced to 90% of the original material mechanical properties, which could meet the room temperature service requirements.

Tab 2 Mechanical properties of room temperature before and after forming

Test subjects	Pre-forming	Post-forming	Reduction rate
Tensile strength/MPa	332	301	9.3%
Yield strength/MPa	155	140	9.7%
Maximum elongation/%	34	33	2.9%

To study the effect of the entire deformation process on the microstructure of the material, EBSD testing of the microstructure before and after deformation was carried out and the test results are shown in Fig.13. After the statistical summary, the average grain size of the original plate was approximately 14.12 μm and the average grain size of the part was approximately 18.97 μm, which was an increase in grain size. The material was deformed in the high-temperature environment at around 4h and the grains grew to reduce the total interface energy. However, the grain shape of the part was approximately equiaxed, without obvious holes, cracks or other microscopic defects, with good overall performance.

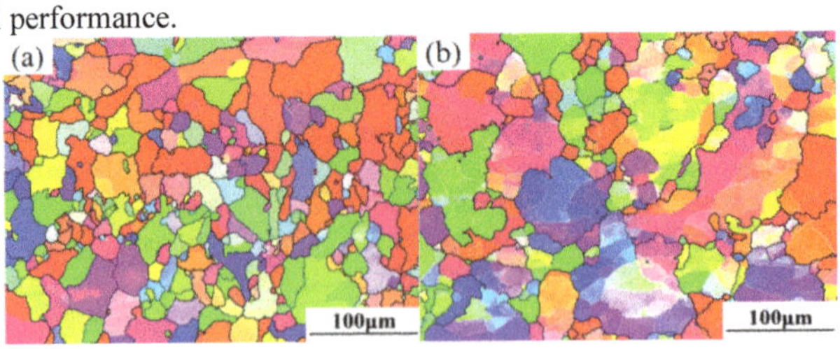

Fig.13 Microstructure before and after forming: (a)before forming; (b)after forming

4. Conclusions

Through the process of preliminary technical research and development, forming simulation, and trial production, the aluminum alloy hatch with good quality was successfully produced. The minimum wall thickness of the part was 3.58mm, which was a reasonable wall thickness distribution; the shape and dimensional deviations of the hatch were less than 1mm, with high shape and dimensional accuracy; the mechanical properties of the formed part were reduced to different degrees compared with the raw material, but were within reasonable values; the microscopic grain size increased slightly, but no obvious microscopic defects appeared. It can be seen that the uniform rapid superplastic forming process can be successfully applied to the production and development of the aluminum alloy hatch body, and combined with cold stamping, argon arc welding, and other processes to obtain the overall lightweight component for opening and closing mechanism hatch of acceptable quality, to promote the development of the rail transit parts manufacturing industry to high-speed, energy-saving, environmental protection and other directions.

References

[1] D. He, J.M. Zhu, Integrated passenger transport rail transitsystem construction model, J. Southwest Jiaotong Univ. 47 (2012) 279-284.

[2] M.H. Sun, Urban mass transit energy-saving technology trends study, Constr. Des. Proj. (2009) 15-19.

[3] G.F. Wang, H.L. Wang, J.Y. Liang, H.H. Jia, Y.B. Gu, Research on rapid superplastic forming technology of AA5083 aluminum alloy for rail transit industry, J. Plast. Eng. 26 (2019) 37-42.

[4] Y. Xu, G.F. Wang, H.P. Lin, Z.P. Wang, H.X. Sun, Research on skin forming of domestic 5083 aluminum alloy high speed maglev cab, J. Plast. Eng. 28 (2021) 25-32.

[5] J.U. Hirsch, Rgen, Aluminium in innovative light-weight car design, Mater. Trans. 52 (2011) 818-824. https://doi.org/10.2320/matertrans.L-MZ201132

[6] W.L. Lu, X.W. Wu, H.J. Liang, Y. Wang, Research on high speed superplastic forming technology of aluminum alloy automotive panel, Hot Work. Technol. 42 (2013) 39-41.

[7] G.F. Wang, C. Sun, S.F. Liu, M. Yang, Research on quick superplastic forming technology of aluminum alloy complex components, Materialwiss. Werkstofftech. 45 (2014) 854-859. https://doi.org/10.1002/mawe.201400294

[8] E.M. Taleff, L.G. Hector, J.R. Bradley, R. Verma, P.E. Krajewski, The effect of stress state on high-temperature deformation of fine-grained aluminum-magnesium alloy AA5083 sheet, Acta. Mater. 57 (2009) 2812-2822. https://doi.org/10.1016/j.actamat.2009.02.036

[9] Z.Q. Li, W.L. Lu, W.L. Wang, Y. Wang, Wall thickness distribution and its control of 5A06 thin-walled shell superplastic bulging, J. Plast. Eng. 24 (2017) 108-113.

[10] M.Y. Sun, G.P. Bi, Research on thickness uniformity of 5A06 aluminum alloy sheet superplastic gas bulging forming and its die design, Hot Work. Technol. 45 (2016) 132-135.

[11] Y.F. Wang, H.P. Lin, L.P. Sui, J.C. Yan, G.F. Wang, H.L. Wang, Application of uniform quick superplastic forming technology on front skin of aluminum alloy for urban vehicles, J. Plast. Eng. 28 (2021) 104-110.

[12] G.Y. Sun, L. Wang, D.D. Shao, L. Zhang, W.K. Li, X.W. Xia, Analysis of adaptability, microstructure and properties of MIG welding process of 5083 aluminum alloy for ship, Electric Weld. Mach. 52 (2022) 75-81.

[13] Y. Liu, W.J. Wang, J.J. Xie, S.G. Sun, L. Wang, Y. Qian, Y. Meng, Y.J. Wei, Microstructure and mechanical properties of aluminum 5083 weldments by gas tungsten arc and gas metal arc welding, Mater. Sci. Eng. A 549 (2012) 7-13. https://doi.org/10.1016/j.msea.2012.03.108

[14] J.Y. Liang, Z.H. Du, H.P. Lin, G.F. Wang, Direct-reverse bulging superplastic forming process of urban railway door of industrial 5083 aluminum alloy, J. Plast. Eng. 29 (2022) 66-72.

[15] Z.F. Chen, C.X. Zhao, H.T. Ren, D.L. Zhang, TIG welding points for Al-Mg alloy pipes, Petro. Chem. Constr. 41 (2019) 89-91.

Superplasticity in Advanced Materials - ICSAM 2023 Materials Research Forum LLC
Materials Research Proceedings **32** (2023) 167-172 https://doi.org/10.21741/9781644902615-18

Low-temperature superplasticity of Si-added medium-Mn steel

Hyun-Bin Jeong[1,a], Seok-Won Choi[1,b], Seok-Hyeon Kang[1,2,c] and Young-Kook Lee[1,d,*]

[1]Department of Materials Science and Engineering, Yonsei University, Seoul, 03722, Republic of Korea

[2]Current address: Research & Development Division, Hyundai-Steel, Dangjin 31719, Republic of Korea

[a]fenixbin@yonsei.ac.kr, [b]csw526@yonsei.ac.kr, [c]ksh546@hyundai-steel.com, [d]yklee@yonsei.ac.kr

Keywords: Low-Temperature Superplasticity, Dynamic Reverse Transformation, Grain Boundary Sliding, High Strength Steel

Abstract. Grain boundary sliding, a deformation mechanism of superplasticity, occurs only at high temperatures, making it difficult for superplastic steels to be practically applied due to high energy consumption and surface oxidation. Therefore, in the present study, we introduce a newly developed Si-added medium-Mn steel, which can be superplastically deformed at such a low temperature of 763 K. The low-temperature superplasticity of this steel was caused by grain boundary sliding between recrystallized α grains and dynamically reverted γ grains. The steel also exhibited high room-temperature tensile strength (1336 MPa) after the superplastic forming simulation.

Introduction

Superplastic materials have drawn significant attention due to high elongation prior to tensile failure, which makes them ideal for making complex-shaped mechanical parts without welding or joining. In particular, superplastic steels have been extensively studied due to low material and production costs, such as high-carbon steel [1-3], duplex stainless steel [4,5], high-specific-strength steel [6], and medium-Mn steel (MMS) [7-12]. However, their practical application is challenging due to their high superplastic deformation temperature (T_s) [7,9]. To overcome the limitation, present authors reported a new Fe-10Mn-3.5Si steel [12], which exhibited superplasticity at 763 K with a strain rate ($\dot{\varepsilon}$) of 1×10^{-3} s^{-1}. The steel has a low material cost and can be produced using a conventional rolling process [12]. Unlike previous superplastic materials with a dual-phase microstructure [7-11], the steel has a single-phase microstructure before plastic deformation [12]. Additionally, after a superplastic forming simulation at 763 K, the steel possesses ultrahigh tensile strength (1336 MPa) at room temperature [12], which far exceeds the tensile strength of Al-added MMS (715 MPa) [9].

Experimental

The ingots of Si-free and Si-added 10Mn steels were produced using an induction melting furnace with an Ar atmosphere. The actual chemical compositions of the steels were Fe-10.5Mn and Fe-10.6Mn-3.5Si (wt%), respectively [12]. After homogenization at 1373 K for 12 h, the ingots were hot-rolled at temperatures ranging from ~1373 K to 1273 K to ~4.0-mm-thick plates, and then air-cooled to room temperature. After surface descaling, the hot-rolled plates were cold-rolled at room temperature to 1.5-mm-thick sheets, which corresponded to thickness reductions of 50%.

For high-temperature tensile testing, tensile specimens with a gauge part of $5.0 \times 10.0 \times 1.5$ mm^3 were machined from the cold-rolled sheets. Uniaxial tensile tests were performed using universal testing machines (MTDI equipped with a high-temperature furnace). Before tensile tests, tensile specimens were heated to various temperatures at a heating rate of 0.5 K s^{-1}, maintained for 5 min at a given tensile temperature, and then deformed at the initial $\dot{\varepsilon}$ of 1 × 10^{-3} s^{-1}. A failed specimen was removed from the furnace, followed by air cooling to room temperature.

For the superplastic forming simulation, high-temperature tensile specimens with the gauge part of 20.0 × 10.0 × 1.5 mm^3 were machined. The specimens were strained by ~100% at the T_s of each specimen with an initial $\dot{\varepsilon}$ of 1 × 10^{-3} s^{-1}. After that, a small-sized tensile specimen was machined from the gauge part of the high-temperature interrupted tensile specimen. The gauge part of the small-sized tensile specimen was 16.0 × 3.0 × 0.8 mm^3. The small-sized tensile specimen was deformed using universal testing machines (Instron 3382) at room temperature with an initial $\dot{\varepsilon}$ of 1 × 10^{-3} s^{-1}.

The microstructures of the specimens were examined at room temperature using an X-ray diffractometer (XRD; Rigaku, Smartlab) and a field-emission scanning electron microscope (FE-SEM; JEOL, JSM-7001F) equipped with an electron backscattered diffractometer (EBSD; EDAX-TSL, Digiview). For XRD analysis, the specimens were mechanically polished with a suspension of 1 μm diamond particles, and then electropolished in a mixed solution of 90% glacial acetic acid (CH_3COOH) and 10% perchloric acid ($HClO_4$) at 15 V for 1 min to remove the damaged layer. For SEM-EBSD observation, the specimens were polished with a suspension of 0.04 μm colloidal silica particles. The SEM-EBSD operation parameters, such as the accelerating voltage, probe current, working distance, and step size, were set to 20 kV, 12 nA, 15 mm, and 25 nm, respectively.

The phase constituents of the present steels were determined using XRD with Cu-Kα_1 radiation, with a scanning rate, step size, and range of 2° min^{-1}, 0.02°, and 40°-100°, respectively. The volume fraction of γ phase was calculated using the integrated intensity method, which takes into account the integrated intensities of all diffracted peaks. The effect of texture on γ fraction was considered by dividing the integrated intensity of each peak by the pole density of each plane [13].

Results and discussion

All hot-rolled plates had a full α'-martensite microstructure without γ phase due to the high hardenability of the steels used (Fig. 1a [12]). Accordingly, cold-rolled sheets had a single phase of deformed α'-martensite (Fig. 1b [12]).

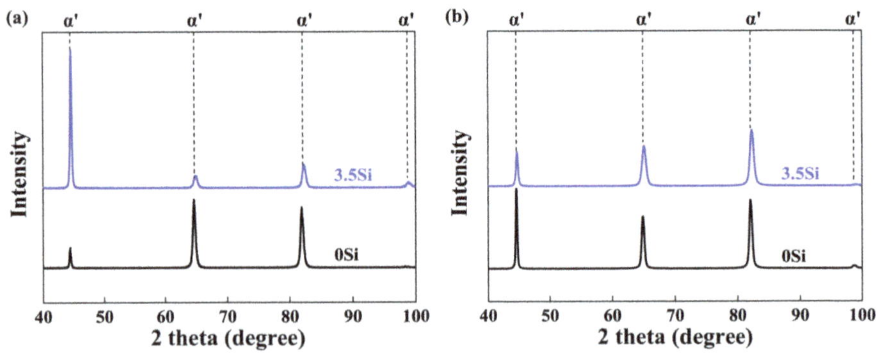

Fig. 1. XRD patterns of (a) hot- and (b) cold-rolled present specimens [12].

To evaluate the low-temperature superplasticity of the present steels, specimens were tensile-deformed at various temperatures with a $\dot{\varepsilon}$ of 1 × 10^{-3} s^{-1} (Fig. 2 [12]). The steels exhibited superplastic elongation above 300% at temperatures below 858 K (Fig. 2a [12]). Notably, the 3.5Si steel exhibited the lowest T_s of 763 K, which was confirmed by deforming the 0Si steel at 763 K

Materials Research Forum LLC
https://doi.org/10.21741/9781644902615-18

(Fig. 2b [12]). The 0Si steel showed lower engineering strain (76%) and higher peak stress than the 3.5Si steel, which indicates that the addition of 3.5wt% Si was effective in lowering the T_s.

Except for our steels, the lowest T_s of superplastic steels, above which the total elongation is 300% or over, is 873 K at a $\dot{\varepsilon}$ of 1×10^{-3} s^{-1} [1,2,7]. The 3.5Si steel exhibited the lowest T_s (763 K) [12], which is lower by 110 K than the lowest T_s (873 K) reported to date.

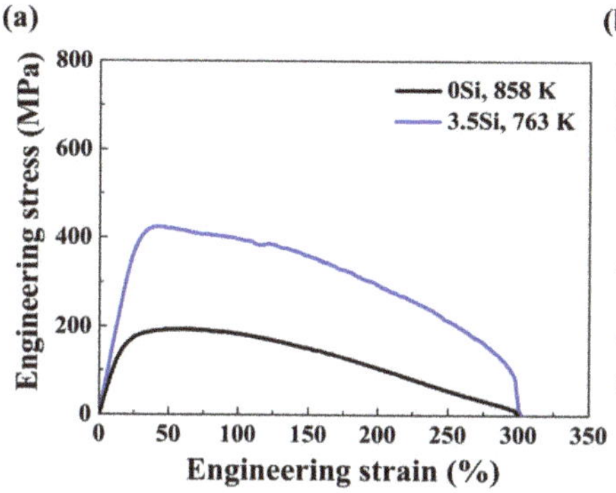

Fig. 2. Engineering stress-strain curves of present steels measured (a) at the lowest temperature (T_s), where total elongation is 300% or over and (b) at 763 K [12].

To elucidate the deformation mechanism responsible for the excellent superplasticity of 3.5Si steel, EBSD phase and grain orientation spread (GOS) maps were taken from the tensile-fractured 3.5Si specimen at 763 K (Fig. 3a). Ultrafine γ grains (~138 nm), which were absent before deformation, were observed on the phase map. This indicates that dynamic reverse transformation partially occurred during tensile deformation at 763 K. By comparing the phase map to the GOS map, fine recrystallized α grains with low GOS values [14] were distinguished. This confirms that the fine equiaxed grains shown in Fig. 3a consist of both recrystallized α and reverted γ grains.

However, the fraction of reverted γ grains was measured using XRD patterns due to the limited probing area of EBSD. For comparison, the fractions of recrystallized α and reverted γ grains were also measured using the 0Si specimen deformed at 763 K with a $\dot{\varepsilon}$ of 1×10^{-3} s^{-1}. As seen in Figs. 3b and c [12], the 3.5Si specimen exhibited remarkably higher fractions of recrystallized α grains and reverted γ grains than the 0Si specimen. In particular, the fraction of reverted γ grains increased linearly in 3.5Si steel during tensile deformation, reaching ~13% after failure.

Materials Research Proceedings **32** (2023) 167-172
https://doi.org/10.21741/9781644902615-18

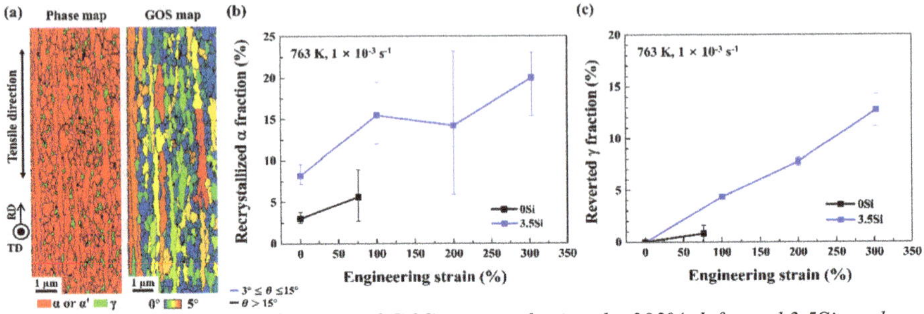

Fig. 3. *(a) EBSD maps of phase and GOS measured using the 303% deformed 3.5Si steel. Variations of the fractions of (b) recrystallized α grains [12] and (c) reverted γ grains with tensile strain in both 0Si and 3.5Si steel [12].*

According to Hashimoto *et al.*'s study [15], the sliding rate of the α/γ phase boundaries was significantly higher than that of α/α and γ/γ boundaries in a shear-deformed DSS bicrystal at 1373 K. Additionally, it is well-known that grain boundary sliding (GBS) actively occurs in superplastic materials with fine equiaxed grains [16]. Based on these, it can be inferred that GBS took place at the phase boundaries between the recrystallized α and reverted γ grains in the deformed 3.5Si steel, which had ultrafine recrystallized α (~307 nm) and reverted γ grains (~138 nm).

To reconfirm the occurrence of GBS, texture analysis was performed using the 3.5Si specimen strained by 303% at 763 K. While the orientation of recrystallized α grains revealed a texture component {114}<110> with an $I_{max.}$ of 5.9 (Fig. 4a), that of reverted γ grains exhibited a random texture with an $I_{max.}$ of 3.0 (Fig. 4b) due to GBS during tensile deformation at 763 K. It is considered that since not all recrystallized α grains were in contact with the reverted γ grains, they still exhibited slightly developed texture with $I_{max.} = 5.9$ (Fig. 4a).

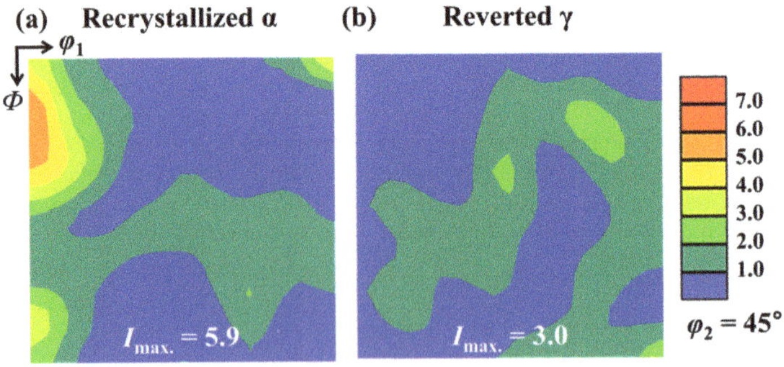

Fig. 4. *ODF maps of (a) recrystallized α and (b) reverted γ grains in the 3.5Si specimen deformed by 303% at 763 K with a $\dot{\varepsilon}$ of 1 × 10^{-3} s^{-1}.*

Meanwhile, the 3.5Si steel demonstrated an exceptional room-temperature tensile strength of 1336 MPa after undergoing superplastic forming simulation at 763 K (Fig. 5 [12]). Considering its numerous advantages, such as the lowest T_s, high room-temperature tensile strength, low material cost, and ease of fabrication, 3.5Si steel is considered an excellent option for producing mechanical parts that require complex shapes and high room-temperature tensile strength simultaneously.

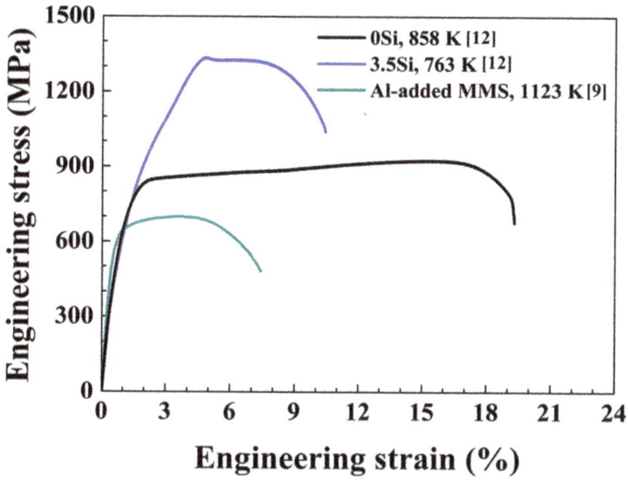

Fig. 5. *Engineering stress-strain curves measured at room temperature with a $\dot{\varepsilon}$ of 1×10^{-3} s^{-1} after the superplastic forming simulation [9,12].*

Summary

In the present study, we introduce a newly developed superplastic Fe-10Mn-3.5Si steel. The steel exhibited the lowest T_s of 763 K, which is 110 K lower than reported to date. The deformation mechanism of excellent superplasticity was GBS occurring at phase boundaries between recrystallized α and reverted γ grains. In addition, the steel showed remarkable room-temperature tensile strength. We suggest that this steel may accelerate the commercialization of superplastic steels.

References

[1] H. Zhang, D. Ponge, D. Raabe, Superplastic Mn-Si-Cr-C duplex and triplex steels: interaction of microstructure and void formation, Mater. Sci. Eng. A 610 (2014) 355-369. https://doi.org/10.1016/j.msea.2014.05.061

[2] H. Zhang, K.G. Pradeep, S. Mandal, D. Ponge, P. Choi, C.C. Tasan, D. Raabe, Enhanced superplasticity in an Al-alloyed multicomponent Mn-Si-Cr-C steel, Acta Mater. 63 (2014) 232-244. https://doi.org/10.1016/j.actamat.2013.10.034

[3] B. Walser, O.D. Sherby, Mechanical behavior of superplastic ultrahigh carbon steels at elevated temperature, Metall. Trans. A 10A (1979) 1461-1471. https://doi.org/10.1007/BF02812011

[4] Y.S. Han, S.H. Hong, The effects of thermos-mechanical treatments on superplasticity of Fe-24Cr-7Ni-3Mo-0.14N duplex stainless steel, Scr. Mater. 36 (1997) 557-563. https://doi.org/10.1016/S1359-6462(96)00421-6

[5] K. Tsuzaki, H. Matsuyama, M. Nagao, T. Maki, High-strain rate superplasticity and role of dynamic recrystallization in a superplastic duplex stainless steel, Mater. Trans., JIM 31 (1990) 983-994. https://doi.org/10.2320/matertrans1989.31.983

[6] W. Wang, M. Yang, D. Yan, P. Jiang, F. Yuan, X. Wu, Deformation mechanisms for superplastic behaviors in a dual-phase high specific strength steel with ultrafine grains, Mater. Sci. Eng. A 702 (2017) 133-141. https://doi.org/10.1016/j.msea.2017.07.011

[7] W. Cao, C. Huang, C. Wang, H. Dong, Y. Weng, Dynamic reverse phase transformation induced high-strain-rate superplasticity in low carbon low alloy steels with commercial potential, Sci. Rep. 7 (2017) 9199. https://doi.org/10.1038/s41598-017-09493-7

[8] Z. Cao, G. Wu, X. Sun, C. Wang, D. Ponge, W. Cao, Revealing the superplastic deformation behaviors of hot rolled 0.10C5Mn2Al steel with an initial martensitic microstructure, Scr. Mater. 152 (2018) 27-30. https://doi.org/10.1016/j.scriptamat.2018.03.046

[9] J. Han, S.-H. Kang, S.-J. Lee, M. Kawasaki, H.-J. Lee, D. Ponge, D. Raabe, Y.-K. Lee, Superplasticity in a lean Fe-Mn-Al steel, Nat. Commun. 8 (2017) 751. https://doi.org/10.1038/s41467-017-00814-y

[10] S.-H. Kang, S.-W. Choi, Y.-D. Im, Y.-K. Lee, Grain boundary sliding during high-temperature tensile deformation in superplastic Fe-6.6Mn-2.3Al steel, Mater. Sci. Eng. A 780 (2020) 139174. https://doi.org/10.1016/j.msea.2020.139174

[11] S.-H. Kang, H.-B. Jeong, J.-S. Hong, Y.-K. Lee, Effect of B on the superplasticity of Fe-6.6Mn-2.0Al alloy, Mater. Sci. Eng. A 822 (2021), 141697. https://doi.org/10.1016/j.msea.2021.141697

[12] H.-B. Jeong, S.-W. Choi, S.-H. Kang, Y.-K. Lee, Ultralow-temperature superplasticity of high strength Fe-10Mn-3.5Si steel, Mater. Sci. Eng. A 848 (2022) 143408. https://doi.org/10.1016/j.msea.2022.143408

[13] Y. Onuki, A. Hoshikawa, S. Sato, T. Ishigaki, T. Tomida, Quantitative phase fraction analysis of steel combined with texture analysis using time-of-flight neutron diffraction, J. Mater. Sci. 52 (2017) 11643-11658. https://doi.org/10.1007/s10853-017-1309-x

[14] V. Torganchuk, I. Vysotskiy, S. Malopheyev, S. Mironov, R. Kaibyshev, Microstructure evolution and strengthening mechanisms in friction-stir welded TWIP steel, Mater. Sci. Eng. A 746 (2019) 248-258. https://doi.org/10.1016/j.msea.2019.01.022

[15] S. Hashimoto, F. Moriwaki, T. Mimaki, S. Miura, Sliding along the interphase boundary in austenite/ferrite duplex stainless steel bicrystals, in: S. Hori (Ed.), Superplasticity in Advanced Materials, The Japan Soc. Res. on Superplasticity, 1991, pp. 23-32.

[16] K.A. Padmanabhan, S.B. Prabu, R.R. Mulyukov, A. Nazarov, R.M. Imayev, S. G. Chowdhury, Superplasticity: Common Basis for a Near-Ubiquitous Phenomenon, first ed., Springer, Berlin, Heidelberg, 2018. https://doi.org/10.1007/978-3-642-31957-0

Materials Research Proceedings **32** (2023) 173-180 https://doi.org/10.21741/9781644902615-19

Friction stir processing to improve grain refinement and superplasticity of Al-Mg-Mn-Cr alloy

Anna A. Kishchik[1,a], Ahmed O. Mosleh[2,b], Olga A. Yakovtseva[1,c], Anton D. Kotov[1,d], Anastasia V. Mikhaylovskaya[1,e*]

[1]Physical Metallurgy of Non-Ferrous Metals, National University of Science and Technology MISIS, Leninsky Prospekt, 4 build 1, 119049 Moscow, Russia

[2]Mechanical Engineering Department, Faculty of Engineering at Shoubra, Benha University

[a]kishchik.aa@misis.ru, [b]ahmed.omar@feng.bu.edu.eg, [c]yakovtsevao@mail.ru, [d]kotov@misis.ru, [e]mihaylovskaya@misis.ru

Keywords: Aluminum Alloys, Friction Stir Processing (FSP), Microstructure, Superplasticity, Grain Size

Abstract. In this study the Al-4.8Mg-1.0Mn-0.15Cr alloy was frictionally stir processed in order to enhance grain refinement effect and superplastic properties. The microstructure after friction stir processing (FSP) was analyzed using light, scanning and transmission electron microscopies. An additional cold rolling (50%) was performed after the FSP, and a comparative study in terms of microstructure and superplasticity was done between the samples processed using hot rolling and further FSP, FSP and cold rolling, and cold rolling without FSP. The superplastic properties were studied at temperatures of 450 and 500 °C with constant strain rates of 1×10^{-2} and 1×10^{-2} s^{-1}. The treatment regime including FSP and further cold rolling improved superplasticity due to fine-grained and homogeneous grain structure with a good thermal stability. At a temperature of 500°C and strain rates of 1×10^{-3} - 1×10^{-2} s^{-1} the alloy demonstrated 330-450 % elongation with a flow stress of 10-20 MPa.

Introduction

Aluminum alloys are widely used in industrial production for urban infrastructure products, due to the combination of high specific strength and corrosion resistance. The possibility of obtaining streamlined shapes is important for the automotive and aircraft industry and it is achievable with the implementation of superplastic forming. Grain boundary sliding is the key superplastic deformation mechanism that accommodated by dislocation and diffusional creep, and grain rotation [1]-[5]. Thus, for a good superplastic properties fine-grained structure and movable high angle grain boundaries are required [6]-[10]. Grain refinement improves superplasticity, strain rates and elongations increase that favor formability [11]-[13]. There are several methods for obtaining an ultrafine-grained structure (below 1 μm) in aluminum-based alloys, among the most effective are severe plastic deformation techniques [5][14][15]. A friction stir processing (FSP) is an approach of solid-state treatment to obtain a ultrafine-grained structure based on the principles of friction stir welding [16]-[19], which is relatively common in a global industry. The method is widely used to improve surface properties [20]-[22]. When processing by friction with stir, severe plastic deformation accumulates, thermal effects occur in the processing zone, which leads to the formation of fine or ultrafine recrystallized grains [17][23][24]. As the result, an increase of strength characteristics at room temperature [25] and an improvement of superplasticity are possible [26]-[30]. FSP is effective in terms of scalability and the result achieved. For a number of conventional aluminum alloys, this method improves superplasticity, including a decrease in temperature and an increase in strain rate [19][31]-[35]. It was shown in [36]-[39] that heating during FSP can lead to the growth of particles of the secondary phases, which negatively affects the mechanical properties.

Undesirably non-uniform distribution of secondary phase particles is noted in [40], with the formation of a depleted central zone, the mechanisms of such undesirable phenomena have not been reliably established at the moment. The authors [41] investigate the influence of the FSP parameters, the transverse speed, rotation and tilt, of the tool on the superplasticity of the conventional high-strength AA7075 alloy. At 1500 rpm, 31.5 mm/min and 2 °, under conditions of low heating in the FSP zone, a defect-free material is obtained and the alloy exhibits superplasticity at a low temperature of 400 °C. Authors of [42] investigated the influence of the processing parameters of the superplasticity of Al–4.5%Mg–0.35%Sc–0.15%Zr alloy. Comparing to a simple thermomechanical treatment, the stress reduces by 2-2.5 times and elongations increase by 2.5-7 times at low temperatures of 350-500 °C and a comparatively high strain rate of 10^{-2} s^{-1}. In recent years, there has been a growing number of scientific papers aimed to improving the application of the FSP method for obtaining semi-finished products with required characteristics of different alloys. In this work, an AA5083-type Al-Mg-Mn-Cr alloy with an increase Mn content (similar to Alnovi-U [43]) was investigated in a purpose to study the FSP influence on the grain structure and superplastic properties.

Materials and methods

For preparing the investigated alloy, technically pure elements 99.85% purity Al and 99.5% purity Mg were used. The other elements were added in the form of Al-master alloys (Al-10wt% Mn and Al-10wt% Cr). The studied alloy was casted at a temperature of 750 ± 20 °C by using a Nabertherm electric furnace. The melt was poured into a water-cooled copper mold with inner dimensions of 15×120×120 mm and casting cooling rate of ~15 K/s. After solidification, the ingot was annealed in two steps, the first step was performed at a temperature of 360 °C for 8 h and the second step was at a temperature of 420 °C for 3 h. The homogenized ingot was hot rolled from an initial thickness of 15 mm to 7 mm in a temperature of 420±20 °C.

Table 1. Chemical composition of investigated alloy (wt%).

Mg	Mn	Cr	Si	Fe	Al
4.85	1.05	0.14	0.05	0.10	Balance

A friction stir processing was performed on the surface of the rolled sheets using an automated milling machine (Bridgeport, Elmira, NY, USA). The FSP was carried out at 40 mm/min. The FSP tool was made of hot work tool steel (W302), and its design can be found in ref [44].

The FSPed samples of alloy were divided into two groups: the first group was cold rolled with a reduction of 75%, from 4 mm (plate was cut from the FSP volume) to 1 mm thickness, along the FSP direction (abbreviated by X). The second group was sliced by 1 mm in thickness sheets (abbreviated by F). Figure 1 shows a schematic diagram for these samples production. The reference treatment included hot rolling and cold rolling to 1 mm (75% reduction) without FSP (abbreviated S). Annealing in a temperature range of 400-550°C for 20 min was performed in order to analyze grain structure before the start of superplastic deformation.

The microstructure was analyzed after casting, two-step homogenization, hot rolling, FSP, cold rolling with final recrystallization annealing for 20 min in a range of 400-550°C. An Axiovert 200 MMat light microscopy (LM, Carl Zeiss Oberkochen, Germany) and a Tescan-VEGA3 scanning electron microscopy (SEM, Tescan Brno s.r.o., Kohoutovice, Czech Republic) equipped with energy dispersive spectrometr (EDS, X-MAX80, Oxford Instruments plc, Abingdon, UK) system were used. The samples were mechanically grinded using SiC papers in a range of 230 to 4000 grits, and then were polished in a solution of colloidal silica-based OPS suspension (Struers) with 20% water. The samples were anodized at a voltage of 20 V for 20 s in a 10% water solution of the H_3BO_4 saturated in the HF and then were captured using polarized light to study grain structure.

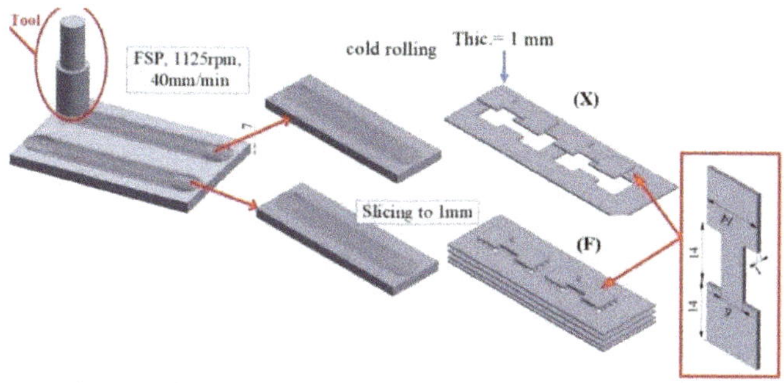

Fig. 1. Schematic diagram for sheets production methods; F is for as-FSP, and X is for FSP with subsequent cold rolling.

Results and discussion

After solidification, the dendrite liquation of Mg at the periphery of the (Al)-solid solution dendritic cells was observed (Fig. 2a). Dark areas enriched with Mg may belong to both Mg segregation and precipitates of non-equilibrium Al_3Mg_2 phase. Bright particles enriched with Mn, and also contained minor Fe and Cr, their size varied between 0.2-5 μm. After homogenization Mg distribution became homogeneous, that confirmed the elimination of the dendrite liquation and dissolution of the Mg-enriched phase (Fig.2d). Mn-bearing particles exhibited the same shape and similar size after homogenization annealing. After hot rolling the particles size decreased to 0.2-2 μm. FSP led further refinement, the size of Mn-bearing particles of solidification origin was varied in a range of 0.1-1.0 μm.

Homogenization annealing of as-cast ingot led to precipitation of elongated and equiaxed shaped dispersoids (Fig. 3a). The precipitates may belong to Al_6Mn and $Al_{18}Mg_3Cr_2$ phases, where Mn and Cr may be partially substituted by Cr, Mn, Fe [45]-[48]. The precipitates of Al_4Mn phase [49] or metastable I-phase [50]-[53] are also possible. The size of precipitates was varied in a range of 30 to 350 nm after homogenization. Hot rolling and subsequent FSP refined dispersoids to 20-140 nm. Precipitates become homogeneously distributed in matrix and their number density increased after FSP compare to hot rolled state (Fig.2c,f).. It is notable, that dispersoids demonstrated the same size after multidirectional forging with cumulative strain of $\sum e=10.5$ for alloy of a similar composition [54]. Fine Al_6Mn dispersoids help to pin grain boundaries and provide stable grain size [25][55].

The morphology of the as-cast grains was characterized by a dendritic structure (Fig. 4a). The average grain size was 150±30 μm. Before the FSP, the microstructure of the hot-rolled alloy (base metal, BM) was non-recrystallized and it was characterized by the large grains with a liner size in a range of 20-300 μm (Fig. 4b) After FSP, grain size was significantly decreased (Fig. 4c,d). The stir zone exhibited a fine-grained structure with a mean size of 4.0±0.3 μm (Fig. 4d).

Materials Research Proceedings **32** (2023) 173-180 https://doi.org/10.21741/9781644902615-19

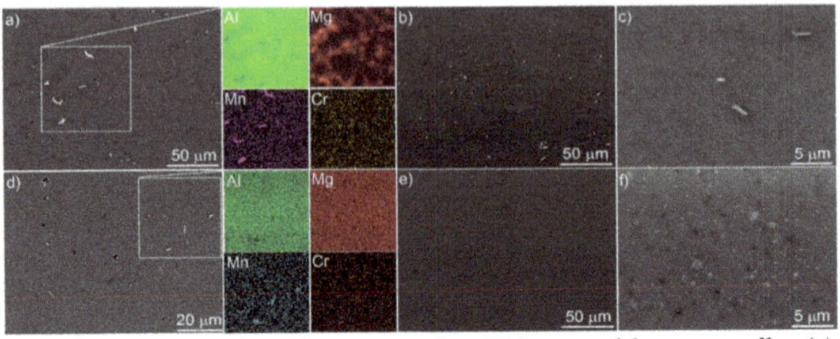

Fig. 2. The SEM micrographs with corresponding EDS maps of the as-cast alloy (a), after homogenization annealing (d), after hot rolling (b,c) and after FSP (e,f) for the Al-Mg-Mn-Cr alloy.

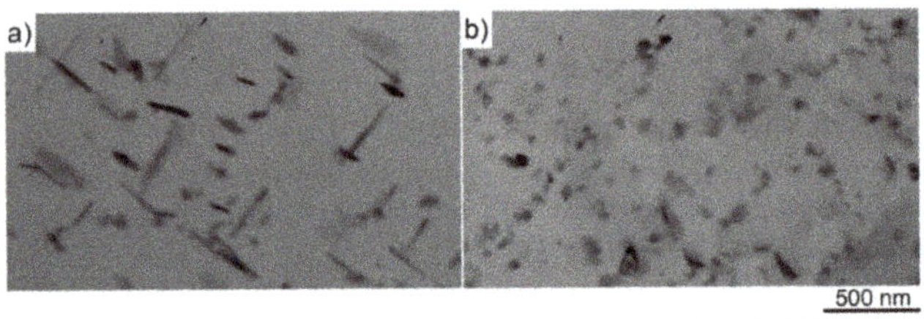

Fig.3. TEM micrographs of the alloy after homogenization (a) and FSP in a stir zone (b)

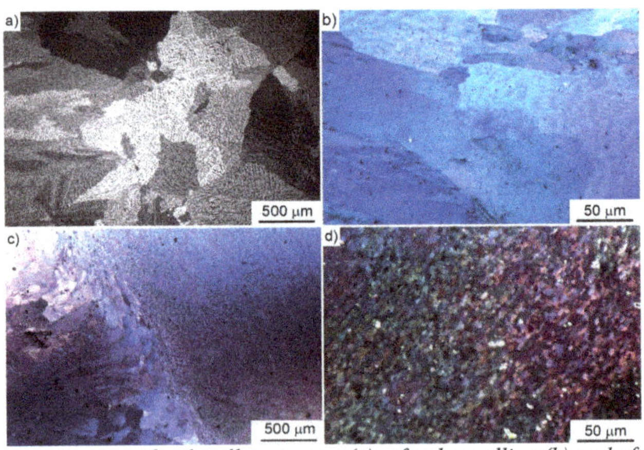

Fig. 4. The microstructures for the alloy: as-cast (a), after hot rolling (b) and after FSP (c,d).

Figure 5 shows the grain structure after annealing for 20 min for the alloys with different treatments. For FSP alloy (regime F), equiaxed grains with a mean grain size of 4.8±0.4 µm were formed at 400 °C (Fig. 5a) and with a mean grain size of 12.5±0.8 µm at 500 °C (Fig. 5c). The sheets treated by X regime with FSP and cold rolling demonstrated grain size of 7.1±0.6 µm at 400 °C (Fig. 5e) and 7.6±0.5 µm at 500 °C (Fig. 5g) and equiaxed grain structure. The alloy after a simple thermomechanical treatment included hot and cold rolling (reference regime S)

 Materials Research Forum LLC
https://doi.org/10.21741/9781644902615-19

demonstrated a mean grain size of 11±1 μm after annealing at both 450 and 500 °C (Fig. 5j,k). Meanwhile, more elongated grains were formed after simple treatment 'S'. The grain aspect ratio was ~1 for F and X treatment and varied between 0.3 to 0.6 for 'S' treatment.

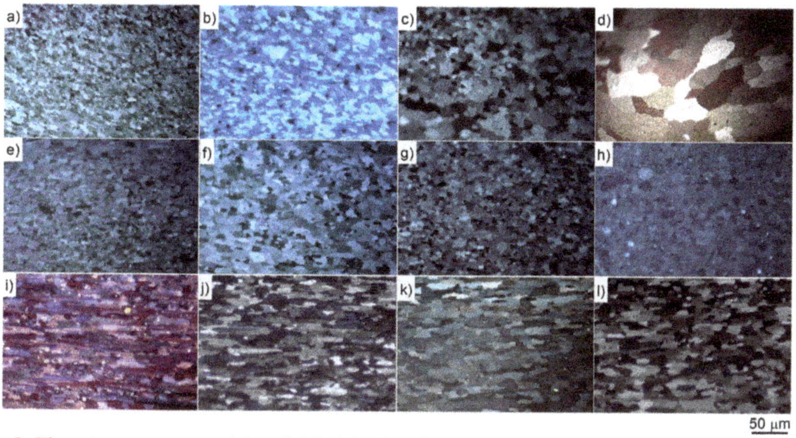

Fig. 5. The microstructure of the Al-Mg-Mn-Cr alloy after annealing at 400 °C (a,e,i); 450 °C (b,f,j), 500 °C (c,g,k) and 550 °C (d,h,l) treated by hot rolling and FSP (F regime) (a-d), hot rolling, FSP and cold rolling (X regime) (e-h), hot rolling and cold rolling (S regime) (i-l).

Figure 6a shows the stress *vs.* strain rate and strain rate sensitivity *m*-coefficient *vs.* strain rate dependences for the studied alloy after different treatment regimes, F, X, and S, at the temperatures of 450 °C and 500 °C. The curves demonstrated sigmoidal shape that is usual for superplastic materials. The maximum *m*-values were corresponded to 1×10^{-3} s^{-1} for the samples processed by the X and S regimes, and near 1×10^{-2} s^{-1} for the FSP mode, the F regime. The constant strain rate tests were perfumed at both strain rates 1×10^{-3} s^{-1} and 1×10^{-2} s^{-1} (Fig. 6b). Stress level decreased with a decrease in a strain rate and an increase in deformation temperature. The F treated and S treated alloys demonstrated small elongations of 160-250% at both strain rates studied. Regime X with FSP and cold rolling demonstrated similar values of 250% at 450° C. Higher temperature of 500°C provided 330% of elongation at 1×10^{-2} s^{-1} and 450% of elongation to failure at 1×10^{-3} s^{-1}. Small elongations of FSP treated F-type alloy is the result of weak grain size stability, meanwhile, rolling after FSP provided a stable fine-grained structure and good superplastic properties.
Thus, FSP as an intermediate processing operation provided a fine grained equiaxed structure and good superplasticity. An equiaxed grained structure and high grain size stability of the FSP treated and cold rolled samples and the resulted large elongations can be the result of a high fraction of fine dispersoids provided a high Zener drag force, uniform distribution and fine size of the solidification-originated particles.

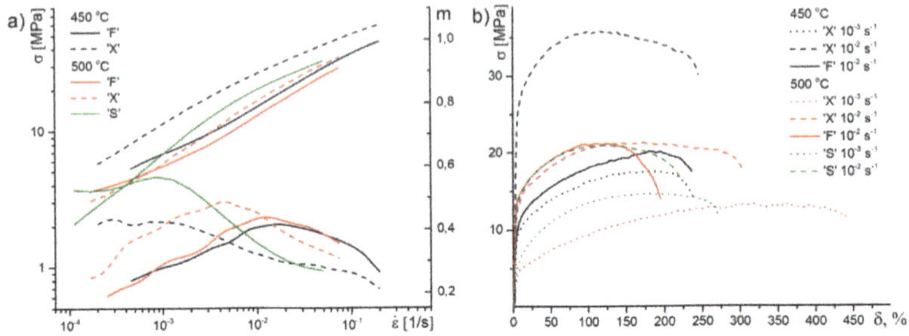

Fig. 6. Strain rate dependences of the flow stress and parameter m for a temperature of 450 °C and 500 °C (a) and dependences of the flow stress on deformation at strain rate of 1×10^{-2} s^{-1} for the investigated alloy for different technologies.

Summary

Three treatment regimes including (F) hot rolling and FSP, (X) hot rolling, FSP and cold rolling, and (S) hot rolling and cold rolling were compared for Al4.8Mg-1Mn-0.1Cr alloy.

Friction stir processing refined phases of solidification origin and dispersoids and provided an eqiaxed recrystallized fine-grained structure with a mean grain size of ~4 μm in a stir zone. Increasing an annealing temperature for the FSP-processed samples in arrange of 400-550°C led a significant grain growth from 4.8 μm to 40 μm, with abnormal grain growth at 550°C. A weak grain structure stability was a reason of low superplastic properties. A simple treatment 'S' with hot and cold rolling forms an elongated grains with a mean longitudinal size of 10.9 μm and aspect ratio of 0.6 after annealing at the same temperature range. An elongated grains of the S-treated alloy also lead a weak superplasticity. Oppositely, the treatment X including hot rolling, FSP and cold rolling provided a fine-grained equiaxed structure after annealing in a temperature range of 400-550 °C with a mean size of 7.1-8.5 μm. This treatment provide a good superplasticity at 500°C with m ~0.5 and 330-450% of elongation to failure at strain rates of 1×10^{-3} - 1×10^{-2} s^{-1}.

Acknowledgment

The study was funded by the RSF Grant # 22-79-00215.

References

[1] A.H. Chokshi, Adv. Eng. Mater. 2020, 22, 1900748. https://doi.org/10.1002/adem.201900748
[2] A.K. Basak, A. Pramanik, C. Prakash, S. Shankar, S.S. Sehgal, Mater. Today Commun. 2023, 35. https://doi.org/10.1016/j.mtcomm.2023.105830
[3] H. Masuda, T. Kanazawa, H. Tobe, E. Sato, Scr. Mater. 2018, 149, 84. https://doi.org/10.1016/j.scriptamat.2018.02.021
[4] H. Masuda, E. Sato, Acta Mater. 2020, 197, 235. https://doi.org/10.1016/j.actamat.2020.07.042
[5] T.G. Langdon, J. Mater. Sci. 2007, 42, 3388. https://doi.org/10.1007/s10853-006-1475-8
[6] E. V. Bobruk, P.D. Dolzhenko, M.Y. Murashkin, R.Z. Valiev, N.A. Enikeev, Materials (Basel). 2022, 15, 6983. https://doi.org/10.3390/ma15196983
[7] E. V. Bobruk, M.Y. Murashkin, I.A. Ramazanov, V.U. Kazykhanov, R.Z. Valiev, Materials (Basel). 2023, 16, 727. https://doi.org/10.3390/ma16020727
[8] M. Kawasaki, T.G. Langdon, Mater. Trans. 2019, 60, 1123. https://doi.org/10.2320/matertrans.MF201915
[9] R.B. Figueiredo, M. Kawasaki, T.G. Langdon, Prog. Mater. Sci. 2023, 137, 101131. https://doi.org/10.1016/j.pmatsci.2023.101131

[10] F.A. Mohamed, Adv. Eng. Mater. 2020, 22, 1900532. https://doi.org/10.1002/adem.201900532
[11] L. Bhatta, A. Pesin, A.P. Zhilyaev, P. Tandon, C. Kong, H. Yu, Metals (Basel). 2020, 10, 77. https://doi.org/10.3390/met10010077
[12] T.G. Nieh, J. Wadsworth, O.D. Sherby, O.D.S.T.G.N. J. Wadsworth, Superplasticity in Metals and Ceramics, Cambridge University Press, New York 1997. https://doi.org/10.1017/CBO9780511525230
[13] T.G. Langdon, Acta Metall. Mater. 1994, 42, 2437. https://doi.org/10.1016/0956-7151(94)90322-0
[14] K. Higashi, Mater. Sci. Technol. 2000, 16, 1320. https://doi.org/10.1179/026708300101507172
[15] T.G. Langdon, Mater. Trans. JIM 1999, 40, 716. https://doi.org/10.2320/matertrans1989.40.716
[16] V. Patel, W. Li, A. Vairis, V. Badheka, Crit. Rev. Solid State Mater. Sci. 2019, 44, 378. https://doi.org/10.1080/10408436.2018.1490251
[17] R.S. Mishra, Z.Y. Ma, Mater. Sci. Eng. R Reports 2005, 50, 1. https://doi.org/10.1016/j.mser.2005.07.001
[18] D. Lohwasser, Z. Chen, in Frict. Stir Weld., Elsevier2010, 1. https://doi.org/10.1533/9781845697716.1
[19] Z.Y. Ma, R.S. Mishra, F.C. Liu, Mater. Sci. Eng. A 2009, 505, 70. https://doi.org/10.1016/j.msea.2008.11.016
[20] K.M. Mehta, V.J. Badheka, Wear 2023, 522, 204719. https://doi.org/10.1016/j.wear.2023.204719
[21] Z. Wei, A. Kulandaivel, T. Hermanto, R. Vaira Vignesh, S. Mehrez, M. Paidar, A. Mohd Zain, V. Mohanavel, Mater. Lett. 2023, 346, 134532. https://doi.org/10.1016/j.matlet.2023.134532
[22] Shalok Bharti, N.D. Ghetiya, K.M. Patel, Phys. Met. Metallogr. 2022, 123, 1387. https://doi.org/10.1134/S0031918X21100586
[23] S. Malopheyev, S. Mironov, R. Kaibyshev, Mater. Charact. 2023, 200, 112909. https://doi.org/10.1016/j.matchar.2023.112909
[24] S. Mironov, Y.S. Sato, H. Kokawa, Acta Mater. 2008, 56, 2602. https://doi.org/10.1016/j.actamat.2008.01.040
[25] I. Vysotskii, S. Malopheyev, S. Mironov, R. Kaibyshev, Mater. Charact. 2022, 185, 111758. https://doi.org/10.1016/j.matchar.2022.111758
[26] H. Mirzadeh, Mater. Sci. Eng. A 2021, 819, 141499. https://doi.org/10.1016/j.msea.2021.141499
[27] Y.B. Sun, X.P. Chen, J. Xie, C. Wang, Y.F. An, Q. Liu, Mater. Today Commun. 2022, 33, 104217. https://doi.org/10.1016/j.mtcomm.2022.104217
[28] F. Cao, C. Xiang, S. Kong, N. Guo, H. Shang, Materials (Basel). 2023, 16, 2345. https://doi.org/10.3390/ma16062345
[29] A. Orozco-Caballero, M. Álvarez-Leal, F. Carreño, O.A. Ruano, Metals (Basel). 2022, 12, 1880. https://doi.org/10.3390/met12111880
[30] J. He, Y. Hu, Y. Sun, W. Li, G. Luo, Mater. Res. Express 2022, 9, 096509. https://doi.org/10.1088/2053-1591/ac8f20
[31] X.C. Luo, D.T. Zhang, G.H. Cao, C. Qiu, D.L. Chen, Mater. Sci. Eng. A 2019, 759, 234. https://doi.org/10.1016/j.msea.2019.05.050

[32] I. Charit, R.S. Mishra, Mater. Sci. Eng. A 2003, 359, 290. https://doi.org/10.1016/S0921-5093(03)00367-8
[33] I. Charit, R.S. Mishra, Acta Mater. 2005, 53, 4211. https://doi.org/10.1016/j.actamat.2005.05.021
[34] Z.. Ma, R.. Mishra, M.. Mahoney, Acta Mater. 2002, 50, 4419. https://doi.org/10.1016/S1359-6454(02)00278-1
[35] F.C. Liu, B.L. Xiao, K. Wang, Z.Y. Ma, Mater. Sci. Eng. A 2010, 527, 4191. https://doi.org/10.1016/j.msea.2010.03.065
[36] A. Denquin, D. Allehaux, M.H. Campagnac, G. Lapasset, Mater. Sci. Forum 2003, 426-432, 2921. https://doi.org/10.4028/www.scientific.net/MSF.426-432.2921
[37] B. Yang, J. Yan, M.A. Sutton, A.P. Reynolds, Mater. Sci. Eng. A 2004, 364, 55. https://doi.org/10.1016/S0921-5093(03)00532-X
[38] H.G. Salem, Scr. Mater. 2003, 49, 1103. https://doi.org/10.1016/j.scriptamat.2003.08.010
[39] V. V. Patel, V. Badheka, A. Kumar, Mater. Manuf. Process. 2016, 31, 1573. https://doi.org/10.1080/10426914.2015.1103868
[40] A. Smolej, D. Klobčar, B. Skaza, A. Nagode, E. Slaček, V. Dragojević, S. Smolej, Mater. Sci. Eng. A 2014, 590, 239. https://doi.org/10.1016/j.msea.2013.10.027
[41] A. Jamali, S.E. Mirsalehi, CIRP J. Manuf. Sci. Technol. 2022, 37, 55. https://doi.org/10.1016/j.cirpj.2021.12.008
[42] S. Pradeep, V. Pancholi, Metall. Mater. Trans. A 2014, 45, 6207. https://doi.org/10.1007/s11661-014-2573-x
[43] T. Kudo, A. Goto, K. Saito, in Mater. Sci. Forum, 2013, 271. https://doi.org/10.4028/www.scientific.net/MSF.735.271
[44] A.O. Mosleh, O.A. Yakovtseva, A.A. Kishchik, A.D. Kotov, E.B. Moustafa, A. V. Mikhaylovskaya, JOM 2023.
[45] O. Engler, S. Miller-Jupp, J. Alloys Compd. 2016, 689, 998. https://doi.org/10.1016/j.jallcom.2016.08.070
[46] B. Yang, M. Gao, Y. Liu, S. Pan, S. Meng, Y. Fu, R. Guan, Mater. Sci. Eng. A 2023, 872, 144952. https://doi.org/10.1016/j.msea.2023.144952
[47] S. Pan, Z. Wang, C. Li, D. Wan, X. Chen, K. Chen, Y. Li, Mater. Des. 2023, 226, 111647. https://doi.org/10.1016/j.matdes.2023.111647
[48] M. Mofarrehi, M. Javidani, X.-G. Chen, Mater. Sci. Eng. A 2022, 845, 143217. https://doi.org/10.1016/j.msea.2022.143217
[49] A.Y. Algendy, K. Liu, X.G. Chen, Mater. Charact. 2021, 181, 111487. https://doi.org/10.1016/j.matchar.2021.111487
[50] A.G. Mochugovskiy, N.Y. Tabachkova, M.E. Ghayoumabadi, V.V. Cheverikin, A.V. Mikhaylovskaya, J. Mater. Sci. Technol. 2021, 87, 196. https://doi.org/10.1016/j.jmst.2021.01.055
[51] A.G. Mochugovskiy, N.Y. Tabachkova, A.V. Mikhaylovskaya, Mater. Lett. 2021, 284, 128945. https://doi.org/10.1016/j.matlet.2020.128945
[52] A. Mochugovskiy, N. Tabachkova, A. Mikhaylovskaya, Mater. Lett. 2019, 247, 200. https://doi.org/10.1016/j.matlet.2019.03.126
[53] F. Zupanič, T. Bončina, Solid State Phenom. 2022, 327, 26. https://doi.org/10.4028/www.scientific.net/SSP.327.26
[54] A.A. Kishchik, M.S. Kishchik, A.D. Kotov, A.V. Mikhaylovskaya, Phys. Met. Metallogr. 2020, 121, 489. https://doi.org/10.1134/S0031918X20050075
[55] I. Nikulin, A. Kipelova, S. Malopheyev, R. Kaibyshev, Acta Mater. 2012, 60, 487. https://doi.org/10.1016/j.actamat.2011.10.023

Effect of Co addition on the microstructure evolution and superplastic behavior of Ti-4Al-3Mo-1V-0.1B alloy

Anton D. Kotov[1,a *], Maria N. Postnikova[1,b], Ahmed O. Mosleh[2,c] and Anastasia V. Mikhaylovskaya[1,d]

[1] Department of Physical Metallurgy of Non-Ferrous Metals, National University of Sciences and Technology "MISIS", 4 Leninskiy ave. 4, 119049 Moscow, Russia

[2] Department of Mechanical Engineering, Shoubra Faculty of Engineering, Benha University, Cairo 11629, Egypt

[a,] kotov@misis.ru, [b] sitkina.m@misis.ru, [c] ahmed.omar@feng.bu.edu.eg, [d]mihaylovskaya@misis.ru

Keywords: Titanium Alloys, Superplasticity, Microstructural Evolution, Dynamic Grain Growth; Flow Stress

Abstract. Temperature reduction of superplastic forming of titanium alloys is currently a significant issue. The present study focused on the effect of modifying the Ti-4Al-3Mo-1V-0.1B alloy with 0.5-2 wt% Co additions on the superplastic behavior and microstructure evolution. The results demonstrated that Co alloying promoted the formation of recrystallized and globular microstructure before the beginning of superplastic deformation due to the acceleration of diffusivity by Co in comparison with the Co-free alloy. The diffusivity acceleration also led to dynamic grain growth during superplastic deformation but promoted a stable superplastic flow of alloys with 0.5-2% Co at temperatures of 700-750 °C. This enhanced the strain rate sensitivity coefficient m from 0.35-0.4 to 0.5-0.65 and the elongation to failure from 200-350% to 500-1000% compared to the Co-free alloy. The 2 wt% Co alloying provided excellent low-temperature superplasticity in the temperature range of 625-775 °C with a coefficient m of 0.5-0.65 and elongation to failure of 800-1000% at a constant strain rate of 1×10^{-3} s^{-1}.

Introduction

Titanium alloys' exceptional mechanical and physical properties make them popular in numerous fields. But the forming process of parts from titanium alloys using traditional methods at low temperatures is a time-consuming and energy-consuming process due to their high strength and relatively low elastic modulus [1,2]. Complex-shaped alloy components can be formed using superplastic forming (SPF) at low gas pressure in a single technological process. The implementation of the SPF method is constrained by a requirement for high temperatures of 800-1000 °C to produce suitable superplasticity performance of titanium alloys [3]. Reducing the forming temperature decreases cycle time, die material cost and wear, and energy consumption, increases die life, and inhibits the formation of an alpha-oxide layer on the surface of manufactured parts [2,4]. There are several effective methods for reducing the superplastic forming temperature of titanium alloys, including grain structure refinement before superplastic forming [5–8] and modification of the alloy chemical composition by alloying with highly diffusive elements [9,10]. Elements with high diffusivity in titanium, such as Fe, Ni, and Co, are of particular interest in this regard [10–13]. Previous studies indicated that alloying with highly diffusive β-stabilizers, such as Fe and Ni, is efficient for obtaining low-temperature superplasticity [10,14–17]. Thus, the study of the effect of Co modification on the superplasticity and microstructure of titanium alloys is an actual issue.

The conventional alloy Ti-4Al-3Mo-1Vdue is widely used in manufacturing applications due to its high strength and corrosion resistance and good creep resistance. The alloy exhibits a fine-grained structure with simple thermomechanical processing that provides excellent superplastic behavior but at high temperatures of 825-875 °C [18]. The purpose of this study is to the influence of 0.5% to 2wt% Co on the microstructure and superplastic behavior of the Ti-4Al-3Mo-1V alloy to reduce the superplastic deformation temperature. In previous studies, it was found that the trace boron addition accelerates recrystallization and globularization of the microstructure [19] which provides a stable flow and reduces the flow stresses at the initial stage of superplastic deformation [20]. In this regard, the investigated alloys were also alloyed with 0.1wt% trace boron addition.

Materials and Methods

Table 1 provides the chemical compositions of the alloys studied. (SEM-EDS data for Al, Mo, V, and Co and nominal for B content). For the preparation of the alloys, technically pure metals of Al (99.99 wt.%), V (99.95 wt.%), Co (99.95 wt.%), and Ti-50wt% Mo master alloy were used. Arc 200 vacuum arc melting crucible furnace with water-cooled copper mold was used to melt and cast 100g ingots with a dimension of $50\times40\times10$ mm^3.

Table 1. Chemical composition of the investigated alloys (wt.%).

Alloy	Al	Mo	V	Co	B	Ti
0Co	3.7	3.0	1.2	0.0	0.1	Bal.
0.5Co	4.1	2.8	0.9	0.5	0.1	Bal.
1Co	4.4	3.1	1.0	1.0	0.1	Bal.
2Co	3.8	3.3	1.1	2.0	0.1	Bal.

The ingots were subjected to a heat treatment process that included homogenization annealing for one hour at 800 °C and subsequent annealing for 0.5 hours in a β-phase field with subsequent water cooling. The quenched ingots were subjected to hot rolling at 750 ± 10 °C in the two-phase ($\alpha+\beta$) field with a total reduction of 90% (from 10 to 1 mm).

The scanning electron microscope (SEM) Tescan Vega 3-LMH with an X-Max 80 energy dispersive detector was used to analyze the microstructure, chemical, and phase composition of alloys. Samples for SEM analysis were mechanically polished on a Struers LaboPol-5 machine with SiC abrasive papers of 300 to 2400 grit and then polished with a 10% H_2O_2 solution in OP-U silica-based suspension.

Thermo-Calc software (TTTI3 database) was used for the construction of a polythermal section of the phase diagram and the determination of β-transus temperatures. The β-transus values of the alloys were also experimentally determined by differential thermal analysis (DTA, Setaram Labsys).

The uniaxial strain rate tensile tests were performed on a Walter Bai LFM-100 machine in an argon atmosphere furnace at temperatures of 625, 700, and 775 °C. The dog-bone samples with dimensions of $14\times6\times1$ mm^3 were held in the furnace for 0.5 h to provide uniform temperature during the tensile test. The superplastic behavior was evaluated using step-by-step strain rate reduction tests ranging from 5×10^{-2} s^{-1} to 5×10^{-5} s^{-1}, constant strain rate tests at 1×10^{-3} s^{-1}, and tests with a change in strain rate (20% from nominal strain rate with a strain step of 0.2) during deformation by ASTM E2448-11.

Materials Research Forum LLC
https://doi.org/10.21741/9781644902615-20

Results and Discussion

Thermo-Calc calculations and DTA analysis

The Thermo-Calc calculated polythermal section of the multicomponent phase diagram for the Ti-4Al-1V-1Fe-1Ni-0.1B-xMo section (Figure 1a) and the DTA spectra for the alloys studied (Figure 1b) showed similar theoretical 920, 909, 901 and 885 °C and experimental 922, 908, 903 and 886 °C temperatures of β-transus for the 0Co, 0.5Co, 1Co and 2Co alloys, respectively. Therefore, an increase in the Co content from 0 to 2wt% leads to a decrease in the β-transus temperature from 922 to 886 °C.

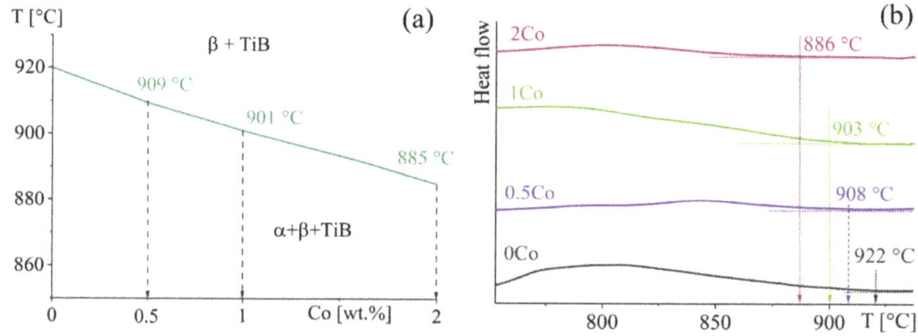

Figure 1. (a) Polythermal section of the Ti-4Al-3Mo-1V-0.1B-xFe diagram constructed by Thermo-Calc software and (b) DTA spectra for the investigated alloys.

Analysis of the microstructure after processing

Figure 2a-c demonstrates the microstructures of the investigated alloys after thermomechanical processing. A similar partially recrystallized microstructure with elongated grains in the rolling direction was observed in all investigated alloys. Table 2 demonstrates the mean grain size measurements both across ($d_{\perp}$) and along ($d_{\parallel}$) the rolling direction and grain's aspect ratio ($d_{\perp}/d_{\parallel}$). Increasing the Co content resulted in insignificant changes in the grain size of the α and β phases; $d_{\parallel}$ ranged from 0.7 to 0.9 μm for the α phase and from 0.8 to 1.0 μm for the β phase, $d_{\perp}$ ranged from 0.4 to 0.5 μm for the α phase and from 0.3 to 0.4 μm for the β phase, respectively. Moreover, when the cobalt content increased from 0 to 2 wt.%, the grain aspect ratio ($d_{\perp}/d_{\parallel}$) changed insignificantly for both phases; $d_{\perp}/d_{\parallel}$ ranged from 0.5 to 0.7 for the α-phase and from 0.3 to 0.4 for the β-phase. Also, alloying of the 0.5-2.0%Co resulted in an increase in β-phase volume fraction to 23-37% compared with 20% for the Co-free alloy.

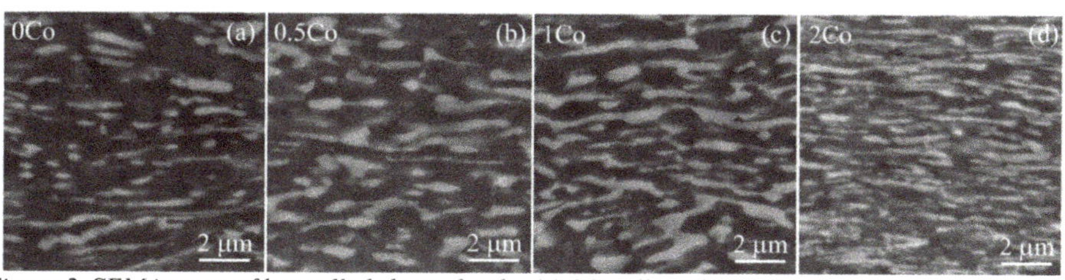

Figure 2. SEM images of hot-rolled sheets for the (a) 0Co alloy, (b) 0.5Co alloy, (c) 1Co alloy, and (d) 2Co alloy.

Table 2. Mean grain size along ($d_{\parallel}$) and across ($d_{\perp}$) the hot deformation direction after processing of the investigated alloys.

Alloy	α [мкм]			β [мкм]		
	$d_{\parallel}$	$d_{\perp}$	$d_{\perp}/d_{\parallel}$	$d_{\parallel}$	$d_{\perp}$	$d_{\perp}/d_{\parallel}$
0Co	0.9 ± 0.1	0.4 ± 0.1	0.5	1.0 ± 0.1	0.3 ± 0.1	0.3
0.5Co	0.7 ± 0.1	0.5 ± 0.1	0.7	0.8 ± 0.2	0.4 ± 0.1	0.4
1Co	0.8 ± 0.1	0.5 ± 0.1	0.6	0.9 ± 0.1	0.3 ± 0.1	0.3
2Co	0.9 ± 0.1	0.4 ± 0.1	0.5	1.0 ± 0.1	0.3 ± 0.1	0.3

Superplastic deformation behavior and microstructure evolution

Figure 3a-c demonstrates the results of the step-by-step decreasing strain rate test at different temperatures. Co alloying resulted in a decrease in flow stress and an increase in the strain rate sensitivity coefficient (*m*) compared to the base 0Co alloy. The base alloy exhibited maximum *m* values of 0.4-0.43 at 775 °C in the range of strain rates 1×10^{-4} - 4×10^{-3} s^{-1}. The 0.5-2% Co alloying increases the coefficient *m* to 0.5-0.65 in the same strain rate range. The 0.5Co,1Co, and 2Co alloys exhibited high *m* values of 0.5-0.6 even at a low temperature of 700 °C. The 2Co alloy demonstrated high *m* values of 0.50-0.53 even at 625 °C in the strain rate range of 5×10^{-5} - 1×10^{-3} s^{-1}.

The strain rate of 1×10^{-3} s^{-1} corresponded to a maximum *m* of 0.55-0.65 at 775 °C and high *m* values of 0.40-0.50 at temperatures of 775-625 °C for Co-containing alloys. Thus, the alloys studied were tested by uniaxial tensile to failure at a constant strain rate of 1×10^{-3} s^{-1} at temperatures of 625, 700, and 775 °C (Figure 3d-f, Figure 4). The alloying with 0.5-2% Co resulted in a decrease in flow stress throughout the investigated temperature range and an increase in elongation to failure from 200-350% to 500-1000% at 700-775 °C. The maximum difference in superplastic behavior was obtained by alloying with 2%Co as compared to the Co-free base alloy at all temperature ranges of 625-775 °C; the flow stress decreased by 2-3 times and the elongation to failure increased to 900-1000% (Figure 5a).

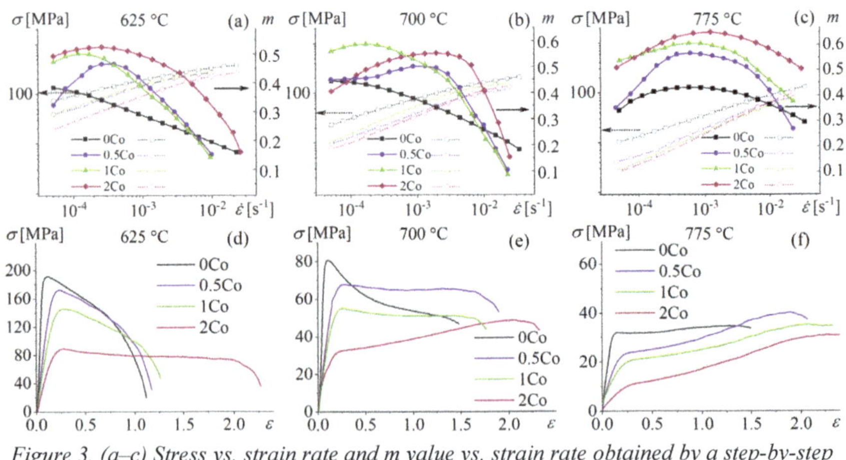

Figure 3. (a–c) Stress vs. strain rate and m value vs. strain rate obtained by a step-by-step decrease in the strain rate and (d–f) stress–strain curves at a constant strain rate of 1×10^{-3} s^{-1} at (a,d) 625 °C, (b,c) 700 °C and (c,f) 775 °C.

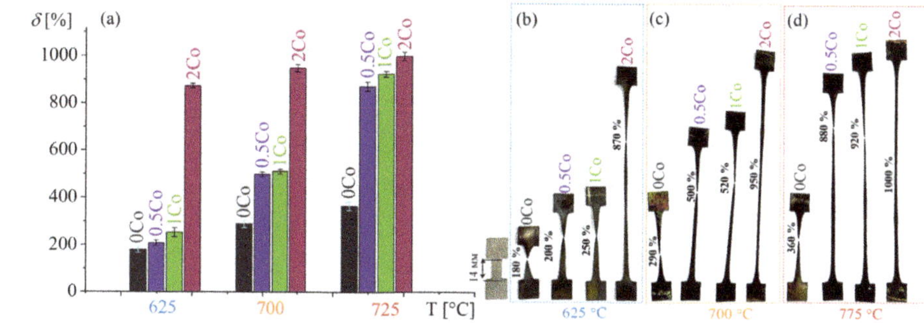

Figure 4. Dependency of elongation to failure (δ) at a constant strain rate of 1 × 10⁻³ s⁻¹ (a) and specimen images after SPD tests at a strain rate of 1 × 10⁻³ s⁻¹ at (a) 625 °C, (b) 700 °C and (c) 775 °C.

The superplastic behavior is strongly influenced by the microstructure evolution during deformation. In this regard, the strain-induced evolution of m with a 20% change in strain rate during deformation to failure was studied by testing at a constant nominal strain rate of 1×10^{-3} s^{-1}. Co-containing alloys exhibit high values of $m \geq 0.5$ at deformation up to 1-1.5 true strain at temperatures of 775 and 700 °C compared to the reference 0Co alloy (m=0.35-0.40). The m-value decreases to 0.4 for true strain exceeding 1-1.5 presumably due to the acceleration of dynamic grain growth. At the same time, an alloy with 2% Co exhibits m values of 0.4-0.5 with a stable superplastic flow up to a strain of 1.8 (Figure 5a).

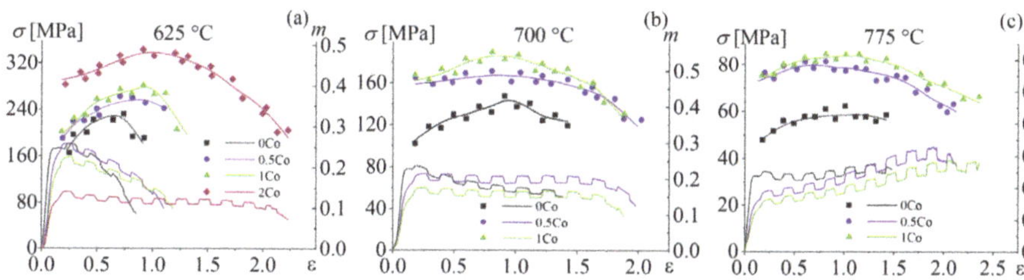

Figure 5. Stress-strain curves and strain-induced evolution of coefficient m at the tests with a 20% change in strain rate according to ASTM standard at (a) 625 °C, (b) 700 °C, and (c) 775 °C at the nominal strain rate of 1×10^{-3} s^{-1}.

The strain-induced microstructural evolution was analyzed at 700 °C for the Co-free base alloy and the alloy with 2% Co with the most difference in deformation behavior. Figure 6 provides microstructures of 0Co and 2Co alloys before (annealing for 30 min) and after superplastic deformation to failure. After annealing at 700 °C the 0Co alloy exhibits a fine grain structure with uncrystallized areas with a 13% β-phase volume fraction and a grain size of 0.6 ± 0.2 μm for the α-phase and 0.8 ± 0.2 μm for the β-phase. The 2Co alloy exhibits a more globular microstructure with a β-phase volume fraction of 27% and a grain size of 1.0 ± 0.2 μm for the α-phase and 1.1 ± 0.2 μm for the β-phase. Superplastic deformation leads to dynamic grain growth to 1.0 ± 0.1 μm for α-phase and to 1.2 ± 0.1 for β-phase in the Co-free alloy and to 1.4 ± 0.2 μm for α-phase and to 2.0 ± 0.2 μm for β-phase in the 2Co alloy.

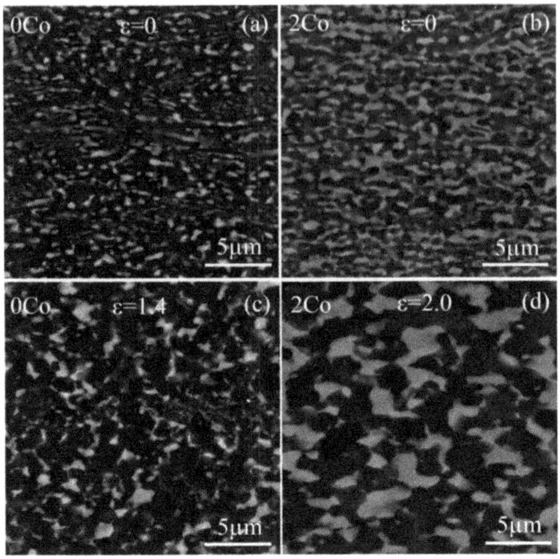

Figure 6. Microstructures of (a,c) 0Co alloy and (b,d) 2Co alloy (a,b) before (annealing for 30 min) and (c,d) after deformation at 700 °C with a strain rate of 1×10^{-3} s^{-1}.

As a result, the 2%Co alloying significantly improved the superplasticity of the base Ti-4Al-3Mo-1V-0.1B alloy despite the acceleration of dynamic grain growth. This effect can be explained for several reasons. The first reason is that Co is a β-stabilizing element and increases the β-phase volume fraction at deformation temperatures. Increasing the β-phase volume fraction simplifies deformation by its high diffusivity and ductility due to the BCC lattice. It is known that most Ti alloys require 20-50% β-phase to provide a stable superplastic flow with limited grain growth [21–23]. Increasing the β-stabilizer content provides the required β-phase fraction at lower temperatures and consequently provides a better superplastic behavior. The second reason is that Co is an element with high diffusivity in Ti and promotes diffusion-controlled mechanisms of superplastic deformation – grain boundary sliding (GBS) and its accommodation [10,16,24,25]. In addition, recrystallization and fragmentation occurred at low temperatures due to accelerated diffusion for the Co-bearing alloys, which provided the required recrystallized microstructure with equiaxial fine grains before superplastic deformation.

Summary

The effect of 0.5-2 wt.% Co addition on the superplastic behavior and microstructure parameters of Ti-4Al-3Mo-1V-0.1B alloy in the temperature range of 625-775 °C was investigated. The main conclusions are the following:

The addition of 0.5-2%Co significantly improved the superplastic behavior of the Ti-4Al-3Mo-1V-0.1B alloy at 700-775 °C: *m* value increased from 0.35-0.4 to 0.5-0.65, elongation to failure increased from 200-350% to 500-1000% with lower flow stress in Co-bearing alloys. The microstructure evolution analysis revealed an acceleration of recrystallization and fragmentation of the microstructure during annealing at superplastic deformation temperature due to diffusivity acceleration by Co addition. Enhanced diffusivity also led to dynamic grain growth during superplastic deformation but promoted stable superplastic flow due to the facilitation of GBS. As a result, the alloying with 2 wt% Co provides excellent superplastic behavior in the temperature range 625-775 °C at a constant strain rate of 1×10^{-3} s^{-1} with a coefficient *m* value of 0.5-0.65 and elongation to failure of 800-1000%.

Acknowledgment
The study was funded by Russian Science Foundation (Grant #21-79-10380).

References

[1] V.N. Moiseyev, Titanium alloys: Russian aircraft and aerospace applications, CRC Press, Boca Raton, FL, USA, 2005.

[2] M. Peters, C. Leyens, Fabrication of Titanium Alloys, in: C. Leyens, M. Peters (Eds.), Titan. Titan. Alloy., Wiley-VCH Verlag GmbH & Co. KGaA, Weinheim, Germany, 2005: pp. 245–261. https://doi.org/10.1002/3527602119.ch8

[3] N. Ridley, Metals for superplastic forming, in: Superplast. Form. Adv. Met. Mater., Woodhead Publishing, Sawston, UK, 2011: pp. 3–33. https://doi.org/10.1533/9780857092779.1.3

[4] M. Jackson, Superplastic forming and diffusion bonding of titanium alloys, in: Superplast. Form. Adv. Met. Mater., Woodhead Publishing Limited, Cambridge, United Kingdom, 2011: pp. 227–246. https://doi.org/10.1533/9780857092779.3.227

[5] T.G. Langdon, Seventy-five years of superplasticity: Historic developments and new opportunities, J. Mater. Sci. 44 (2009) 5998–6010. https://doi.org/10.1007/s10853-009-3780-5

[6] S.V. Zherebtsov, E.A. Kudryavtsev, G.A. Salishchev, B.B. Straumal, S.L. Semiatin, Microstructure evolution and mechanical behavior of ultrafine Ti 6Al 4V during low-temperature superplastic deformation, Acta Mater. 121 (2016) 152–163. https://doi.org/10.1016/j.actamat.2016.09.003

[7] R.B. Figueiredo, M. Kawasaki, T.G. Langdon, Seventy years of Hall-Petch, ninety years of superplasticity and a generalized approach to the effect of grain size on flow stress, Prog. Mater. Sci. 137 (2023) 101131. https://doi.org/10.1016/j.pmatsci.2023.101131

[8] A.V. Sergueeva, V.V. Stolyarov, R.Z. Valiev, A.K. Mukherjee, Enhanced superplasticity in a Ti-6Al-4V alloy processed by severe plastic deformation, Scr. Mater. 43 (2000) 819–824. https://doi.org/10.1016/S1359-6462(00)00496-6

[9] J.A. Wert, N.E. Paton, Enhanced superplasticity and strength in modified Ti-6AI-4V alloys, Metall. Trans. A. 14 (1983) 2535–2544. https://doi.org/10.1007/BF02668895

[10] E. Alabort, D. Barba, M.R. Shagiev, M.A. Murzinova, R.M. Galeyev, O.R. Valiakhmetov, A.F. Aletdinov, R.C. Reed, Alloys-by-design: Application to titanium alloys for optimal superplasticity, Acta Mater. 178 (2019) 275–287. https://doi.org/10.1016/j.actamat.2019.07.026

[11] J. Koike, Y. Shimoyama, H. Fujii, K. Maruyama, Characterization of superplasticity in Ti-5.5Al-1Fe alloys, Scr. Mater. 39 (1998) 1009–1014. https://doi.org/10.1016/S1359-6462(98)00286-3

[12] B.H. Prada, J. Mukhopadhyay, A.K. Mukherjee, Effect of Strain and Temperature in a Superplastic Ni-Modified Ti-6A1-4V Alloy, Mater. Trans. JIM. 31 (1990) 200–206. https://doi.org/10.2320/matertrans1989.31.200

[13] D. Klimenko, M. Ozerov, S. Suresh, N. Stepanov, M.A. Tikhonovsky, G. Salishchev, S. Zherebtsov, Microstructure Evolution and Properties of Ti-6Al-4V Alloy Doped with Fe and Mo during Deformation at 800°C, Defect Diffus. Forum. 385 (2018) 144–149. https://doi.org/10.4028/www.scientific.net/DDF.385.144

[14] M.L. Meier, D.R. Lesuer, A.K. Mukherjee, The effects of the α/β phase proportion on the superplasticity of Ti-6Al-4V and iron-modified Ti-6Al-4V, Mater. Sci. Eng. A. 154 (1992) 165–173. https://doi.org/10.1016/0921-5093(92)90342-X

[15] A. V. Mikhaylovskaya, A.O. Mosleh, P. Mestre-Rinn, A.D. Kotov, M.N. Sitkina, A.I. Bazlov, D. V. Louzguine-Luzgin, High-Strength Titanium-Based Alloy for Low-Temperature Superplastic Forming, Metall. Mater. Trans. A Phys. Metall. Mater. Sci. 52 (2021) 293–302. https://doi.org/10.1007/s11661-020-06058-8

[16] A.D. Kotov, M.N. Postnikova, A.O. Mosleh, A. V. Mikhaylovskaya, Influence of Fe on the microstructure, superplasticity and room-temperature mechanical properties of Ti–4Al–3Mo–1V-0.1B alloy, Mater. Sci. Eng. A. 845 (2022) 143245. https://doi.org/10.1016/j.msea.2022.143245

[17] A.D. Kotov, M.N. Postnikova, A.O. Mosleh, V. V. Cheverikin, A. V. Mikhaylovskaya, Microstructure and Superplastic Behavior of Ni-Modified Ti-Al-Mo-V Alloys, Metals (Basel). 12 (2022) 741. https://doi.org/10.3390/met12050741

[18] A.D. Kotov, A. V. Mikhailovskaya, A.O. Mosleh, T.P. Pourcelot, A.S. Prosviryakov, V.K. Portnoi, Superplasticity of an Ultrafine-Grained Ti–4% Al–1% V–3% Mo Alloy, Phys. Met. Metallogr. 120 (2019) 60–68. https://doi.org/10.1134/S0031918X18100083

[19] D. Hill, R. Banerjee, D. Huber, J. Tiley, H.L. Fraser, Formation of equiaxed alpha in TiB reinforced Ti alloy composites, Scr. Mater. 52 (2005) 387–392. https://doi.org/10.1016/j.scriptamat.2004.10.019

[20] M.N. Postnikova, A.D. Kotov, A.I. Bazlov, A.O. Mosleh, S. V. Medvedeva, A. V. Mikhaylovskaya, Effect of Boron on the Microstructure, Superplastic Behavior, and Mechanical Properties of Ti-4Al-3Mo-1V Alloy, Materials (Basel). 16 (2023) 3714. https://doi.org/10.3390/ma16103714

[21] M. Jain, M.C. Chaturvedi, N.L. Richards, N.C. Goel, Microstructural characteristics in α phase during superplastic deformation of Ti6Al4V, Mater. Sci. Eng. A. 145 (1991) 205–214. https://doi.org/10.1016/0921-5093(91)90250-Q

[22] H. Imai, G. Yamane, H. Matsumoto, V. Vidal, V. Velay, Superplasticity of metastable ultrafine-grained Ti-6242S alloy: Mechanical flow behavior and microstructural evolution, Mater. Sci. Eng. A. 754 (2019) 569–580. https://doi.org/10.1016/j.msea.2019.03.085

[23] M.L. Meier, D.R. Lesuer, A.K. Mukherjee, α Grain size and β volume fraction aspects of the superplasticity of Ti-6Al-4V, Mater. Sci. Eng. A. 136 (1991) 71–78. https://doi.org/10.1016/0921-5093(91)90442-P

[24] J. Koike, Y. Shimoyama, I. Ohnuma, T. Okamura, R. Kainuma, K. Ishida, K. Maruyama, Stress-induced phase transformation during superplastic deformation in two-phase Ti–Al–Fe alloy, Acta Mater. 48 (2000) 2059–2069. https://doi.org/10.1016/S1359-6454(00)00049-5

[25] J.S. Kim, D.M. Li, C.S. Lee, Alloying effects on superplastic behaviour of Ti-Fe-Al-Ni alloys, Mater. Sci. Technol. 14 (1998) 676–682. https://doi.org/10.1179/mst.1998.14.7.676

Predicting ECAP misorientation evolution and its influence on superplasticity for an Al-Zn-Mg-Cu alloy

Fernando Carreño *

Dpt. Physical Metallurgy, CENIM, CSIC, Av. Gregorio del Amo 8, 28040 Madrid, Spain

carreno@cenim.csic.es

Keywords: Aluminium SPD Grain Refinement, ECAP Misorientation Evolution, New Misorientation Expression, GBS Superplastic Constitutive Equation

Abstract. Extensive research on severe plastic deformation (SPD) of metallic materials has been performed so far in order to improve mechanical properties, both at low and high temperatures. As a result of the extensive grain refinement obtained, increased superplastic behaviour has been attained at higher strains rates and lower temperatures than usual. This is due to finer grain sizes and higher average misorientations obtained, which enhance grain boundary sliding (GBS). However, the misorientation effect on superplasticity, and the prediction of misorientation evolution during SPD has remained so far, qualitative. In this research, especial attention has been given to the quantitative misorientation evolution with increasing equal channel angular pressing (ECAP) deformation, in order to propose an expression useful to describe and predict the influence of the different processing parameters on the misorientation evolution with strain. This is exemplified for ECAPed Al-Zn-Mg-Cu and other aluminium alloys, and it could serve as a basis for predicting the misorientation evolution, and its influence on superplasticity, of other SPD processes and metallic alloys.

Introduction

Widespread research on severe plastic deformation (SPD) of metallic materials has been performed in order to improve mechanical properties, both at low and high temperatures as a result of the extensive grain refinement obtained [1-15]. At low temperatures, high yield and ultimate tensile stresses, as well as higher than usual ductility is found, increasing toughness. At high temperatures, increased superplastic behaviour has been attained at higher strain rates and lower temperatures than usual due to finer grain sizes, L, and higher average misorientations, $\bar{\delta}$, which enhance grain boundary sliding (GBS) mechanism [8-9,11-16]. However, the misorientation effect on superplasticity has not been studied as much as the effect of grain size. In this research, a tool for description, analysis and prediction of the misorientation evolution with increasing applied strain is proposed, which will be useful to predict superplastic behaviour with higher accuracy.

Equal channel angular pressing (ECAP) is a severe plastic deformation processing technique consisting in a die having two channels, of equal cross section, intersecting at an angle, typically 90°, or 120°. It was first described by Segal et al. [1]. In an ECAP, ideally, a billet experiences simple shear without a change in cross-sectional area while it is pressed through, and so, the pressing may be repeated many times. As a consequence, large cumulative plastic strains can be introduced to a billet in multi-pass ECAP processing. This processing leads in metals and alloys to microstructure refinement and the development of ultra-fine and nanoscale grains. Additionally, during multi-pass ECAP the strain-path can be varied by choice of processing route, for instance, route A, B_A, B_C or C, etc., differing in the sense of billet rotation in between successive pressing passes. B_C (rotating 90°) and A (no rotations) are the most common routes.

An interesting feature of ECAP processing is that, for a 90° or 120° ECAP, one pass is approximately equivalent to a true deformation (ε) value of 1 or 0.7, respectively, being necessary at least 12 or 16 passes to obtain highly misoriented microstructures. Therefore, ECAP allows convenient study of microstructure evolution as a function of applied strain. Additionally, ECAP allows proper measuring of processing stress [7-8, 11] and adequate temperature control. Generally, the lower the ECAP temperature, the finer the microstructure and the slower the microstructural evolution towards highly misoriented ultrafine grains. Another factor affecting the microstructure evolution during ECAP processing is the composition of the alloy, which may have precipitates and/or solutes in solid solution.

Although the mechanisms of microstructure evolution are an important subject, they are not in the scope of this article. This work will focus in the description of the misorientation evolution irrespective of the underlying microstructural mechanism. Preferably, it is convenient not to have discontinuous behaviour such as given by discontinuous recrystallization, or abnormal grain growth between ECAP passes. The ideal situation would be that corresponding, approximately, to a continuous dynamic recovery (CDRV), for which all the deformation will be employed in increasing progressively the misorientation of the ultrafine subgrains, which were obtained in the initial passes. It must be noted that this is not the general situation, because usually, certain progressive refinement is occurring even for a high number of passes.

Nevertheless, the objective of this work is to offer an adequate tool for the description, analysis and prediction of the misorientation evolution during increasing number of ECAP passes, i.e., as a function of applied strain.

Proposed tool for misorientation evolution study

For metallic materials being processed by SPD, the misorientation evolution with applied strain can be divided in three stages: i) the initial vast grain subdivision from coarse grain to fine or ultrafine subgrains with low angle grain boundaries, LAGBs (Stage I), ii) the approximately linear increase of average subgrain misorientation, gradually, towards high angle grain boundaries, HAGBs (Stage II), and iii) the final slow and asymptotic behaviour towards a maximum average misorientation (Stage III). The three stages have already been reported by other researchers [5]. This maximum average misorientation, $\bar{\delta}_{max}$, should correspond, ideally, to the average misorientation of a random distribution, $\bar{\delta}_{max} = \bar{\delta}_{random}$, as described by Mackenzie, which is $\bar{\delta}_{random} = 41°$ for cubic metals and alloys [17].

An important idea to highlight in this work is that the final slow misorientation increment with applied deformation is not attributed, in principle, to any recrystallization mechanism, but just the result of being close to a total misorientation randomization. Of course, any possible recrystallization process could accelerate the randomization process or even change the final result with unexpected misorientation distributions. Anyway, the tool proposed below will help in the description and analysis of the underlying mechanisms. In this work, it won't be considered, nor predicted, the initial grain subdivision, just characterized by a "zero point" parameter, designated k0, as it will be shown in the following.

The tool takes into account two facts: i) the initial misorientation increments of the newly formed subgrains behave approximately linear with applied deformation, and ii) the final misorientation increments of the grains are small for large applied strains, trending to an asymptotic value. For these reasons, it is proposed the use of the arctan function ($\bar{\delta} \approx \arctan(f(\varepsilon))$), which meets these features. In this way, we can concentrate our future analysis in the argument, $f(\varepsilon)$, to get insights about the influence of different microstructural and processing parameters, such as grain size, precipitates, solid solution, stacking fault energy (SFE), ECAP geometry (90°, 120°), processing route (A, B_C), temperature (RT, HT), etc.

The arctan function presents an initial positive linear slope from zero value, and asymptotic behaviour for high argument values. With just two parameters, k1 and k2, it can describe quite different Stages II (linear) and III (asymptotic). It can be added another parameter, k0, characterizing a starting point depending on the Stage I (initial subgrain subdivision). This function, in terms of average misorientation vs. applied strain is given by:

$$\bar{\delta} = \bar{\delta}_{random} \left(\frac{2}{\pi}\right) arctan\left(\left(\frac{\pi}{2\,\bar{\delta}_{random}}\right) k1\,(\epsilon - k0)^{k2}\right) \quad (1)$$

where $\bar{\delta}$ is the average misorientation, $\bar{\delta}_{random}$ = 41° for cubic metals and alloys, as derived by Mackenzie [17], ε is the total applied true strain, k0 is the "zero point" parameter (to obtain the fine initial subgrains, the "zero point" for counting the strain in Stage II), k1 can be regarded as a misorientation rate, and k2 as an effectiveness parameter. We put inside arctan the value "($\pi/2\bar{\delta}_{random}$)" because we had to put outside its inverse to get the asymptotic value of $\bar{\delta}_{random}$ at high strains, and the initial slope = k1. In principle, the expected values of the parameters are around k0 ≈ 1-2, k1 ≈ 6 and k2 ≈ 1. This means that we expect about one or two ECAP passes to obtain the initial ultrafine subgrains with minimum misorientation (k0 ≈ 1-2), an increase of average misorientation of 6° per pass (k1 ≈ 6), and a "typical" effectiveness (k2 ≈ 1). The value k2 = 1 means that the average misorientation increases linearly with applied strain at small-medium misorientations. This is expected for a constant (sub)grain size. However, there is a tendency for decreasing grain sizes with increasing strain that may influence the experimental values of Eq.1 parameters.

Relationship between %HAGB and $\bar{\delta}$

The misorientation can be described in several ways. Histograms can be complicated when performing comparisons. Although not very precise, the simplest and most common misorientation parameters are the fraction of HAGBs, %HAGB, and average misorientation, $\bar{\delta}$. The author prefers the latter, $\bar{\delta}$, because there is only one average, $\bar{\delta}$, value from a given histogram. On the contrary, for the %HAGB, an arbitrary threshold should be stablished between low and high angle boundaries to calculate its value (usually 15°). Nevertheless, an approximately linear relation can be found to relate both parameters in a wide range of %HAGB or $\bar{\delta}$ [6, 9]. Therefore, to compare the different alloys data, from distinct research works, and show them in useful plots, it is convenient to make explicit the relation between $\bar{\delta}$ and %HAGB. The simplest translation should follow the expression:

$$\bar{\delta} = a + b\ \%HAGB \quad (2)$$

where approximate values of *a* and *b* are about a ≈ 6°, b ≈ 0.36, for cubic metals and alloys. Other approximate values are a ≈ 4°, b ≈ 0.395, fitted from [6], or a ≈ 5.5°, b ≈ 0.352, fitted from [9]. Following the Mackenzie random distribution and having in mind that for %HAGB = 100%, $\bar{\delta} > \bar{\delta}_{random}$, values of a = 6° and b = 0.36 have been chosen as quite reasonable. However, each experimental misorientation histogram is different and the extreme values could show some dispersion.

Microstructure evolution during ECAP of aluminium alloys

Several examples of misorientation evolution have been taken from the literature. Some of them use $\bar{\delta}$ values, but most use the fraction of high angle boundaries, %HAGB. An example of ECAP processing at room temperature (RT) for pure Al (99.99%) up to 12 passes is given by Kawasaki et al. [6] with a 90° ECAP route B_C, rod of 10 mm diameter, 60 mm length. Each pass applied a true strain of 1. Initial grain size, L, was about 1 mm. The values of grain size, L, and misorientation, $\bar{\delta}$, are plotted together in Fig.1a up to a total strain ε = 12.

Fig.1b shows data from Apps et al. [3] using a 120° ECAP at room temperature following Route A (without rotation) to process Al-0.13Mg and AA8079 alloys up to a strain ε = 10 (upper axis).

In Fig.1b, it is shown that the aluminium alloy containing some coarse precipitates needs less ECAP passes to obtain similar microstructure evolution.

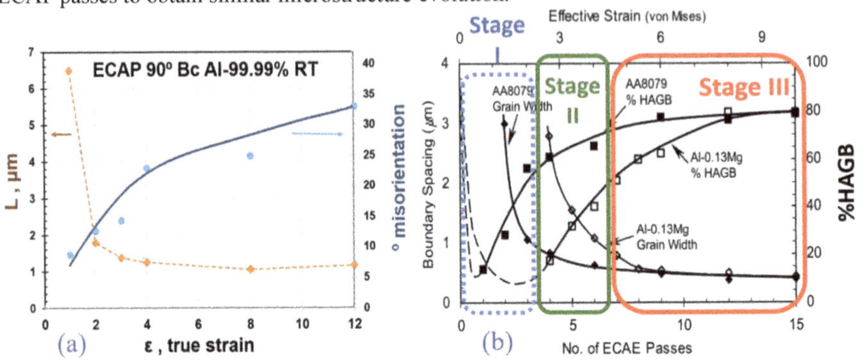

Fig.1. L and $\bar{\delta}$ vs. ε for a) ECAPed 90° RT pure Al (99.99%) [6] and b) ECAPed 120° RT Al-0.13Mg and AA8079 alloys [3]. Stages I, II and III are also highlighted.

Additionally, in this Fig.1b, it can be clearly shown the three stages for the misorientation evolution during ECAP. The first, Stage I, having different span depending on the alloy composition, relates to the fine subgrain initial subdivision. Following, Stage II is associated to the approximately linear misorientation increase of the fine subgrains (LAGBs) towards HAGBs. Finally, Stage III shows the asymptotic behaviour of the average misorientation as a function of applied strain. In these two alloys, the final grain size is smaller than that for the Al 99.99% alloy.

For other Al alloys including precipitates and/or elements in solid solution, an approximately similar shape of curves is obtained, although the misorientation rate is different. Fig.2 shows data for alloy AA6060-T6 processed under ECAP 90° following Route B_C at room temperature [10]. As expected, composition can make a difference in the misorientation evolution. In this case also, the final grain size is smaller than for the Al 99.99% alloy.

Regarding the behaviour of Al-Zn-Mg-Cu alloys, similar to AA7075, Goloborodko et al. [4] have performed ECAP 90° on alloy AA7475 following route A, at 523 and 673K (HT), and compared to other Al alloys, as shown in Fig.3a. The misorientation evolution of AA7475 follows the three stages, as other Al alloys (Al-0.1Mg, Al-0.13 Mg, Al-0.3Mn, Al-3Mg and AA5052). Some variations may arise by the use of a different ECAP Route, or temperature, as

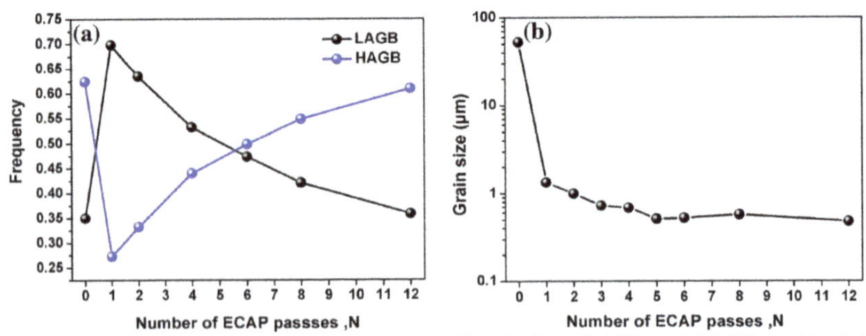

Fig.2. ECAPed 90° B_C RT AA6060-T6 Al-Mg-Si alloy [10]: a) LAGB and HAGB, and b) L (µm) vs. N (true strain).

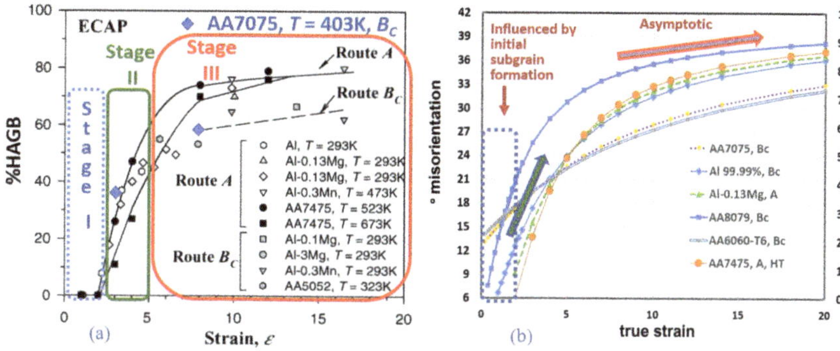

Fig.3. a) AA7475 %HAGB vs. strain compared to other Al alloys [4], and b) calculations from Eq.1 with parameters from Table 1.

Table 1. Eq.1 parameter values associated to various Al alloys, and corresponding calculations of $\bar{\delta}$ and %HAGB (Eq.2) for ε = 8.

Alloy	k0	k1	k2	ε	$\bar{\delta}$	%HAGB
Al 99.99%	0.00	**6.8**	1	8	29.3	65
Al-0.13Mg	0.80	**8.0**	1	8	29.9	67
AA8079	-0.25	**12.0**	1	8	34.3	79
AA6060-T6	-5.00	**3.0**	1	8	25.6	55
AA7475 HT	1.35	**9.0**	1	8	30.3	68
AA7075	-4.00	**3.4**	1	8	26.1	56

well as composition. Additionally, Fig.3a shows data from an overaged AA7075 ECAPed 90° at 130 °C (403K) using Route B_C (37% and 56% HAGB at 3 and 8 passes, respectively, shown as the large blue coloured diamonds) [7-8]. For this AA7075 overaged alloy, grain size values diminished as ECAP passes were applied, from about 200 to 163 nm, in the shortest grain axis (transverse), for 3 and 8 ECAP passes, respectively, in line with the decline observed also by Goloborodko et al. [4]. In the case of AA7075 alloy, although only two points are available, a tentative prediction and comparison with other alloys will be performed.

Fig.3b shows the approximate misorientation evolution curves of the analysed ECAPed alloys as described using Eq.1. They will be explained in the next paragraph.

Description and predictions of misorientation evolution

To have easy and clear comparisons among alloys, the arctan function of Eq.1 will be used with only two parameters, setting the parameter k2 to 1. This is equivalent to say that misorientation increases linearly with applied strain in Stage II. This should be the ideal situation if grain size remains stable. Although it is not the real case, it is useful for approximate descriptions and comparison among different alloys. The parameter k0 corresponds to the “zero point” of Stage II and k1 gives the initial misorientation rate, as shown in Table 1 for the six alloys considered. Additionally, $\bar{\delta}$ and %HAGB values given by Eq.1 and Eq.2, respectively, for ε = 8 are included.

Observing Table 1 and Fig.3b it can be noticed that Al 99.99% behaviour is in the middle. Its k1 = 6.8 is between 3 (AA6060-T6, full of small precipitates) and 12 (AA8079, with some coarse precipitates). Similar behaviour to AA6060-T6 shows the AA7075 alloy, having the slowest misorientation rates, probably due to their ultrafine grains still evolving. Close to the average behaviour, that of Al 99.99%, are the dilute alloy Al-0.13Mg, and the AA7475 HT ECAPed at high temperature by Route A. The processing at high temperature helps also in the formation of HAGBs. Additionally, Route A may be more effective than Route B_C because B_C is cyclic every 4 passes. From this comparison it is clear that having a clean matrix with some fraction of non-fine precipitates helps in the development of a highly misoriented microstructure, as pointed out by Apps et al. [3].

Regarding the expected superplastic behaviour of the ECAPed alloys, interesting questions arise: When a fine, but non-randomly misoriented, grain structure will be superplastic? And, if so, in what extent?

Influence of misorientation on superplastic behaviour

Superplasticity at intermediate-high temperatures is obtained thanks to the operation of the grain boundary sliding (GBS) mechanism. As a consequence, it is necessary a fine, equiaxed and highly misoriented grain structure. In fact, the finer, the more equiaxed and the more misoriented the grain, the better. During decades, several phenomenological, although similar, constitutive equations have been proposed, but they do not contemplate the misorientation as an explicit parameter [8, 14, 16]. The only attempts to obtain a superplastic equation including explicitly the fraction of HAGBs (%HAGB) or average misorientation ($\bar{\delta}$) are recent [9, 12-13]. In [12-13] it was proposed the idea that if the average misorientation is less than random, the material behaves as if the "randomly misoriented equivalent grain size for GBS", L_{GBSeq}, is larger than the grain size, L. This is conveniently described by expression [13]:

$$L_{GBSeq} = L\,(\bar{\delta}_{random}/\bar{\delta})^2 \tag{3}$$

where $\bar{\delta}$ is the experimental average misorientation corresponding to the experimental L value. Therefore, the smaller the alloy misorientation, the less superplastic. The full expression for the new superplastic GBS constitutive equation is [12-13]:

$$\dot{\varepsilon} = A\left(\frac{\sigma}{E}\right)^2\left[\left(\frac{\bar{\delta}}{\bar{\delta}_{random}}\right)^2\left(\frac{b}{L}\right)\right]^2\frac{D_L}{b^2} \tag{4}$$

where σ is the applied stress, E is the Young Modulus, b is the Burger's vector, D_L is the coefficient of lattice self-diffusion and A is a material constant.

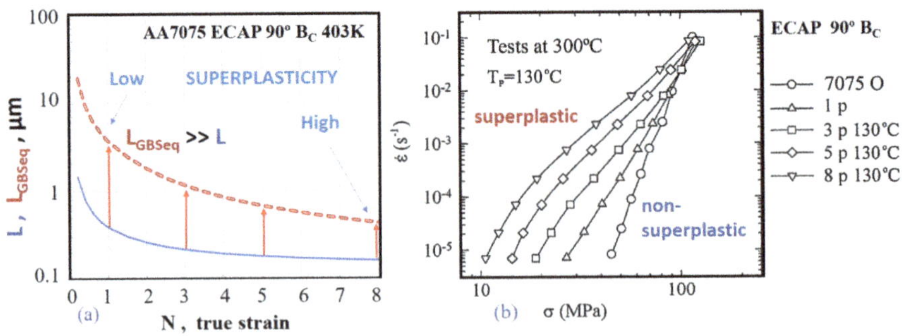

Fig.4. a) L and L_{GBSeq} vs. ε for ECAPed AA7075 explaining b) the lesser or greater superplasticity after various ECAP passes. L and tensile test data taken from [7-8].

Now, having data of both grain size, L, and average misorientation, $\bar{\delta}$, we are able to predict the effective grain size for GBS superplastic behaviour, L_{GBSeq}. For instance, for alloy AA7075 [7] taking into account the misorientation evolution given by Eq. 1 with parameters of Table 1, shown in Fig.3b, and L evolution with ECAP passes, shown in Fig.4a, the L_{GBSeq} values predicted by Eq.3 are shown also in Fig.4a. Additionally, Fig.4b shows strain rate-stress data at 300 °C after various ECAP passes at 130°C (403K) [8] presenting the large effect in the superplastic deformation of the GBS-equivalent grain size.

Not surprisingly, L_{GBSeq} values can be between two and ten times larger than L values, being L_{GBSeq} larger the lower the applied strain, and thus, making an important difference in superplastic behaviour from one to eight passes, as shown in Fig.4b. These predicted L_{GBSeq} values can be very useful for technical applications, such as superplastic forming in the aeronautical and aerospace industries.

Summary

- Average grain misorientation, $\bar{\delta}$, increases continuously with applied strain, ε, in SPD processes, such as ECAP, which is particularly useful to model the misorientation evolution because the misorientation saturation is usually found after about 12 or 16 ECAP passes.
- Initially, a network of low-misoriented subgrains should be formed (Stage I). This step could take a variable amount of strain depending on initial grain size and composition. For instance, solid solution may delay this step, whereas precipitates may favour it.
- Next, for increasing ECAP passes, generally, misorientation should increase linearly with increasing applied deformation, transforming LAGBs into HAGBs (Stage II).
- Finally, for large strains (Stage III), the average misorientation values tend to saturate asymptotically towards the Mackenzie distribution average misorientation value, $\bar{\delta}_{random}$. For cubic metals and alloys $\bar{\delta}_{random} = 41°$.
- To describe and analyse the misorientation evolution with applied strain, it is proposed the use of the arctan function to take into account both the initial linear behaviour and the final asymptotic behaviour towards the misorientation saturation value. In its simpler form, it is proposed the equation: $\bar{\boldsymbol{\delta}} = \bar{\delta}_{random}\,(2/\pi)\,\mathbf{arctan}(\pi/2\bar{\delta}_{random} \cdot \mathbf{k1} \cdot (\varepsilon - \mathbf{k0})^{\mathbf{k2}})$, where k0, k1 and k2 are constants depending on the material and processing variables. k0 relates to the strain needed to start Stage II, k1 relates to the initial misorientation rate respect to the applied strain (in Stage II), and the exponent k2 is about 1, for a constant grain size, related to the linearity of misorientation evolution in Stage II. These parameters depend on composition, microstructure and processing parameters.
- Once the alloys have been processed, and knowing their values of grain size, L, and average misorientation, $\bar{\delta}$, "GBS-equivalent grain size" values can be obtained so that their superplastic behaviour can be adequately compared. The GBS-equivalent grain size is given by: $\mathbf{L_{GBSeq}} = \mathbf{L} \cdot (\bar{\boldsymbol{\delta}}_{\mathbf{random}} / \bar{\boldsymbol{\delta}})^2$ [12-13]. Finally, L_{GBSeq} substitutes L in the superplastic GBS constitutive equation, thus providing more accurate predictions, having taken into account the average misorientation values.

Acknowledgements

Financial support from MINECO (Spain), Project MAT2015-68919-C3-1-R (MINECO/FEDER) is gratefully acknowledged. Funding from the project PID2020-118626RB-I00 awarded by MCIN/AEI/10.13039/501100011033 is also acknowledged. O.A. Ruano, J.M. García-Infanta, C.M. Cepeda-Jiménez and A. Orozco-Caballero are greatly acknowledged for making this work possible.

References

[1] V. M. Segal, Plastic Working of Metals by Simple Shear, Russian Metallurgy, Vol. 1, 1981, pp. 99-105

[2] R. Valiev, R. Islamgaliev, I.V. Alexandrov, Bulk Nanostructured Materials from Severe Plastic Deformation. Prog. Mater. Sci. 45 (2000) 103-189. https://doi.org/10.1016/S0079-6425(99)00007-9
[3] P.J. Apps, J.R. Bowen, P.B. Prangnell, The effect of coarse second-phase particles on the rate of grain refinement during severe deformation processing, Acta Mater. 51 (2003) 2811-2822. https://doi.org/10.1016/S1359-6454(03)00086-7
[4] A. Goloborodko, O. Sitdikov, R. Kaibyshev, H. Miura, T. Sakai, Effect of pressing temperature on fine-grained structure formation in 7475 aluminum alloy during ECAP, Mater. Sci. Eng. A 381 (2004) 121–128. https://doi.org/10.1016/j.msea.2004.04.049
[5] I. Mazurina, T. Sakai, H. Miura, O. Sitdikov, R. Kaibyshev, Grain refinement in aluminum alloy 2219 during ECAP at 250 °C, Mater. Sci. Eng. A 473 (2008) 297-305. https://doi.org/10.1016/j.msea.2007.04.112
[6] M. Kawasaki, Z. Horita, T.G. Langdon, Microstructural Evolution in High Purity Aluminum Processed by ECAP, Mater. Sci. Eng. A 524 (2009) 143-150. https://doi.org/10.1016/j.msea.2009.06.032
[7] C.M. Cepeda-Jiménez, J.M. García-Infanta, O.A. Ruano, F. Carreño, Mechanical properties at room temperature of an Al-Zn-Mg-Cu alloy processed by equal channel angular pressing, J. Alloys Compounds 509 (2011) 8649–8656. https://doi.org/10.1016/j.jallcom.2011.06.070
[8] C.M. Cepeda-Jiménez, J.M. García-Infanta, O.A. Ruano, F. Carreño, High strain rate superplasticity at intermediate temperatures of the Al 7075 alloy severely processed by equal channel angular pressing, J. Alloys Compounds 509 (2011) 9589-9597. https://doi.org/10.1016/j.jallcom.2011.07.076
[9] K. Wang, F.C. Liu, P. Xue, D. Wang, B.L. Xiao, Z.Y. Ma, Superplastic Constitutive Equation Including Percentage of High-Angle Grain Boundaries as a Microstructural Parameter, Metall. Mater. Trans. A 47A (2016) 546-559. https://doi.org/10.1007/s11661-015-3230-8
[10] T. Khelfa, M. A. Rekik, J. A. Muñoz-Bolaños, J. M. Cabrera-Marrero, M. Khitouni, Microstructure and strengthening mechanisms in an Al-Mg-Si alloy processed by equal channel angular pressing (ECAP), The Int. J. Adv. Manuf. Techn. 95 (2018) 1165–1177. https://doi.org/10.1007/s00170-017-1310-1
[11] F. Carreño, O.A. Ruano, Superplasticity of Aerospace 7075 (Al-Zn-Mg-Cu) Aluminium Alloy Obtained by Severe Plastic Deformation, Defect Diff. Forum 385 (2018) 39-44. https://doi.org/10.4028/www.scientific.net/DDF.385.39
[12] A. Orozco-Caballero, O.A. Ruano, E. F. Rauch, F. Carreño, Severe friction stir processing of an Al-Zn-Mg-Cu alloy: Misorientation and its influence on superplasticity, Mater. Design 137 (2018) 128-139. https://doi.org/10.1016/j.matdes.2017.10.008
[13] F. Carreño, A. Orozco-Caballero, Superplastic GBS Constitutive Equation Incorporating Average Grain Misorientation Dependence, Mater. Sci. Forum 941 (2018) 1501-1506. https://doi.org/10.4028/www.scientific.net/MSF.941.1501
[14] R.B. Figueiredo, M. Kawasaki, T.G. Langdon, Seventy years of Hall-Petch, ninety years of superplasticity and a generalized approach to the effect of grain size on flow stress, Prog. Mater. Sci. 137 (2023) 101131. https://doi.org/10.1016/j.pmatsci.2023.101131
[15] K. Edalati, Superfunctional Materials by Ultra-Severe Plastic Deformation, Materials 16 (2023) 587. https://doi.org/10.3390/ma16020587
[16] O.A. Ruano, O.D. Sherby, On constitutive equations for various diffusion-controlled creep mechanisms, Rev. Phys. Appl. 23 (1988) 625–637. https://doi.org/10.1051/rphysap:01988002304062500
[17] J.K. Mackenzie, Second Paper on Statistics Associated with the Random Disorientation of Cubes, Biometrika 45 (1958) 229-240. http://www.jstor.org/stable/2333059

Microstructural characteristics, mechanical and corrosion properties of a low-alloyed Mg alloy after different deformation processing

Zahra Abbasi[1,2 a*], Jose Maria Cabrera[3,b], Ramin Ebrahimi[2,c], and Erhard Schafler[1,d]

[1]Physics of Nanostructured Materials, Dynamics of Condensed Systems, Faculty of Physics, University of Vienna, Vienna, Austria

[2]Department of Materials Science and Engineering, School of Engineering, Shiraz University, Shiraz, Iran

[3]Department of Materials Science and Engineering EEBE, Universidad Politécnica de Catalunya, Barcelona, Spain

[a]zahraabbassi@shirazu.ac.ir, [b]jose.maria.cabrera@upc.edu, [c]ebrahimy@shirazu.ac.ir, [d]erhard.schafler@univie.ac.at

Keywords: Mg Alloy, ECAP, Microstructure, Mechanical Properties, Corrosion

Abstract. In the evolution of characteristics in Mg alloys, the combined influence of grain refinement by severe plastic deformation (SPD) and alloying elements usually plays a crucial role. Rare earth elements (Y, Gd, and Nd) in combination with Zn have a substantial impact on Mg characteristics in various compositions. In this study, a new dilute extruded Mg-Zn-Gd-Y-Nd alloy was exposed to 5 passes of equal channel angular pressing (ECAP), in a die with a 90° channel angle following route Bc. The initial deformation temperature was 300°C, and it dropped to 200°C with a 25°C step until the fifth pass. Initial and deformed samples were subjected to hardness testing, optical and scanning electron microscopy (SEM) examinations and corrosion tests. After the fifth run of ECAP at 200°C, necklaces of fine recrystallized grains along grain boundaries of elongated unrecrystallized grains in extruded samples transformed to an ultrafine grained microstructure. SEM images reveal the presence of very fine nanoscale dynamic recrystallization (DRX) nuclei in the context of the ECAPed alloy. Furthermore, measures of hardness show the increase in hardness from the starting state to the fifth pass of ECAP. The increase in hardness was caused by dynamic recrystallization, which resulted in a higher percentage of freshly produced grains and grain boundaries. Furthermore, the inclusion of rare earth elements increased grain refinement and controlled the rate of dynamic recrystallization (DRX) during ECAP. On the other hand, severe plastic deformation cause changes in the density and distribution of grain boundaries and defects, which affect the corrosion behavior of magnesium alloys. Additionally, in comparison to the as cast condition the extruded-annealed, ECAPed and as-extruded samples have better corrosion resistance, respectively. It can be concluded that grain refinement has positive effect on decreasing the corrosion rate while homogenization of the extruded microstructure is more effective.

Introduction

Known as the lightest structural material with combined properties of low density (1.7 g/cm^3) and high specific strength, magnesium alloys are potential candidates for the pioneering generation of structural materials arranged for many applications, from the aerospace and automotive industry to biomedical implants [1-3]. However, there are still some limitations in the wide application of magnesium alloys due to the poor formability of their hexagonal closed pack crystal structure at room temperature as well as poor corrosion resistance in most environmental area [4]. Taking the advantage of alloying elements accompanied by reducing the grain size is an effective solution to conquer these limitations to a wide extent. Among different alloying elements, rare earth elements alongside zinc have attracted great attention in the case of improving magnesium properties [5, 6]. Amongst more common RE elements, Gd, Nd and Y have been found to bring about advantageous effects on casting characteristics, as well as electrochemical, mechanical and physical properties of magnesium alloys at ambient and high temperatures [4]. From the grain refinement aspect, severe plastic deformation (SPD) techniques have been extensively used to prepare ultrafine grained materials [7]. Equal channel angular pressing (ECAP) is one of the most recognized methods to prepare bulk ultrafine-grain (UFG) materials [8, 9]. The evolution of the material's structure during ECAP processing highly depends on the material, the strain introduced (number of passes), temperature, the processing rate, route variations and other conditions [10, 11]. The most common channel angle in ECAP is 90˚; in case of magnesium deformation, the larger the channel angle than 90˚ the lower the deformation temperature as well as the smaller the shear deformation of a single pass [12]. The route Bc (the sample rotated 90° clockwise after each pass) is generally the most efficient way to develop the UFG structure when using a die with a channel angle of = 90° [13]. Recent researches proved that ECAP could change the corrosion mode from pitting corrosion to uniform corrosion [14]. However, the complexity of influencing factors, such as grain size, second phase distribution and crystallographic orientation, make the effect of ECAP on the corrosion resistance of magnesium alloys debatable [15, 16].

The main object of this research is to compare the microstructural changing and corrosion behavior of the novel Mg-Zn-Gd-Y-Nd alloy after extrusion and severe plastic deformation by ECAP process. The ECAP process can produce intense and uniform deformation by simple shear and provides a convenient procedure for introducing an ultrafine grain size into a material. The samples were prepared by using hot extrusion methods. Hardness and polarization tests were carried out on the as cast, extruded and ECAPed rods, and the microstructure was examined using optical and scanning electron microscopy (SEM).

Experimental details

The initial material is a novel magnesium-based alloys obtained from die casting of pure Mg (99.99%) and alloying elements (Zn, Y, Gd and Nd). The Mg based alloy chemical composition was analysed by inductively coupled plasma (Arian ICP-OES spectrometer 730-ES) and known as Mg-0.5Zn-<0.1Gd-0.5Y- <0.1Nd [17].

Samples of circular cross section of 30 mm in diameter and 70 mm in height were machined from the as cast billets and were homogenized in Ar atmosphere at 520°C for 12 hours. The samples were extruded at the temperature of 480°C with an extrusion ratio of 8 (d_0=34 and d=12mm) using a Zwick/Roell 250 testing machine with the extrusion speed of 1 mm/min. Afterwards, round billets of 5 mm in diameter and 30 mm in height were extracted from the as-extruded rods and annealed at 220°C in an electric furnace and quenched immediately prior to ECAP. The samples were preheated in an ECAP die with channel angle of 90° for 5 minutes. ECAP was carried out following the route B_C up to five passes with the aim of MoS_2 spray as lubricant. Experiments were carried out at 300 °C for the first pass and continued by a 25°C step decrease of temperature up to the fifth pass.

A cross section of the as-cast, as-extruded, as-extruded annealed, fourth pass ECAPed sample and fifth pass ECAPed sample were extracted for microstructural analysis. Specimens for optical microscopy were prepared by conventional grinding and diamond polishing (down to 0.25 μm), followed by etching in an acetic-picral solution. In addition, micro-hardness measurement was conducted on samples with the applied load of 50g and dwell time of 15s. The open circuit potential (OCP) and potentiodynamic polarization tests, adopting a standard three-electrode cell set-up, in a simulated body fluid solution (SBF) with the PH of 7.4, at 37°C. Samples were cut from either as-cast, extruded, extruded-annealed or ECAP treated billets (5 pass) and mounted in epoxy resin and grounded successively with 1000~3000 grit SiC paper, washed with distilled water, ethanol and dried by a blower. Potentiodynamic polarization curve measurement were performed on IVIUM potentiostat with the conventional three compartment cell; magnesium alloys as working electrode, a platinum rod as counter electrode and a silver chloride electrode (Ag/AgCl, sat. KCl) as reference electrode. Potentiodynamic polarization curves are performed over the range E_{ocp} (open circuit potential) ±500mV, with scanning rate of $1mVs^{-1}$ and measuring the OCP for 3000s.

Results and discussion

The microstructural observation of as-cast samples revealed a coarse-grained inhomogeneous microstructure. Applying extrusion at temperature of 480°C created necklaces of fine grains at grain boundaries. This bimodal microstructure of fine recrystallized grains and large unrecrystallized grains stretched along the extrusion direction is a clear demonstration of dynamic recrystallization (DRX). Layers of nuclei formed across the grain boundaries during deformation and progress by further deformation in expense of the deformed domain by means of forming new layers of the DRXed grains. But it should be noted that the recrystallization rate was not high enough to produce an equiaxed microstructure. Annealing after extrusion at the temperature of 220°C for 2h converted the bimodal microstructure to almost an equiaxed microstructure being an indication of static recrystallization (SRX) (fully completed) and, most probably, some slight grain growth took place during annealing (Figure 1).

Fig. 1 The microstructure of the samples in different conditions of a) as-cast, b) as extruded and c) extruded-annealed.

Performing ECAP at the temperature of 300°C for the first pass via route Bc and continuing up to fifth pass by decreasing the temperature to 200°C (with 25°C step) resulted in samples without any cracks. Figure 2 depicts the ECAPed samples after the first, fourth and the fifth passes of processing by route Bc.

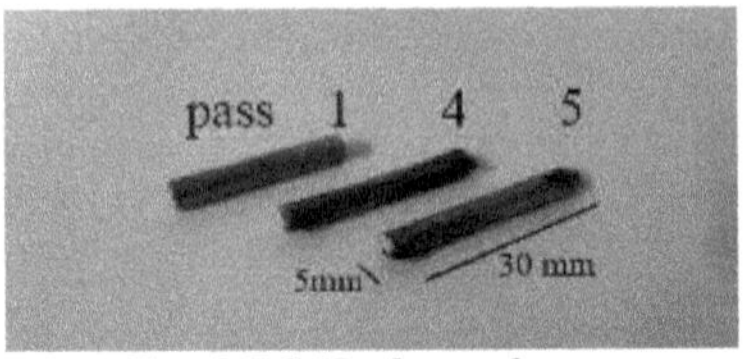

Fig. 2 ECAPed samples.

Figure 3 shows the microstructure of the as ECAPed samples after the first, fourth and fifth passes via route Bc in the cross sections perpendicular to extrusion direction. Generally, by applying ECAP and increasing the number of ECAP passes, dislocation density increases as well which highten the energy of system [18]. When the dislocation density is high enoughinside the grains, they will rearrange to form low angle grain boundaries, by further deformation high angle grain boundaries (HAGB) form. These HAGBs are suitable places for DRX (figure 3(d)) [19]. As depicted in Figure 1(c), the extruded-annealed samples are the initial samples for ECAP process. The coarse grain microstructure is refined and also the dislocations rearrangement consumed the stored energy induced by SPD as a term of softening mechanism. The grain size after 5 ECAP passes is about 5 μm with some fine DRXed grains of 0.4 μm (determined by white rings in figure 3(d)).

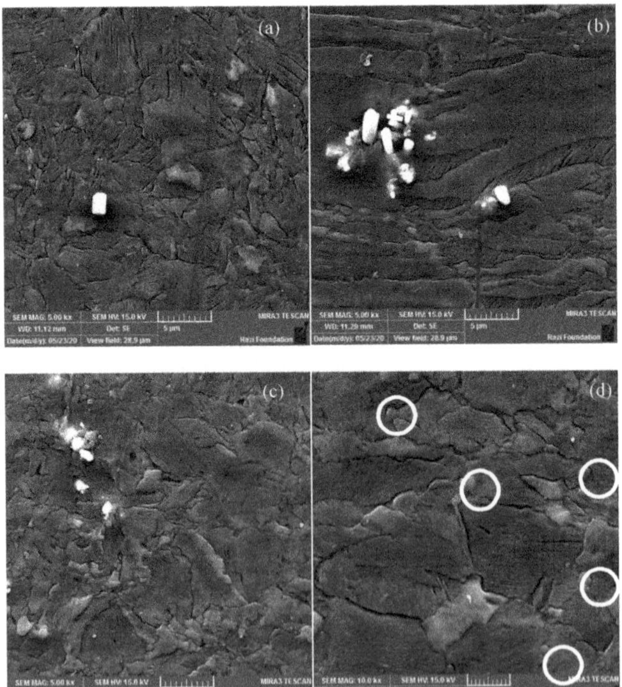

Fig. 3 Microstructure of samples after the (a) first, (b) fourth and (c and d) the fifth passes of ECAP (route Bc).

Materials Research Forum LLC
https://doi.org/10.21741/9781644902615-22

According to hardness results (Figure 4), there is a decreasing trend from as cast condition to as extruded one, while annealing after extrusion decreased the hardness value to some extent but the microstructure is more equiaxed. On the other hand, by the increase of ECAP pass number from the first to the fourth one the hardness number did not change while the grain morphology is completely different in both samples. It can be deducted that dynamic recrystallization during the ECAP process was incomplete, and still preserve plenty of energy in the grain boundaries and there was a balance between the hardening and softening processes [20]. Proceeding to the fifth pass, generation of more dislocation density accompanied by lower temperature of 200°C raise the hardness value.

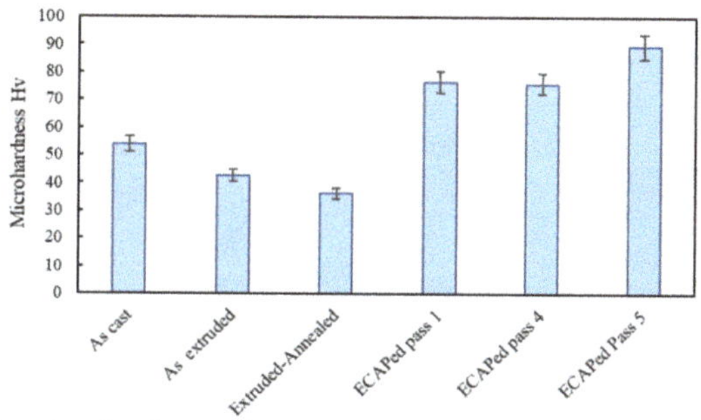

Fig. 4 The microhardness of samples processed by different conditions.

The corrosion behavior of Mg alloys is highly affected by the microstructure. Grain refinements lead to changes in the density and distribution of grain boundaries which influence the mechanical and corrosion behavior of Mg alloys. Zeng et al. [21] reported that the corrosion rate decreases by the decrease in the grain size while the occurrence of filiform corrosion can be retarded at finer grains. Filiform corrosion originates from corrosion pits and extends forward along the active area. One of the explanations for the enhanced corrosion resistance with finer grains is that the stress on surface film due to the mismatch between Mg oxide and underlying Mg metal substrate, can be relieved by the aid of grain refinement. However, there are some studies also reported that the corrosion resistance deteriorates as the grain size decreases. Song et al. [22] reported that a grain refinement obtained by ECAP has weakened corrosion resistance of ECAPed magnesium alloy, resulting in more corrosion pits and obvious filiform-like corrosion. Here, extrusion as well as ECAP decreased the grain size in the sample, but regarding the corrosion current densities (i_{corr}) estimated from the Tafel extrapolation method, the as cast and as extruded samples were found to show similar i_{corr} values, 24 and 27.2 μA cm^{-2}, respectively. While this number for ECAPed sample is 72.9 μA cm^{-2}. Therefore, it can be concluded that the permanent protection is not provided and also a faster localized corrosion would happen upon the onset of pitting. Many studies indicated that the mechanism by which grain size influences the corrosion behavior of Mg alloys is still under debate [23]. Additionally, the variation in grain morphology affect the corrosion process because the anodic and cathodic reactions are to be maintained in equilibrium which is controlled by the potential of the metal. In the case of anodic reaction with more electrons proposed into the metal, the excess electrons then shift the potential to more negative values, thus, slows down the anodic reaction and increases the cathodic reaction [13].

Table 1. Fitting results of the polarization curves

	As cast	Extruded	Extruded-Annealed	ECAPed 5th pass
E. corr [V]	-1.5453	-1.5373	-1.59125	-1.48835
i_{corr} [μA/cm^2]	24	27.2	18.6	72.9
C. Rate [mm/y]	0.5476	0.62195	0.4239	1.6665

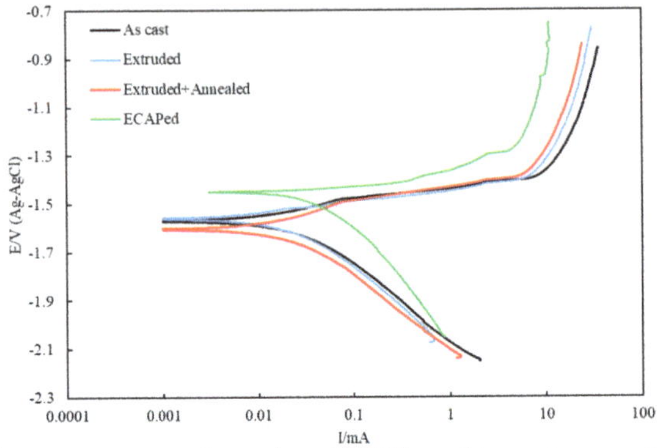

Fig. 5 Polarization curves of samples with different history immersed in SBF.

Therefore, it can be concluded the extruded-annealed sample has the lowest corrosion rate among all while the ECAPed sample has the highest. Although ECAP had increase the hardness of the sample and improve the mechanical property by decreasing the grain size, it reduced the corrosion resistance in the sample.

Conclusions

The significant effects of ECAP on the mechanical properties and corrosion resistance of the Mg-0.5Zn-<0.1Gd-0.5Y- <0.1Nd alloy are as follows:

- By converting the bimodal microstructure to an equiaxed ultrafine grain one after five passes of ECAP, the grain size of Mg-Gd-Nd-Zn-Zr alloy was refined to about 5 μm with some 0.4 μm DRXed grains.
- ECAP enhanced the mechanical properties of the magnesium alloy as compared with the as-extruded sample. The hardness number increased from 35 in the as extuded- annealed sample to 90 after 5 passes of ECAP.
- An adverse effect was observed on the corrosion resistance as compared with the extruded sample. The extruded-annealed sample showed the higher corrosion resistance as compared to the ECAPed sample which showed the highest corrosion rate.

References

[1] B. Mordike and T. Ebert, Magnesium: properties-applications-potential, Materials Science and Engineering: A 302, no. 1 (2001) 37-45. https://doi.org/10.1016/S0921-5093(00)01351-4

[2] A. A. Luo and A. K. Sachdev, Applications of magnesium alloys in automotive engineering, Fundam. Process. Prop. Appl. (2012) 393-426. https://doi.org/10.1533/9780857093844.3.393

[3] N. Hort, Y. Huang, D. Fechner, M. Störmer, C. Blawert, F. Witte, C. Vogt, H. Drücker, R. Willumeit, K. U. Kainer and F. Feyerabend, Magnesium alloys as implant materials – Principles

of property design for Mg–RE alloys, Acta Biomater. 6 (2010) 1714-1725. https://doi.org/10.1016/j.actbio.2009.09.010

[4] S. Tekumalla, S. Seetharaman, A. Almajid and M. Gupta, Mechanical Properties of Magnesium-Rare Earth Alloy Systems: A Review, Metals, 5(1), (2014) 1-39. https://doi.org/10.3390/met5010001

[5] Y. Huang, W. Gan, K.U. Kainer and N. Hort, Role of multi-microalloying by rare earth elements in ductilization of magnesium alloys, J. of Magnesium and Alloys 2(1) (2014) 1-7. https://doi.org/10.1016/j.jma.2014.01.005

[6] S. M. Zhu, M. A. Gibson, M. A. Easton and J. F. Nie, Evaluation of Magnesium Die-Casting Alloys for Elevated Temperature Applications: Microstructure, Tensile Properties, and Creep Resistance, Scr. Mater. 63, (2010) 698-703. https://doi.org/10.1007/s11661-015-2946-9

[7] R. B. Figueiredo, M. T. P. Aguilar, P. R. Cetlin and T. G. Langdon, Processing magnesium alloys by severe plastic deformation, Materials Science and Engineering 63 (2014). https://doi.org/10.1088/1757-899X/63/1/012171

[8] T. G. Langdon, The principles of grain refinement in equal-channel angular pressing, Materials Science and Engineering A, 462 (2007) 3-11. https://doi.org/10.1016/j.msea.2006.02.473

[9] R. Z. Valiev and T.G. Langdon, Principles of equal-channel angular pressing as a processing tool for grain refinement, Prog. Mater. Sci. 51 (2006) 881-981. https://doi.org/10.1016/j.pmatsci.2006.02.003

[10] T. G. Langdon, Twenty-five years of ultrafine-grained materials: Achieving exceptional properties through grain refinement, Acta Materialia, 61(19) (2013) 7035-7059. https://doi.org/10.1016/j.actamat.2013.08.018

[11] F. Kang, J. T. Wang and Y. Peng, The effect of hydrostatic pressure on the activation of non-basal slip in a magnesium alloy, Materials Science and Engineering: A, 487 no.1-2 (2008) 68-73. https://doi.org/10.1016/j.scriptamat.2009.07.011

[12] Y. Iwahashi, Z. Horita, M. Nemoto and T. G. Langdon, Principle of equal-channel angular pressing for the processing of ultra-fine grained materials, Acta materialia, 46 (1998) 3317-3331. https://doi.org/10.1016/1359-6462(96)00107-8

[13] K. R. Gopi, H. S. Nayaka and S. Sahu, Corrosion Behavior of ECAP-Processed AM90 Magnesium Alloy, Arabian Journal for Science & Engineering 43(9) (2018). https://doi.org/10.1007/s13369-018-3203-5

[14] D. Song, A. B. Ma, J. H. Jiang, P. H. Lin, D. H. Yang, J. F. Fan, Corrosion behaviour of bulk ultra-fine grained AZ91D magnesium alloy fabricated by equal-channel angular pressing, Corros. Sci. 53 (2011) 362-373. https://doi.org/10.1016/j.corsci.2010.09.044

[15] H. Wang, Y. Estrin, H. M. Fu, G. L. Song and Z. Zúberová, Mechanical properties and biocorrosion resistance of the Mg-Gd-Nd-Zn-Zr alloy processed by equal channel angular pressing, Adv. Eng. Mater. (2007) 967-972. https://doi.org/10.1016/j.msec.2016.05.118

[16] G.R. Argade, S.K. Panigrahi, R.S. Mishra, Effects of grain size on the corrosion resistance of wrought magnesium alloys containing neodymium, Corros. Sci. 58 (2012) 145-151. https://doi.org/10.1016/j.corsci.2012.01.021

[17] Z. Abbasi, R. Ebrahimi and J. M. Cabrera, Investigation on Texture Evolution and Recrystallization Aspects of Novel Mg–Zn–Gd–Y–Nd Alloys, Metals and Materials International 27 (2021) 3983–3992. https://doi.org/10.1007/s12540-020-00729-2

[18] M. Gholami-Kermanshahi, V. D. Neubert, M. Tavakoli, F. Pastorek, B. Smola and V. Neubert, Effect of ECAP Processing on Corrosion Behavior and Mechanical Properties of the ZFW MP Magnesium Alloy as a Biodegradable Implant Material, Advanced Engineering Materials, 20 (2018) 1800121. https://doi.org/10.1002/adem.201800121

[19] S. Khani, M. R. Aboutalebi, M. T. Salehi, H. R. Samim and H. Palkowski, Microstructural development during equal channel angular pressing of as-cast AZ91 alloy, Materials Science and Engineering: A, 678 (2016) 44-56. https://doi.org/10.1016/j.msea.2016.09.066

[20] A. Galiyev, R. Kaibyshev and G. Gottstein, Correlation of plastic deformation and dynamic recrystallization in magnesium alloy ZK60, Acta materialia., 49(7) (2001) 1199-207. https://doi.org/10.1016/S1359-6454(01)00020-9

[21] R. C. Zeng, Z. Z. Yin, X. B. Chen, D. K. Xu, Corrosion types of magnesium alloys, Magnesium Alloys-Selected Issue, 2018. https://dx.doi.org/10.5772/intechopen.80083

[22] D. Song, A. Ma, J. Jiang, P. Lin, D. Yang and J. Fan, Corrosion behavior of equal-channel-angular-pressed pure magnesium in NaCl aqueous solution, Corrosion Science, 52(2) (2010) 481-490. https://doi.org/10.1016/j.corsci.2009.10.004

[23] Y. Ding, C. Wen, P. Hodgson and Y. Li, Effects of alloying elements on the corrosion behavior and biocompatibility of biodegradable magnesium alloys: a review, J. of materials chemistry B, 2(14) (2014) 1912-1933. https://doi.org/10.1039/C3TB21746A

Effect of severe plastic deformation on microstructure and properties of duplex stainless steel

Claudio Gennari[1a], Luca Pezzato[1b], Nuria Llorca-Isern[2c], Alisiya Biserova-Tahchieva[2d], Irene Calliari[1e]

[1] Department of Industrial Engineering, University of Padova, Italy

[2] Dept. Ciència de Materials i Química Física, Universitat de Barcelona, Spain

[a]claudio.gennari@unipd.it, [b]luca.pezzato@unipd.it, [c]nullorca@ub.edu, [d]abiserova.tahchieva@ub.edu, [e]irene.calliari@unipd.it

Abstract. The excellent performance of duplex and super duplex stainless steels due to their high corrosion resistance and high mechanical strength is directly related to their austeno-ferritic microstructure. However, these steels may suffer the formation of secondary brittle phases when they reach temperatures between 600°C and 950°C causing catastrophic service failure of components. In order to understand the influence of the mechanical history of the steel, the equal channel angular pressing was applied on UNS S32750 bar samples. ECAP is processed by using a hydraulic press (DE-80). Microstructural characterization was carried out on the severe plastic deformed samples by means of OM, SEM and EBSD. Ultimate tensile strength has improved with a low decreasing of elongation after the first pass. The ECAP process resulted to produce a faster precipitation of secondary phases if compared with the non-ECAP samples.

Keywords: Duplex Stainless Steels, Severe Plastic Deformation, ECAP, Grain Refinement

Introduction

Severe plastic deformation (SPD) is a technique used to process metallic materials with the aim of achieving enhanced mechanical properties through the creation of a structural state characterized by ultrafine grains on a sub-micrometer scale (ranging from 1 μm to less than 100 nm). Furthermore, if these ultrafine grain materials exhibit reasonable stability at elevated temperatures, they can demonstrate superplastic forming capabilities.

The structural changes induced by SPD, which lead to the development of sub microcrystalline states, primarily occur through the formation of dense dislocation walls known as dislocation sub-boundaries. As deformation increases, the misorientation of these sub-boundaries also rises, eventually reaching values typical of conventional high-angle grain boundaries, thereby resulting in the formation of a microcrystalline state [1-3].

Studying the structural modifications that occur during SPD of dual-phase alloys, such as duplex stainless steels (DSS) [4], is intriguing for several reasons. In these alloys, consisting of ferritic and austenitic phases, interphase boundaries can serve as markers for different mechanisms operating in the transformations of the two phases during SPD. This provides valuable insights into changes in phase shapes, their mutual arrangement, and the distinct plastic flow exhibited by the two phases during deformation. Furthermore, such studies can shed light on other potential transformations during and after SPD, including martensite formation from the austenite phase, precipitation of detrimental phases (such as sigma and chi), and the behavior of the alloy during severe extrusion processing [5-8].

This paper examines the SPD achieved through Equal Channel Angular Pressing (ECAP) in the 2507 SDSS grade. This specific grade was chosen due to the stability of its austenitic phase against martensitic transformation. The objective is to analyze the formation and evolution of the sub microcrystalline structure during ECAP processing in both the starting ferrite and austenite phases.

Materials and experimental procedures

The material tested shows a composition summarized in Table 1. It was delivered as a wrought rod of 20 mm in diameter in the solution-annealed condition at 1120°C for 1 hour and water quenched. The samples for the experiment had a diameter of 5mm and a length of 30 mm.

Table 1. Chemical composition of SDSS steel.

Cr	Mo	N	C	Ni	Mn	P	S	Si	Co
24.90	3.623	0.27	0.0463	6.55	0.530	0.0254	0.0113	0.624	0.079

The ECAP processes was performed on a hydraulic press DE-80 with 80 tons capacity, electronically controlled (speed of the plunger, pressure, travel length of the plunger). As a lubricant Molybdenum disulfide (MoS2) was used inside the channel.

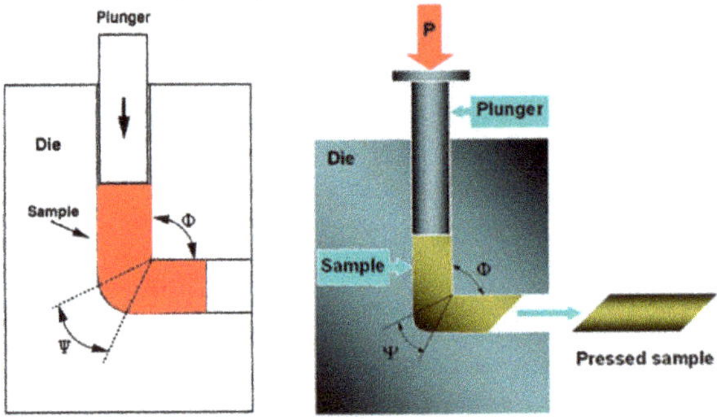

Fig. 1. ECAP worling

Samples for metallographic examination were obtained longitudinally and transversely, grinded with standard metallographic procedures (Silicon carbides paper up to 1200 grit size) and then polished with diamond suspension down to 1 micrometers. Etching was performed with 30wt% KOH solution under 2V for 15 sec. EBSD analysis was additionally performed but the samples had to undergo a further polishing step using colloidal silica (50 nm) for approximately 2 hours. EBSD was performed using and EBSD probe by Oxford Instrument in a field emission SEM FESEM JEOL J-71000F.

Mechanical testing was carried out on undersized specimen from ECAP processed samples. Moreover microhardness were performed with 1 Kg of weight for 20 s in a Vickers microhardness tester.

Results and discussion

Hardness:

As expected due to the extensive coldworking, hardness increased after each pass. The hardness was increased by 23% after the first pass, and only by 10% after the second pass (Fig. 2). This is because hardness will get saturated to a particular value after subsequent passes.

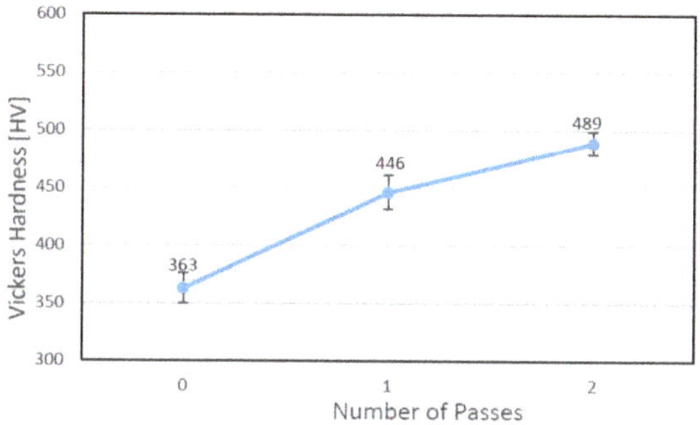

Fig. 2. Hardness improvement with respect to number of passes

Tensile test:

Compared to the coarse-graine counterpart, ECAP-processed SDSSs show significantly higher strength (Fig.3). A reduction in ductility is noted but generally is less compared to the conventional processing techniques such as drawing, rolling or extrusion. The ductility retention of ECAP processed specimens is greater compared to the cold rolled material [7].

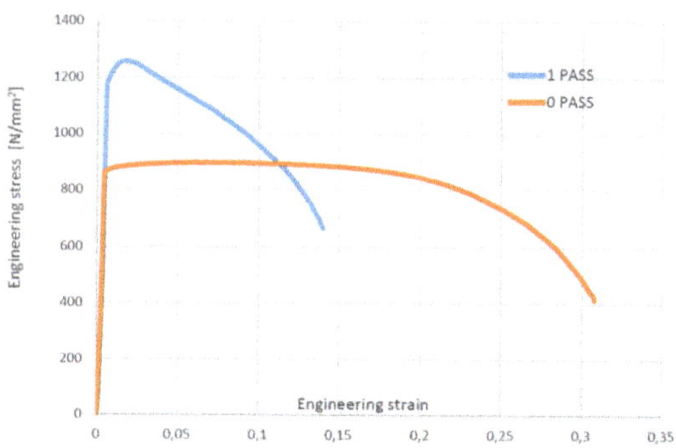

Fig. 3. Engineering stress-strain curve for 0 and 1 pass ECAPed samples

Table 2. Micro tensile test results

	0 Pass	1 Pass
Yield stress (YS) [N/mm^2]	864.62	1219.17
Ultimate tensile strength (UTS) [N/mm^2]	900.02	1310.06
Breaking stress [N/mm^2]	414.99	667.44
Percentage elongation (%)	43.37	28.47

Optical microscopy:
The phase balance of the as received material is almost equal (55±4% austenite and 45±3% ferrite). As can be seen from Fig.4 the microstructure shows elongated grains along the rolling direction, while the transverse section shows mainly equiaxed grains.

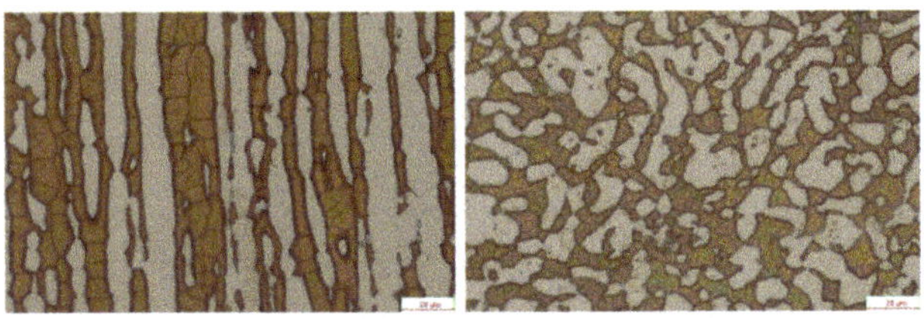

(a) (b)

Fig.4. Microstructure of studied DSS in (a) longitudinal section (b) transverse section

From Fig.5 it can be seen that deformations bands are present both after the first and the second pass of ECAP deformation. Those bands are common in multiphase metals deformed to high strain. This is due to the saturation of work hardening that causes local plastic instability. Ferrite and austenite have different plastic deformation behaviors due to their very different high stacking fault energy. Moreover, strain hardening is quite different in both ferrite and austenite which leads to the appearance of deformation bands. Optical microscope images (Fig. 5) clearly show that the spacing of deformation bands decreases with increasing deformation, indicating that grain subdivision continues to refine at high strain rate.

Materials Research Forum LLC
https://doi.org/10.21741/9781644902615-23

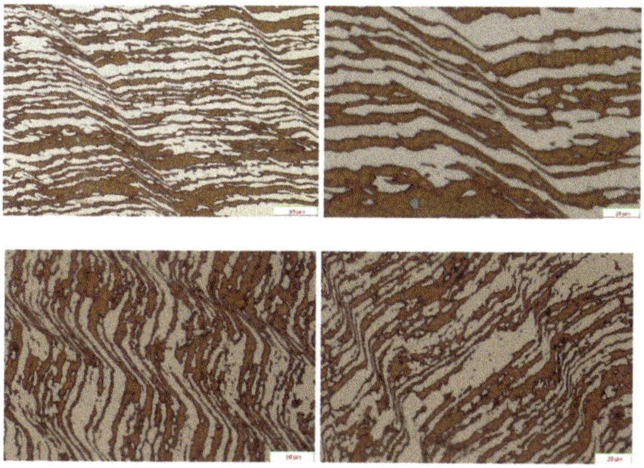

Fig. 5. Deformation bands observed after 1and 2 pass in optical microscope

EBSD analysis:
EBSD investigations is a very powerful tool in order to investigate the evolution of microstructure after each pass in ECAP processed materials. The EBSD maps have been performed along the longitudinal direction with respect the ECAP deformation (Fig.6).

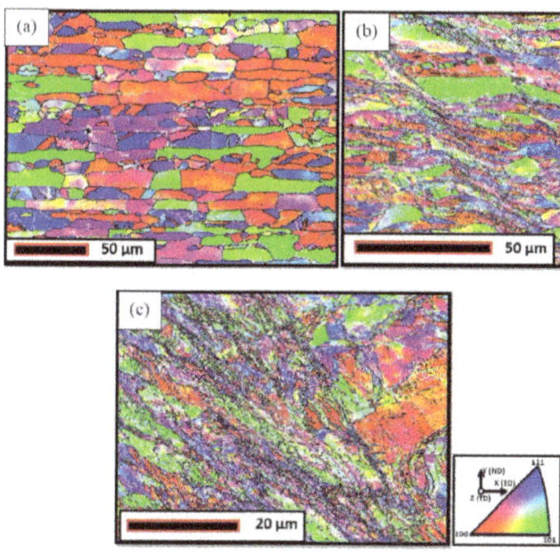

Fig. 6. IPF map with misorientations greater than 3°. Elongated as well as equi-axed grains are seen in the image. Also stereographic triangle, which indicates the orientatons of the grains in the IPF map. (a) 0 Pass (b) 1Pass (c) 2 Pass.

From the inverse pole figure map (FIG6.) it can be observed very extensive grain refinement near the deformation bands. After the two ECAP passes, the microstructure shows mainly equiaxed grains near the deformation bands and grains elongated near them.

The grain size just after the first pass is very small (0.9 μm). After the second pass the avarege grain size is reduced down to 0.4 μm.

Distribution of grain size in the non ECAPed material and after each pass is shown in Fig.7. The average grain size of the undeformed material is approximately 8.5 μm which get reduced to 0.9 μm after the first pass and to 0.4 μm after the second pass. Even though the as received material shows a very fine grain microstructure due to the effect of austenite and ferrite, the grain refinement of the ECAP processed is very strong.

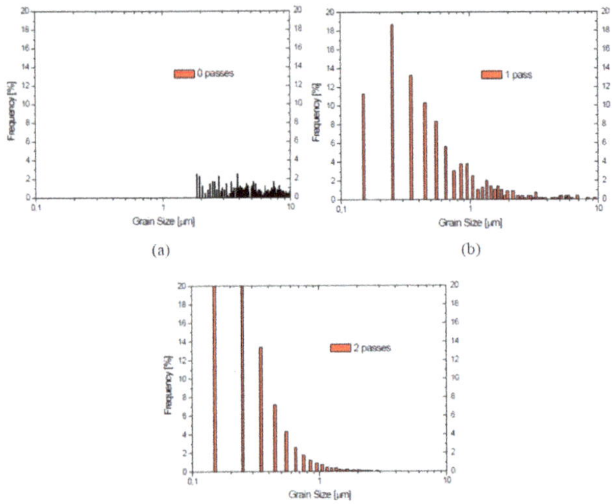

Fig. 7. Grain size distribution (a) 0 Pass (b) 1 Pass (c) 2 Pass

Grain refinement occurring during Equal Channel Angular Pressing (ECAP) can be attributed to the formation of deformation bands, which are an inevitable result of the process and have two potential origins. The first concept suggests that deformation bands originate due to the uncertainty involved in selecting the operative slip systems. In many cases, the applied strain can be accommodated by multiple sets of slip systems, resulting in different lattice rotations. Another origin is associated with the non-uniform straining, where different regions of a grain experience varying levels of strain. If the work done within the bands is insufficient for homogeneous deformation and if the bands can be arranged in a way that their net strain matches the overall deformation, they contribute to grain refinement. These deformation bands are separated by Geometrically Necessary Boundaries (GNBs). Additionally, boundaries can also form through the statistical trapping of dislocations, leading to the creation of Incidental Dislocation Boundaries (IDBs). The misorientation of both types of boundaries increases as the strain progresses. Furthermore, as deformation increases, the spacing between both types of boundaries decreases, indicating that grain subdivision continues to refine at higher strain rates.

Fig.8 shows the corresponding grain boundary maps, where the high angle grain boundaries (HAGBs), with a misorientation angle above 15°, are shown as black lines and boundaries 3° and 15° are shown as red lines.

The microstructure observed after the initial pass reveals a prevalence of low-angle boundaries, primarily in the austenite phase. Upon the second pass, the microstructure exhibits low angle misorientation boundaries in both the ferrite (BCC) and austenite (FCC) phases. Additionally, as

Materials Research Proceedings **32** (2023) 205-212 https://doi.org/10.21741/9781644902615-23

the number of passes in ECAP is increased, a gradual transition occurs from an array predominantly composed of low-angle grain boundaries (LAGBs) to a predominantly high-angle configuration.

In the examined Dual-Phase Stainless Steel (DSS), the application of ECAP initially generates grain boundary misorientations with an abundance of low-angle boundaries. This pattern of misorientation distribution is consistently observed after ECAP due to the significant number of excess dislocations introduced into the billet during each pass through the die. However, there is an observed trend of increasing average boundaries misorientation with an increase in the number of passes. This presents an opportunity to utilize ECAP or other severe plastic deformation techniques to obtain fine-grained materials with varying but controlled boundary misorientation distributions.

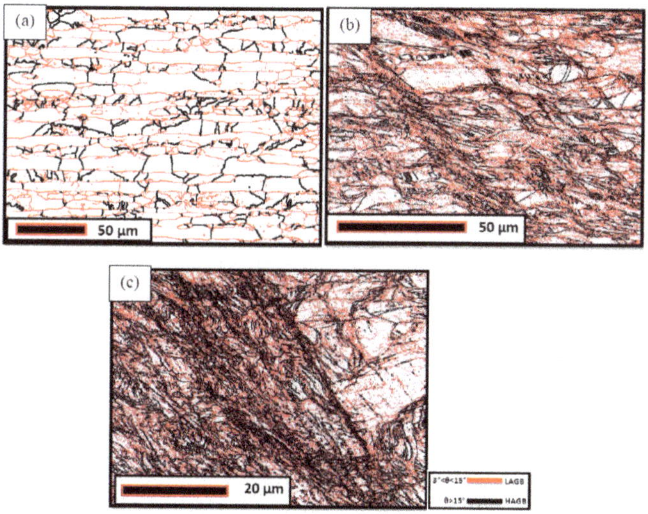

Fig.8. Boundary map which shows low angle grain boundaries and high angle grain boundaries (a) 0 Pass (b) 1 Pass (c) 2 Pass

After undergoing ECAP, the analysis of grain boundary misorientations indicates a predominance of low-angle boundaries. The consistent occurrence of a high proportion of low-angle boundaries in the misorientation distributions after ECAP is a direct result of the substantial number of excess dislocations introduced into the billet during each individual pass through the ECAP die. However, there is an observable trend of the average boundary misorientation increasing as the number of passes is increased. This presents a valuable opportunity to employ ECAP processing in the creation of materials with distinct yet controlled distributions of boundary misorientations.

Conclusions

The main results of the mechanical and microstructural characterization of the DSS steel processed by ECAP can be summarized:

1. Significant increment in hardness after the first pass about 23%. After the second pass the hardness has improved to 489HV, as compared to 363HV for its coarser part.

2. Ultimate tensile strength has improved to 45% after first pass. Percentage elongation decreased from 43% to 28% after the first pass. Engineering stress – strain curves showed increment in strength, reduction in ductility.
3. There is a significant grain refinement, the average grain size after the 2 pass is about 0.4μm, as compared to 8.5μm for its coarser part of 0 pass.
4. EBSD analysis and different maps showed the grain refinement especially near the deformation bands. Spacing of these deformation bands decreases with increase in deformation, indicating that grain subdivision continues to refine at high strain rate. IPF maps of ECAPed samples contains equiaxed grains at deformation bands and elongated grains next to the deformation bands.

References

[1] Y. Estrin, A. Vinogradov, Extreme grain refinement by severe plastic deformation: A wealth of challenging science, Acta Mater. 61, (2013), p. 782 https://doi.org/10.1016/j.actamat.2012.10.038

[2] R.Z. Valiev, T.G. Langdon, Principles of equal-channel angular pressing as a processing tool for grain refinement, J. Progr. Mat. Sci. 51, (2006), p. 881 https://doi.org/10.1016/j.pmatsci.2006.02.003

[3] R.Z. Valiev, Nanostructuring metals by severe plastic deformation for advanced properties, Nature Materials, 3, (2004) p. 511 https://doi.org/10.1038/nmat1180

[4] I. Alvarez-Armas, S. Degallaix-Moreuil, Duplex Stainless Steels. Wiley-ISTE (2009)

[5] A. Belyakov, Y. Kimura, K. Tsuzaki. Microstructure evolution in dual-phase stainless steels during severe deformation. Acta Mater. 54, (2006) p. 2521 https://doi.org/10.1016/j.actamat.2006.01.035

[6] S. Wronskia, J. TAarasiuk, B: Bacroix, A. Baczmanski, C. Braham, Investigation of plastic deformation heterogeneities in duplex stainless steels by EBSD, Mat. Charact. 73, (2012), p. 52. https://doi.org/10.1016/j.matchar.2012.07.016

[7] L. Chen, F.P. Yuan, P. Jiang, X.L. Wu, Mechanical properties and nanostructures in duplex stainless steels. J. Mat. Sci. Eng. A551, (2012) p. 154 https://doi.org/10.1016/j.msea.2012.04.112

[8] A. Momeni, K. Dehgani, Effect of Hot Working on Secondary Phase Formation in 2205 Duplex Stainless Steel, J. Mat. Sci. Technol., 26(9), /2010) p. 851 https://doi.org/10.1016/S1005-0302(10)60136-3

Consolidation of AZ31 magnesium chips using equal channel angular pressing

Majid Al-Maharbi[1,a *] and Almundher Al-Namani[1,b]

[1]Department of Mechanical and Industrial Engineering, Sultan Qaboos University, Muscat, Sultanate of Oman

[a]majidm@squ.edu.om, [b]almundher.spro6@gmail.com

Keywords: Magnesium Alloy, AZ31, Consolidation, Solid-State Recycling, Equal Channel Angular Pressing, Severe Plastic Deformation

Abstract. Equal channel angular pressing (ECAP) as a severe plastic deformation (SPD) technique was employed for consolidation of AZ31 magnesium chips. The chips were consolidated at a temperature of 300 °C following ECAP route A up to two passes. The optical and scanning electron micrographs reveal that consolidation have taken place and the quality of consolidation improves with number of ECAP passes and hence the total imposed strain. The Vickers hardness of the processed samples increased from 41 of the as cast used to generate the chips to 47.7 ad 58.5 of consolidated samples after one and two passes respectively. The consolidation of chips was accompanied by the breakage and dispersion of the oxide layer developed on the surface of the chips during and after machining.

Introduction

During the last two decades, magnesium (Mg) and its alloys have attracted great interests as structural materials for many engineering applications because of their high strength-to-weight ratio. However, frequent observations on the relatively low strength and ductility of these materials presents great concerns and have limits their applications. The consolidation or solid state recycling of chips is considered as a possible option to improve the mechanical properties of materials especially the lightweight materials like magnesium alloys. In the recent years many studies have been conducted to consolidate lightweight materials like magnesium, aluminum and titanium alloys. Mabuchi et al., [1] solid-state recycled (SSR) AZ91 Mg alloy using hot extrusion and were able to obtain a good combination of mechanical properties of the consolidated material. The solid-state recycling of Mg alloys using hot extrusion followed after the success of this study [2–8].

Wan et al [9] and Peng et al [10] suggested the consolidation mechanism of Mg alloys chips during solid-state recycling. Mg alloy chips are usually covered with a layer of oxide that may prevent the consolidation process. The oxide layers are broken down due to the large shear forces and high pressure at elevated temperature during the deformation process [11,12]. This break of the oxide layer helps meet the first requirements for the perfect bonding of chips, where the broken oxides particles are moved to the boundary between the recrystallized grains [89,101]. Therefore, direct contact between fresh metal surfaces, and interface bonding of chips can be accomplished. Cracks and pores are also expected to be distributed around the interface between chips [3,6]. However, the pores or voids can be reduced due to the plastic flow of the matrix under severe deformation and hence enhancing the compactness of the recycled Mg alloys [14]. Moreover, the physical consolidation may be fulfilled when the interfaces of the chips get dissolved because of the metallurgical bonding by atom diffusion under severe plastic deformation at elevated temperature [13]. The interfaces and crystal defects such as dislocation and grain boundaries caused by shear deformation act as diffusion paths of atoms [10,14]. Eventually, a fully dense material is obtained with uniform dissipation of oxide particles in the Mg alloy matrix.

The AZ31 alloy, as the most commonly used Mg alloys, has been solid-state recycled using hot extrusion [8], hot rolling [3], spark plasma sintering [15], and friction stir extrusion [16]. ECAP-based SSR method, to the best of the authors' knowledge has not been used to recycle Mg alloy chips before the current study. Wu et al. [8] found that double extrusion of AZ31 chips is necessary for the breakage and dispersion of the oxide particles into the structure and hence improving the consolidation of the recycled material. Similarly, Chino et al. [17] reported that high extrusion ratio of 1600:1 produced more uniform dispersion of oxide contaminant than lower extrusion ratio of 45:1. One of the advantages of ECAP over other SPD techniques is the possibility of continuous deformation without affecting the shape of the workpiece and hence accumulating imposed strain with more number of passes. This technique of solid-state recycling metals has been used before for solid-state recycle several titanium and aluminum alloys. In this study, ECAP is investigated as a means of consolidating the Mg AZ31 alloy chips into the bulk material.

Experimental Procedures

The raw material used in this investigation was cast AZ31 Mg alloy. The required AZ31 chips were generated using milling machine. Prior to production of the chips, all surfaces of AZ31 ingot have been machined to remove oxide layers. The milling process was carried out with a cutting tool speed of 900 rpm, a feed rate 75 (mm/min) and a depth of cut of 1 mm. No coolant or lubricant has been used to avoid any contamination with the chips. The produced chips have a curly shape with various sizes as shown in Figure 1.

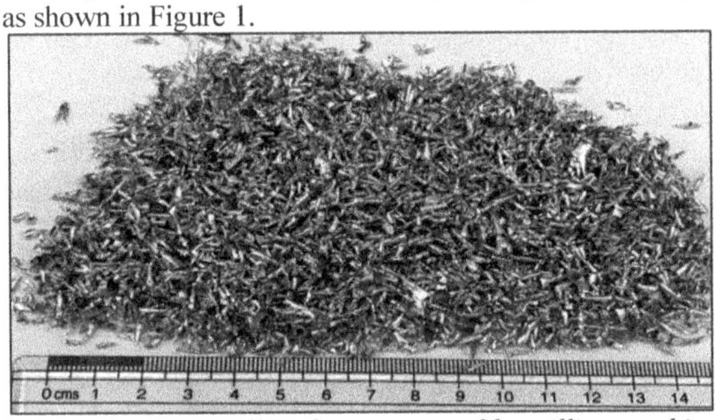

Figure 1 *AZ31 Mg alloy chips generated by milling machine.*

Prior to ECAP process, chips were pre-compacted using a locally fabricated steel mold. The compaction process was carried out at room temperature using 100 MPa pressure. Compacted billets with an approximate diameter of 14.5 mm and a length of 80 mm were produced. The compacted AZ31 billets were then inserted into aluminum cans for the ECAP process. The extrusion process was carried out using locally designed and fabricated ECAP die with a 90° angle and a 24.5 mm by 24.5 mm channels' cross-section.

Optical and electron microscopy were carried out on the consolidated specimens extracted from the fully consolidated region of the processed workpiece. All measurements and investigations were carried out on the flow plane normal to the flow direction FD. Samples were cut from 1-A and 2-A specimens were prepared following the standard metallographic procedures. For optical microscopy (OM), the samples were etched using Acetic Glycol etchant solution which is composed of 20 mL acetic acid, 1ml HNO_3, 20 ml H_2O and 60 ml of ethylene glycol. The samples were immersed in the etching solution for 25-30 seconds. VHX 1000E Keyence digital microscope was used for optical microscopy (OM) investigation. The same metallographic procedures were followed to prepare samples for scanning electron microscopy (SEM) investigation. Back-scattered electron (BSE) and secondary electron (SE) micrographs of the samples were taken to check the quality of the consolidation after 1 and 2 passes of route A. Elemental mapping of both samples was done using energy-dispersive X-ray spectroscopy (EDS) attached to the SEM. Vickers hardness (HV) with a diamond indenter was carried out on the flow plane of the recycled AZ31 samples. The sample face was subjected to a load of 10 kgf by indenter having a shape of right pyramid with a square base and an angle of 136 degrees between opposite faces. Several hardness measurements were taken at different regions per sample to evaluate the homogeneity of the recycled materials.

Experimental Results and Discussion

The optical micrographs (OM) of the consolidated AZ31 alloy samples taken from flow plane after one pass (1-A) and two passes (2-A) are shown in Figure 2.a and 2.b, respectively. It is well known that Mg and its alloy are easily oxidized during milling even at room temperature and hence a thick oxide layer will be formed on the chips surface. The thick dark boundaries seen in the micrographs of the consolidated AZ31 alloy samples are oxide layers [18–21]. It is expected that the consolidated chips after experiencing simple shear deformation during severe plastic deformation ECAP process will be distorted and rearranged from random directions to specific inclination angle. This angle changes with the number of passes and processing route used. In the current study the chip boundaries in the 1-A sample are oriented along the principal strain direction with an inclination angle of about 24-27° from the extrusion direction (ED). In previous chips consolidation studies using ECAP, inclination angles of 21°[18]and 22.5° [20] from the ED were measured.

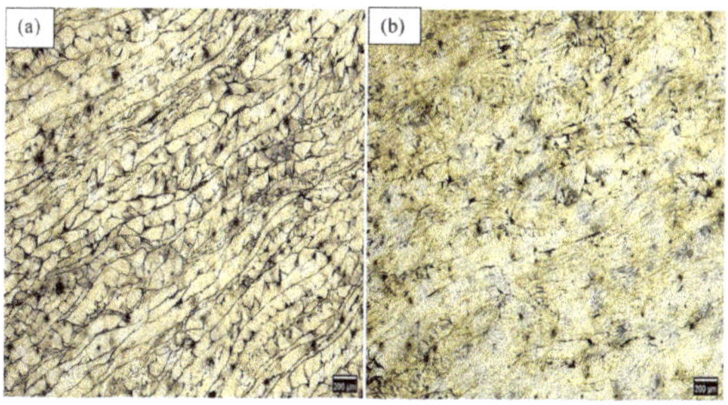

Figure 2 *Optical micrographs of consolidated AZ31 samples using ECAP at 300 C after (a) 1-A (one pass) and (b) 2-A (two passes). Extrusion direction (ED) is to the right.*

Figure 3 shows scanning electron micrographs (SEM micrographs) of 1-A and 2-A samples. These micrographs were taken from as polished samples without etching. The chips boundaries are visible as dark lines in 1-A sample as shown in Figure 3.a. These chips boundaries are clearly seen at high magnification SEM micrograph of 1-A (Figure 3.c). In contrast, 2-A sample shows less and thinner chip boundaries (see Figure 3.b. and 3.d) .This suggests that a higher number of passes may induce more breakdown and dispersion of chip boundaries, and hence achieve better consolidation between the chips. Similar chips boundaries appearance of SSR Mg alloy have also been reported in earlier studies utilizing different consolidation techniques like hot extrusion [6,22–24] and cyclic extrusion compression(CEC) [10,25].

Materials Research Forum LLC
https://doi.org/10.21741/9781644902615-24

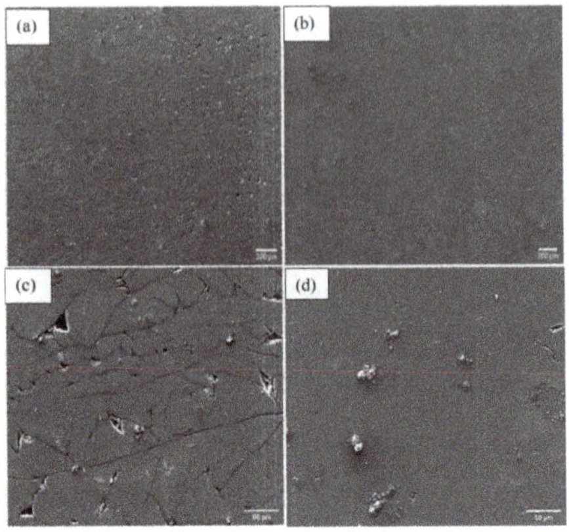

Figure 3 *SEM micrographs of 1-A and 2-A samples: (a) low magnification 1-A, (b) low magnification image 2-A, (c) high magnification 1-A and (d) high magnification 2-A.*

Figure 4 shows the elemental mapping of oxygen measured using EDS mounted in SEM. The mapping reveals more clearly the difference in oxide distribution between 1-A and 2-A samples. The oxide is concentrated at the chips surface and formed boundaries between the chips. After one pass shear deformation (1-A) of ECAP as shown in figure 4.a the oxide is still concentrated at the chips surfaces and formed boundaries between them. Oxide layers work as a barrier to fulfil the requirement of consolidation between the chips. In contrast to 1-A sample, it is clearly observed that most of the oxide layers between the chips are no longer visible in 2-A sample. The accumulation of shear deformation after two ECAP passes is capable of breaking up oxide layers into particles distributed in the structure of the recycled 2-A sample as shown in Figure 4..b. The destruction of oxide layers allows stronger bonding between chips.

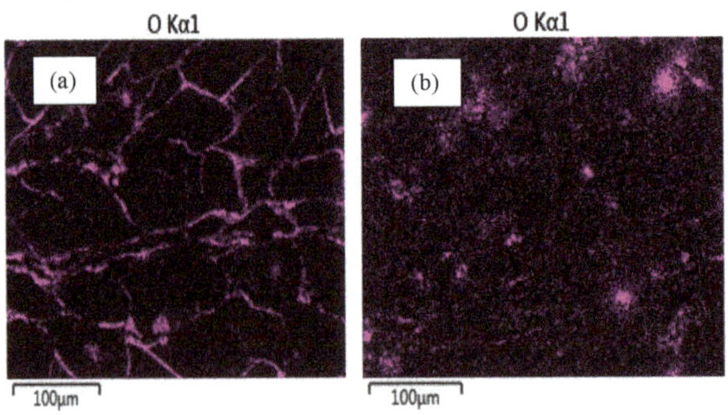

Figure 4. *EDS elemental mapping of oxygen in (a) 1-A (b) 2-A samples.*

In order to evaluate the mechanical properties of the recycled materials, hardness measurements were carried out on the flow plane of the recycled samples. The Vickers hardness (HV) of recycled material after one pass 1-A, two passes 2-A and as cast AZ31 were measured. Table 1 lists the HV values of 1-A, 2-A and as cast AZ31 samples taken from different locations. For easy comparison, Figure 5 shows the average HV value of recycled material 1-A, 2-A and as cast AZ31 samples. It can be clearly noticed that the hardness of the SSR materials is higher than the as cast material used to generate the chips. The average hardness values increased from 41± 1HV for as cast to 47.7±1.15 HV (~15% increase) after the first ECAP pass and further increased to 58.5±1.4 HV after the second pass about (~23% increase).

Table 1 *Vickers hardness values of as-cast and ECAP SSR AZ31 samples after one and two passes.*

Location	Cast AZ31 (0 pass) [HV]	1-A (One pass) [HV]	2-A (Two passes) [HV]
1	41.5	44	61.1
2	40.4	49.5	61.3
3	38.5	49.8	56.4
4	42.8	50.3	60.4
5	43.2	47.3	55.4
6	39.7	45.7	56.7
Average	**41.0**	**47.7**	**58.5**

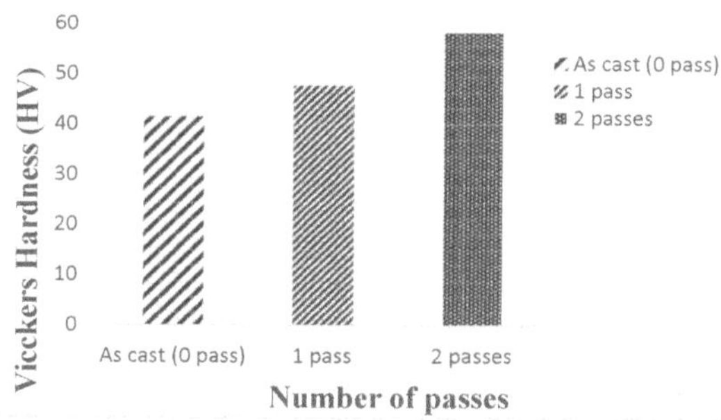

Figure 5 *The average values of Vickers hardness of as cast and ECAP SSR AZ31 samples.*

Conclusions

- Machining chips of AZ31 have been successfully consolidated at moderate a temperature below melting temperature using equal channel angular pressing (ECAP).
- The consolidation quality is significantly improved with higher number of passes, wherein the accumulation of shear deformation leads to easier destruction of chip boundaries (oxide layers), dissolve the interfaces between chips and hence better solid bonding occurred.
- Oxide contaminants were dispersed more uniformly in the recycled AZ31 material after multiple passes ECAP.
- The hardness of recycled AZ31 alloy is higher than that of the starting as cast AZ31 and it is further increased with increasing number of passes as a result of the greatly improved physical contact of chips and more uniform dispersion of the fine oxide particles.

References

[1] M. Mabuchi, K. Kubota, K. Higashi, New recycling process by extrusion for machined chips of AZ91 magnesium and mechanical properties of extruded bars, Mater. Trans. JIM. 36 (1995) 1249–1254. https://doi.org/10.2320/matertrans1989.36.1249

[2] Y. Chino, M. Kobata, K. Shimojima, H. Hosokawa, Y. Yamada, H. Iwasaki, M. Mabuchi, Blow Forming of Mg Alloy Recycled by Solid-State Recycling, Mater. Trans. 45

(2004) 361–364. https://doi.org/10.2320/matertrans.45.361

[3] Y. Chino, T. Hoshika, M. Mabuchi, Enhanced corrosion properties of pure Mg and AZ31 Mg alloy recycled by solid-state process, Mater. Sci. Eng. A. 435–436 (2006) 275–281. https://doi.org/10.1016/j.msea.2006.07.019

[4] M. Hu, Z. Ji, X. Chen, Z. Zhang, Effect of chip size on mechanical property and microstructure of AZ91D magnesium alloy prepared by solid state recycling, Mater. Charact. 59 (2008) 385–389. https://doi.org/10.1016/j.matchar.2007.02.002

[5] S. Wu, Z. Ji, T. Zhang, Microstructure and mechanical properties of AZ31B magnesium alloy recycled by solid-state process from different size chips, J. Mater. Process. Technol. 209 (2009) 5319–5324. https://doi.org/10.1016/j.jmatprotec.2009.04.002

[6] M.L. Hu, Z.S. Ji, X.Y. Chen, Q.D. Wang, W.J. Ding, Solid-state recycling of AZ91D magnesium alloy chips, Trans. Nonferrous Met. Soc. China (English Ed. 22 (2012) s68--s73. https://doi.org/10.1016/S1003-6326(12)61685-9

[7] S. Wu, Z. Ji, M. Hu, Z. Huang, C. Tian, M. Wu, Microstructure and Mechanical Properties of AZ31B Magnesium Alloy Prepared by Solid State Recycling, Xiyou Jinshu Cailiao Yu Gongcheng/Rare Met. Mater. Eng. 47 (2018) 736–741. https://doi.org/10.1016/s1875-5372(18)30101-2

[8] S. yan Wu, Z. sheng Ji, S. fan Rong, M. liang Hu, Microstructure and mechanical properties of AZ31B magnesium alloy prepared by solid-state recycling process from chips, Trans. Nonferrous Met. Soc. China (English Ed. 20 (2010) 783–788. https://doi.org/10.1016/S1003-6326(09)60214-4.

[9] B. Wan, W. Chen, T. Lu, F. Liu, Z. Jiang, M. Mao, Review of solid state recycling of aluminum chips, Resour. Conserv. Recycl. 125 (2017) 37–47. https://doi.org/10.1016/j.resconrec.2017.06.004

[10] T. Peng, Q.D. Wang, Y.K. Han, J. Zheng, W. Guo, Consolidation behavior of Mg-10Gd-2Y-0.5Zr chips during solid-state recycling, J. Alloys Compd. 503 (2010) 253–259. https://doi.org/10.1016/j.jallcom.2010.05.011

[11] D. Baffari, G. Buffa, L. Fratini, A numerical model for Wire integrity prediction in Friction Stir Extrusion of magnesium alloys, J. Mater. Process. Technol. 247 (2017) 1–10. https://doi.org/10.1016/j.jmatprotec.2017.04.007

[12] S. WU, Z. JI, S. RONG, M. HU, Microstructure and mechanical properties of AZ31B magnesium alloy prepared by solid-state recycling process from chips, Trans. Nonferrous Met. Soc. China. (2010). https://doi.org/10.1016/S1003-6326(09)60214-4

[13] T. Ying, M.Y. Zheng, X.S. Hu, K. Wu, Recycling of AZ91 Mg alloy through consolidation of machined chips by extrusion and ECAP, Trans. Nonferrous Met. Soc. China (English Ed. 20 (2010). https://doi.org/10.1016/S1003-6326(10)60547-X

[14] M. Moss, R. Lapovok, C.J. Bettles, The equal channel angular pressing of magnesium and magnesium alloy powders, Jom. 59 (2007) 54–57. https://doi.org/10.1007/s11837-007-0105-5

[15] D. Paraskevas, S. Dadbakhsh, J. Vleugels, K. Vanmeensel, W. Dewulf, J.R. Duflou, Solid state recycling of pure Mg and AZ31 Mg machining chips via spark plasma sintering, Mater. Des. 109 (2016) 520–529. https://doi.org/10.1016/j.matdes.2016.07.082

[16] D. Baffari, G. Buffa, D. Campanella, L. Fratini, A.P. Reynolds, Process mechanics in Friction Stir Extrusion of magnesium alloys chips through experiments and numerical simulation, J. Manuf. Process. 29 (2017) 41–49. https://doi.org/10.1016/j.jmapro.2017.07.010

[17] Y. Chino, T. Hoshika, J.S. Lee, M. Mabuchi, Mechanical properties of AZ31 Mg alloy recycled by severe deformation, J. Mater. Res. 21 (2006) 754–760. https://doi.org/10.1557/jmr.2006.0090

[18] Q. Shi, Y.Y. Tse, R.L. Higginson, Effects of processing parameters on relative density, microhardness and microstructure of recycled Ti-6Al-4V from machining chips produced by equal channel angular pressing, Mater. Sci. Eng. A. 651 (2016) 248–258. https://doi.org/10.1016/j.msea.2015.11.002

[19] P. Luo, H. Xie, M. Paladugu, S. Palanisamy, M.S. Dargusch, K. Xia, Recycling of titanium machining chips by severe plastic deformation consolidation, J. Mater. Sci. 45 (2010) 4606–4612. https://doi.org/10.1007/s10853-010-4443-2

[20] P. Luo, D.T. McDonald, S.M. Zhu, S. Palanisamy, M.S. Dargusch, K. Xia, Analysis of microstructure and strengthening in pure titanium recycled from machining chips by equal channel angular pressing using electron backscatter diffraction, Mater. Sci. Eng. A. 538 (2012) 252–258. https://doi.org/10.1016/j.msea.2012.01.039

[21] D.T. McDonald, P. Luo, S. Palanisamy, M.S. Dargusch, K. Xia, Ti-6Al-4V recycled from machining chips by equal channel angular pressing, Key Eng. Mater. 520 (2012) 295–300. https://doi.org/10.4028/www.scientific.net/KEM.520.295

[22] S. Wu, Z. Ji, T. Zhang, Microstructure and mechanical properties of AZ31B magnesium alloy recycled by solid-state process from different size chips, J. Mater. Process. Technol. 209 (2009) 5319–5324. https://doi.org/10.1016/j.jmatprotec.2009.04.002

[23] L. Wen, Z. Ji, X. Li, Effect of extrusion ratio on microstructure and mechanical properties of Mg – Nd – Zn – Zr alloys prepared by a solid recycling process, 9 (2008) 2–7. https://doi.org/10.1016/j.matchar.2008.03.009

[24] Z. Zu-de, C. Qiang, Y. Lin, S.H.U. Da-yu, Z. Zhi-xiang, Microstructure and mechanical properties of Mg-Zn-Y-Zr alloy prepared by solid state recycling, Trans. Nonferrous Met. Soc. China. 21 (2010) 265–271. https://doi.org/10.1016/S1003-6326(11)60708-5

[25] T. Peng, Q.D. Wang, J.B. Lin, Microstructure and mechanical properties of Mg–10Gd–2Y–0.5Zr alloy recycled by cyclic extrusion compression, Mater. Sci. Eng. A. 516 (2009) 23–30. https://doi.org/10.1016/j.msea.2009.04.024.

Secondary phase precipitation in ultrafine-grained superduplex stainless steels

Alisiya Biserova-Tahchieva[1,a*], Núria Llorca-Isern[1,b] Jose Maria Cabrera[2,c]

[1]Universitat de Barcelona, Departament de Ciència de Materials i Química Física, Martí i Franquès 1-10, E-08028 Barcelona

[2]Universitat Politècnica de Catalunya, Dept. Ciència i Enginyeria de Materials EEBE, c/Eduard Maristany 10-14, E-08019 Barcelona

[a]abiserova.tahchieva@ub.edu, [b]nullorca@ub.edu, [c]jose.maria.cabrera@upc.edu

Keywords: Duplex Stainless Steel, HPT, Severe Plastic Deformation, Secondary Phases

Abstract. Ultrafine-grained S322750 superduplex stainless steel has been obtained through high-pressure torsion (HPT) process. Microstructural evolution after the severe plastic deformation (SPD) has shown no phase transformation. Isothermal heat treatment was applied after the deformation process in order to promote secondary phase precipitation. Precipitation was enhanced in the ultrafine-grained stainless steel in comparison to the coarse-grained stainless steel at the same isothermal heat treatment conditions. Mainly sigma phase and chromium nitrides were found as a product of the local diffusion process. The morphology of the secondary phases has been analyzed by using field emission scanning electron microscopy (FESEM) and transmission electron microscopy (TEM).

Introduction

The increasing global demand for high-performance stainless steels is driven by their significant environmental and economic impact. While renewable energies and alternative energy sources are on the rise, the oil and gas industries still require materials that can withstand geologically challenging reservoir conditions and operational complexities [1]

Duplex stainless steels continue to be utilized in various industries, including chemical engineering, offshore structures, and construction, due to their exceptional properties. These steels exhibit superior toughness, corrosion resistance, and higher strength compared to austenitic stainless steels [2]. Additionally, their cost is lower due to the reduced nickel content. The combination of ferrite (δ) and austenite (γ) phases in equal proportion meets the increased demands for yield strength and corrosion resistance, which cannot be achieved by ferritic or austenitic stainless steels alone.

Furthermore, diverse applications, such as nuclear plant reactors, require careful examination of steel microstructure evolution and the formation of secondary phases. The presence of secondary phases, including sigma (σ) phase, chromium nitrides (CrN, Cr_2N), and carbides ($M_{23}C_6$, M_7C_3), is of significant importance [3]–[6] .Sigma phase is particularly studied for its impact on toughness and corrosion susceptibility due to Cr-depleted zones [7]–[9]. Precipitation of chromium nitrides or carbides, especially in grades with higher nitrogen content, can affect the mechanical properties of the material. Consequently, critical alloying elements such as Cr, Mo, and Ni decompose, leading to a decrease in mechanical and corrosion properties. Extensive research has been conducted on the morphology, size, precipitation mechanism, and orientation relationship between precipitates and the matrix, which are documented in the literature [10]–[13]. However, most of these studies focus on the microstructure of steels immediately after conventional deformation processes or after industrial processing and aging processes, neglecting any previous procedures that may impact microstructural changes. For instance, processing the steel to achieve an ultrafine-grained microstructure can significantly enhance mechanical properties such as toughness, strength, and fatigue resistance [14]–[16]. Severe plastic deformation (SPD) methods, including high-pressure torsion (HPT), are considered more effective than conventional plastic deformation methods. HPT, a deformation process that minimally affects the dimensions of the object, enables the formation of nanostructures that exhibit substantial improvements in mechanical properties, including superplasticity [17], [18]. It is essential to investigate the combined effects of SPD and thermodynamically induced changes in microstructure.

This study aims to examine the microstructural evolution of UNS S32750 superduplex stainless steel subjected to HPT deformation at room temperature and its impact on the precipitation process. The ultrafine grains and precipitates are analyzed using field-emission scanning electron microscopy (FE-SEM) and transmission electron microscopy (TEM).

Experimental procedure

The material chosen was superduplex stainless steel grade 2507 (UNS S32750), with the following chemical composition: C: 0,018; Si: 0,26; Mn: 0,84; P: 0,019; S: 0,001; Cr: 25,08; Ni: 6,88; Mo: 3,82; N: 0,29; Cu: 0,17, balanced in Fe. For the experiments, disc-shaped samples (10mm diameter and 1mm thick) were prepared by cutting a hot extruded tube. A solution annealing process (1080°C and held for 30 minutes) was conducted previously, aimed to eliminate any undesirable phases and promote structural homogenization. Subsequently, the samples were quenched in water to rapidly cool them at. The high-pressure torsion (HPT) process was performed at room temperature with an unconstrained setup. The pressure applied during the deformation was 6 GPa. Different numbers of full revolutions, referred to as "turns," were applied: N=1, 5, 10 and 12. The rotation speed of 1 revolution per minute was maintained during the process. The samples were individually placed between a pair of anvils and strained by pressing and rotating the lower anvil. Following the HPT deformation, the samples underwent an isothermal treatment to induce the precipitation of secondary phases (at 830 ±5 °C for around 3 minutes and rapidly quenched in water). Figure 1 provides a schematic representation of the appearance of the samples, and they were characterized.

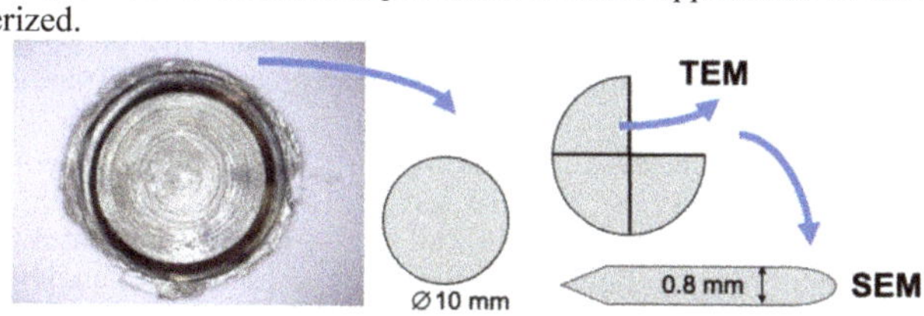

Figure 1. Aspect and dimensions and schematic representation of the characterization of a HPT-deformed sample.

Field-emission scanning electron microscope (FE-SEM) - JEOL J-7100F (JEOL Ltd., Tokyo, Japan) with a coupled Robinson back scattered electron (BSE) detector was used. A JEOL JEM 2100 (JEOL Ltd., Tokyo, Japan) operated at acceleration voltage of 200 kV, equipped with EDS system (80 mm^2 SDD, Oxford Instruments AZtec).

Results and Discussion

Micrographs of the samples deformed by 1, 5, 10 and 12 turns of HPT near the center of the disc samples confirm the deformation of the microstructure. It can be seen the increase of homogeneity from 1 to 12 turns and a similar microstructure with elongated and twisted austenite grains (brighter phase).

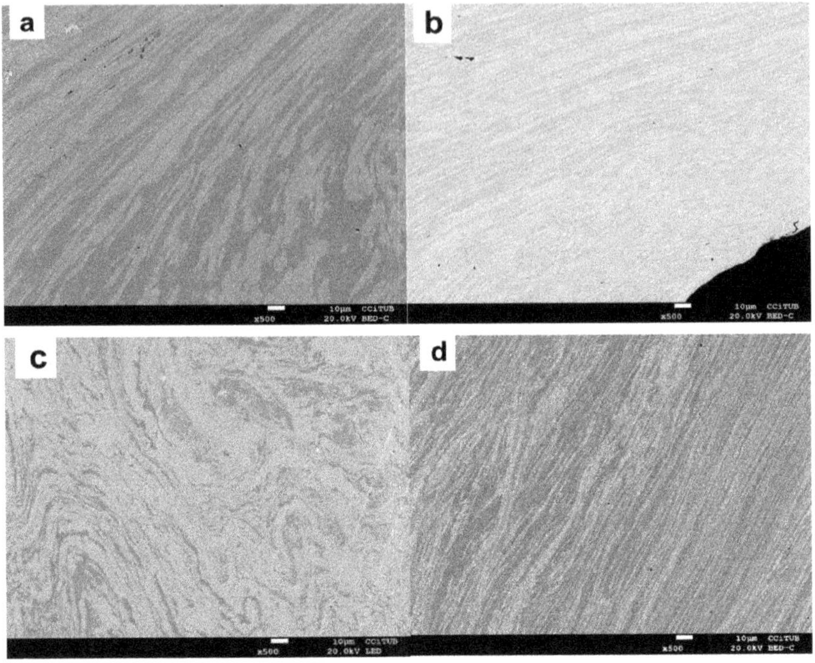

Figure 2. FE-SEM micrographs of SDSS samples near the center of the discs after different number of HPT turns: (a) N=1, (b) N=5, (c) N=10 and (d) N=12.

As observed in Figure 2, there is no evidence of phase transformation or precipitation of secondary phases after HPT. Grains morphology and microstructure analysis has been carried out through TEM.

Bright field (BF) images and selected area diffraction (SAED) patterns have been obtained from the four samples after deformation by 1, 5, 10 and 12 HPT turns. Figure 3 shows the BF micrographs and the SAED patterns, respectively. In Figure 3 (a), the two phases can be seen, in the area highlighted with a dashed yellow line, where the bcc-ferritic phase with less defects is indicated, unlike the fcc-austenite with significant deformation and accumulated dislocations. From the diffraction pattern of the selected zone (marked in Figure 3 (a) with a white circle), the crystallinity of the sample and its random orientation can be appreciated. The fuzzy appearance of grain boundaries is characteristic of non-equilibrium ultrafine grains (UFG) [19]. These, obtained after severe plastic deformation, contain a high density of extrinsic and intrinsic dislocations, as well as elastically distorted layers along the grain boundaries that lead to diffuse images. An example of this aspect can be seen in Figure 3 (b) of the sample after 5 turns. In addition, the accumulation of dislocations forming walls is observed. From the rings of the diffraction patterns, the polycrystallization and the selected zone corresponding to the fcc-ferrite phase are highlighted. As the deformation increases, a greater homogenization is seen, which is reflected in the nanostructure, where after 12 turns, there is a distribution of equiaxed grains and less differentiation between phases, as well as their corresponding diffraction of electrons in the form of diffuse rings, indicating a greater polycrystallization, as can be seen in the image of the diffraction patterns in Figure 3 (c), highlighting the selected zone corresponding to the fcc-austenite phase.

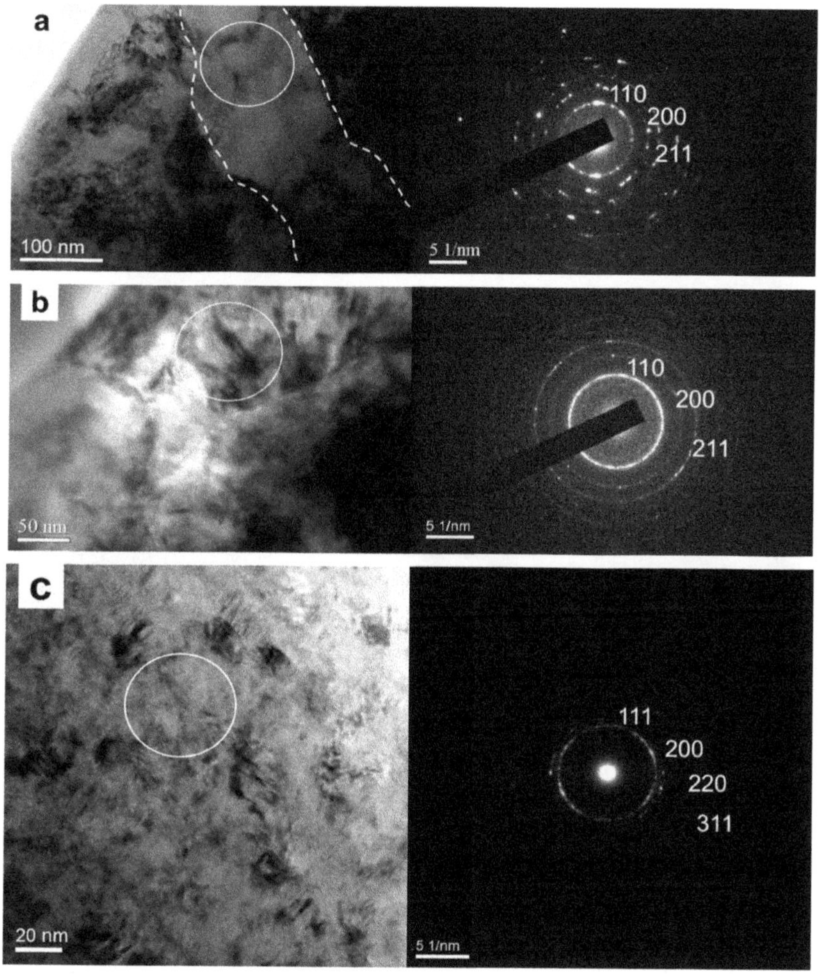

Figure 3. Bright-field TEM micrographs of SDSS deformed samples by HPT: (a) N=1, (b) N=5, (c) N=12 and their corresponding SAED patterns, respectively.

With the short thermal treatment carried out after the HPT process, the precipitation of secondary phases has been easily distinguished in the grains stressed by the deformation, as can be seen in Figure 4. The bright field images of the samples after (a) 1 turn, (b) 5 turns, (c) 10 turns and (d) 12 turns can be observed. The grain size in the different microstructures is relatively comparable, grains between 50 and 100 nm correlated to chromium nitride precipitates and larger grains ~500 nm for the austenite grains and some sigma phase grains. On the other hand, stressed grains are observed in which the accumulation of dislocations stands out (marked with yellow arrows in Figure 4 (a) and (b). In addition, twinning due to mechanical deformation (marked with red arrows in Figure 4 (a), (b), and (d) can be detected in some grains, characterized by a minimal distortion, sometimes difficult to observe, which can confuse them with twins of thermal origin, but their smaller size, around 100 nm or less, as well as their lenticular shape, are some of the indications to distinguish them from these.

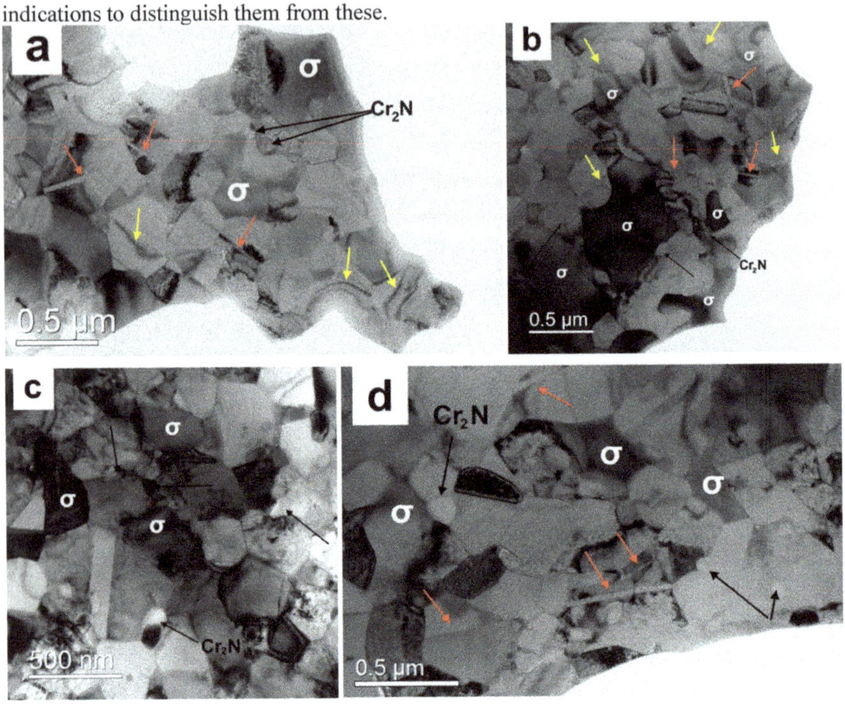

Figure 4. Bright-field TEM micrographs of SDSS deformed samples by HPT and thermally treated: (a) N=1, (b) N=5, (c) N=10 and (d) N=12. Red arrows point out twinning, yellow arrows mark dislocations accumulation and black arrows show some of the Cr_2N.

Smaller grains, around 100 nm, are associated with the precipitates of chromium nitrides (marked with black arrows in Figure 4). The precipitation of Cr_2N has been observed at the grain boundaries of γ/γ, considered as intergranular, as well as in the austenitic grains. The high nitrogen content increases the degree of undercooling, reduces the free energy, and accelerates the precipitation of Cr_2N. The substantial plastic deformation, combined with subsequent temperature exposure, provides favorable conditions for the rapid mobility of nitrogen atoms with a high diffusion coefficient. The precipitation of the sigma phase occurs after the nucleation of the chi phase, as the precipitation temperature of the latter is lower than that of the sigma phase, a phenomenon observed in various studies by multiple authors. Subsequently, the chi phase is gradually consumed by the sigma phase until complete disappearance. In this study, the precipitation of the chi phase was not observed in the severely deformed microstructure after the HPT process. Comparatively with previous studies, secondary phase precipitation was much greater in the deformed microstructures, and the evolution of the sigma phase was significant enough to completely consume the ferrite phase, as ferrite was not detected.

Due to the continuous diffusion of elements during the precipitation process, starting with the diffusion of nitrogen towards the grain boundaries, the formation of initial intermetallic phases occurs, and chromium nitrides configure zones where some elements decrease while others are in excess. Specifically, after the nucleation of the sigma phase, the diffusion of elements such as Fe, Cr, and Mo from the ferrite phase towards the sigma phase leads to the generation of these depleted zones in Cr and Mo. Moreover, the diffusion of elements from the austenite phase, such as Ni, also results in an enriched zone of Ni. Many authors in literature consider this specific zone generated due to the diffusion of elements and the precipitation of secondary phases as a new phase, referred to as secondary austenite or $\gamma 2$. It is considered that the precipitation of the sigma phase together with the "secondary austenite" as a eutectoid reaction is not appropriate. Therefore, the same logic applies to the precipitation in superduplex stainless steel, considering it as an invariant transformation.

Conclusions

Superduplex stainless steel has shown no secondary precipitation after HPT deformation. Subsequent isothermal treatment has shown increased precipitation of intermetallic secondary phases. Sigma phase and Cr_2N are the main precipitates found in the microstructure after a short thermal treatment on the 1, 5, 10 and 12 turns of HPT deformed samples. There is a similar trend and homogeneous distribution, as well as size and morphology in the four samples. The sigma phase has precipitated mainly, but not exclusively, at the austenite/chromium nitride/austenite and austenite/austenite/austenite triple point. The results confirmed that chromium nitrides appear to be the preferred nucleation sites for the sigma phase in the most strained samples.

References

[1] B. D. Craig and L. Smith, “Corrosion Resistant Alloys (CRAs) in the oil and gas industry – selection guidelines update,” *Nickel Institute Technical Series*, no. 0073, 2011.

[2] P. Kangas and G. Chai, “Use of advanced austenitic and duplex stainless steels for applications in Oil & Gas and Process industry,” *Century of Stainless Steels*, vol. 794, pp. 645–669, 2013. https://doi.org/10.4028/www.scientific.net/AMR.794.645

[3] L. Pezzato, M. Lago, K. Brunelli, M. Breda, E. Piva, and I. Calliari, “Effect of secondary phases precipitation on corrosion resistance of duplex stainless steels,” in *Materials Science Forum*, Trans Tech Publications Ltd, 2017, pp. 1495–1500. https://doi.org/10.4028/www.scientific.net/MSF.879.1495

[4] G. Argandona, M. V. Biezma, J. M. Berrueta, C. Berlanga, and A. Ruiz, “Detection of Secondary Phases in UNS S32760 Superduplex Stainless Steel by Destructive and Non-destructive Techniques,” *J Mater Eng Perform*, vol. 25, no. 12, pp. 5269–5279, Dec. 2016. https://doi.org/10.1007/s11665-016-2395-7

[5] Y. J. Kim, S. W. Kim, H. B. Kim, C. N. Park, Y. Il Choi, and C. J. Park, “Effects of the precipitation of secondary phases on the erosion-corrosion of 25% Cr duplex stainless steel,” *Corros Sci*, vol. 152, no. March, pp. 202–210, 2019. https://doi.org/10.1016/j.corsci.2019.03.006

[6] K. W. Chan and S. C. Tjong, “Effect of secondary phase precipitation on the corrosion behavior of duplex stainless steels,” *Materials*, vol. 7, no. 7, pp. 5268–5304, 2014. https://doi.org/10.3390/ma7075268

[7] C.-C. Hsieh and W. Wu, “Overview of Intermetallic Sigma () Phase Precipitation in Stainless Steels,” *ISRN Metallurgy*, vol. 2012, pp. 1–16, Mar. 2012. https://doi.org/10.5402/2012/732471

[8] M. B. Cortie and E. M. L. E. M. Jackson, "Simulation of the precipitation of sigma phase in duplex stainless steels," *Metall Mater Trans A Phys Metall Mater Sci*, vol. 28, no. 12, pp. 2477–2484, 1997. https://doi.org/10.1007/s11661-997-0005-x
[9] W. Yongqiang, S. Hao, N. Li, X. Yanhao, and J. Hemin, "Effect of Sigma Phase Precipitation on the Pitting Corrosion Mechanism of Duplex Stainless Steels," *Int. J. Electrochem. Sci*, vol. 13, pp. 9868–9887, 2018. https://doi.org/10.20964/2018.10.38
[10] A. D. Warren, R. L. Harniman, Z. Guo, C. M. Younes, P. E. J. Flewitt, and T. B. Scott, "Quantification of sigma-phase evolution in thermally aged 2205 duplex stainless steel," *J Mater Sci*, vol. 51, no. 2, pp. 694–707, Jan. 2016. https://doi.org/10.1007/s10853-015-9131-9
[11] N. Llorca-Isern, H. López-Luque, I. López-Jiménez, and M. V. Biezma, "Identification of sigma and chi phases in duplex stainless steels," *Mater Charact*, vol. 112, pp. 20–29, Feb. 2016. https://doi.org/10.1016/j.matchar.2015.12.004
[12] J. S. Li *et al.*, "Thermal cycling induced stress–assisted sigma phase formation in super duplex stainless steel," *Mater Des*, vol. 182, 2019. https://doi.org/10.1016/j.matdes.2019.108003
[13] R. O. Sousa, P. Lacerda, P. J. Ferreira, and L. M. M. Ribeiro, "On the Precipitation of Sigma and Chi Phases in a Cast Super Duplex Stainless Steel," *Metallurgical and Materials Transactions A*, vol. 50, no. 10, pp. 4758–4778, 2019. https://doi.org/10.1007/s11661-019-05396-6
[14] R. Z. Valiev, Y. Estrin, Z. Horita, T. G. Langdon, M. J. Zehetbauer, and Y. Zhu, "Producing Bulk Ultrafine-Grained Materials by Severe Plastic Deformation: Ten Years Later," *JOM*, vol. 68, no. 4, pp. 1216–1226, Apr. 2016. https://doi.org/10.1007/s11837-016-1820-6
[15] M. Kawasaki, H.-J. Lee, B. Ahn, A. P. Zhilyaev, and T. G. Langdon, "Evolution of hardness in ultrafine-grained metals processed by high-pressure torsion," *Journal of Materials Research and Technology*, vol. 3, no. 4, pp. 311–318, Oct. 2014. https://doi.org/10.1016/J.JMRT.2014.06.002
[16] T. G. Langdon, "Twenty-five years of ultrafine-grained materials: Achieving exceptional properties through grain refinement," *Acta Mater*, vol. 61, no. 19, pp. 7035–7059, Nov. 2013. https://doi.org/10.1016/j.actamat.2013.08.018
[17] A. P. Zhilyaev and T. G. Langdon, "Using high-pressure torsion for metal processing: Fundamentals and applications," *Prog Mater Sci*, vol. 53, no. 6, pp. 893–979, Aug. 2008. https://doi.org/10.1016/j.pmatsci.2008.03.002
[18] K. Edalati and Z. Horita, "A review on high-pressure torsion (HPT) from 1935 to 1988," *Materials Science and Engineering: A*, vol. 652, pp. 325–352, Jan. 2016. https://doi.org/10.1016/j.msea.2015.11.074
[19] R. Z. Valiev, V. Yu. Gertsman, and O. A. Kaibyshev, "Grain boundary structure and properties under external influences," *physica status solidi (a)*, vol. 97, no. 1. John Wiley & Sons, Ltd, pp. 11–56, Sep. 16, 1986. doi: 10.1002/pssa.2210970102

Grain boundary sliding at low temperatures

Roberto B. Figueiredo

Department of Metallurgical and Materials Engineering,
Universidade Federal de Minas Gerais, Brazil

figueiredo@demet.ufmg.br

Keywords: Deformation Mechanisms, Grain Boundary Sliding, Superplasticity

Abstract. Grain boundary sliding plays a key role on the high temperature deformation of fine grained materials. This mechanism is related to a high strain rate sensitivity of approximately 0.5 and usually gives rise to high superplastic elongations. The rate controlling equation for the mechanism of grain boundary sliding has shown good agreement with experimental data for multiple materials, with different grain sizes and tested at different strain rates. However, the predictive ability of the rate controlling equation seems to deteriorate at low temperatures. Although there are experimental evidences of high strain-rate sensitivities in ultrafine grained materials tested at low temperatures, this parameter does not reach values near 0.5 and also there seems to be disagreement in stress level in many conditions. The present overview evaluates the occurrence of grain boundary sliding in ultrafine grained materials at low temperatures considering an adapted rate controlling equation which display good agreement with experimental data. A gradual transition from grain refinement softening at high temperature to grain refinement hardening at low temperatures and a gradual increase in strain rate sensitivity with increasing temperature are observed.

Introduction

The occurrence of grain boundary sliding as a deformation mechanism at high homologous temperatures is now well established. This mechanism is observed in materials with fine grains, typically less than 10 μm, at temperatures typically above half of the material melting point. There is also a strain rate range, or stress range, in which this mechanism is rate controlling. The rate controlling equations for high temperature creep mechanisms are usually in the format in eq. 1 [1]

$$\dot{\varepsilon} = \frac{ADGb}{kT}\left(\frac{\sigma}{G}\right)^n \left(\frac{b}{d}\right)^p \tag{1}$$

in which $\dot{\varepsilon}$ is the effective strain rate, A is a dimensionless constant, D is the diffusion coefficient, G is the shear modulus, b is the Burgers vector, k is the Boltzmann constant, T is the absolute temperature for creep, σ is the effective stress, n is the stress exponent and p the inverse of the grain size exponent. Different creep mechanisms have different parameters A, D, n and p and so each is rate controlling at specific conditions of stress, temperature and grain size. A common way to visualize the rate controlling mechanism for the different conditions is through deformation mechanism maps and an example is given in Fig. 1 for the AZ31 magnesium alloy tested at different temperatures [2]. Thus, the range for each of the deformation mechanisms is delineated and experimental data from the literature [2-13] is also shown. A different color is used for each deformation mechanism and the same color is used to indicate the deformation mechanism reported in each experiment. It is apparent there is a good correlation between the theoretical deformation mechanism maps and the experimental data. Thus, the high temperature deformation of metallic materials is fairly well predicted in terms of creep mechanisms.

https://doi.org/10.21741/9781644902615-26

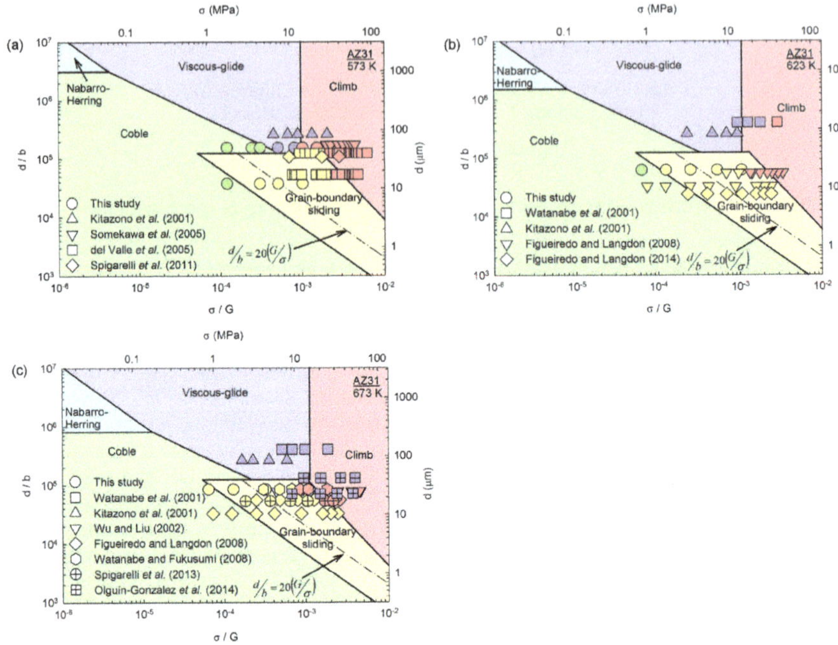

Figure 1 – Deformation mechanism maps for high temperature creep of a Mg alloy AZ31 [2].

The occurrence of grain boundary sliding (GBS) in fine grained materials at temperatures above half the melting point is well established. However, there has been great interest in the development of ultrafine and nanocrystalline materials [14] and in the evaluation of the occurrence of GBS at lower temperatures. Early studies showed that some features of GBS were observed at lower temperatures in these materials such as grain boundary offsets after deformation and increased strain rate sensitivity. However, the experimental data of flow stress and strain rate do not agree with the predictions from the rate equation for high temperature GBS. A recent paper [15] showed that a simplification considered in the model for high temperature GBS is not valid for lower temperature deformation and a modified version of the rate equation was suggested. The present overview evaluates this modified model and show that the mechanism of grain boundary sliding agrees with experimental data for lower temperature deformation of many metallic materials.

The model for grain boundary sliding and data for room temperature strength

The model [15] considers that grain boundary sliding takes place by the movement of dislocations in the vicinity of grain boundaries. These dislocations pile up at triple junctions increasing the stress and trigger dislocation motion in the neighboring grains. These dislocations then pile up at the opposite boundaries where they undergo climbing assisted by the high stresses. The equation for the deformation rate is available elsewhere [15] and an equation for the flow stress is given below [15].

Materials Research Forum LLC
https://doi.org/10.21741/9781644902615-26

$$\sigma \approx \sqrt{\frac{3GkT}{2d_s b^2} ln\left(\frac{\dot{\varepsilon} d_s^{\ 3}}{10\delta D_{gb}} + 1\right)} \tag{2}$$

where d_s is the spatial grain size and δ is the grain boundary width which is usually considered as $2b$. It was shown that the predictions of this model agrees with experimental data for multiple materials tested at different temperatures and strain rates [15]. For instance, Fig. 2 shows a good agreement between the flow stress determined in experiments at room temperature and the predictions from eq. 2 for 27 different materials processed by high-pressure torsion [16].

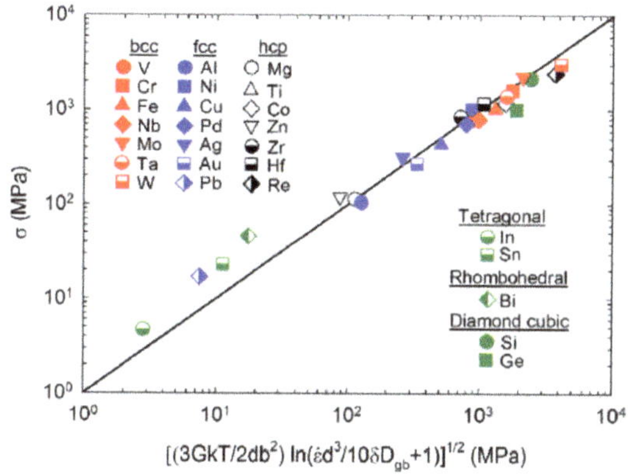

Figure 2 – Flow stress determined in experiments at room temperature plotted as a function of the flow stress predicted by the model of grain boundary sliding for 27 different materials [16].

The model of grain boundary sliding only predicts the contribution of the grain size to the flow stress and other strengthening mechanisms such as solid solution hardening must be incorporated in order to evaluate the overall strength of materials. Thus, the contribution of other strengthening mechanisms might be introduced by the incorporation of a parameter σ_0 in addition to the right side of eq. 2. Also, many studies evaluate the grain size using the mean linear intercept length method and this affects the prediction of the GBS model. Thus, eq. 3 shows the prediction of the flow stress considering a threshold stress, σ_0, and the mean linear intercept grain size, d_l.

$$\sigma = \sigma_0 + \sqrt{\frac{\sqrt{3}GkT}{2d_l b^2} ln\left(\frac{\dot{\varepsilon} d_l^{\ 3}}{2\delta D_{gb}} + 1\right)} \tag{3}$$

It has been shown that the model of GBS might predict a Hall-Petch relationship between the flow stress and the grain size at low temperatures and a good agreement has been reported with experimental data for multiple materials and a broad range of grain sizes [15, 17-21]. The transition between the high temperature and low temperature behavior is examined next.

Transition between high- and low-temperature behavior

A recent review [21] evaluated the relationship between the grain size and the flow stress for different temperature ranges considering the grain boundary sliding (GBS) model. A Hall-Petch relationship is predicted at low temperatures and superplasticity is predicted at high temperatures for fine grained materials. Low temperature is considered as $T<0.3T_M$ and high temperature is considered as $T>0.5T_M$ where T_M is the melting temperature of the material. Moderate temperatures are considered as the range 0.3 $T_M<T<0.5$ T_M and a gradual transition is predicted in the moderate temperature range. Figure 3 shows representative curves of the prediction of the GBS model for the relationship between flow stress and strain rate for two grain sizes at the different temperature ranges [21]. The behavior of an ultrafine grained material is depicted by the dashed lines and d_1 while the behavior of a fine grained material is depicted by continuous lines and d_2. The ultrafine grained material display higher flow stress than the fine grained counterpart at low temperatures and the opposite is observed at high temperature. At moderate temperatures the ultrafine grained material might display higher strength at high strain rates and lower strength at low strain rates. The slope of the curve of the ultrafine grained material is larger than the slope of the fine grained counterpart.

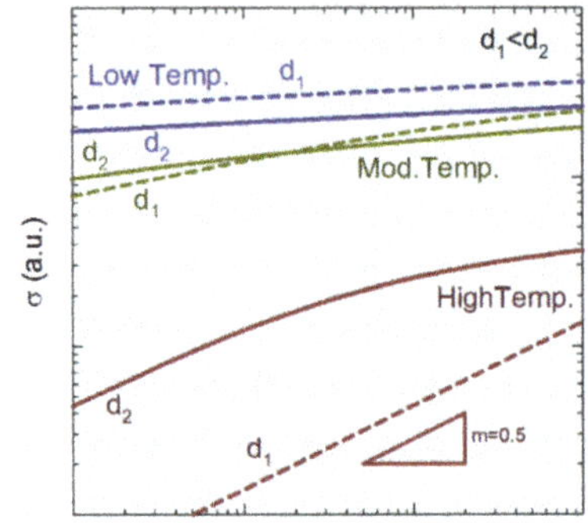

Figure 3 – General trends for the relationship between flow stress and strain rate for materials with different grain sizes and tested at different temperature ranges [21].

The strain rate sensitivity is estimated by the slope in the plots of flow stress vs strain rate. Thus, the predictions of the GBS model depicted in Fig. 3 suggests that the strain rate sensitivity of ultrafine grained materials increases with increasing temperature. This effect has been reported in the literature for multiple materials and Fig. 4 shows experimental data [22-24] of strain rate sensitivity of a CrMnFeCoNi multi component alloy tested at different temperatures. The predictions from the GBS model are also shown. The experimental data shows an increase in the strain rate sensitivity values with increasing temperature in agreement with the predictions from the model.

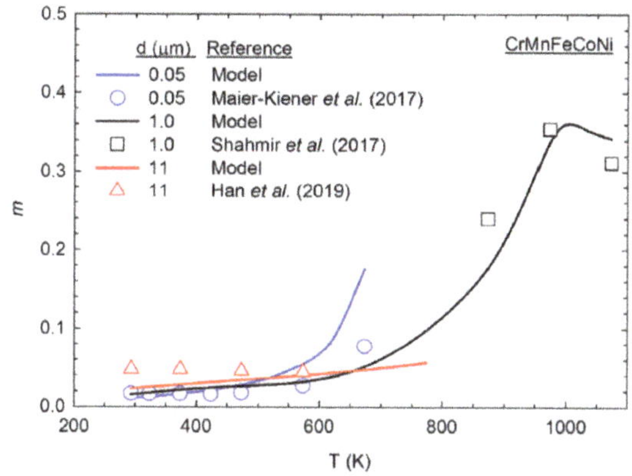

Figure 4 – Strain rate sensitivity of CrMnFeCoNi alloy at different temperatures [20]

The gradual transition between low temperature behavior, with low strain rate sensitivity, and high temperature behavior, with high strain rate sensitivity, is also confirmed by experimental data from ultrafine grained Al-Mg alloys tested at different temperatures. Figure 5 shows the predictions from the model of grain boundary sliding for the relationship between flow stress and strain rate for temperatures of 298 K, 403 K, 523 K and 673 K. It is important to note that the threshold stress is considered as a thermally activated process in the prediction depicted in Fig. 5 [21]. Experimental data from the literature is also shown for comparison and display good agreement with the predictions. Thus, a gradual increase in slope in the relationship between flow stress and strain rate is observed with increasing temperature. This analysis confirms that the model of grain boundary sliding [15] display good agreement with experimental data and predicts a gradual increase in strain rate sensitivity of fine grained material with increasing temperature.

Summary and conclusions

Grain boundary sliding is an established deformation mechanism for high temperature deformation of fine grained materials. A recent paper [15] showed that the rate controlling equation for the mechanism of grain boundary sliding can be adapted to account for the higher stresses of low temperature deformation.

The present overview shows that the adapted equation for grain boundary sliding predicts a gradual transition from the typical behavior observed at high temperature in fine grained materials to a Hall-Petch type of behavior at low temperatures.

The transition from low temperature to high temperature changes the relationship between flow stress and grain size in a way that grain refinement hardening is observed at low temperatures and grain refinement softening at high temperatures.

A gradual increase in strain rate sensitivity with increasing temperature is predicted and confirmed by experiments in ultrafine grained materials.

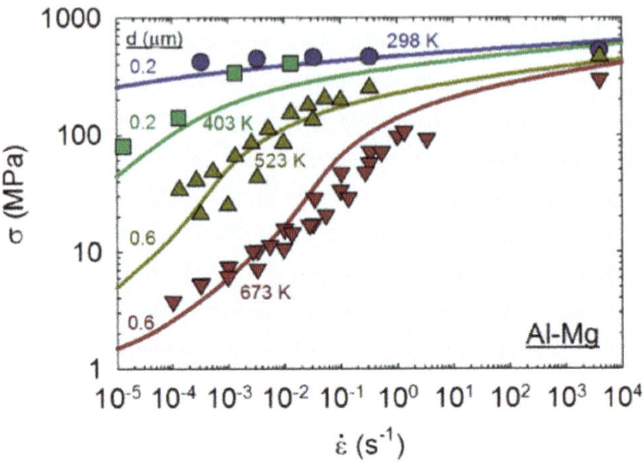

Figure 5 – Experimental data and model prediction of flow stress of ultrafine Al-Mg alloys tested at different temperatures plotted as a function of the strain rate [21].

Acknowledgement
The author acknowledge support from CNPq (grant 302832/2022-0) and FAPEMIG (grants BPD-00228-22 and PPM-00324-17).

References

[1] J.E. Bird, A.K. Mukherjee, J.E. Dorn, Quantitative Relation Between Properties and Microstructure, Israel Universities Press, Jerusalem, Israel, 1969.

[2] R.B. Figueiredo, T.G. Langdon, Analysis of the creep behavior of fine-grained AZ31 magnesium alloy, Materials Science and Engineering: A 787 (2020) 139489. https://doi.org/10.1016/j.msea.2020.139489

[3] K. Kitazono, E. Sato, K. Kuribayashi, Internal stress superplasticity in polycrystalline AZ31 magnesium alloy, Scripta Materialia 44(12) (2001) 2695-2702. https://doi.org/10.1016/S1359-6462(01)00967-8

[4] H. Somekawa, K. Hirai, H. Watanabe, Y. Takigawa, K. Higashi, Dislocation creep behavior in Mg-Al-Zn alloys, Materials Science and Engineering: A 407(1) (2005) 53-61. https://doi.org/10.1016/j.msea.2005.06.059

[5] J.A. del Valle, M.T. Pérez-Prado, O.A. Ruano, Deformation mechanisms responsible for the high ductility in a Mg AZ31 alloy analyzed by electron backscattered diffraction, Metallurgical and Materials Transactions A 36(6) (2005) 1427-1438. https://doi.org/10.1007/s11661-005-0235-8

[6] S. Spigarelli, M. El Mehtedi, D. Ciccarelli, M. Regev, Effect of grain size on high temperature deformation of AZ31 alloy, Materials Science and Engineering: A 528(22) (2011) 6919-6926. https://doi.org/10.1016/j.msea.2011.05.032

[7] H. Watanabe, H. Tsutsui, T. Mukai, M. Kohzu, S. Tanabe, K. Higashi, Deformation mechanism in a coarse-grained Mg-Al-Zn alloy at elevated temperatures, International Journal of Plasticity 17(3) (2001) 387-397. https://doi.org/10.1016/S0749-6419(00)00042-5

[8] R.B. Figueiredo, T.G. Langdon, Evaluating the superplastic flow of a magnesium AZ31 alloy processed by equal-channel angular pressing, Metallurgical and Materials Transactions A 45(8) (2014) 3197-3204. https://doi.org/10.1007/s11661-013-1920-7

[9] R.B. Figueiredo, T.G. Langdon, Developing superplasticity in a magnesium AZ31 alloy by ECAP, Journal of Materials Science 43(23-24) (2008) 7366-7371. https://doi.org/10.1007/s10853-008-2846-0

[10] H. Watanabe, M. Fukusumi, Mechanical properties and texture of a superplastically deformed AZ31 magnesium alloy, Materials Science and Engineering: A 477(1) (2008) 153-161. https://doi.org/10.1016/j.msea.2007.05.012

[11] M.L. Olguín-González, D. Hernández-Silva, M.A. García-Bernal, V.M. Sauce-Rangel, Hot deformation behavior of hot-rolled AZ31 and AZ61 magnesium alloys, Materials Science and Engineering: A 597 (2014) 82-88. https://doi.org/10.1016/j.msea.2013.12.027

[12] S. Spigarelli, O.A. Ruano, M. El Mehtedi, J.A. del Valle, High temperature deformation and microstructural instability in AZ31 magnesium alloy, Materials Science and Engineering: A 570 (2013) 135-148. https://doi.org/10.1016/j.msea.2013.01.060

[13] X. Wu, Y. Liu, Superplasticity of coarse-grained magnesium alloy, Scripta Materialia 46(4) (2002) 269-274. https://doi.org/10.1016/S1359-6462(01)01234-9

[14] K. Edalati, A. Bachmaier, V.A. Beloshenko, Y. Beygelzimer, V.D. Blank, W.J. Botta, K. Bryła, J. Čížek, S. Divinski, N.A. Enikeev, Y. Estrin, G. Faraji, R.B. Figueiredo, M. Fuji, T. Furuta, T. Grosdidier, J. Gubicza, A. Hohenwarter, Z. Horita, J. Huot, Y. Ikoma, M. Janeček, M. Kawasaki, P. Král, S. Kuramoto, T.G. Langdon, D.R. Leiva, V.I. Levitas, A. Mazilkin, M. Mito, H. Miyamoto, T. Nishizaki, R. Pippan, V.V. Popov, E.N. Popova, G. Purcek, O. Renk, Á. Révész, X. Sauvage, V. Sklenicka, W. Skrotzki, B.B. Straumal, S. Suwas, L.S. Toth, N. Tsuji, R.Z. Valiev, G. Wilde, M.J. Zehetbauer, X. Zhu, Nanomaterials by severe plastic deformation: review of historical developments and recent advances, Materials Research Letters 10(4) (2022) 163-256. https://doi.org/10.1080/21663831.2022.2029779

[15] R.B. Figueiredo, T.G. Langdon, Deformation mechanisms in ultrafine-grained metals with an emphasis on the Hall-Petch relationship and strain rate sensitivity, Journal of Materials Research and Technology 14 (2021) 137-159. https://doi.org/10.1016/j.jmrt.2021.06.016

[16] R.B. Figueiredo, K. Edalati, T.G. Langdon, Effect of creep parameters on the steady-state flow stress of pure metals processed by high-pressure torsion, Materials Science & Engineering A 835 (2022) 142666. https://doi.org/10.1016/j.msea.2022.142666

[17] R.B. Figueiredo, T.G. Langdon, Effect of grain size on strength and strain rate sensitivity in metals, Journal of Materials Science 57 (2022) 5210-5229. https://doi.org/10.1007/s10853-022-06919-0

[18] A.P. Carvalho, R.B. Figueiredo, An Overview of the Effect of Grain Size on Mechanical Properties of Magnesium and Its Alloys, Materials Transactions advpub (2023). https://doi.org/10.2320/matertrans.MT-MF2022005

[19] A.P. Carvalho, R.B. Figueiredo, The Effect of Ultragrain Refinement on the Strength and Strain Rate Sensitivity of a ZK60 Magnesium Alloy, Advanced Engineering Materials 24(3) (2022) 2100846. https://doi.org/10.1002/adem.202100846

[20] R.B. Figueiredo, W. Wolf, T.G. Langdon, Effect of grain size on strength and strain rate sensitivity in the CrMnFeCoNi high-entropy alloy, Journal of Materials Research and Technology 20 (2022) 2358. https://doi.org/10.1016/j.jmrt.2022.07.181

[21] R.B. Figueiredo, M. Kawasaki, T.G. Langdon, Seventy years of Hall-Petch, ninety years of superplasticity and a generalized approach to the effect of grain size on flow stress, Progress in Materials Science 137 (2023) 101131. https://doi.org/10.1016/j.pmatsci.2023.101131

[22] V. Maier-Kiener, B. Schuh, E.P. George, H. Clemens, A. Hohenwarter, Insights into the deformation behavior of the CrMnFeCoNi high-entropy alloy revealed by elevated temperature nanoindentation, Journal of Materials Research 32(14) (2017) 2658-2667. https://doi.org/10.1557/jmr.2017.260

[23] H. Shahmir, J. He, Z. Lu, M. Kawasaki, T.G. Langdon, Evidence for superplasticity in a CoCrFeNiMn high-entropy alloy processed by high-pressure torsion, Materials Science and Engineering: A 685 (2017) 342-348. https://doi.org/10.1016/j.msea.2017.01.016

[24] Z. Han, W. Ren, J. Yang, Y. Du, R. Wei, C. Zhang, Y. Chen, G. Zhang, The deformation behavior and strain rate sensitivity of ultra-fine grained CoNiFeCrMn high-entropy alloys at temperatures ranging from 77 K to 573 K, Journal of Alloys and Compounds 791 (2019) 962-970. https://doi.org/10.1016/j.jallcom.2019.03.373

In-situ heating observations on microstructure relaxation of ultrafine-grained high-entropy alloys using neutron diffraction and laser-scanning confocal microscopy

Megumi Kawasaki[1,a*], Jae-Kyung Han[1,b], Suk-Chun Moon[2,c], Klaus-Dieter Liss[2,d]

[1] School of Mechanical, Industrial and Manufacturing Engineering, Oregon State University, Corvallis, OR 97331, U.S.A.

[2] School of Mechanical, Materials, Mechatronic and Biomedical Engineering, University of Wollongong, NSW 2522, Australia

[a]megumi.kawasaki@oregonstate.edu, [b]hanja@oregonstate.edu, [c]scmoon@uow.edu.au, [d]kdl@uow.edu.au

Keywords: High-Pressure Torsion, Laser-Scanning Confocal Microscopy, Nanostructure, Neutron Diffraction, X-Ray Diffraction

Abstract. The thermal stability of ultrafine-grained metals can be fully understood when observing time-resolved microstructural changes over multiple-length scales. The global microstructural relaxation behavior upon heating of an ultrafine-grained (UFG) CoCrFeNi high-entropy alloy (HEA) was characterized by in-situ heating neutron diffraction measurements. Before heating, the nanocrystalline microstructure was introduced by applying high-pressure torsion (HPT), leading to severe lattice distortion by excess dislocations and defects. The sequential information on the structural relaxation of recovery, recrystallization, and grain growth are identified by in-situ heating neutron diffraction analysis defining the texture development, linear thermal lattice expansion, and stress relaxation behaviors of the UFG HEA with increasing temperature up to 1300K. By contrast, nanocrystalline metals processed by HPT are often inhomogeneous microstructurally and compositionally. The influence of such inhomogeneity on the macro-scale microstructural relaxation is monitored using an HPT-processed CoCrFeNiMn high-entropy alloy through in-situ heating laser-scanning confocal microscopy. This study emphasizes the importance of characterization techniques for further in-depth exploration of the SPD-processed ultrafine-grained structure.

Introduction

The thermal stability of metallic materials is a critical ability and design characteristic that can decide the specific applications of engineering metals. It is especially important for ultrafine-grained (UFG) metals and bulk nanostructured materials, whose stored free energy at the grain boundaries is high leading to accelerated microstructural recovery and grain growth [1,2]. Such changes in microstructure directly influence the mechanical properties, i.e., deterioration of strength and hardness, of UFG metals. Thus, the understanding of thermal behavior including microstructural evolution and changes in mechanical properties is significantly important along with the successful development of severe plastic deformation (SPD) techniques for microstructural refinement of numerous metals including multiple-principal element alloys often called high entropy alloy (HEA) [3]. This study describes a recent result on the *in-situ* heating microstructural relaxation behavior of a nanostructured CoCrFeNi alloy examined using neutron diffraction. Moreover, the direct observation of the relaxation microstructure at large length scales is demonstrated on a nanocrystalline CoCrFeNiMn alloy through the application of high-temperature laser-scanning confocal microscopy.

Formation of nanostructure

The present study uses the CoCrFeNi alloy produced by additive manufacturing (AM) of the powder bed fusion technology. The detailed conditions for AM are described earlier [4]. After machining to have disk samples having 10 mm diameter and 0.85 mm thickness, where the disk surfaces are perpendicular to the building direction, high-pressure torsion (HPT) processing was applied to the first batch of the as-built alloy disks for 1/2, 1, 2, 4 and 8 turns under 6 GPa at room temperature. Figure 1 shows the micrographs of the alloy after AM at (a) normal to the built direction and (b) the mid-radius of the disk sample after HPT for 8 turns [5]. The as-built microstructure shows a coarse, columnar grain microstructure. The HPT process refined the microstructure to have reasonably equiaxed grains with an average grain size of d = 90 nm. The microstructural changes during HPT processing occur gradually with increasing numbers of turns, and it is confirmed by taking the hardness contour maps across the disk cross-sections as shown in Fig. 1(c). The initial Vickers microhardness values before HPT was H_V = 260. Due to the imposed shear strain by torsion being proportional to the disk radius [6], high hardness, thus grain refinement, is developed from the disk periphery towards the disk center with increasing HPT turns, while the hardness is homogeneously developed across the disk height. After 8 turns, the hardness distribution became homogeneous in the nanostructured AM CoCrFeNi alloy.

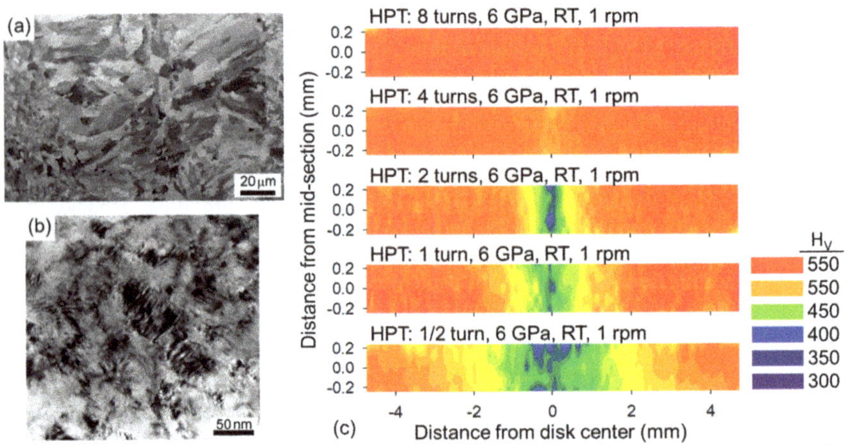

Figure 1. Micrographs of the alloy (a) in as-built and (b) after HPT for 8 turns, and (c) color-coded hardness contour maps after HPT for 1/2 turn through 8 turns of AM CoCrFeNi alloy.

A series of samples including AM powder before AM, after AM (as-built), and after HPT for 1/2 turn through 8 turns were examined by the X-ray diffraction (XRD) method. All sample conditions showed diffraction peaks of only face-centered cubic (f.c.c.) structure, and the estimated microstructure parameters including crystallite size, microstrain, and lattice constant are listed in Table 1 [5]. While there is a slight change before and after AM, significant changes in these microstructural parameters are initiated by the application of HPT even for 1/2 turn. The changes in the parameters are maintained reasonably through 8 turns. It is common to observe a smaller domain size of 75 nm by XRD after 8 turns than by TEM (Fig. 1(b)) due to the high populations of subgrains and dislocation walls as well as crystallographic orientation [7].

Materials Research Forum LLC
https://doi.org/10.21741/9781644902615-27

Table 1. Summary of estimated crystallite size, microstrain, and lattice parameters for the AM CoCrFeNi alloy in the conditions of powder, after AM, and HPT for up to 8 turns [5].

	Crystallite size (nm)	Microstrain	Lattice constant (Å)
Powder	462.54	0.00495	3.5718 ± 0.0007
As-built	652.27	0.00410	3.5744 ± 0.0009
HPT: 1/2 turn	101.25	0.00810	3.5742 ± 0.0048
HPT: 1 turn	116.11	0.00881	3.5728 ± 0.0051
HPT: 2 turns	80.94	0.00858	3.5752 ± 0.0059
HPT: 4 turns	89.65	0.00878	3.5745 ± 0.0059
HPT: 8 turns	75.07	0.00834	3.5771 ± 0.0069

In-situ heating neutron diffraction analysis

Microstructural evolution upon heating at 4 K/min from 300 K to 1273 K followed by quenching was examined for the as-built and HPT-processed CoCrFeNi alloy through neutron diffraction experiments at the iMATERIA, beamline BL20, at the Japan Proton Accelerator Research Complex (J-PARC) [8]. The obtained time-resolved neutron diffraction patterns from Banks 1-3 are plotted into diffraction contour maps with temperature changes as shown in Fig. 2 for the CoCrFeNi alloy in (a) as-built and (b) AM and HPT for 15 turns [8]. HPT for 15 turns was applied to ensure the homogeneous nanostructure of the alloy. The peak profiles at $t = 0$ min of heating imply the textures of the alloy conditions immediately after AM (a) and HPT (b).

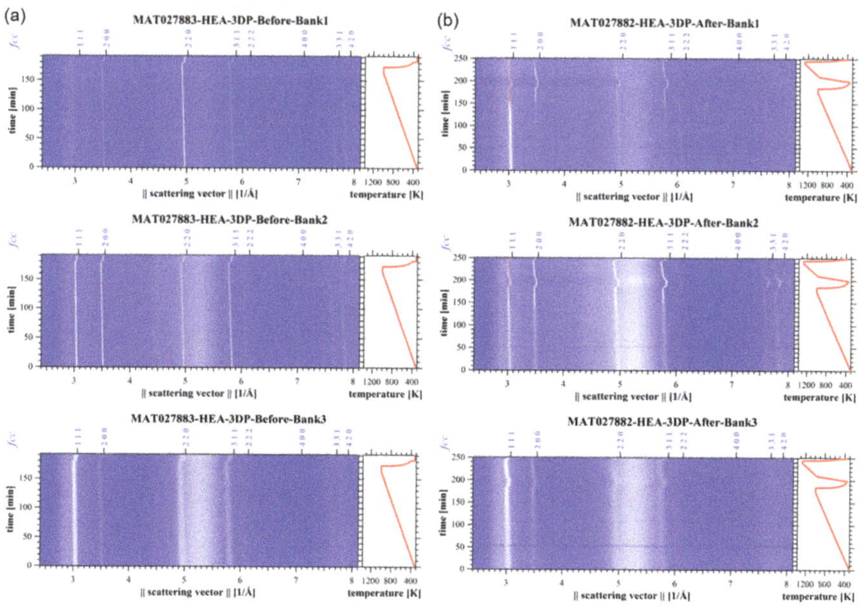

Figure 2. Contour plots showing variation in neutron diffraction profiles from three detector banks 1-3 with time and temperature for the AM CoCrFeNi HEA after (a) printing and (b) HPT for 15 turns [8].

Both materials showed diffraction peaks of a single-phase f.c.c. crystal structure throughout heating followed by cooling, thereby proving no phase transformation. In addition, two features are noticeable in the contour plots of neutron diffraction profiles with time and temperature for the CoCrFeNi alloy: the continuous and concurrent changes in the peak shifting and peak width. The former indicates lattice expansion and contraction with heating and cooling, respectively, and the latter involves relaxation in microstrain and growth in crystallite size, indicating microstructural relaxation. These parameters are further analyzed, and Fig. 3 shows the evolution of (a) peak lattice expansion and (b) relative peak width at the plane coordinate of 111 of the neutron diffraction peaks with temperature for both as-built (AM) and HPT-processed (AM + HPT) CoCrFeNi alloy [8]. The relative lattice expansion and contraction in Fig. 3(a) were estimated using the base lattice parameter of $a_0 = 3.60$ Å taken at the minimum strain condition after the completion of a heating-cooling cycle in the diffraction measurement. Ex-situ hardness measurements after heating to the specific target temperatures followed by cooling were conducted for the samples, where the results are shown in Fig. 3(c).

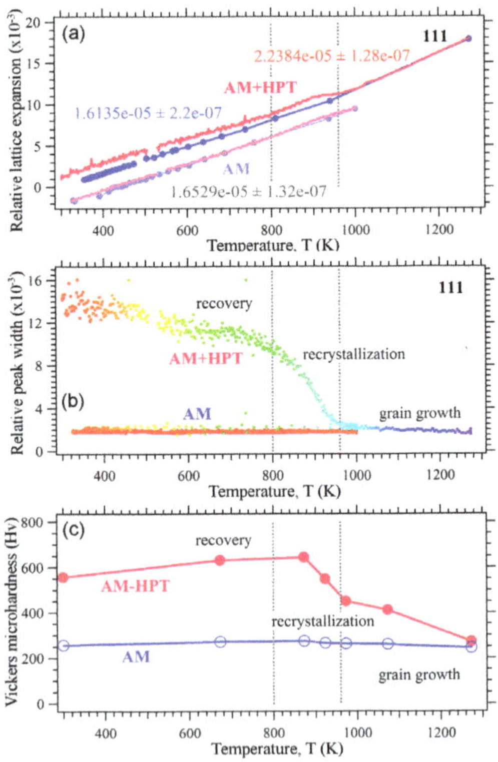

Figure 3. Evolution of (a) relative peak width, (b) relative lattice expansion, and (c) Vickers microhardness upon heating for the as-printed (AM) and HPT-processed (AM + HPT) CoCrFeNi alloy [8].

For both AM and AM + HPT samples, there is a consistent linear lattice contraction of ~1.6 × 10^{-5} K^{-1} at <900 K, while a high lattice expansion of 2.2 × 10^{-5} K^{-1} is observed over 960 K with some plateau at 900-960 K for AM + HPT. Such inconsistency in relative lattice expansion without phase transformation can be explained by the inhomogeneous distribution of thermal vacancies leading to heterogeneous lattice strain in CoCrFeNi and CoCrFeNiMn high-entropy alloys at >973 K [9]. The changes in relative peak width and hardness values with heating towards 1273 K in Fig. 3(b) and (c) provide microstructural relaxation behavior of the nanostructured CoCrFeNi alloy, which are not visible in the AM sample. In practice, the gradual reduction in relative peak width and a hardness increase from H_V of 550 to 650 are shown up to 873 K, followed by a dramatic decrease in both relative peak width and hardness to $H_V \approx 450$ through 973 K, and these achieved the minimum saturated values of peak width and hardness of $H_V \approx 370$ with further heating thought 1273 K. These three stages of microstructural relaxation are considered as recovery, recrystallization, and grain growth.

Annealing-induced hardening (AH) [10] is achieved upon heating to 873 K in the present UFG CoCrFeNi alloy. AH has been observed in other SPD-processed UFG metals, including stainless steel [11], single-phase CoCrNi-based medium entropy alloys [12], Ni-Mo alloy [13], and Ti [14]. AH must be differentiated from transformation hardening, including the nucleation of Guinier-Preston zones in Al alloys and the formation of NiMn-, FeCo-, and Cr-rich phases in the CoCrFeNiMn Cantor alloy. Moreover, AH is achievable without solute segregation strengthening [15]. Specifically, UFG metals processed through a top-down strategy, such as SPD, achieved AH by defect relaxation involving (i) relaxation of mobile dislocations inside the grains by clustering into low energy configurations without any significant loss in dislocation density; (ii) relaxation of non-equilibrium grain boundaries leading to a difficult emission of dislocations; and (iii) clustering of excess vacancies into planer defects, impeding the motion of dislocations. The ductility of UFG metals after AH is anticipated to be improved because of the microstructural relaxation during annealing. Accordingly, the AH effect on the UFG metals differs from the "yield strength anomaly", which strengthens materials at elevated temperatures by dislocation locking and dynamic strain aging strengthening leading to lower ductility [16].

The in-situ neutron diffraction experiments provide a knowledge base to minutely design crystalline microstructures and improve the desired properties of a UFG material by an accurate post-processing heating treatment. A recent report summarized currently available examples of SPD-processed nanostructured materials which are characterized by synchrotron X-ray and neutron diffraction techniques [17].

In-situ heating laser scanning confocal microscopy

The present neutron diffraction analysis as well as lab-scale X-ray and synchrotron diffraction analyses can capture the non-monotonous distribution of diffraction peak broadening, indicating strain anisotropy, due to the heterogeneous distributions of excess numbers of dislocations, vacancies, and other defects within crystals of SPD-processed metals [18]. The microstructural inhomogeneity and heterogeneous evolution of a UFG microstructure can be observed by microscopy techniques that are capable of scanning over large-length scales. A recent experiment demonstrated a heterogeneous microstructural relaxation behavior of an HPT-processed nanocrystalline CoCrFeNiMn alloy having saturated grain sizes of 40-50 nm through laser scanning confocal microscopy (LSCM) (Model VL2000DX-SVF18SP) with a heating facility [17]. The principles of LSCM including how the observed surface information represents the bulk behavior of materials are explained elsewhere [19]. Figure 4 shows a series of the LSCM micrographs taken at cross-sections of the alloy disk edges upon heating from 373 K to 923 K at 10 K/min.

Materials Research Proceedings **32** (2023) 235-243 https://doi.org/10.21741/9781644902615-27

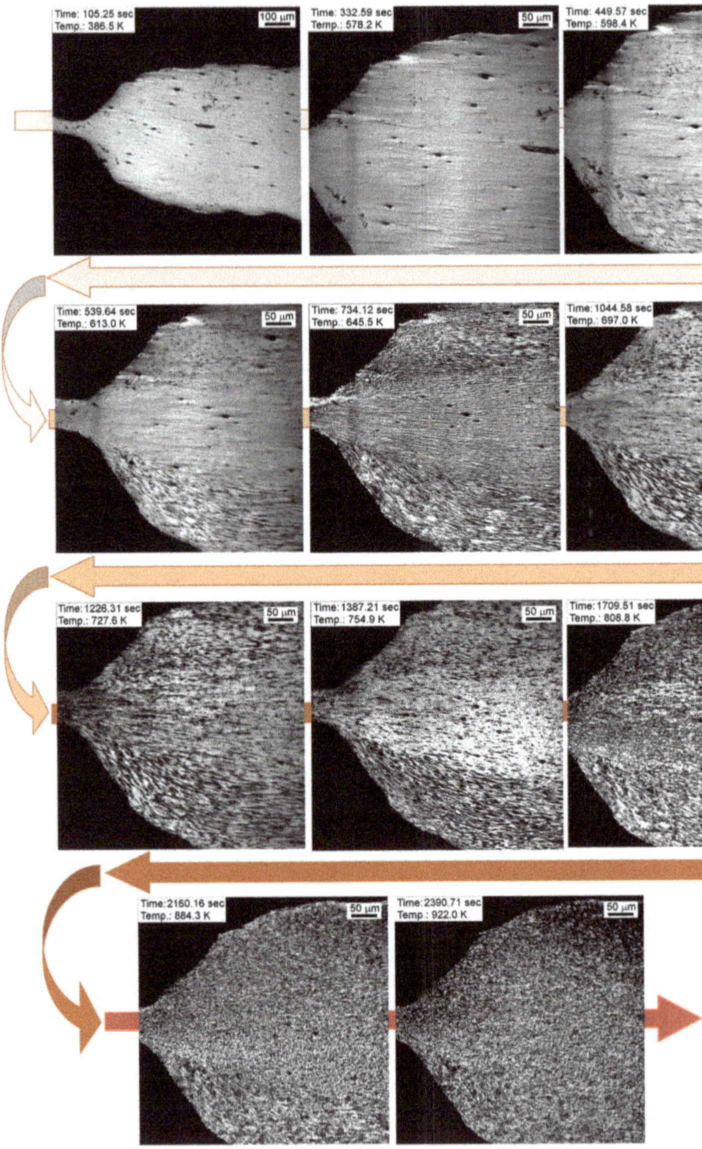

Figure 4. A series of LSCM micrographs taken at a disk edge region of the HPT-processed CoCrFeNiMn HEA at arbitrarily selected temperatures between 373 K and 923 K. The micrographs are adopted from [17].

Overall homogeneous microstructure until 580 K started to show coarser multi-phase elongated phases aligning shear directions only at the sample surfaces except for the middle-thickness without coarsening through 650 K. The separated phase evolution with phase coarsening was maintained until 800 K. Above this temperature, all regions are coarsened, and the equiaxed phases are homogeneously distributed across the sample thickness. While it is a preliminary result without characterization of the coarsened phases, the heterogeneous microstructure evolution during heating was visualized within the HPT-processed CoCrFeNiMn HEA disk sample by the application of high-temperature LSCM for the first time. The phase transformation and agglomerations of the nanostructured CoCrFeNiMn alloy were observed at 723 K [20], but careful consideration is required for the local measurements for microstructural analysis and small-scale mechanical testing, i.e. nanoindentation, due to the possible heterogeneous microstructure and texture introduced by SPD processing. These diffraction and microscopy techniques which analyze the large sample volumes and areas are essential and complementary to the other local measurements to fully understand the evolution of microstructure and mechanical properties of the SPD-processed UFG metals.

Summary

The novel characterization techniques of in-situ heating neutron diffraction and high-temperature laser scanning confocal microscopy provided a comprehensive view of the continuous microstructural relaxation of bulk nanocrystalline HEAs. With the application of different techniques, the heterogeneous UFG microstructures developed during heating can be revealed and understood precisely.

Acknowledgments

This study was supported in part by the National Science Foundation of the United States under Grant No. CMMI-2051205 (M.K.). The authors greatly acknowledge the Ibaraki prefectural government and the J-PARC facility for granting access to iMATERIA under the 2019 Overseas Academic User Program of Ibaraki Neutron Beamline BL20, proposal number 2019PM2014. The authors acknowledge the University of Wollongong for granting access to facilities in the High-temperature Microscopy Laboratory.

References

[1] A. Kumpmann, B. Gunther, H.D. Kunze, Thermal-stability of ultrafine-grained metals and alloys, Mater. Sci. Eng. A 168 (1993) 165-169. https://doi.org/10.1016/0921-5093(93)90722-Q

[2] C.C. Koch, R.O. Scattergood, K.A. Darling, J.E. Semones, Stabilization of nanocrystalline grain sizes by solute additions, J. Mater. Sci. 43 (2008) 7264-7272. https://doi.org/10.1007/s10853-008-2870-0

[3] D.-H. Lee, I.-C. Choi, M.-Y. Seok, J. He, Z. Lu, J.-Y. Suh, M. Kawasaki, T.G. Langdon, J.-i. Jang, Nanomechanical behavior and structural stability of a nanocrystalline CoCrFeNiMn high-entropy alloy processed by high-pressure torsion, J. Mater. Res. 30 (2015) 2804-2815. https://doi.org/10.1557/jmr.2015.239

[4] Y. Kuzminova, D. Firsov, A. Dudin, S. Sergeev, A. Zhilyaev, A. Dyakov, A. Chupeeva, A. Alekseev, D. Martynov, I. Akhatov, S. Evlashin, The effect of the parameters of the powder bed fusion process on the microstructure and mechanical properties of CrFeCoNi medium-entropy alloys, Intermetallics 116 (2020) 106651. https://doi.org/10.1016/j.intermet.2019.106651

[5] W. Zhao, J.-K. Han, Y.O. Kuzminova, S.A. Evlashin, A.P. Zhilyaev, A.M. Pesin, J.-i. Jang, K.-D. Liss, M. Kawasaki, Significance of grain refinement on micro-mechanical properties and

structures of additively-manufactured CoCrFeNi high-entropy alloy, Mater. Sci. Eng. A 807 (2021) 140898. https://doi.org/10.1016/j.msea.2021.140898

[6] A.P. Zhilyaev, T.G. Langdon, Using high-pressure torsion for metal processing: Fundamentals and applications, Prog. Mater. Sci. 53 (2008) 893-979. https://doi.org/10.1016/j.pmatsci.2008.03.002

[7] J. Gubicza, P.T. Hung, M. Kawasaki, J.-K. Han, Y. Zhao, Y. Xue, J.L. Lábár, Influence of severe plastic deformation on the microstructure and hardness of a CoCrFeNi high-entropy alloy: A comparison with CoCrFeNiMn, Mater. Charact. 154 (2019) 304-314. https://doi.org/10.1016/j.matchar.2019.06.015

[8] X. Liu, J.-K. Han, Y. Onuki, Y.O. Kuzminova, S.A. Evlashin, M. Kawasaki, K.-D. Liss, In situ neutron diffraction investigating microstructure and texture evolution upon heating of nanostructured CoCrFeNi high-entropy alloy, Adv. Eng. Mater. 25 (2023) 2201256. https://doi.org/10.1002/adem.202201256

[9] E.W. Huang, H.-S. Chou, K.N. Tu, W.-S. Hung, T.-N. Lam, C.-W. Tsai, C.-Y. Chiang, B.-H. Lin, A.-C. Yeh, S.-H. Chang, Y.-J. Chang, J.-J. Yang, X.-Y. Li, C.-S. Ku, K. An, Y.-W. Chang, Y.-L. Jao, Element effects on high-entropy alloy vacancy and heterogeneous lattice distortion subjected to quasi-equilibrium heating, Sci. Rep. 9 (2019) 14788. https://doi.org/10.1038/s41598-019-51297-4

[10] J. Gubicza, Annealing-induced hardening in ultrafine-grained and nanocrystalline materials, Adv. Eng. Mater. 22 (2020) 1900507. https://doi.org/10.1002/adem.201900507

[11] M. Kawasaki, J.-K. Han, X. Liu, Y. Onuki, Y.O. Kuzminova, S.A. Evlashin, A.M. Pesin, A.P. Zhilyaev, K.-D. Liss, In situ heating neutron and X-ray diffraction analyses for revealing structural evolution during post-printing treatments of additive-manufactured 316L stainless steel, Adv. Eng. Mater. 24 (2021) 2100968. https://doi.org/10.1002/adem.202100968

[12] P.T. Hung, M. Kawasaki, J.K. Han, A. Szabo, J.L. Labar, Z. Hegedus, J. Gubicza, Thermal stability of nanocrystalline CoCrFeNi multi-principal element alloy: Effect of the degree of severe plastic deformation, Intermetallics 142 (2022) 107445. https://doi.org/10.1016/j.intermet.2021.107445

[13] J. Gubicza, P.H.R. Pereira, G. Kapoor, Y. Huang, S.S. Vadlamani, T.G. Langdon, Annealing-induced hardening in ultrafine-grained Ni-Mo alloys, Adv. Eng. Mater. 20 (2018) 1800184. https://doi.org/10.1002/adem.201800184

[14] R.Z. Valiev, A.V. Sergueeva, A.K. Mukherjee, The effect of annealing on tensile deformation behavior of nanostructured SPD titanium, Scr. Mater. 49 (2003) 669-674. https://doi.org/10.1016/S1359-6462(03)00395-6

[15] O. Renk, A. Hohenwarter, K. Eder, K.S. Kormout, J.M. Cairney, R. Pippan, Increasing the strength of nanocrystalline steels by annealing: Is segregation necessary?, Scr. Mater. 95 (2015) 27-30. https://doi.org/10.1016/j.scriptamat.2014.09.023

[16] H. Shang, Q. Ma, Q. Gao, H. Zhang, H. Li, H. Zhang, L. Sun, Yield strength anomaly evaluation of W-free Co-Ni-Al-based superalloys during high temperature tensile tests, Mater. Charact. 192 (2022) 112242. https://doi.org/10.1016/j.matchar.2022.112242

[17] M. Kawasaki, J.-K. Han, X. Liu, S.-C. Moon, K.-D. Liss, Synchrotron high-energy X-ray & neutron diffraction, and laser-scanning confocal microscopy: in-situ characterization techniques for bulk nanocrystalline metals, Mater. Trans. (2023) in press. https://doi.org/10.2320/matertrans.MT-MF2022022

[18] T. Ungár, Characterization of nanocrystalline materials by X-ray line profile analysis, J. Mater. Sci. 42 (2006) 1584-1593. https://doi.org/10.1007/s10853-006-0696-1

[19] S.-C. Moon, D. Phelan, R. Dippenaar, New insights of the peritectic phase transition in steel through in-situ measurement of thermal response in a high-temperature confocal microscope, Mater. Charact. 172 (2021) 110841. https://doi.org/10.1016/j.matchar.2020.110841

[20] D.H. Lee, J.A. Lee, Y.K. Zhao, Z.P. Lu, J.Y. Suh, J.Y. Kim, U. Ramamurty, M. Kawasaki, T.G. Langdon, J.I. Jang, Annealing effect on plastic flow in nanocrystalline CoCrFeMnNi high-entropy alloy: A nanomechanical analysis, Acta Mater. 140 (2017) 443-451. https://doi.org/10.1016/j.actamat.2017.08.057

Hot deformation of ultrafine-grained TiC-reinforced AlCrFeNbMoTiV refractory high entropy alloys

P. Martin[1,a*], C. Aguilar[2,b], B. Ferrari[3,c], A.J. Sánchez-Herencia[3,d], J.M. Cabrera[1,4,e]

[1]Deparment of Materials Science and Engineering EEBE, Technical University of Catalonia, c/Eduard Maristany 10-14, 08019 Barcelona, Spain

[2]Department of Materials and Metallurgical Engineering, Technical University Federico Santa María, Av. España 1680, 2340000 Valparaíso, Chile

[3]Institute of Glass and Ceramics ICV-CSIC, c/Kelsen 5, 28049 Madrid, Spain

[4]Fundació CIM-UPC, c/Llorens i Artigas 12, 08028 Barcelona, Spain

[a]pablo.martin.saint-laurence@upc.edu, [b]claudio.aguilar@usm.cl, [c]bferrari@icv.csic.es, [d]ajsanchez@icv.csic.es, [e]jose.maria.cabrera@upc.edu

Keywords: Constitutive Equation, High-Entropy Alloys, Complex Concentrated Alloys, Ultrafine-Grained, Mechanical Properties

Abstract. In the present work, the hot deformation behavior of the $AlCr_{0.2}FeNbMoTiV_2$ refractory high-entropy alloy was studied by means of its constitutive modeling. The samples were fabricated using mechanical alloying and spark plasma sintering, resulting in an ultrafine-grained microstructure consisting of a main bcc solid solution, Laves phase, and TiC and Al_2O_3 precipitates. Compression tests between 950 and 1100 °C and $5 \cdot 10^{-4}$ and 10^{-2} s^{-1} of strain rate were performed in order to determine the constitutive equations. Additionally, the samples were subjected to a heat treatment at 1350 °C for 16 h, presenting a considerable increment in the specific yield strength (from 33 to 94 $MPa \cdot g^{-1} \cdot cm^3$ at 1000 °C) accompanied by a slight increase of the grain size (from 0.35 to 1.2 μm). Overall, the heat-treated $AlCr_{0.2}FeNbMoTiV_2$ alloy presented one of the highest specific yield strength values among refractory high entropy alloys.

Introduction

The current challenges of the aerospace industry require advanced materials that can overcome the performance of current materials in terms of high-temperature strength and oxidation resistance, among others. Among the potential candidates to fulfill these requirements, refractory high-entropy alloys (RHEAs), first proposed in 2010 [1,2], have attracted the interest of the scientific community because of their outstanding performance in terms mechanical properties at elevated temperatures (even superior to that of commercial Ni-based superalloys) and elevated oxidation resistance [3]. As part of the high-entropy alloys (HEAs) family, these materials correspond to multicomponent alloys (typically over five) with contents between 5 and 35 at. pct., resulting in complex compositions far away from the boundaries of the traditional alloy design concept. RHEAs, particularly, are mainly constituted by refractory elements (like Mo, Nb, Ta, or W) that provide elevated hot softening resistance, alloyed with non-refractory elements (like Al, Si, or Fe) that reduce the density or enhance oxidation resistance properties. Even though RHEAs are commonly fabricated by casting techniques, the mechanical alloying (MA) followed by spark plasma sintering (SPS) route has proven to produce the strongest RHEAs because of the combination of fine or ultrafine grains and the formation of finely and homogeneously dispersed oxides and nitrides [4,5]. This particular microstructure results in superior properties at high temperatures and a superior balance of strength and ductility at room temperature [6,7].

The constitutive analysis of the high-temperature deformation of metallic materials can result not only in an accurate description of the effect of temperature and strain rate over the flow stress but it also provides insightful details about the deformation and softening mechanisms involved. Several expressions have been proposed for these purposes. One of them, the power law (Eq. 1), relates the Zenner-Hollomon (Z) parameter (defined as the temperature-compensated strain rate) with the stress, particularly in low-stress conditions (high temperatures and low strain rates) [8].

$$Z = \dot{\varepsilon} \exp\left(\frac{Q_{app}}{RT}\right) = A'\sigma^{n'} \tag{1}$$

where Q_{app} corresponds to the apparent activation energy, $\dot{\varepsilon}$ to the strain rate, R to the gas constant, T to the temperature, and A' and n' are parameters of the model.
To the authors' knowledge, this is the first work that addresses the constitutive modeling of a PM-fabricated RHEA, resulting in an expression that accurately predicts the peak stress (σ_P) between 1000 – 1100 °C and 0.0005 – 0.01 s^{-1}. Additionally, the physical meaning of n' and Q_{app} was discussed in order to contribute to the comprehension of the deformation mechanisms of RHEAs with typical microstructural features of powder metallurgy processing. Lastly, the effect of annealing on the alloy's microstructure and high-temperature mechanical properties was evaluated.

Materials and methodologies

The powder alloy (denominated R2) was produced mixing pure commercial powders of Al (< 45 µm, 99.5%), Cr (< 45 µm, 99%), Mo (3-7 µm, 99.95%), Nb (< 45 µm, 99.8%), Ti (< 45 µm, 99%), and V (<45 µm, 99.5%), employing hardened steel vials and balls which can act as a source of Fe. The milling process was conducted in a high-energy planetary ball mill Fritsch Pulverisette 5, for 50 h, using a ball-to-powder ratio (BPR) of 20:1 and 300 rpm of rotational speed under an Ar atmosphere. Additionally, 2.6 wt. pct. of ethylene bis-stereamide was added as a Process Control Agent (PCA), and on/off cycles of 30 min/30 min were utilized to avoid severe cold welding. Table 1 summarizes the composition of the 50 h-milled powder used in the sintering stage. The SPS was conducted in a Dr. Sinter SPS-1050-CE press, using 1150 °C and 50 MPa of stress, with a dwell time of 10 min in vacuum. Disks of 30 mm in diameter and between 8 and 9 mm in height were sintered using graphite die and punches. As-sintered samples (R2AS) were subjected to an annealing treatment at 1350 °C for 16 h under an Ar atmosphere (R2HT).

Table 1: Composition of the Mechanically Alloyed (MA-ed) R2 powder.

Al (at. pct.)	Cr (at. pct.)	Fe (at. pct.)	Mo (at. pct.)	Nb (at. pct.)	Ti (at. pct.)	V (at. pct.)
12.49	4.67	13.63	14.48	14.36	13.66	26.72

X-ray diffraction (XRD) analysis was performed in a Bruker D8 Advance diffractometer, using $Cu_{k\alpha 1}$-radiation (λ = 0.15406 nm) equipped with a Ge monochromator. Besides, transmission electron microscopy (TEM) was performed employing a J2100F microscope equipped with energy-dispersive X-ray spectroscopy (EDS) detector. For these purposes, a lamella was prepared in a Focus Ion Beam (FIB)-SEM Neon40 Crossbeam™ workstation. A field-emission gun SEM JEOL JSM-7001F equipped with EDS and electron backscattered diffraction (EBSD) detectors was also used, utilizing the MTEX package to process the EBSD data. Lastly, 6 mm in diameter and 8 mm in height cylindrical samples were cut from the sintered samples for compression testing. Tests at different temperatures (950, 1000, 1050, and 1100 °C) and different strain rates (0.0005, 0.001, 0.005, and 0.01 s^{-1}) were performed in an Instron 4507 universal testing machine equipped with a load cell of 100 kN, using Ar flow to avoid oxidation of the samples, and a layer of mica and boron nitride to diminish the friction between the samples and the punches.

Results and discussion

According to the XRD analysis (Figure 1), both R2AS and R2HT samples are mainly constituted by a bcc phase with lattice parameters similar to that of V, TiC, and Laves phases (C14 and C36 polytypes observed in both samples, while minor C15 was observed in R2AS too). Additionally, in the case of the R2HT sample, Al_2O_3 peaks were observed. Figure 2.a and Figure 2.d illustrates the microstructure of R2AS and R2HT, respectively, evidencing fully-dense microstructures. According to the respective EDS mapping images (Figure 2.b and Figure 2.e), both samples exhibited the same elemental distribution: a V, Mo-rich solid solution, the Ti, Nb-rich carbide, the Fe, Nb-rich Laves phases, and minor Al_2O_3 particles. According to the image analysis performed over the microstructure images, the bcc phase reaches over 45 pct. of the area, while carbides are around 25 pct. In the same way, Figure 2.c and Figure 2.f depicts the Inverse Pole Figure (IPF) map of the respective samples, showing the orientation of the bcc and fcc grains. The average grain size of the bcc and carbide phases of the R2AS sample is around 0.35 and 0.25 μm which increases to 1.3 and 1.1 μm in the R2HT sample, respectively. Apart from that, no remarkable changes in phase equilibrium or composition were observed.

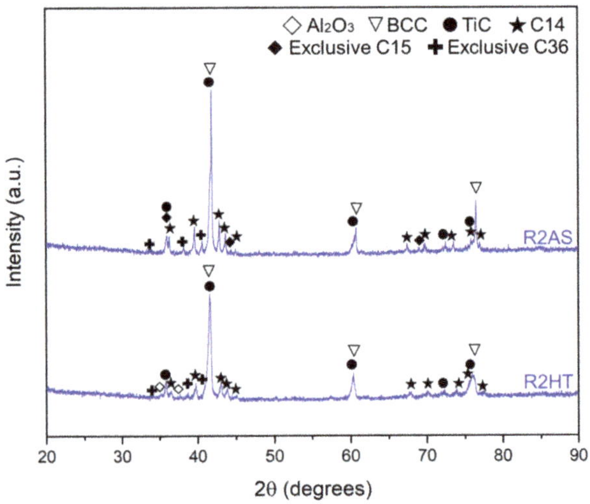

Figure 1: XRD pattern of R2AS and R2HT.

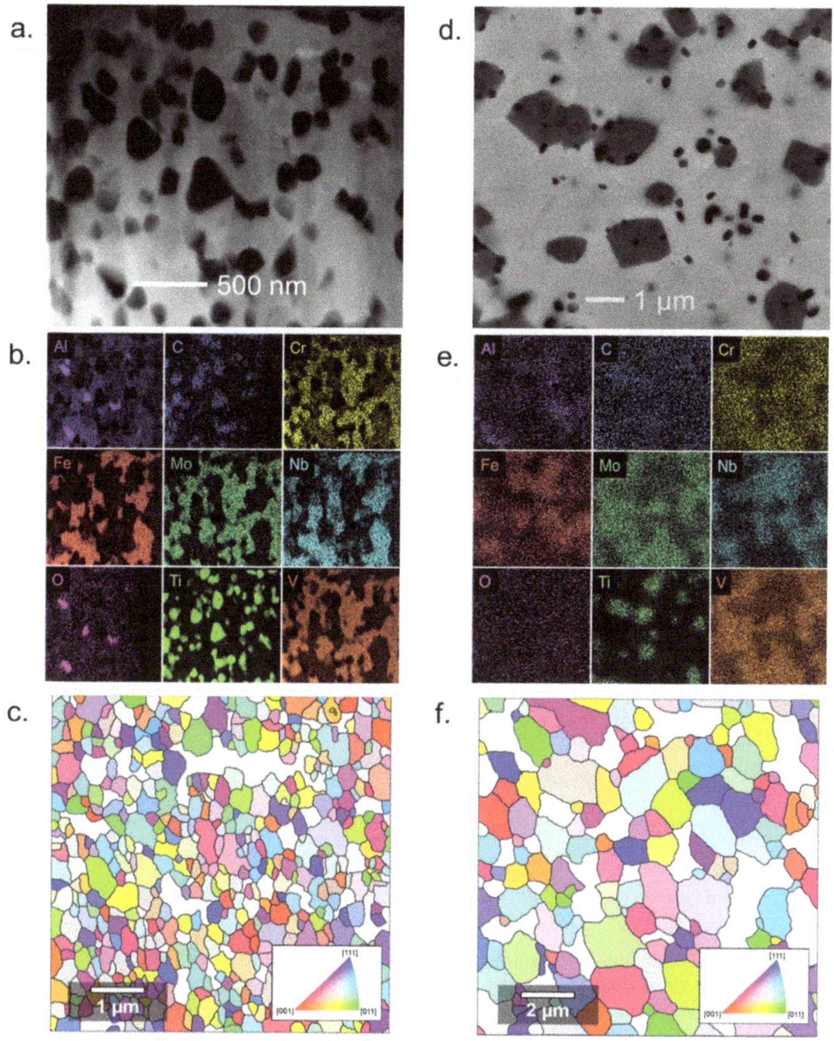

Figure 2: a) TEM image of R2AS microstructure; b) EDS mapping images of the region showed in a; c) IPF map of bcc and fcc grains subset of R2AS; d) SEM image of R2HT microstructure; e) EDS mapping images of the region showed in d; and f) IPF map of bcc and fcc grains subset of R2HT.

Figure 3 presents the compression curves of the R2AS samples at different temperatures and strain rates. The material exhibited the typical behavior of metallic materials: increasing the temperature or decreasing the strain rate reduces the flow stress. Apart from that, the shape of the curves obtained at 1100 and 1050 °C indicates the predominance of dynamic recovery (DRV) as the softening mechanism. In opposition, curves obtained at high strain rates at 1000 °C and at low strain rates at 950 °C suggest that dynamic recrystallization (DRX) may have predominated instead. It is also important to highlight that a strain of 0.7 was achieved in most cases, except for those tests at 950 °C using strain rates of 0.005 and 0.01 s^{-1}. In those samples, some cracks appeared during the elastic deformation region, indicating that the material was already brittle, which resulted in scarce ductility. Additionally, there is a considerable difference in the peak stresses (σ_P) obtained at 1000 °C and 950 °C at 0.001 and 0.0005 s^{-1}. All these findings suggest that the mechanical behavior changes in that temperature interval, either by dissolution/precipitation phenomena or a change in the controlling deformation mechanism. For the previous reasons, the constitutive analysis will be performed only considering the 1000-1100 °C range.

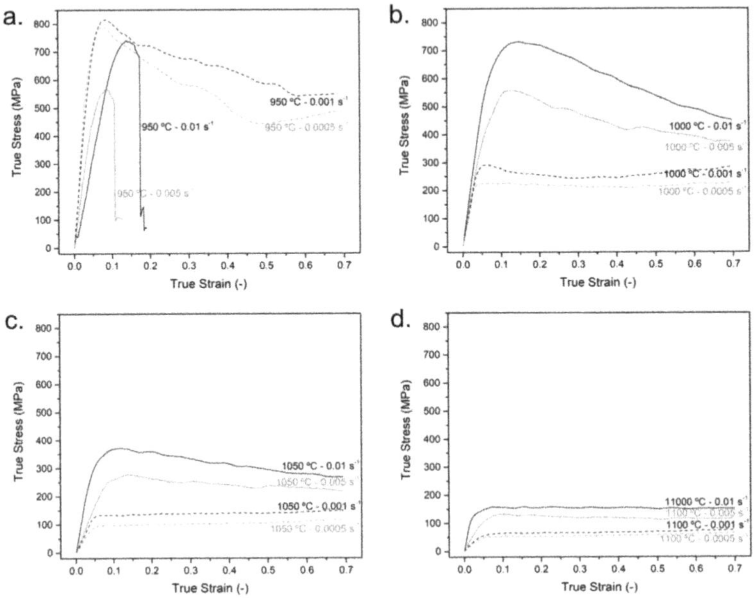

Figure 3: Compression curves of the R2AS samples at different strain rates and at a) 950 °C, b) 1000 °C, c) 1050 °C, and d) 1100 °C.

Figure 4.a illustrates the linear relationship between ln $\dot{\varepsilon}$ and ln σ_P as described in Eq. 2. The slopes of the curves of each plot correspond to n', resulting in values of 2.60, 2.23, and 2.54, at 1100, 1050, and 1000 °C, respectively, resulting in an average n' value of 2.46. The excellent R^2 values, as well as the similarity between the calculated n' values, confirm that there is no change in the deformation mechanism in the range of study. According to the literature, n' = 2 is associated with grain-boundary sliding (GBS) as the governing deformation mechanism. This is the typical deformation mechanism of ultrafine-grained materials, and it has been reported in several works contributing to the hot deformation of PM-fabricated RHEAs or in the recrystallized grains of coarse-grained RHEAs [9,10]. In the present analysis, the obtained n' value is slightly superior to the theoretical one of GBS. Nevertheless, it has been described that in multi-phase materials, the n' value obtained by means of the constitutive analysis is superior to the actual value; the load is actually distributed between two or more phases, and therefore, the stress actually exerted over the matrix phase is inferior to the experimental measured one [11]. Hence, the rest of the analysis considers n' = 2.

$$\ln \dot{\varepsilon} + Q_{app}/R(1/T) = \ln A' + n' \ln \sigma_P \quad (2)$$

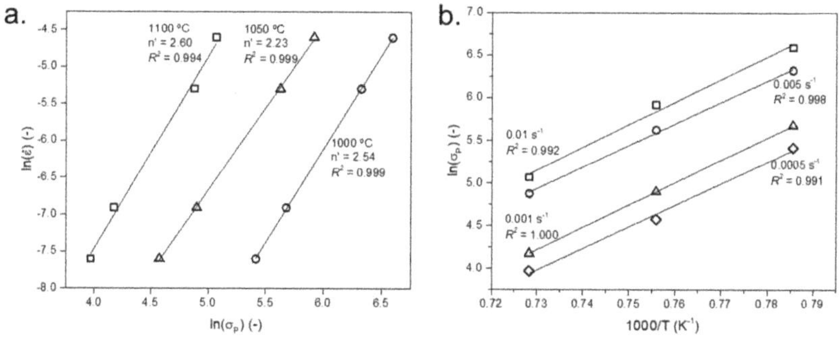

Figure 4: Plots used to determine the values of a) n' and Q_{app} according to the power law.

Figure 4.b shows the linear relationship between ln σ_P and 1000/T in order to calculate Q_{app} as Eq. 3 indicates. An average R^2 coefficient of 0.995 was obtained, evidencing that a single Q_{app} value can represent the experimental data (at least, at each strain rate). Considering n' = 2, an average Q_{app} value of 429.8 kJ·mol^{-1} was calculated. When GBS governs the deformation, Q_{app} corresponds either to the activation energy of grain boundary or of lattice diffusion (depending on which one of them controls it). Nevertheless, the evaluation of Q_{app} in the constitutive analysis of HEAs is still complex and unclear. Because of the compositional complexity of these alloys, there is usually no clear solute-solvent distinction, therefore is not easy to identify which elements play the main role in controlling the diffusion, indistinctly of the mechanism involved. In the present study, the obtained Q_{app} value is in the range of the self-diffusion activation energies of Mo or V [12]. Even though this would agree with the idea that lattice-controlled GBS is the governing deformation mechanism, an in-depth characterization of the microstructure and texture of the deformed samples is necessary to confirm it.

$$Q_{app} = Rn'[\partial \ln \sigma_P / \partial(1/T)]_{\dot{\varepsilon}} \quad (3)$$

Lastly, Figure 5 illustrates the linear relationship between ln Z and ln σ_P. An elevated R^2 coefficient (0.993) was obtained, confirming the excellent agreement of the experimental data with the power law. Thus, a value of A' equals to $2.38 \cdot 10^9$ was calculated as the intercept of this curve. Therefore, the resulting relationship between Z and σ_P is represented in Eq. 4.

$$Z = \dot{\varepsilon} \exp(429.8/RT) = 2.38 \cdot 10^9 \sigma_P^2 \quad (4)$$

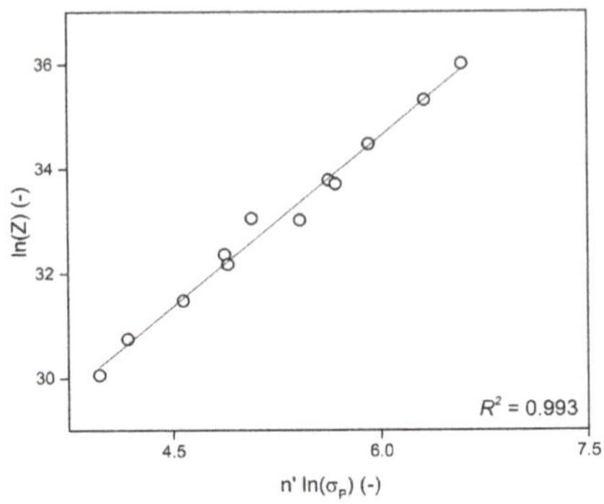

Figure 5: Plot used to the obtention of the value of A' and the corresponding R^2 coefficient of the power law.

On the other hand, Figure 6 illustrates the compression curves of R2AS and R2HT at 1000, 1050, and 1100 °C employing a strain rate of 0.001 s^{-1}. As can be observed, the R2HT samples presented a considerably superior σ_P than the R2AS samples. Since the only considerable difference between R2AS and R2HT microstructures is the grain size of the constituents, this was attributed as the responsible of this superior mechanical behavior. Despite the increase in strength, R2HT samples still presented elevated ductility. These findings suggest that adjusting the microstructural features of PM RHEAs can result in an improvement in their high-temperature mechanical performance.

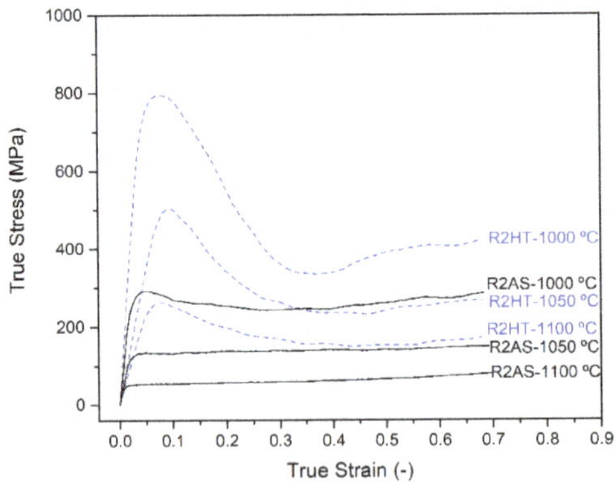

Figure 6: High-temperature compression curves of R2AS and R2HT samples, obtained using a strain rate of 10^{-3} s^{-1}.

Regarding the controlling deformation mechanism, two possibilities are on the table. Firstly, the R2AS and R2HT samples presented different deformation mechanisms, which would be responsible for this considerable difference. Senkov et al. [13] found that the strength remarkably dropped when GBS governs deformation; this agrees with the lower strength of the R2AS, which deformation is controlled by GBS according to the constitutive analysis. The second option is that GBS is also the controlling deformation mechanism of R2HT samples. This is plausible since GBS is hindered by any condition that restricts the movement of grain boundaries, such as lower temperatures, higher strain rates, second phases in grain boundaries, and, as may be in the present case, coarser grain size; all these actions would produce an increase of the flow stress. Nevertheless, further microstructural analysis is required to determine the governing deformation mechanism.

Figure 7 illustrates the specific yield strength (SYS) of PM RHEAs, casting RHEAs, and the alloys of the present study. As can be observed, the R2AS alloy presented a considerably inferior SYS than the R2HT samples, increasing from 33 to 98 MPa/g·cm^{-3} at 1000 °C. On the other hand, the alloys of the present study presented a poorer performance compared to $Al_xCrNbMoV$ (x = 0, 0.1, 0.5, and 1) RHEAs produced by MA+SPS [4,5], which ranges between 155 and 188 MPa/g·cm^{-3} at 1000 °C. Nevertheless, except for the AlCrNbMoV alloy, none of them achieved a strain of 0.3, revealing that this superior strength was obtained by sacrificing their ductility. Apart from that, the SYS of the R2HT samples exceeds most of that of RHEAs that present considerable ductility at 1000 °C.

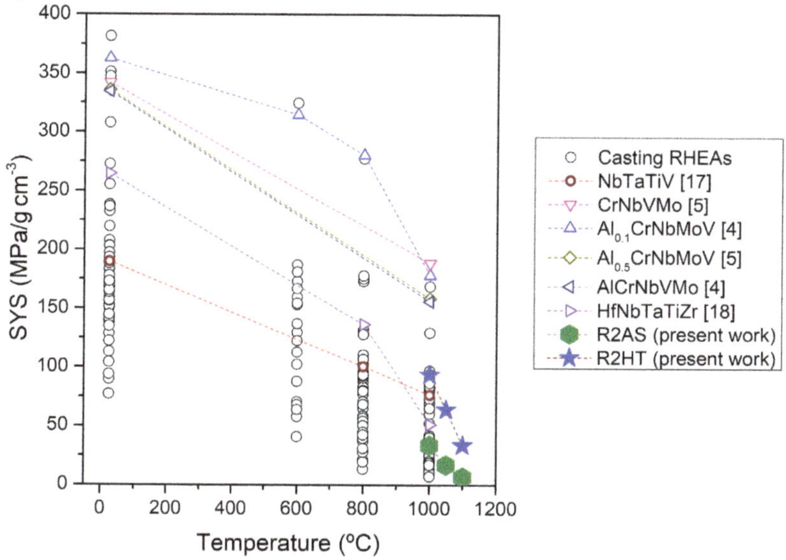

Figure 7: SYS of RHEAs produced by powder metallurgy [4,5,14,15], R2AS, R2HT, and casting RHEAs at different temperatures ($\dot{\varepsilon}$ = 0.001 s^{-1}).

Conclusions

The high-temperature compression tests of the R2 RHEA fabricated by powder metallurgy were effectively performed. Hence, the main conclusions of this work are:

i) A change in the mechanical behavior between 950 and 1000 °C was observed that can be attributed to the precipitation of a phase, the dissolution of the Laves phase at high temperatures, or a change in the controlling deformation mechanism. Further studies are required to address this aspect.

ii) The expression $\dot{\varepsilon}\exp(429.8/RT) = 2.38 \cdot 10^9\, \sigma_P^2$ was obtained to model the σ_P in the studied temperature and strain rate range. An R^2 coefficient of 0.99 was obtained, evidencing the good agreement of the experimental data with the model.

iii) n' and Q_{app} values of 2 and 429.8 kJ·mol^{-1} were obtained with the conventional constitutive analysis considering the load partitioning effect produced by the presence of second phases. Hence, it was assumed that grain boundary sliding controlled the hot deformation of the alloy, while Q_{app} is in the range of the activation energies of self-diffusion of Mo and V.
iv) An increase of the grain size (from 0.35 to 1.2 μm in the case of the bcc phase) increased the strength while the ductility was not reduced. This can be attributed to a change in the controlling deformation mechanism, or well, that GBS is still the governing deformation mechanism but is hindered by the coarser grain size.

Acknowledgments
P. Martin gratefully acknowledges ANID/Doctorado Becas Chile/2019-72200338. JMC is grateful to the funding of project RTI2018-097885-B-C31 by the Agencia Estatal de Investigación (AEI) of Spain.

References

[1] O.N. Senkov, G.B. Wilks, D.B. Miracle, C.P. Chuang, P.K. Liaw, Intermetallics 18 (2010) 1758-1765. https://doi.org/10.1016/j.intermet.2010.05.014

[2] O.N. Senkov, G.B. Wilks, J.M. Scott, D.B. Miracle, Intermetallics 19 (2011) 698-706. https://doi.org/10.1016/j.intermet.2011.01.004

[3] O.N. Senkov, D.B. Miracle, K.J. Chaput, J.P. Couzinie, J. Mater. Res. 33 (2018) 3092-3128. https://doi.org/10.1557/jmr.2018.153

[4] B. Kang, J. Lee, H.J. Ryu, S.H. Hong, J. Alloys Compd. 767 (2018) 1012-1021. https://doi.org/10.1016/j.jallcom.2018.07.145

[5] B. Kang, T. Kong, H.J. Ryu, S.H. Hong, J. Mater. Sci. Technol. 69 (2021) 32-41. https://doi.org/10.1016/j.jmst.2020.07.012

[6] J. Pan, T. Dai, T. Lu, X. Ni, J. Dai, M. Li, Mater. Sci. Eng. A 738 (2018) 362-366. https://doi.org/10.1016/j.msea.2018.09.089

[7] B. Kang, J. Lee, H.J. Ryu, S.H. Hong, Mater. Sci. Eng. A 712 (2018) 616-624. https://doi.org/10.1016/j.msea.2017.12.021

[8] J.M. Cabrera, A. Al Omar, J.J. Jonas, J.M. Prado, Metall. Mater. Trans. A Phys. Metall. Mater. Sci. 28 (1997) 2233-2244. https://doi.org/10.1007/s11661-997-0181-8

[9] Q. Liu, G. Wang, Y. Liu, X. Sui, Y. Chen, S. Luo, Mater. Sci. Eng. A 809 (2021) 140922. https://doi.org/10.1016/j.msea.2021.140922

[10] R.R. Eleti, A.H. Chokshi, A. Shibata, N. Tsuji, Acta Mater. 183 (2020) 64-77. https://doi.org/10.1016/j.actamat.2019.11.001

[11] Z. Savaedi, R. Motallebi, H. Mirzadeh, J. Alloys Compd. 903 (2022) 163964. https://doi.org/10.1016/j.jallcom.2022.163964

[12] G. Neumann, C. Tuijn, Self-Diffusion and Impurity Diffusion in Pure Metals, First Edit, Elsevier Ltd., 2009. https://doi.org/10.1016/S1470-1804(08)00010-2

[13] O.N. Senkov, J.M. Scott, S. V. Senkova, F. Meisenkothen, D.B. Miracle, C.F. Woodward, J. Mater. Sci. 47 (2012) 4062-4074. https://doi.org/10.1007/s10853-012-6260-2

[14] A. Fu, W. Guo, B. Liu, Y. Cao, L. Xu, Q. Fang, H. Yang, Y. Liu, J. Alloys Compd. 815 (2020). https://doi.org/10.1016/j.jallcom.2019.152466

[15] X. Li, A. Fu, Y. Cao, S. Xu, S. Gao, J. Alloys Compd. 894 (2022). https://doi.org/10.1016/j.jallcom.2021.162414

Deformation behaviour and microstructural evolution of high-entropy CoFeMnNi alloy at hot deformation condition with low strain rate

Lihua Du[1,2,3,a*], Jiaai Shi[4,b], Jiayu Liang[4,c], Kailun Zheng[4,d]

[1] AVIC Manufacturing Technology Institute, Beijing 100024, China

[2]Aviation Key Laboratory of Science and Technology on Plastic Forming, Beijing 100024, China

[3]Beijing Key Laboratory on Digital Plastic Forming and Equipment, Beijing 100024, China

[4]Dalian University of Technology, Dalian 116024, China

[a]quietness2007@126.com, [b]homura_zero@126.com, [c]1262976726@qq.com,[d]zhengkailun@dlut.edu.cn

Keywords: High-Entropy Alloy, Deformation Behaviour, Super-Plastic Forming, Deformation

Abstract. High-entropy alloy is a promising structural material for high-temperature service applications due to the designable superior properties. Superplasticity, normally achieved at relatively high temperature and low strain rate, enables difficult to formation materials to manufacture complex-shaped parts. This paper presents a feasibility experimental study on the deformation and microstructure of CoFeMnNi high-entropy alloy under hot deformation conditions with low strain rate. A series of hot uniaxial tests were conducted at the temperature range between 800-1000°C with typical strain rates of 10^{-4} and 10^{-5}/s to determine the stress-strain behaviours. Additionally, to understand the deformation mechanism, microstructure evolution after deformation was characterized. Finally, a mechanism based constitutive model of high-entropy alloy was developed, which facilities the process optimization by finite element simulations.

1. Introduction

The fabrication of High-entropy alloys (HEAs), featured by containing more than four to five major elements typically, is considered to be one of the greatest breakthroughs of alloy design [1]. Due to excellent properties such as high strength [2], excellent toughness [3] and wear resistance [4] for various extreme service environments, HEAs have attracted intensive research work in recent years. For further attempt to manufacture engineering parts, the investigation of deformation behaviour of HEAs is essential. HEAs are basically brittle at room temperature, with poor ductility and limited formability for further processing [5]. However, the ductility of HEA increases with the increase of processing temperature, which enable the various forming processes becomes possible.

Research on the formability of HEAs at elevated temperature has been carried out. Zhao et al. [6] manufactured CoCrFeNi HEA by laser powder-bed fusion technique and refined its grains, and the results show that the additive manufactured HEA has high strain hardening capability and increased plasticity. Stepanov et al. [7] investigated the characteristics of mechanical behaviour and microstructure evolution of AlCoCrCuFeNi HEA during superplastic rheology and found that the elongation of the HEA can reach 1240% at the temperatures of 800°C-1000°C and strain rates of 10^{-4}/s to 10^{-1}/s. Shahmir et al. [8] obtained CoCrFeNiMn HEA with a grain size of approximately 10nm by high-pressure torsion method, and found that the maximum elongation of HEA can exceed 600% at temperature of 973K. Li et al. [9] investigated the superplastic behaviour of CoCrFeNiCu HEA prepared by stir friction at the temperature of 900~950°C and the strain rate of 3×10^{-4}/s~3×10^{-2}/s and found that the friction stir processed CoCrFeNiCu HEA exhibited a maximum elongation of 620% and a flow stress of 5 MPa at the temperature of 950°C and the strain of 3×10^{-3}/s. Li et al. [10] investigated the effects of annealing temperatures from 573 to 1373 K on the microstructure and mechanical properties of FeCoCrNiMn HEA produced by severe cold rolling with ultrafine structure, and found that the severe cold rolled HEA annealed at 923K and 973K has high strength–ductility combinations because of the formation of the fully recrystallized ultrafine-grained microstructure. However, most of the above investigations are based on highly cost consuming fabrication methods of HEAs. There are few investigations on deformation behaviour of HEAs fabricated by traditional process methods, which are more cost-effective for engineering application.

This paper presents a preliminary study on the new HEA CoFeMnNi under typical superplastic deformation conditions. The as-received material was extruded tube. Tensile specimens were machined from the initial extruded tubes, and a series of tensile tests at hot deformation condition with low strain rate were conducted. To reveal the underlying mechanisms, microstructural evolutions, including the average grain size, dislocation density and fracture morphology were also characterized. Finally, a mechanism-based material model of CoFeMnNi was established, which enables to predict both the macroscopic and microscopic variable evolutions.

2. Experimental method

The as-received material was extruded tube, from which the specimens used for characterizing the hot deformation behaviour of HEA were machined. Sub-sized specimen was machined along the extrusion direction from the tube, as schematically shown in Fig. 1. In the practical machining process, the extruded tube was firstly cut into columns with a fan ring cross-section and flattened to achieve rectangular sheets, and dog-bone shaped specimen was cut from the sheet.

A series of high temperature tensile tests were conducted investigate the formability of the CoFeMnNi alloy at the elevated temperature and low strain rate. Four deformation temperatures ranging from 850°C to 1000°C with a gap of 50°C and two strain rates, 0.0001/s and 0.00001/s were chosen for the tests.

After tensile tests, the fracture region A at the finer part of the specimen and the undeformed region B at the gripped part of the specimen were machined for EBSD observation to investigate the influence of the deformation on initial microstructure.

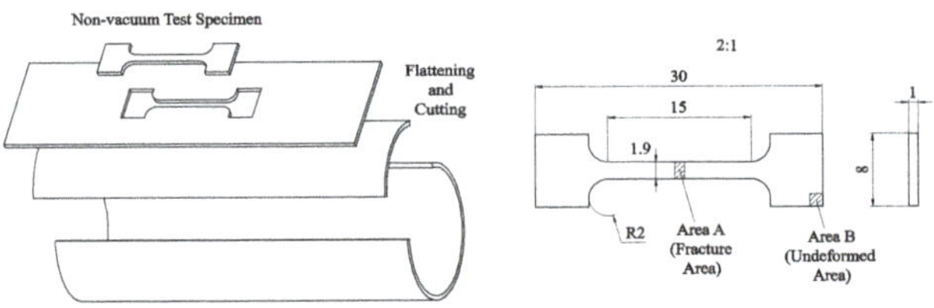

***Fig. 1** The schematic representation of specimen machining and dimensions of the specimen*

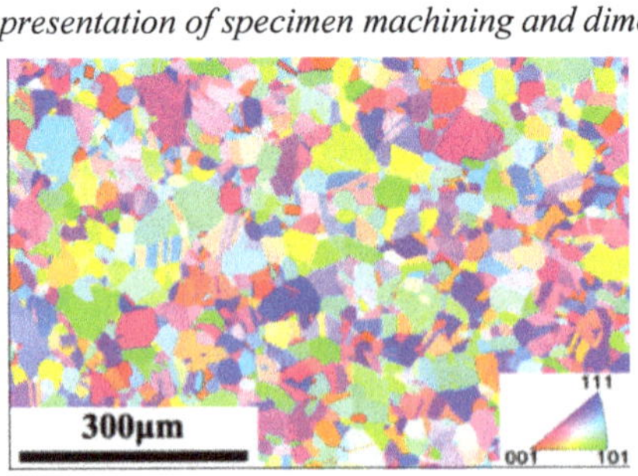

***Fig. 2** The as-received microstructure (with grain size of 6.13μm)*

3. Results of discussion

3.1 Stress-strain behaviours at hot deformation condition

Fig. 3 shows the true stress-strain curve of the CoFeMnNi HEA sheet specimen at different deformation conditions. Fig.3a showed that the true stress obviously increases by more than one-third with the strain rate increasing from 0.00001/s to 0.0001/s. Fig. 2b shows the temperature effect on the visco-plastic behavior at the strain rate of 0.0001/s, the value of the true stress obviously decreases as the temperature increases from 800°C to 1000°C. In addition, the plasticity of HEA is not uniform as the temperature increases evenly. When the temperature increases form 800°C to 850°C, the maximum true strain of the true stress-strain curve only slightly increases. As the temperature increases further, the maximum true strain of the true stress-strain curve increases sharply. When the temperature is 1000°C, the maximum true strain of the true stress-strain curve can be greater than 0.3, which means that the CoFeMnNi HEA has a relatively good plasticity at the temperature of 1000°C and may have acceptable formability to manufacture simple engineering parts.

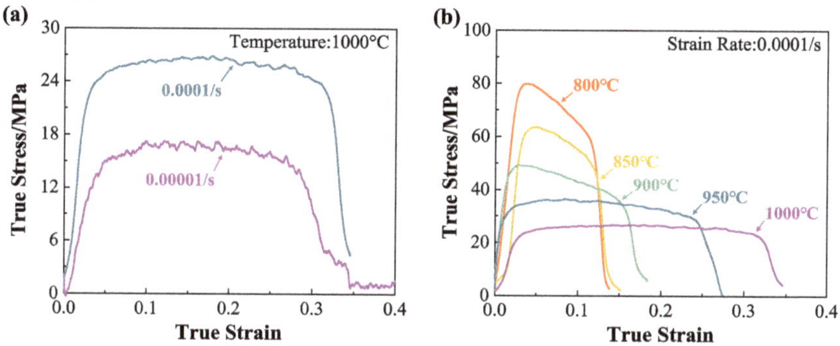

(a) Temperature of 1000°C (b) Strain Rate of 0.0001/s.

Fig. 3 *The true stress-strain curve of the CoFeMnNi HEA at the temperature of 1000°C and strain rate of 0.0001/s.*

3.2 EBSD Microstructure

3.2.1 Grain size and boundaries

Fig. 4 shows EBSD maps at various temperatures for fracture region A and undeformed region B, respectively. When deformation temperature ranged from 850°C to 950°C, the grain size of region A is slightly smaller than region B. Until the temperature was set to 1000°C, the grain size increased from 6.13 to 7.06μm. This phenomenon was attributed to refinement by the dynamic recrystallization during hot deformation. Regarding the same zones, as temperature increases, the grains tend to grow up for both the grains of region A and region B. However, the grain morphology at the region A have no obvious sign of uniaxial elongation and broadly remains isometric. In addition, as temperature increases from 850°C to 1000°C, the difference of the amount of twinning becomes more obvious. When the temperature is 1000°C, the amount of twinning is obviously less than that at the region B.

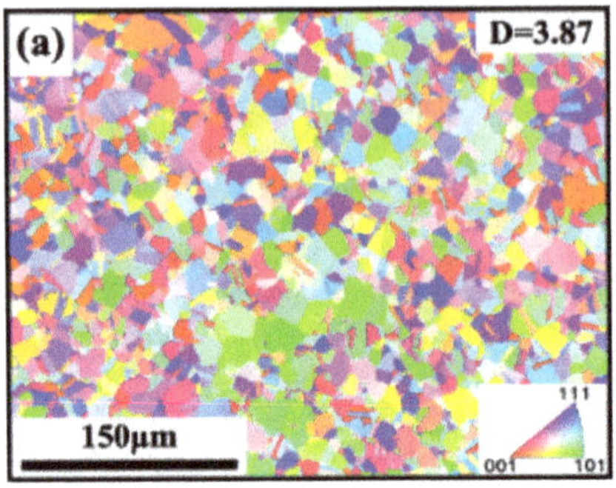

(a) 850°C-0.0001/s-region A

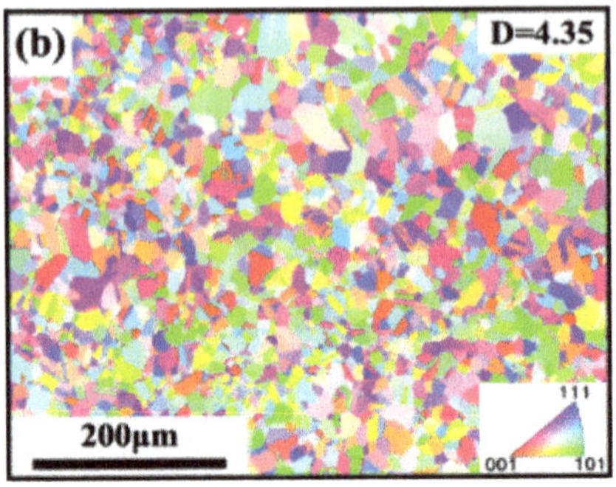

(b) 850°C-0.0001/s-region B

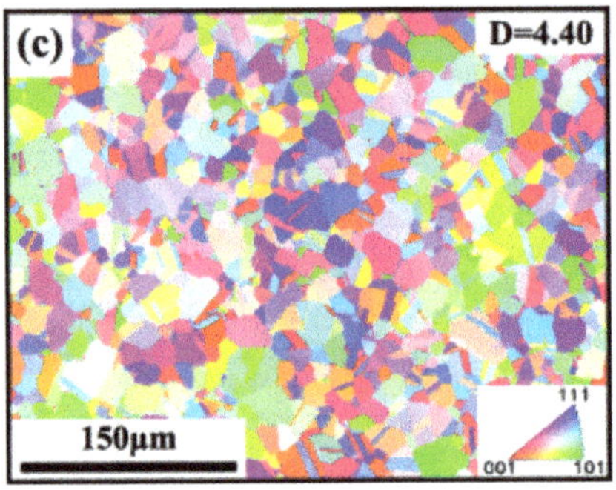

(c) 900°C-0.0001/s-region A

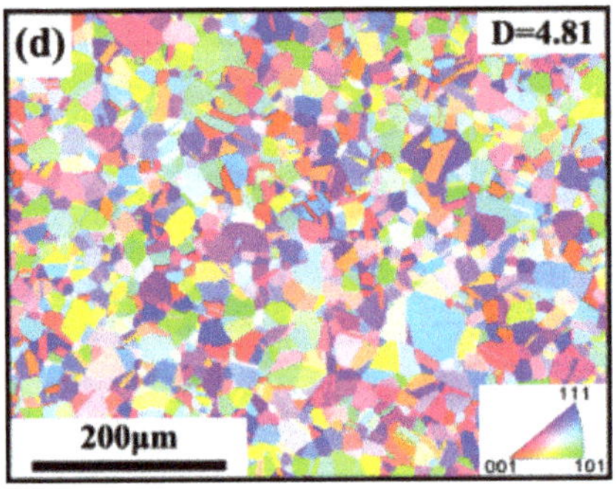

(d) 900°C-0.0001/s-region B

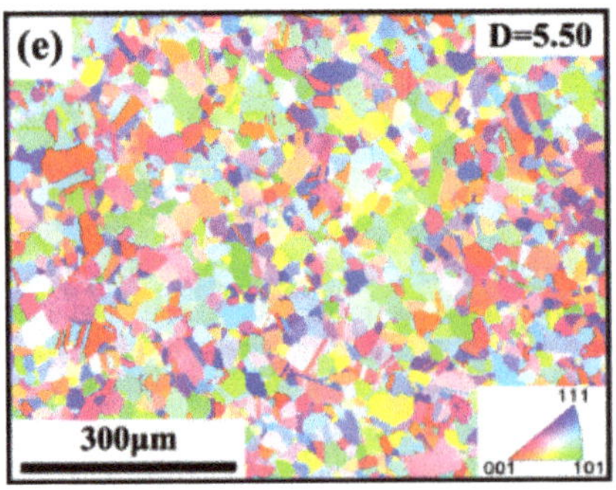

(e) 950°C-0.0001/s-region A

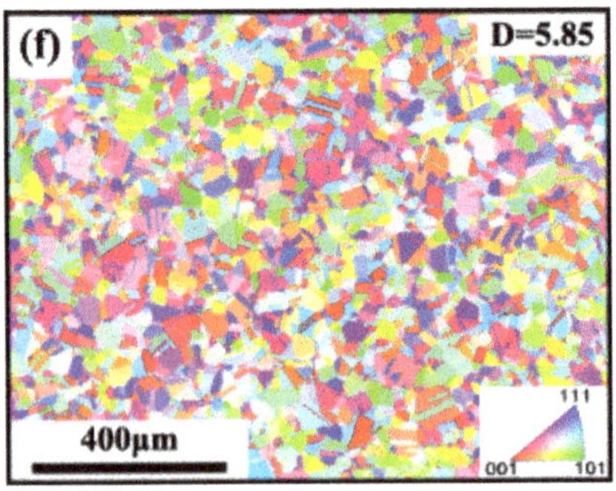

(f) 950°C-0.0001/s-region B

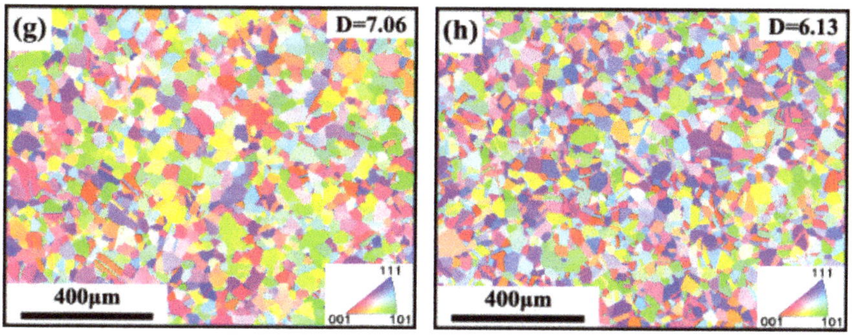

(g) 1000°C-0.0001/s-region A (h) 1000°C-0.0001/s-region B

Fig. 4 *The EBSD observations at different conditions and regions.*

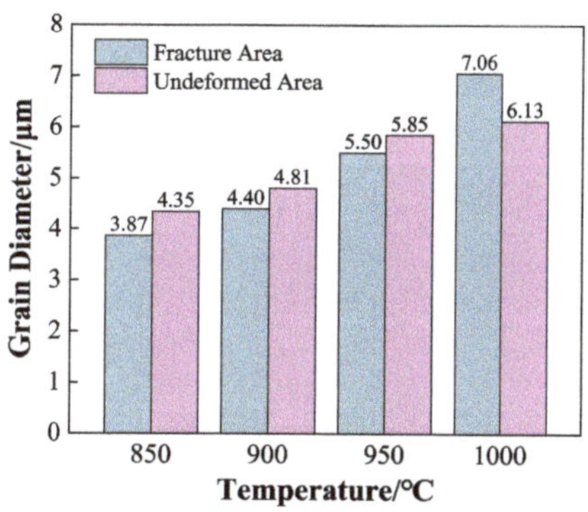

Fig. 5 *The calculated grain diameters under various temperatures and at different regions.*

The obtained EBSD grain boundary maps at various temperatures are shown in Fig. 6. The grain boundaries, indicated by red, green and blue relatively means the angle of the grain boundaries are relatively ranged from 2° to 5°, from 5° to 15° and from 15° to 180°. The results shows that the angle of majority of the grain boundaries ranges from 15° to 180° for whole temperature range and both two regions. It is indicated that grain boundary sliding dominates the deformation rather than the dislocation motion mechanism.

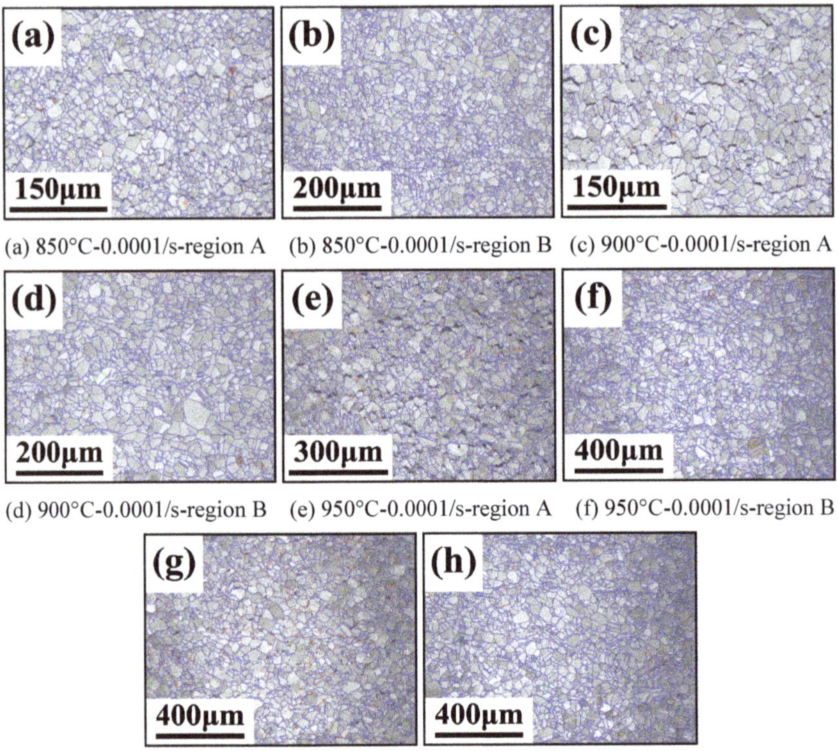

(a) 850°C-0.0001/s-region A (b) 850°C-0.0001/s-region B (c) 900°C-0.0001/s-region A

(d) 900°C-0.0001/s-region B (e) 950°C-0.0001/s-region A (f) 950°C-0.0001/s-region B

(g) 1000°C-0.0001/s-region A (h) 1000°C-0.0001/s-region B

***Fig. 6** The grain boundary map at different condition and different region.*

Fig. 7 summarises the quantification of the fractions of the grain boundaries with different angles at various temperatures for two regions. For the high-angle grain boundary fraction, as temperature increases from 850°C to 1000°C, the fraction of the grain boundaries (>15°) gradually decreases in both fracture region and undeformed region. In addition, the decline magnitude of the fraction of the grain boundaries (>15°) is obviously larger than that at the undeformed region as the temperature increases from 900°C to 1000°C. However, when the temperature is 1000°C, the fraction of the grain boundaries (>15°) in both fracture region and undeformed region can still be larger than 0.85. The fraction of the grain boundaries (5°-15°) has no significant changes as the temperature increases from 850°C to 1000°C. However, the fraction of the grain boundaries (2°-5°) has different variation trends as the temperature from 850°C to 1000°C. In the fracture region, the fraction of the grain boundaries (2°-5°) obviously increases. However, for the undeformed region, the fraction of the grain boundaries (2°-5°) has no significant change as the temperature increases from 850°C to 1000°C.

Materials Research Forum LLC
https://doi.org/10.21741/9781644902615-29

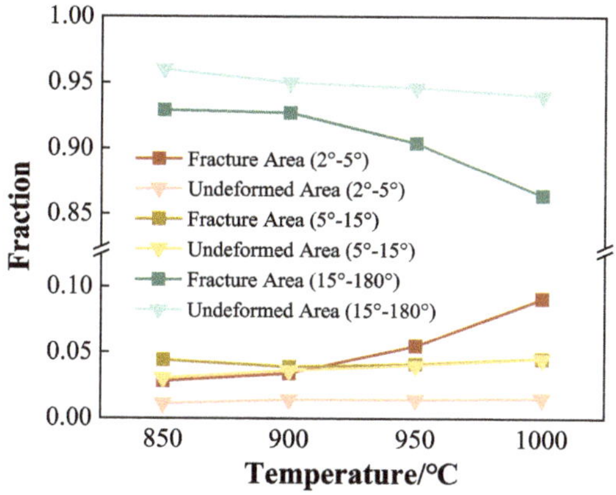

Fig. *7 The fraction of grain boundaries with different angles at various temperatures in fracture and undeformed regions, respectively*

3.3 Visco-plastic constitutive modelling

3.3.1 Material modelling

In order to characterize the hot flow behaviour of HEA, constitutive modelling to model the stress-strain at hot deformation behaviour with low strain rate is essential. In addition, to realize the overall prediction of microstructure evolutions, mechanism-based modelling scheme is used in this paper. The complete equation set is summarized using Eqs. (1) to (8):

Equation (1) is the generalized visco-plastic flow law expressed in the form of rate. For visco-plastic deformation, the strain rate equation is established considering the strain-hardening behavior at the beginning of the deformation and the microstructure evolution, e.g. grain size evolution, during the hot straining.

$$\dot{\varepsilon}^P = \left(\frac{\frac{\sigma}{(1-\omega_{1D})} - R - k}{K} \right)^{n_1} (\bar{d})^{-u} \tag{1}$$

where, $\dot{\varepsilon}^P$ is the plastic strain rate, σ is the equivalent force, R is the hardening-related hardening state variable, k is the initial yield stress (critical stress), which is temperature-dependent, K and n_1 are temperature-dependent coefficients, μ is a constant, and $\bar{d}$ is the average grain size. In addition, the equations take into account the damage and grain size evolution during thermal deformation by introducing the normalized grain size $\bar{d}$ and the one-dimensional damage ω_{1D}.

Equation (2) is the dislocation density rate equation composed by three terms. The first term describes the dislocation density accumulation and dynamic recovery in plastic deformation, while the second and third terms consider the effect of static recovery and dynamic recrystallization on the evolution of dislocation density, respectively. Usually, the dislocation density tends to increase, then decrease, and then becomes a saturated state. At the beginning of the material deformation, strain hardening causes the dislocation density to increase, and after the deformation proceeds to a certain extent, dynamic reversion and other effects cause the dislocation density to decrease, and at the end of the deformation, the dislocation density tends to stabilize.

$$\dot{\rho} = A(1-\bar{\rho})|\dot{\varepsilon}^P| - C_1\bar{\rho}^{n_2} - [C_2\bar{\rho}/(1-S)]\dot{S} \tag{2}$$

where, A, C_1 are temperature dependent coefficients and C_2 is a constant.

Equation (3) is the critical dislocation density equation, which can be used to determine the moment when recrystallization occurs during the deformation of a high-entropy alloy, i.e., when the dislocation density value is higher than the critical dislocation density, the material recrystallizes.

$$\bar{\rho}_c = q_3 \dot{\varepsilon}^{q_4} \quad (3)$$

where, q_3 and q_4 are temperature-dependent coefficients.

Equation (4) is the hardening rate equation, which describes the evolution of the hardening behavior of the material. The hardening of the material is determined by the dislocation density, which increases monotonically to saturation, and is directly related to the plastic strain because recrystallization and annealing are not considered in the effect on hardening at the beginning of the high temperature deformation process.

$$\dot{R} = 0.5b\bar{\rho}^{-0.5}\dot{\bar{\rho}} \quad (4)$$

where, B is the temperature-dependent coefficient and $\bar{\rho}$ is the normalized dislocation density of the material, which is 0 at the beginning of the deformation and increases with the deepening of the deformation until it reaches 1 at saturation.

Equation (5) is the recrystallization evolution equation, the recrystallization coefficient is maximum from 0 to 1. During recrystallization, new grains are nucleated and the total number of grains increases. During thermal deformation, dislocation accumulation leads to slip and shear at grain boundaries. In addition, the occurrence of dynamic recrystallization leads to the creation of new grains, which reduces the average grain size and can even reduce the rate of dislocation accumulation and dislocation density.

$$\dot{S} = \frac{q_1(0.1+S)^{q_2}(1-S)\bar{\rho}^2}{\bar{d}} \quad (5)$$

where, q_1 and q_2 are both temperature-dependent equation coefficients.

Equation (6) is the grain evolution equation, which consists of three terms, where the first term represents the static grain growth during thermal deformation, which is determined by the migration and energy density of grain boundaries; the second term describes the plastic strain leading to grain evolution, and the third term represents the combined effect of recrystallization leading to grain refinement and strain rate.

$$\dot{d} = w_1 d^{-\gamma_1} + w_2 \dot{\varepsilon}^{-\gamma_2} - w_3 \dot{S}^{\gamma_3} (\bar{d})^{\gamma_4} \quad (6)$$

where, w_1 and w_2 are material constants related to grain mobility and grain boundary energy density. w_3 is a temperature dependent coefficient, andγ_1, γ_2, γ_3 and γ_4 are constants.

Equation (7) is the damage rate equation. The damage equation was established based on the continuous damage mechanism and the superplastic forming model.

$$\dot{\omega} = \left[\frac{\eta_1}{(1-\omega)^{\eta_3}}(\dot{\varepsilon}^P)^{\eta_2}\right](1-\delta) + \left[\eta(1-\xi)\dot{\varepsilon}_P + \frac{F\varepsilon^p}{(1-\xi)}\sigma\dot{\varepsilon}_P\right] \quad (7)$$

where, ω denotes the region fraction of the damaged material and ranges from 0 to 1. the damage value $\dot{w} = 0$ is the initial state of the material, and failure is assumed to occur when $\dot{w}$ reaches 0.7. η_1, η_2, η_3 are temperature-dependent coefficients.

Equation (8) is a one-dimensional stress equation considering damage.

$$\sigma = E(1-\omega)(\varepsilon^T - \varepsilon^P) \quad (8)$$

where, E is Young's modulus, ε^T is the total strain, and ε^P is the plastic strain.

For the temperature dependent coefficients, the equations are summarized in Table 2.

***Table 2** Temperature dependent material coefficient equations*

$K = K_0 exp\,(\frac{QK_0}{RT})$	$k = k_0 exp\,(\frac{Qk_0}{RT})$	$n_1 = n_{10} \exp\,(\frac{Qn_{10}}{RT})$
$A = A_0 exp\,(\frac{QA_0}{RT})$	$B = B_0 exp\,(-\frac{QB_0}{RT})$	$W_1 = W_{10} exp\,(-\frac{QW_{10}}{RT})$
$q_1 = q_{10} exp\,(\frac{Qq_{10}}{RT})$	$q_2 = q_{20} exp\,(\frac{Qq_{20}}{RT})$	$W_2 = W_{20} exp\,(-\frac{QW_{20}}{RT})$
$q_4 = q_{40} \exp\,(\frac{Qq_{40}}{RT})$	$q_3 = q_{30} \exp\,(\frac{Qq_{30}}{RT})$	$W_3 = W_{30} exp\,(-\frac{QW_{30}}{RT})$
$\gamma_1 = \gamma_{10} exp\,(\frac{-Q\gamma_{10}}{RT})$	$\gamma_2 = \gamma_{20} exp\,(\frac{Q\gamma_{20}}{RT})$	$\eta_1 = \eta_{10} exp\,(\frac{Q\eta_{10}}{RT})$
$\gamma_3 = \gamma_{30} exp\,(\frac{-Q\gamma_{10}}{RT})$	$\gamma_4 = \gamma_{40} exp\,(\frac{Q\gamma_{40}}{RT})$	$C_1 = C_{10} exp\,(\frac{-QC_{10}}{RT})$
$C_2 = C_{20} exp\,(\frac{-QC_{20}}{RT})$	$E = E_0 exp\,(\frac{QE_0}{RT})$	$P = P_1 \cdot \dot{\varepsilon}_P^2 + P_2 \cdot \dot{\varepsilon}_P + P_3$
$F = \dot{\varepsilon}_p^{0.4}$		

3.3.2 Model prediction accuracy

Figure 7 compares the hot deformation behaviour between the model predictions (scattered points) and experimental results (solid line). The material constants were determined by trial & error fitting, which are given in Table 3.

***Table 3** Coefficient of CoFeMnNi high entropy alloy material model equation*

Parameter	K_0	k_0	n_{10}	E_0	B_0
Value	800	0.002265	2.56	258.5	0.5
Parameter	QK_0	Qk_0	Qn_{10}	QE_0	QB_0
Value	-1.84e-07	6.78E+04	0.0002122	1.67E+04	9.43E-05
Parameter	A_0	C_{10}	C_{20}	W_{10}	W_{20}
Value	4.43	0.0982	6.8	4.62E-4	0.0053
Parameter	QA_0	QC_{10}	QC_{20}	QW_{10}	QW_{20}
Value	0.0002424	-4.29E-05	0.000241	0.1622	4.0107E-04
Parameter	W_{30}	q_{10}	q_{20}	q_{30}	q_{40}
Value	2860	0.3	0.3	0.154	0.268
Parameter	QW_{30}	Qq_{10}	Qq_{20}	Qq_{30}	Qq_{40}
Value	1.14E-0	6.26E-05	6.26E-05	-2.09E-06	2.92E-05
Parameter	γ_1	γ_2	γ_3	γ_4	η_1
Value	3.073126	2.088153	2	2	0.01523
Parameter	$Q\gamma_{10}$	$Q\gamma_{20}$	$Q\gamma_{30}$	$Q\gamma_{40}$	$Q\eta_{10}$
Value	-4.73E-06	1.22E-04	3.11E-04	3.11E-04	1.27E+04
Parameter	η_2	η_3	R	U	
Value	1.452	24.9	8.314	6	
Parameter	$Q\eta_{20}$	$Q\eta_{30}$			
Value	-3537	1.48E-05			

Materials Research Forum LLC
https://doi.org/10.21741/9781644902615-29

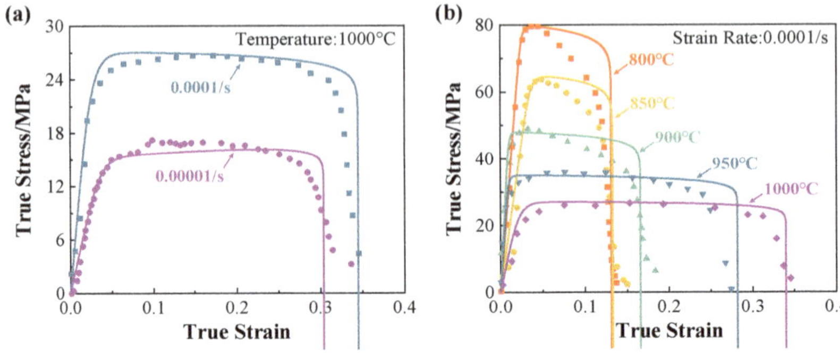

(a) Temperature of 1000° C (b) Strain Rate of 0.0001/s

Fig.7 *Comparison of experimental results and model prediction of hot flow behaviour of CoFeMnNi high entropy alloy under compression*

The overall prediction of the fitted stress-strain curves for the hardening (dislocation accumulation) and softening (dynamic recrystallization) under different conditions is in relatively good agreement. The accuracy of the predictions of the developed high-temperature alloy model was verified by Pearson's correlation coefficient (Person's R), and the equation is shown in (9):

$$R = \frac{\sum_{i=1}^{N}(\sigma_E - \bar{\sigma}_E)(\sigma_p - \bar{\sigma}_p)}{\sqrt{\sum_{i=1}^{N}(\sigma_E - \bar{\sigma}_E)^2 \sum_{i=1}^{N}(\sigma_p - \bar{\sigma}_p)^2}} \tag{9}$$

where, σ_E and σ_P are the experimental stress value and predicted stress value, $\overline{\sigma}_E$ and $\overline{\sigma}_P$ are the mean experimental stress value and the mean predicted stress value, and N is the number of data points. The calculated R value indicates the comparison accuracy, and the R value ranges from 0 to 1, where the closer the R value is to 1, the higher the prediction accuracy of the model. From the calculations, it can be concluded that the average R-values are 0.99689 and 0.9904 for different temperatures and different strain rates, respectively, which means that the prediction accuracy of the model is 99%.

In addition, the prediction error of the model can be quantified by the absolute mean error AARE. The equation is given in (10):

$$AARE = \frac{1}{N}\sum_{i=1}^{N}\left|\frac{\sigma_E - \sigma_p}{\sigma_E}\right| \times 100\% \tag{10}$$

where E is the experimental stress value, P is the stress value predicted by the model, and N is the number of experiments used in the calculation. The AARE values shown in figures (a) and (b) are 0.03789 and 0.08453, respectively, which implies a maximum average relative error of 8.45%.

Conclusion

1) The stress-strain curves of CoFeMnNi HEA under hot conditions with low strain rate exhibited typical visco-plastic deformation behaviour. The ductility increase was not uniform with the deformation temperature increases from 850°C to 1000°C. When the temperature was set to 1000

°C further, the maximum of the true strain (slightly larger than 0.3) was achieved, which means that the material exhibited relatively good feasibility at such deformation condition.
2) EBSD observation results showed that the average grain size of HEAs slightly decreased after hot deformation at the temperatures ranging from 850°C to 950°C, which means the grain size was much more stable compared to typical titanium alloys deformed at hot deformation with low strain rate. Until the deformation temperature reached 1000 °C, the grain size increased by 15% approximately.
3) A mechanism based hot deformation constitutive model of high entropy alloy was developed. Acceptable agreements between experimental and calculation results were achieved.

References

[1] Yeh J.W. Alloy design strategies and future trends in high-entropy alloys[J]. Jom, 2013, 65: 1759-1771. https://doi.org/10.1007/s11837-013-0761-6

[2] Ye Y.F. Wang Q, Lu J, et al. High-entropy alloy: challenges and prospects[J]. Materials Today, 2015, 19(6). https://doi.org/10.1016/j.mattod.2015.11.026

[3] Zhao S., Li Z., Zhu C. et al. Amorphization in extreme deformation of the CrMnFeCoNi high-entropy alloy[J]. Science advances, 2021, 7(5): eabb3108 https://doi.org/10.1126/sciadv.abb3108

[4] Zhang Y., Zuo T. T, Tang Z., et al. Microstructures and properties of high-entropy alloys[J]. Progress in materials science, 2014, 61: 1-93. https://doi.org/10.1016/j.pmatsci.2013.10.001

[5] D. B. Miracle and O. N. Senkov. A critical review of high entropy alloys and related concepts[J]. Acta Materialia, vol. 122, pp. 448-511, 2017/01/01/ 2017. https://doi.org/10.1016/j.actamat.2016.08.081

[6] W. Zhao et al. Significance of grain refinement on micro-mechanical properties and structures of additively-manufactured CoCrFeNi high-entropy alloy[J]. Materials Science and Engineering A, vol. 807, 202 https://doi.org/10.1016/j.msea.2021.140898

[7] N.D. Stepanov, D.G. Shaysultanov, G.A. Salishchev, O.N. Senkov, Mechanical behavior and microstructure evolution during superplastic deformation of the fine-grained AlCoCrCuFeNi high entropy alloy[J]. Materials Science Forum, 1662-9752, Vols. 838-839, pp 302-30 https://doi.org/10.4028/www.scientific.net/MSF.838-839.302

[8] H. Shahmir, J. He, Z. Lu, M. Kawasaki, and T. G. Langdon. Evidence for superplasticity in a CoCrFeNiMn high-entropy alloy processed by high-pressure torsion[J]. Materials Science and Engineering: A, vol. 685, pp. 342-348, 2017/02/08/ 2017. https://doi.org/10.1016/j.msea.2017.01.016

[9] N. Li et al. Achieving superior superplasticity in CoCrFeNiCu high entropy alloy via friction stir processing with an improved convex tool[J]. Materials Science and Engineering A, vol. 873, 2023. https://doi.org/10.1016/j.msea.2023.145034

[10] Z. Li et al. Effect of Annealing Temperature on Microstructure and Mechanical Properties of a Severe Cold-Rolled FeCoCrNiMn High-Entropy Alloy[J]. Metallurgical and Materials Transactions A: Physical Metallurgy and Materials Science, vol. 50, no. 7, pp. 3223-3237, 2019. https://doi.org/10.1007/s11661-019-05231-y

[11] J.G. Lin. Fundamentals of materials modelling for metals processing technologies: theories and applications[M]. Changsha, China: Central South University Press, 2019.

Superplasticity in Advanced Materials - ICSAM 2023 Materials Research Forum LLC
Materials Research Proceedings **32** (2023) 264-273 https://doi.org/10.21741/9781644902615-30

Comparative studies of hyperbolic sine constitutive models for constant pressure superplastic tests

Luis García-Barrachina[1,a,*] and Antonio J. Gámez[1,b]

[1]School of Engineering, University of Cadiz, Av. Universidad de Cadiz, 10, E-11519 Puerto Real (Cadiz), Spain

[a]luis.barrachina@uca.es, [b]antoniojuan.gamez@uca.es

Keywords: Modelling, Constitutive Parameters, Hyperbolic Sine Model, Free-Inflation Test

Abstract. The constant pressure free-inflation test is a very versatile and simple tool for analyzing different features of superplastic forming. In recent years, different methods have been proposed to measure constitutive parameters of the stress–strain relationship for the superplastic material, specifically the K and m parameters of the power law model. However, this law is restricted to be used in narrow strain-rate ranges, and poor results are obtained when applied in a broader spectrum. To overcome this problem, numerous constitutive models covering the full strain-rate range applicable in superplastic forming have been proposed historically, including the hyperbolic sine equation. However, there is no clear consensus on the type of hyperbolic sine function to use. Some authors include a sensitivity parameter while others do not. This article aims to study the characteristics of the hyperbolic sine constitutive model, checking which of the historically proposed models achieves better results in the test at free deformation and constant pressure.

Introduction

It is well known the multi-variable complexity of the superplastic behavior and its dependency on so many factors, such as temperature, microstructure, deformation, strain rate or degradation of the material itself [1]. This makes modelling a complex task and continuous studies are carried out to improve the accuracy and consistency [2–4].

The behavior of a material in its superplastic regime can be extracted, as is usual, from tensile tests where the sample is subjected to a constant strain rate and the stress versus strain is contrasted [5]. Thus, the flow stress dependence of the strain rate is normally characterized by the sensitivity index m.

$$m = \frac{\partial \log \sigma}{\partial \log \dot{\varepsilon}} \tag{1}$$

This is taken as a parameter (between 0.3 and 0.7) for the quality of the superplastic behavior, providing shorter forming times and smaller final thickness gradients for higher values [6]. Thus, it is also common to express the behavior of the material as a flow stress function of the strain rate. Where the parameter m is related as the slope in the logarithmic variant of this graph. So it can be expressed as a potential law adding two material constants (K, m)

$$\sigma = K\dot{\varepsilon}^m \tag{2}$$

Alternatively, the importance of m can be understood changing (2) by introducing an apparent viscosity variable μ_a [7], such as

$$\sigma = \mu_a \dot{\varepsilon} \tag{3}$$

Thus, it is straightforward to associate the behavior of the material and its strain-rate dependence with the apparent viscosity variable

$$\mu_a = K \dot{\varepsilon}^{m-1} \tag{4}$$

such that if it is plotted versus the strain rate, Fig. 1, it is observed that the higher the value of m, the more homogeneous the behavior in the strain-rate range.

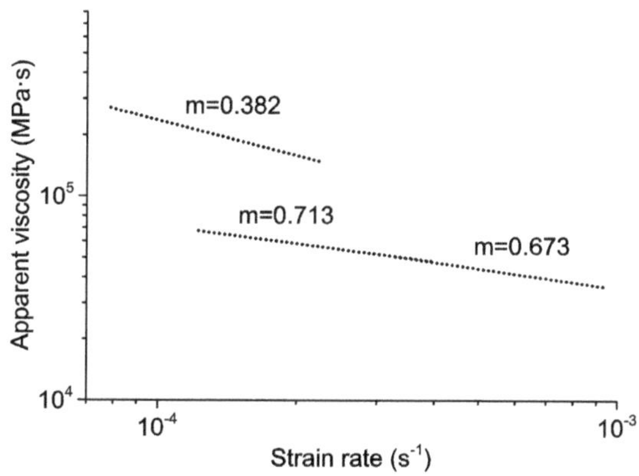

Fig 1. Effect of the m parameter on the apparent viscosity material variable. A higher value of m provides a more homogeneous behavior in the strain-rate range. Values for a titanium alloy at 800 (m=0.328) and 850 °C (m=0.713 and m=0.673)

However, as it was previously mentioned, the material is highly dependent on the strain rate. Thus, the ability to model the behavior of the material through a constant-parameter potential equation is only valid for a given range of strain rate, where the behavior can be fitted to a straight line on the logarithmic scale. As the constitutive equation moves away from the actual behavior of the material, the ability to replicate real examples in computational models is reduced, increasing the discrepancy between the two.

In order to match the material behavior over a wider strain-rate range, different constitutive models have historically been proposed: polynomial forms [8], unified constitutive equations including microstructure evolution models [2,9] or anisotropy plus grain size models [10]. [11] can be consulted for a summary of different constitutive models.

Of all these models, the hyperbolic sine model illustrates the natural step of development from the potential law, as it allows the strain-rate range to be widened without introducing further parameters into the model. It represents an essentially phenomenological model which tries to replicate by means of an equation the behavior of the material collected on the S-curve.

Sellars in 1966 [12] proposed a constitutive model for hot deformation where the strain rate is related to a hyperbolic sine function of the flow stress, where in addition the effect of temperature is introduced by an exponential term of the activation energy.

$$\dot{\varepsilon} = A(\sinh \alpha\sigma)^n e^{\left(-\frac{Q}{RT}\right)} \tag{5}$$

Lin in 1993 [13] proposed a similar model, excluding the exponential term, for creep including damage variables ω. Additionally, his model is extended to a multi-axial formulation by defining a potential dissipation energy hyperbolic cosine function.

$$\dot{\varepsilon} = A \sinh(B\sigma) f(\omega) \tag{6}$$

In 1997, Dunne [14] points out the need to move from the potential model to the hyperbolic sine model if the strain rate range of the constitutive model is to be extended. He also proposes the inclusion of a grain size evolution term.

$$\dot{\varepsilon} = \frac{\alpha}{d^{\gamma}} \sinh(\beta\sigma) \tag{7}$$

This same model is then used by Lin in 2001 [15] to characterize the titanium alloy Ti-6Al-4V with determination of constants by genetic algorithm. The microstructural part of this model was further developed in 2005 [9]. Zhang in 2004 [16] returned to Sellar's model and related the values of the exponential parameter to the α and *m* parameters. In 2006, Bonet [17] proposed a slight modification of the previous model where the exponential affects the hyperbolic sine function and not its argument.

$$\sigma = \sigma_o \left[\sinh^{-1}\left(\frac{\dot{\varepsilon}}{\dot{\varepsilon}_o}\right)\right]^m \tag{8}$$

More recently Yang in 2020 [18] and Mosleh in 2021 [19] have incorporated the hyperbolic sine model into more complex models where potential, exponential and hyperbolic sine laws are coupled to characterize the material over a wide range of strain rate. This leads to an increase in the required number of parameters and their subsequent determination.

$$\dot{\varepsilon} = A[\sinh(\alpha\sigma)]^{\eta} e^{\left(-\frac{Q}{RT}\right)} \tag{9}$$

The aim of this work is to carry out a comparative study of the different hyperbolic sine models historically proposed to describe the superplastic behavior over a wide strain rate range with a number of parameters equal to or less than three. More complex multifunctional models or models with a description of the evolution of the microstructure are discarded. In order to do so, a brief theoretical assessment of the models studied is first made. Then, from a series of tests on AZ31 magnesium alloy at 450 °C, the parameters for each model are extracted. These values are introduced into the constitutive model of the material in a finite element code to computationally obtain the forming time. This value is then compared with the experimental values to verify the model.

Methodology

a) Theoretical framework. The different constitutive models previously presented are summarized in only three options. The terms relating to temperature are not taken into account, as single temperature tests are studied, and those relating to microstructural evolution are also omitted since, due to the characteristics of the method for extracting the behavior of the material [7], this parameter is implicitly included.

Thus, the models to be compared are:

$$\sigma = \sigma_o \left[\sinh^{-1}\left(\frac{\dot{\varepsilon}}{\dot{\varepsilon}_o}\right)\right]^{m^*} \tag{10}$$

$$\sigma = \sigma_o \sinh^{-1}\left(\left[\frac{\dot{\varepsilon}}{\dot{\varepsilon}_o}\right]^{m^*}\right) \tag{11}$$

$$\sigma = \sigma_o \sinh^{-1}\left(\frac{\dot{\varepsilon}}{\dot{\varepsilon}_o}\right) \tag{12}$$

The different models have been combined into only three constitutive equations (10), (11) and (12) including a stress-related parameter σ_o, a strain-rate-related parameter $\dot{\varepsilon}_o$, and a third exponential parameter m^*, where m^* is used to differentiate from the parameter used for the potential function (2) and which is well defined in (1). To obtain the relationship between m^* and the sensitivity index, the expression (11) can be changed to

$$\sinh\left(\frac{\sigma}{\sigma_o}\right) = \left[\frac{\dot{\varepsilon}}{\dot{\varepsilon}_o}\right]^{m^*} \tag{13}$$

Given that the Taylor expansion of the hyperbolic sine function is

$$\sinh(x) = x + \frac{x^3}{3!} + \frac{x^5}{5!} + \cdots \tag{14}$$

Equation (13) can be simplified for small stress values as

$$\sinh\left(\frac{\sigma}{\sigma_o}\right) \cong \frac{\sigma}{\sigma_o} = \left[\frac{\dot{\varepsilon}}{\dot{\varepsilon}_o}\right]^{m^*} \tag{15}$$

arriving at the same potential function with value $K = {}^{\sigma_o}/_{\dot{\varepsilon}_o^{m^*}}$. Therefore, it can be concluded that $m^* = m$ when $\sigma \to 0$. In other words, m^* is the slope of the hyperbolic sine function for low stress values and can be then related to the sensitivity index. This same development can be followed with identical result for (10).

b) Experimental procedure. Experimental tests were carried out on disc specimens on a cylindrical mold of 22.5 mm radius and 3 mm entry radius. The sheets, with an initial thickness of 0.75 mm, were preheated at 450 °C and subjected to 0.2, 0.25, 0.35, 0.5 and 0.75 MPa constant pressure [7,20], Table 1. The output variable to be studied is the evolution of the height at the central point and specifically the time when the height equals the mold radius. A final dome configuration is expected at this point. Forming time experimental values are included in Table 1 together with the values of K and m. A progressive decrease of m is observed as higher strain-rate values are reached.

Table 1. Summary table of trials including estimative values of K and m

External Pressure [MPa]	*Forming time [s]*	$K\ [MPa \cdot s^m]$	m
0.2	3403	565.2	0.544
0.25	2435	565.2	0.544
0.35	1185	565.2	0.544
0.5	423	171.2	0.391
0.75	87	70.6	0.267

With these five tests and following the methodology described in [7], five corresponding values of stress and strain rate are obtained to characterize the material by plotting the hyperbolic sine law that best match the material behavior. Higher pressure from [20] were neglected from the study since they are relative fast processes where transient periods are important and the aforementioned method is less accurate. Using the *Matlab* [21] curve-fitting application, the following parameter values are obtained for the hyperbolic sine models, where the R^2 value is attached, Table 2. Fig. 2 shows how these three models adjust to experimental values of stress and strain-rate.

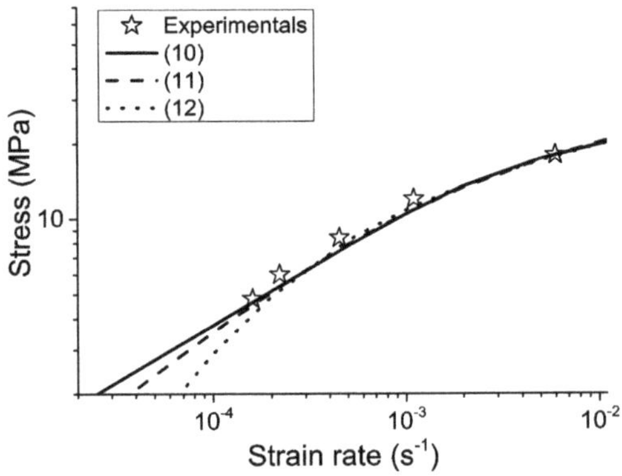

Fig 2. Adjustment of hyperbolic sine models to experimental values

Table 2. Numerical values for the three models' parameters

Model	$\sigma_o\ [MPa]$	$\dot{\varepsilon}_o\ [s^{-1}]$	m^*	R^2
(10)	12.72	0.00144	0.4567	0.998
(11)	7.416	0.00033	0.5932	0.999
(12)	3.872	0.00012		0.996

The reason for using a hyperbolic sine function is illustrated in Fig. 2, in order to accommodate processes with a wide strain-rate range, from ~10^{-4} to ~10^{-2}. For narrower ranges, e.g. those obtained for [0.2-0.5] MPa, the hyperbolic sine closely matches a typical potential function and its results, so its use is less justified.

c) FEM simulations. The three hyperbolic sine models are then tested under *Marc Mentat* FE code using an axisymmetric finite element model, with constant pressure and static analysis, considering that the dynamic effects are negligible [22], and with a termination criterion when the central node is displaced by 22.5 mm. 0.25 mm length quadratic elements are used through the blank thickness.

Finally, the material model is added externally by means of a simple Fortran routine. Fig. 3 depicts an example of the strain-rate evolution during five simulation times for 0.5 MPa.

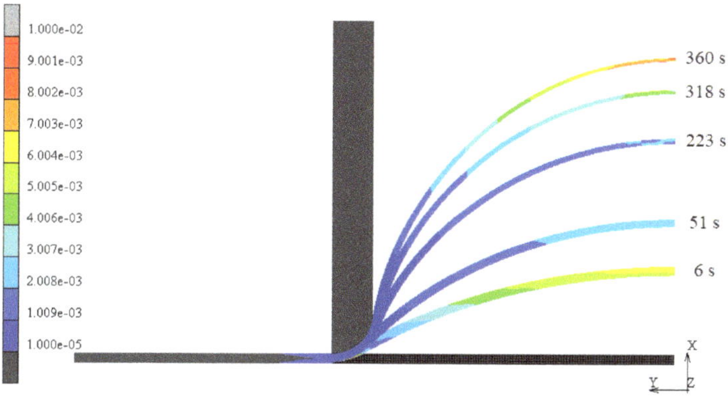

Fig. 3. Strain-rate evolution at five stages during a 0.5 MPa test

Results and Discussions

Table 3 shows the comparative results, for the five pressure values, between the experimental forming time and the three computational forming times using models (10), (11) and (12) together with the parameters in Table 2.

Table 3. Forming time results

External Pressure [MPa]	*Forming Time [s]*			
	Experimental	(10)	(11)	(12)
0.2	3403	3035	3226	2776
0.25	2435	1835	2051	1900
0.35	1185	828	865	914
0.5	423	296	201	295
0.75	87	37	11	44

These values are shown graphically in Fig. 4. A number of remarks can be drawn from these results:

- None of the three models performs significantly better than the others. Even the 2-parameter model presents time estimations of the same order as the models where the exponent term is included.
- The relative deviation from the experimental values shows a positive trend as the external pressure increases.
- All three models present an underestimation of the forming time, i.e. they predict a lower time than the real one. This may be due, as shown in Fig. 2, to the fact that all three models predict a higher strain rate for the reference stress values for the five tests.

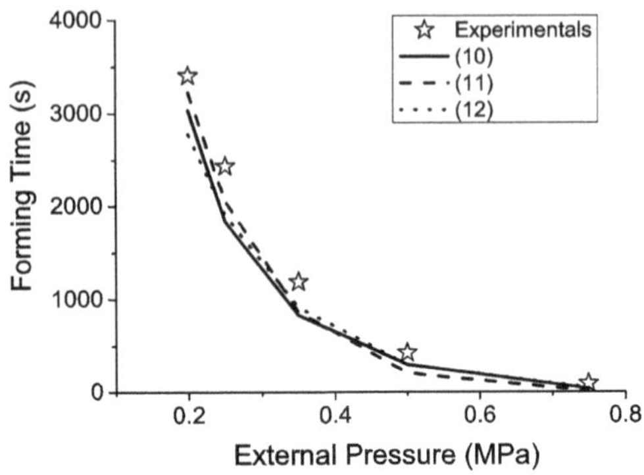

Fig. 4. Forming time versus external pressure

On this last observation, it should be added that this behavior is mainly seen in the pressure range from 0.2 to 0.5 MPa, Fig. 5. However, for the 0.75 MPa test, a sudden softening in the final stage of forming must be included, Fig. 6. This unexpected behavior can also be seen in Fig. 7 where the final shape is compared for pressures of 0.5 and 0.75 MPa. For the latter test, the final shape departs from a dome pattern.

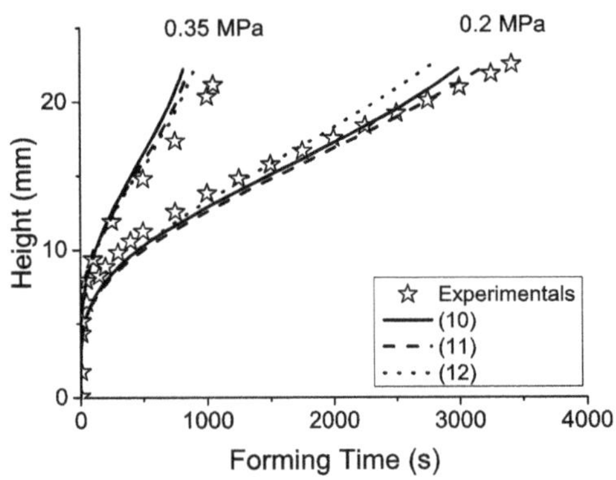

Fig. 5. Center sample height evolution for 0.2 and 0.35 MPa

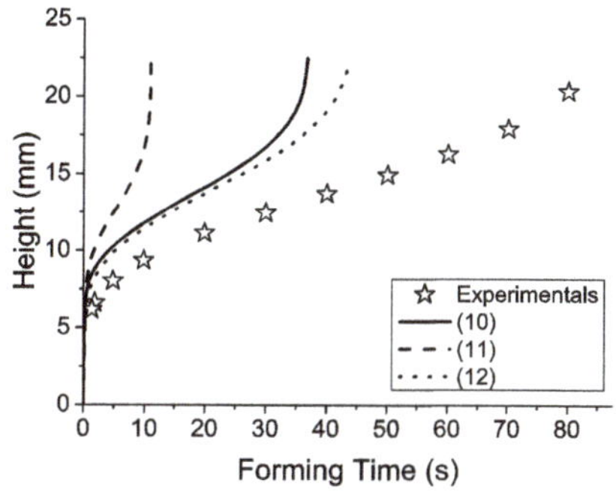

Fig. 6. Center sample height evolution for 0.75 MPa

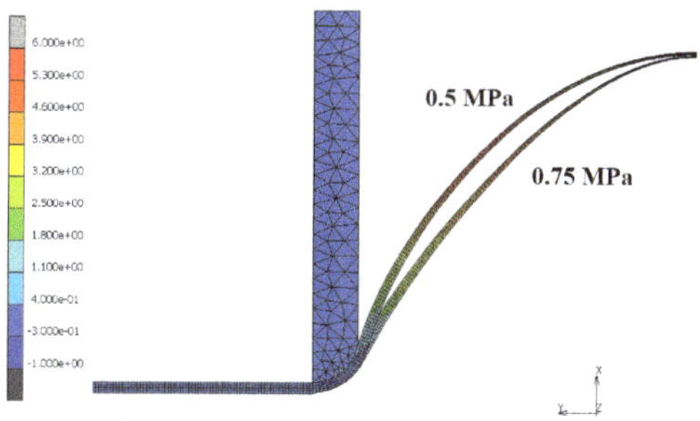

Fig 7. Final configuration for tests at 0.5 and 0.75 MPa. The color scheme represents the horizontal displacements.

Conclusions

In view of the results obtained following the methodology described above, the next conclusions can be drawn:

- The three proposed models give similar results. In other words, the inclusion of a third parameter in the form of an exponent does not provide significant improvements.
- In case such an exponent is used, it seems desirable to avoid confusion with the sensitivity index *m*, with a different nomenclature, since, as previously developed, its relationship is limited to stress values tending to 0, where SPF processes are not common.
- All three models overestimate the strain-rate values over the flow stress. It seems necessary to corroborate whether this trend is repeated for other materials in future work.
- The values of forming time obtained computationally present a first estimation with values in the order of the experimental time with errors ranging from 10% to 50%. It is also observed that these values increase progressively as the external pressure increases.
- It seems also necessary, in future work, to study if the inclusion of an additional parameter allows correcting this overestimation of the strain rate in order to improve the model.

- For higher strain rate values, i.e., approximately 10^{-2} s^{-1} and an approximate forming time of 100 s, anomalous behavior is observed in the FEM model, which requires a separate study to correct the finite element model and achieve more stable behavior for larger strain-rate ranges.

References

[1] I.C. And, R.S. Mishra, Materials Processing Handbook, Taylor & Francis Group, 2007.

[2] E. Alabort, D. Putman, R.C. Reed, Superplasticity in Ti-6Al-4V: Characterisation, modelling and applications, Acta Mater., 95 (2015) 428-442. https://doi.org/10.1016/j.actamat.2015.04.056

[3] T. Yasmeen, Z. Shao, L. Zhao, P. Gao, J. Lin, J. Jiang, Constitutive modeling for the simulation of the superplastic forming of TA15 titanium alloy, Int. J. Mech. Sci., 164 (2019). https://doi.org/10.1016/j.ijmecsci.2019.105178

[4] W. Zhou, J. Lin, D.S. Balint, T.A. Dean, Clarification of the effect of temperature and strain rate on workpiece deformation behaviour in metal forming processes, Int. J. Mach. Tools Manuf., 171 (2021) 103815. https://doi.org/10.1016/j.ijmachtools.2021.103815

[5] P.N. Comley, ASTM E2448-A Unified Test for Determining SPF Properties, J. Mater. Eng. Perform., 17 (2008) 183-186. https://doi.org/10.1007/s11665-007-9181-5

[6] G.C. Cornfield, R.H. Johnson, The forming of superplastic sheet metal, Int. J. Mech. Sci., 12 (1970) 479-490. https://doi.org/10.1016/0020-7403(70)90075-5

[7] L. García-Barrachina, D. Sorgente, L. Tricarico, A.J. Gámez, A method for estimating superplastic material parameters via free-inflation tests, J. Mater. Res. Technol., 11 (2021) 1387-1395. https://doi.org/10.1016/j.jmrt.2021.01.116

[8] N. Chandra, Constitutive behaviour of superplastic materials, Int. J. Non. Linear. Mech., 37 (2002) 461-484. https://doi.org/10.1016/S0020-7462(01)00021-X

[9] J. Lin, T.A. Dean, Modelling of microstructure evolution in hot forming using unified constitutive equations, J. Mater. Process. Technol., 167 (2005) 354-362. https://doi.org/10.1016/j.jmatprotec.2005.06.026

[10] A.J. Carpenter, A.R. Antoniswamy, J.T. Carter, L.G. Hector, E.M. Taleff, A mechanism-dependent material model for the effects of grain growth and anisotropy on plastic deformation of magnesium alloy AZ31 sheet at 450 C, Acta Mater., 68 (2014) 254-266. https://doi.org/10.1016/j.actamat.2014.01.043

[11] O. Majidi, M. Jahazi, N. Bombardier, A viscoplastic model based on a variable strain rate sensitivity index for superplastic sheet metals, Int. J. Mater. Form., 12 (2019) 693-702. https://doi.org/10.1007/s12289-018-1443-2

[12] C.M. Sellars, W.J. McTegart, On the mechanism of hot deformation, Acta Metall., 14 (1966) 1136-1138. https://doi.org/10.1016/0001-6160(66)90207-0

[13] H.D.R. Lin J., D. B.F., A new design of uniaxial testpiece with slit extensometer ridges for improved accuracy of strain measurement, Int. J. Mech. Sci., 35 (1993) 63-78. https://doi.org/10.1016/0020-7403(93)90065-3

[14] T.W. Kim, F.P.E. Dünne, Determination of superplastic constitutive equations and strain rate sensitivities for aerospace alloys, Proc. Inst. Mech. Eng. Part G J. Aerosp. Eng., 211 (1997) 367-380. https://doi.org/10.1243/0954410971532730

[15] J. Lin, Selection of material models for predicting necking in superplastic forming, Int. J. Plast., 19 (2002) 469-481. https://doi.org/10.1016/S0749-6419(01)00059-6

[16] B. Zhang, D.J. Mynors, A. Mugarra, K. Ostolaza, Representing the superplasticity of Inconel 718, J. Mater. Process. Technol., 153-154 (2004) 694-698. https://doi.org/10.1016/j.jmatprotec.2004.04.138

[17] J. Bonet, A. Gil, R.D.R.D. Wood, R. Said, R. V Curtis, Simulating superplastic forming, Comput. Methods Appl. Mech. Eng., 195 (2006) 6580-6603. https://doi.org/10.1016/j.cma.2005.03.012

[18] J. Yang, J. Wu, Q. Zhang, R. Han, K. Wang, The simple hyperbolic-sine equation for superplastic deformation and parameters optimization, J. Mater. Res. Technol., 9 (2020) 10819-10829. https://doi.org/10.1016/j.jmrt.2020.07.076

[19] A.O. Mosleh, A.D. Kotov, A.A. Kishchik, O. V. Rofman, A. V. Mikhaylovskaya, Characterization of superplastic deformation behavior for a novel al-mg-fe-ni-zr-sc-based alloy: Arrhenius-based modeling and artificial neural network approach, Appl. Sci., 11 (2021) 1-18. https://doi.org/10.3390/app11052208

[20] D. Sorgente, G. Palumbo, L.D. Scintilla, L. Tricarico, Gas forming of an AZ31 magnesium alloy at elevated strain rates, Int. J. Adv. Manuf. Technol., 83 (2016) 861-872. https://doi.org/10.1007/s00170-015-7614-0

[21] The MathWorks Inc., MATLAB Version: 9901592791 (R2020b), (2020).

[22] L. García-Barrachina, A.J. Gámez, A forming time estimator of superplastic free bulge tests based on dimensional analysis, Int. J. Mater. Form., 14 (2021) 499-506. https://doi.org/10.1007/s12289-019-01527-x

Warm tensile properties of low-alloy TRIP steel warm-rolled just prior to isothermal holding treatment

Katsumi Hattori[1,a], Genta Kojima[1,b], Junya Kobayashi[2c], Goroh Itoh[2,d,*], Shigeru Kuramoto[2,e] and Tomohiko Hojo[3,f]

[1] Major in Mechanical Engineering, School of Science and Engineering, Graduate School, Ibaraki University, 4-12-1 Naka-Narusawa-cho, Hitachi, 316-8511 Japan

[2] Department of Mechanical System Engineering, Faculty of Engineering, Ibaraki University, 4-12-1 Naka-Narusawa-cho, Hitachi, 316-8511 Japan

[3]Department of Mechanical Engineering and Intelligent Systems, Tohoku Gakuin University, 3-1 Shimizu-koji, Wakabayashi-ku, Sendai, Miyagi, 984-8588 Japan

[a] 23nm479f@vc.ibaraki.ac.jp, [b] 22nm441f@vc.ibaraki.ac.jp,
[c] junya.kobayashi.jkoba@vc.ibaraki.ac.jp, [d] goroh.itoh.ibaraki@vc.ibaraki.ac.jp,
[e] shigeru.kuramoto.11@vc.ibaraki.ac.jp, [f] tomohiko.hojo@mail.tohoku-gakuin.ac.jp

Keywords: TRIP-Aided Steel, High-Strength Steel, Warm Rolling, Retained Austenite, Martensite

Abstract. Low-alloy TRIP steel sheets with ultra-high-strength that utilize transformation-induced plasticity of retained austenite have high strength and ductility. In this study, low-alloy TRIP steel sheets with thermo-mechanical treatment with warm rolling were prepared, and the effects of warm rolling on tensile properties were investigated. A hot-rolled sheet of the steel with Ms of 418.5°C was austenized and then quenched in a salt bath kept at 450°C. Two specimens of the sheet were warm-rolled by 40 and 60% and then isothermally held at 350°C for 1ks. For comparison, a specimen was directly isothermally heat-treated at 350°C for 1ks. The three specimens were subjected to microstructural characterizations and tensile tests at temperatures ranging from 25 to 300°C. It was confirmed that the warm rolling increases the ductility/strength balance. The X-ray diffraction showed that the warm rolling causes an increase in volume fraction of the retained austenite, while its stability against the tensile deformation was somewhat complicated. The effect of the temperature and rolling reduction on the uniform elongation, which is the parameter mostly affected by TRIP, was successfully understood in terms of the volume fraction and the stability of the retained austenite.

Introduction

Low-alloy TRIP steel sheets, which are ultra-high strength steel sheets utilizing transformation-induced plasticity of retained austenite, are attracting attention as automotive steel sheets that contribute to carbon neutrality due to their high strength and high ductility. Low-alloy TRIP steel sheet is known to exhibit high ductility at room temperature, and it has been reported that high ductility is also exhibited during warm deformation by controlling stress or strain-induced transformation of retained austenite [1]. Previous studies [2-6] have reported the properties of low-alloy TRIP steels prepared by heat treatment only, and there are no reports on the warm tensile properties of low-alloy TRIP steels subjected to thermos-mechanical treatment.

In this study, low-alloy TRIP steel sheet specimens were fabricated by thermo-mechanical treatment including warm rolling and their tensile properties were assessed at temperatures ranging from 25 to 300°C. Then discussion on the effect of warm rolling was made based on the experimental evaluation of the volume fraction of the retained austenite and its stability against the tensile deformation.

Table 1 Chemical composition of the sample in mass %.

C	Si	Mn	Nb	Mo	Al	Fe
0.20	1.50	1.50	0.050	0.20	<0.005	Bal.

Experimental procedure

Form a hot-rolled steel sheet of 5 mm thickness with a chemical composition shown in Table 1 (Ms: 418.5°C), three specimens of 1, 1.67 and 2.5 mm thickness were cut out. The latter two specimens were thermo-mechanically treated as shown in Fig. 1, i.e., austenized in a salt bath at 950°C for 1.2 ks, rapidly cooled to 450°C, below the A_1 temperature, warm-rolled with reductions of 40 and 60%, respectively, to get the same final thickness of 1 mm, iso-thermally heat-treated in a salt bath at 350°C for 1 ks, and finally oil-quenched to room temperature. For comparison, the other specimen of 1 mm thickness was austenized and rapidly cooled to 450°C in the same way as above, and then iso-thermally heat-treated at 350°C for 1 ks without warm rolling (0% reduction).

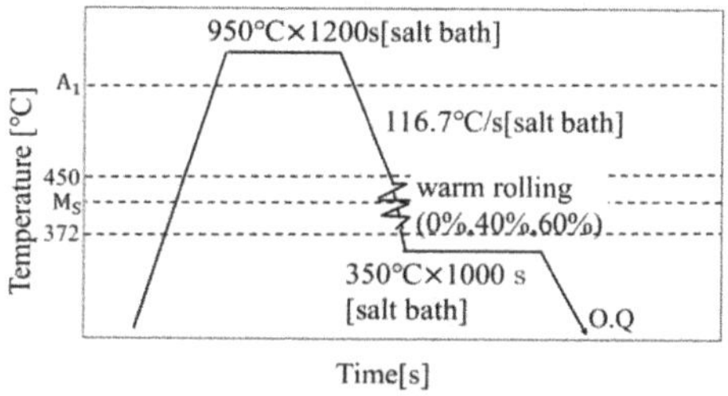

Fig. 1 Schematic of thermo-mechanical processing used in this study.

Tensile-test pieces with a gauge length of 6 mm and width of 3 mm were cut out of the thermo-mechanically treated sheets of the thickness of 1 mm with three reductions (0, 40 and 60%). Tensile tests were conducted using a screw-driven testing machine (Shimadzu AG-G) with an electric furnace at a crosshead speed of 1.0 mm/min at temperatures of 25, 200 and 300°C. Tests were started 300 s after the target specimen temperature was reached. Two tests were carried out for each condition.

To observe the microstructure after the thermo-mechanical treatment, the specimens were wet-ground up to #2000 grit, buffed to a mirror finish, and etched with a nital reagent (ethanol with 5% nitric acid). Microstructural observation and fracture surface observation after the tensile tests were conducted using a scanning electron microscope (SEM, Hitachi S-3400N).

To assess the amount of retained austenite in the three specimens with warm reduction of 0, 40 and 60%, 10 mm square test pieces were cut, electropolished to remove the surface deformed layer , and subjected to X-ray five-peak diffraction method using X-ray diffractometer (Rigaku Ultima IV) with Cu-Kα radiation. The retained austenite volume fraction was determined from the obtained measurement results by combining the diffraction planes, and the average of these values was used as the final retained austenite volume fraction.

Using the k value defined by eq. (1), stability of the retained austenite against tensile deformation was evaluated.

$$k = \frac{\log f_{r0} - \log f_r}{\varepsilon}, \quad (1)$$

where f_{r0}, f_r and ε are volume fractions of retained austenite before and after the tensile deformation and tensile true strain, respectively. Lower k value means higher stability of the retained austenite. To measure the f_{r0} and f_r, samples were cut out from grip portion and portion adjacent to the fracture point, respectively, after the tensile test, while was estimated from thickness and the width before and after the test at the latter portion.

Results and discussion

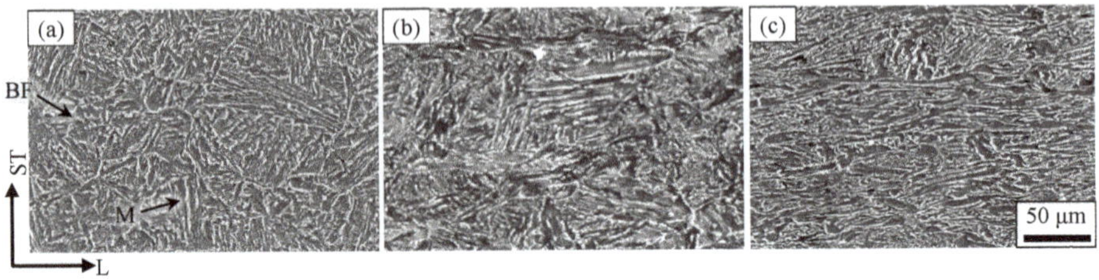

Fig. 2 Microstructures of the three specimens. (a) 0%, (b) 40%, (c) 60%.

Microstructures of the three specimens subjected to the thermo-mechanical processing are shown in Figs. 2 (a) to 2(c). The microstructures of all the three specimens are composed of bainitic ferrite and martensite, examples of which are indicated by labels of BF and M, respectively, and they come to elongate along the rolling direction as the rolling reduction increases. As will be mentioned later, retained austenite should be present but not visible in these images.

Figures 3(a) to 3(c) show stress-strain curves of the three specimens tested at 25, 200 and 300°C, while Figures 4(a) to 4(e) show five mechanical properties obtained from the tests. From these figures, it is clear that warm rolling increases the strength at all the temperatures and this effect is higher in the reduction of 40% than 60%. The effect of warm rolling on the ductility is somewhat complicated: ductility is decreased by 40% rolling but increased by 60% rolling. Temperature dependency of the product of ultimate tensile strength (σ_U) with total elongation (ε_T) is demonstrated in Fig. 4(e) as a measure of toughness and the balance of strength with ductility. In terms of the $\sigma_U \times \varepsilon_T$ value or the balance between strength and ductility, the relationship between unrolled and 40% rolled specimens seems to be normal, which can be frequently seen in many metallic materials: the higher the strength, the lower the ductility. This tendency is seen in Fig. 4(e): almost the same values of $\sigma_U \times \varepsilon_T$ for the 40% rolled and unrolled

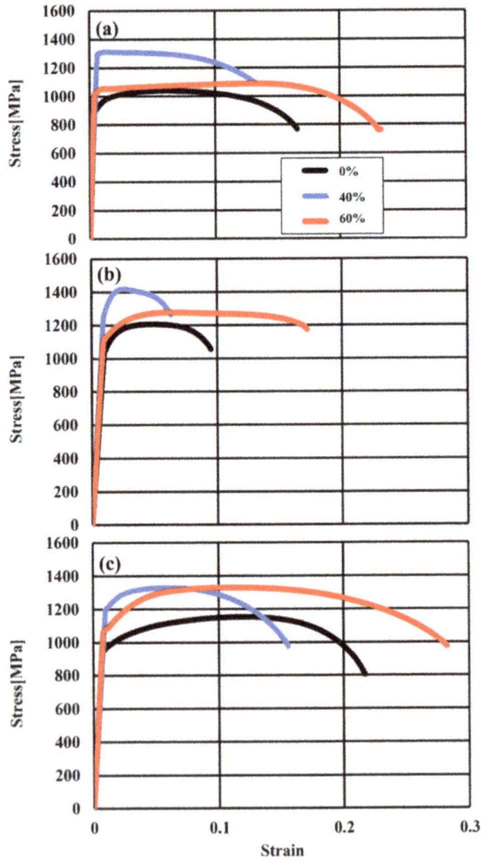

Fig. 3 Stress-strain curves of the three specimens tested at 25°C (a), 200°C (b) and 300°C (c).

Materials Research Forum LLC
https://doi.org/10.21741/9781644902615-31

specimens at all the temperatures. In contrast, the 60% rolled specimen shows both higher strength and higher ductility than unrolled specimen at all the temperatures.

The higher strength and ductility of the 60% rolled specimen will be discussed here. The uniform elongation, ε_U, is the parameter most affected by TRIP [2,7] since the martensite newly formed by the local deformation causes strain hardening, leading to diffusion of the local necking. In Fig. 4(c), ε_U of the 60% rolled specimen is higher than that of the unrolled specimen but the difference decreases as the temperature increases, resulting in almost the same values at 300°C for the two specimens. The cause of this phenomenon will be discussed further in terms of the amount and stability of retained austenite.

Figures 5(a) and 5(b) show temperature dependency of f_{r0} and k. It is obvious that f_{r0} increases with increasing rolling reduction. This can be attributed to the refinement of the microstructure, probably dislocation cell size, by warm rolling, which causes an increase in the number density of the nucleation sites for martensite, resulting in the increase in the number density of film-like retained austenite that forms on the periphery of the martensite. Another probable cause for the increase in f_{r0} with increasing rolling reduction is a decrease in the cooling rate from the austenization temperature to the isothermal holding temperature caused by adding the warm rolling process. According to the previous study [3], a decrease in the cooling rate promotes the refinement of the matrix microstructure in a block, leading to the same effect as above. The higher f_{r0} indicates the larger potential capability to exert the TRIP effect that enhances the strain hardening during tensile deformation, leading to larger uniform elongation. However, lower stability of the austenite, higher k value, is also needed for the TRIP to actually take place. From Fig. 5(b), k value of the 60% rolled specimen is higher than the other two specimens at 25°C, but it decreases with increasing temperature, and becomes lower at 300°C. Thus, it can be deduced that the higher ε_U value of the 60% rolled specimen than unrolled specimen at 25°C was caused both by the higher f_{r0} and k values, while the effect of the higher f_{r0} was alleviated by the lower k at 300°C.

Next, the cause of the change of k value with temperature will be discussed. Sugimoto et al. [4] proposed that a trough should appear in the k value vs. temperature curve when strain-induced martensite transformation (SIMT) becomes stagnant and strain-

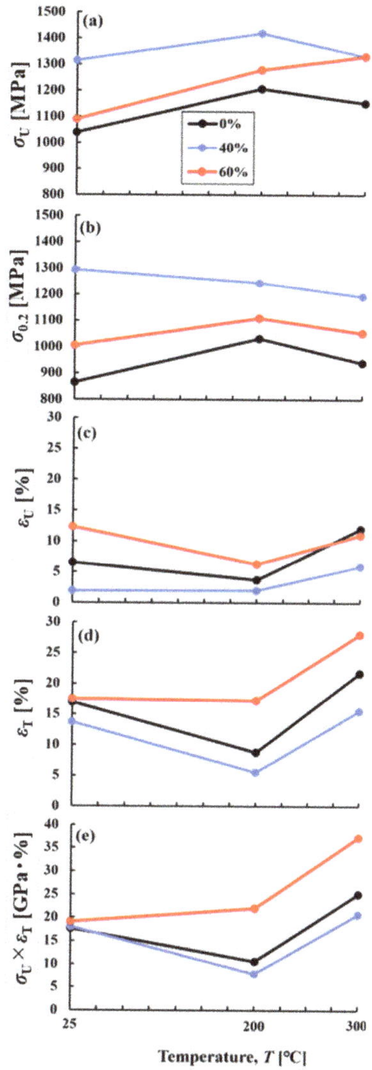

Fig. 4 Temperature dependence of mechanical properties for each specimen. (a) ultimate tensile strength (σ_U), (b) 0.2% proof stress ($\sigma_{0.2}$), (c) uniform elongation (ε_U), (d) total elongation (ε_T), (e) $\sigma_U \times \varepsilon_T$

induced bainitic transformation (SIBT) starts but not abundantly: SIMT occurs at a lower temperature range while SIBT at a higher temperature range. They suggested that the trough temperature corresponds to the temperature where the ductility becomes maximum. In the present study, from Fig. 5(b), the trough of *k* values seem to be present at around 25°C and over 300°C for unrolled and 60% rolled specimens, respectively. Therefore, the temperature range for SIMT has been spread toward higher temperature side in 60% rolled specimen compared to the unrolled specimen.

Finally, the cause for the shift of the *k*-trough temperature in the rolled specimens will be discussed. As shown in Fig. 5(a), f_{r0} increases with rolling reduction, resulting in the decrease of carbon concentration in the retained austenite, which will then raise Ms and Bs temperatures for the retained austenite [8-11]. Hence, the transition temperature from SIMT to SIBT (trough temperature in *k* value) is deduced to be raised by the rolling.

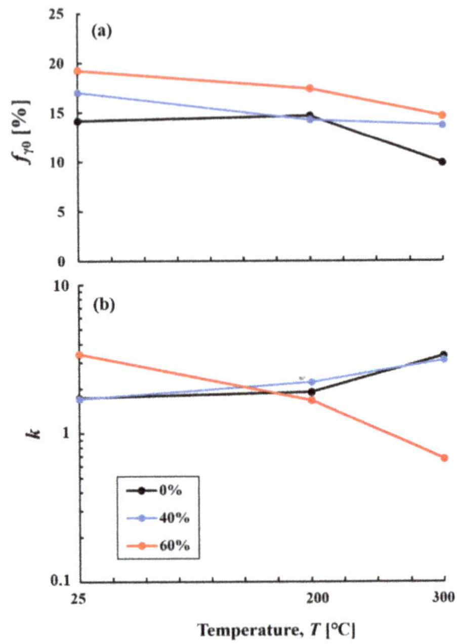

Fig. 5 Temperature dependence of f_r (a) and k valueu (b).

Summary
Tensile properties of a low-alloy TRIP steel affected by warm-rolling prior to isothermal holding treatment were investigated, and the results were discussed in terms of volume fraction of retained austenite and its stability against tensile deformation. It was confirmed that the strength/ductility balance was almost the same at the reduction of 40% and that it was significantly increased by the 60% warm rolling. The latter phenomenon was attributed to the increase in the volume fraction of the retained austenite with the rolling reduction as well as the stability of the retained austenite against the tensile deformation.

Acknowledgement
The authors would like to thank the Grants from the AMADA foundation and a part of JSPS KAKENHI Grant for its financial support. This work was performed under the GIMRT Program of the Institute for Materials Research, Tohoku University (Proposal No. 202112-RDKGE-0071).

References
[1] K. Sugimoto, S-M. Song, J. Sakaguchi, A. Nagasaka, T. Kashima, Warm formability of ultra high-strength low alloy TRIP-aided sheet steels with bainitic ferrite matrix, Tetsu-to-Hagané 91 (2005) 278-284. https://doi.org/10.2355/tetsutohagane1955.91.2_278

[2] P. P. Suikkanen, A-J. Ristola, A. M. Hirvi, P. Sahu, M. C. Somani, D. A. Porter, L. P. Karjalainen, Effects of carbon content and cooling path on the microstructure and properties of trip-aided ultra-high strength steels, ISIJ Int. 53 (2013) 337-346. https://doi.org/10.2355/isijinternational.53.337

[3] N. Kakefuda, Effect of microstructure on hydrogen embrittlement resistance in TRIP-aided martensitic steel with different cooling rate below Ms temperature, Master Thesis, Major in

Mechanical Systems Engineering, Graduate School of Science and Engineering, Ibaraki University, 2021. https://doi.org/10.4028/www.scientific.net/MSF.1016.654

[4] K. Sugimoto, M. Tsunezawa, T. Hojo, S. Ikeda, Ductility of 0.1-0.6C-1.5Si-1.5Mn ultra high-strength TRIP-aided sheet steels with bainitic ferrite matrix, ISIJ Int. 44 (2004) 1608-1614. https://doi.org/10.2355/isijinternational.44.1608

[5] K. Sugimoto, J. Kobayashi, T. Hojo, Microstructure and mechanical properties of ultrahigh-strength TRIP-aided steels, Tetsu-to-Hagané 103 (2017) 1-5. https://doi.org/10.2355/tetsutohagane.TETSU-2016-064

[6] T. Hojo, J. Kobayashi, T. Kochi, K. Sugimoto, Effects of thermomechanical processing on microstructure and shear properties of 22SiMnCrMoB TRIP-aided martensitic steel, The Iron & Steel Technology Conference and Exposition, 2015, pp. 102-110.

[7] A. Barbacki, The role of bainite in shaping mechanical properties of steels, J. Mater. Process. Technol. 53 (1995) 57-63. https://doi.org/10.1016/0924-0136(95)01961-D

[8] Z. Dai, H. Chen, R. Ding, Q. Lu, C. Zhang, Z. Yang, S. van der Zwaag, Fundamentals and application of solid-state phase transformations for advanced high strength steels containing metastable retained austenite, Mater. Sci. Eng. R, 143 (2021) 100590. https://doi.org/10.1016/j.mser.2020.100590

[9] S. van der Zwaag, L. Zhao, S.O. Kruijver, J. Sietsma, Thermal and mechanical stability of retained austenite in aluminum-containing multiphase TRIP steels, ISIJ Int. 42 (2002) 1565-1570. https://doi.org/10.2355/isijinternational.42.1565

[10] R. Blonde, E. Jimenez-Melero, L. Zhao, J. P. Wright, E. Bruck, S. van der Zwaag, N. H. van Dijk, Acta Mater. 60 (2012) 565-577. https://doi.org/10.1016/j.actamat.2011.10.019

[11] R. Blonde, E. Jimenez-Melero, L. Zhao, N. Schell, E. Bruck, S. van der Zwaag, N. H. van Dijk, Mater. Sci. Eng. A 594 (2014) 125-134. https://doi.org/10.1016/j.msea.2013.11.001

Mechanical behavior and microstructure evolution of Ti-6.5Al-2Zr-1Mo-1V/TiB metal-matrix composites during deformation

OZEROV Maxim[1,a], SOKOLOVSKY Vitaly[1,b], GALTSEV Alexander[1,c], STEPANOV Nikita[1,d], ZHEREBTSOV Sergey[1,e]

[1]Belgorod State University, Pobeda 85, Belgorod 308015, Russia

[a]ozerov@bsu.edu.ru, [b]sokolovskiy@bsu.edu.ru, [c]galtsev@bsu.edu.ru, [d]stepanov@bsu.edu.ru, [e]zherebtsov@bsu.edu.ru

Keywords: Titanium Alloy, Metal-Matrix Composite, Boride Fibers, Microstructure, Mechanical Properties, Mechanical Behavior, Aspect Ratio

Abstract. The Ti-6.5Al-2Zr-1Mo-1V/TiB metal-matrix composites with different amounts of TiB reinforcements (~2.0, 6.0 and 10.0 vol. %) were produced by vacuum arc melting. The initial microstructure of the synthesized composites composed of two-phase $\alpha+\beta$ matrix with embedded boride particles. The addition of borides resulted in 15-35% increase in strength without a visible drop in ductility. The alloy with the highest amount of the reinforcement attained yield strength of 1100 MPa, peak strength of 1670 MPa and compression ductility of 10 % at room temperature. Microstructure evolution during hot compression was associated with dynamic recrystallization of the matrix and rearrangement/shortening of TiB fibers. The composite with 10.0 vol. % of TiB demonstrated noticeably higher strength at elevated temperatures in comparison with non-reinforced alloy.

Introduction

Titanium alloys occupy a special place in aviation and shipbuilding industries due to their high specific strength, technological ductility and corrosion resistance [1]. Titanium alloys can also be used at elevated temperatures, for example, in aircraft engine turbine compressors; however, their maximum operating temperature does not exceed 550-600 °C due to a noticeable decrease in strength at these temperatures. Ti-6.5Al-2Zr-1Mo-1V belongs to the pseudo-alpha class; this alloy is used for manufacturing hull structures operated up to 500 °C [2]. However, increasing the high-temperature strength (mainly up to ~600°C; limitation of long-term operation at higher temperatures are rather associated with poor oxidation resistance) of this alloy would significantly expand the area of application of these materials due to replacing heavier steels and nickel alloys [1-3].

One of the promising approaches to increase the high-temperature strength of titanium alloys is the creation of composites by introducing refractory ceramic phases into the titanium matrix [4,5]. The best choice is TiB particles which (i) are well matched with the titanium matrix without the formation of a transition region (ii) have a close coefficient of thermal expansion (iii) due to its good thermal stability TiB provides increased strength even at elevated temperatures [4-6].

Metal-matrix composites (MMCs) based on titanium and its alloys manufactured by various methods have usually rather limited plasticity at room and warm temperatures [4-8]. The properties of MMCs can be noticeably changed by thermomechanical treatment, which affects the structure of both the matrix and the reinforcements, to improve the strength/ductility balance at room and elevated temperatures [8,9]. For example thermomechanical processing, namely multiaxial forging, hot rolling, hot extrusion and others, can be used for MMCs to close defects retained after melting/sintering and further enhance their mechanical properties [8–12].

Meanwhile there is lack of information on mechanical properties on Ti-6.5Al-2Zr-1Mo-1V alloy based composites at room and elevated temperatures and on the influence of

Materials Research Forum LLC
https://doi.org/10.21741/9781644902615-32

thermomechanical treatment on structure and properties of these materials. Thus, within the framework of the current study the effect of TiB particles on the high-temperature properties and microstructure evolution of the near-alpha 6.5Al-2Zr-1Mo-1V alloy was examined.

Materials and Methods

The 60 g ingots of Ti-6.5Al-2Zr-1Mo-1V alloy based composites were produced by vacuum arc melting of pure (≥ 99.9 wt. %) elements of Ti, Al, Zr, Mo, V and commercial purity powder of TiB_2 (99.9 % purity) in a high purity argon atmosphere. 0.7, 2 and 3 wt. % of TiB_2 was added to the master alloy to obtain Ti-6.5Al-2Zr-1Mo-1V/TiB composites with different amounts of borides; hereafter these states are denoted as Alloy A, Alloy B and Alloy C, respectively.

The rectangular samples measured 4x4x6 mm^3 were isothermally strained in compression in air at 20, 600, 800, 900 or 950 °C at a nominal strain rate 10^{-3} s^{-1}. Microstructures of the initial and deformed conditions of the composites were determined using a FEI Quanta 600 scanning electron microscope (SEM) operated at an accelerating voltage of 20 kV. Specimens for SEM were prepared by careful mechanical polishing; deep etching was performed using the Kroll's reagent (95 % H_2O, 3 % HNO_3, 2 % HF). Digimizer software was used for the average length, diameter and volume fraction analysis. XRD was carried out on a RIGAKU diffractometer with Cu-Kα radiation.

Results and Discussion

In the as-cast condition the composites had a two-phase matrix with embedded boride particles (Fig. 1). The matrix consisted of α and β phases, the volume fraction of the β phase did not exceed 4% in all states. The boride particles had a needle-like shape with the average cross-section size of ~ 1.0, ~ 1.2 and ~ 2.6 μm for the Alloys A, B and C, respectively. The values of the average apparent length, as well as the diameter, did not change in the structure of both Alloys A and B and tends to increase significantly in the Alloy C (the measured values of the average apparent length of borides were ~ 10, ~ 11 and ~ 26 μm, respectively). The volume fraction of borides in the structure of three conditions of composites was found to be ~ 2.0, ~ 6.0 and ~ 10.0 % for Alloy A, Alloy B and Alloy C, respectively.

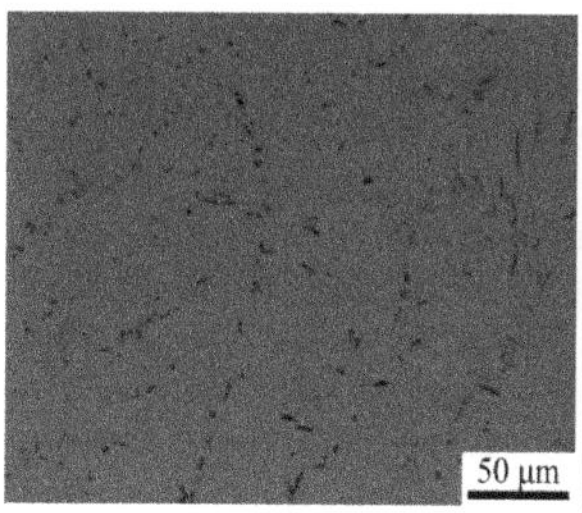

a

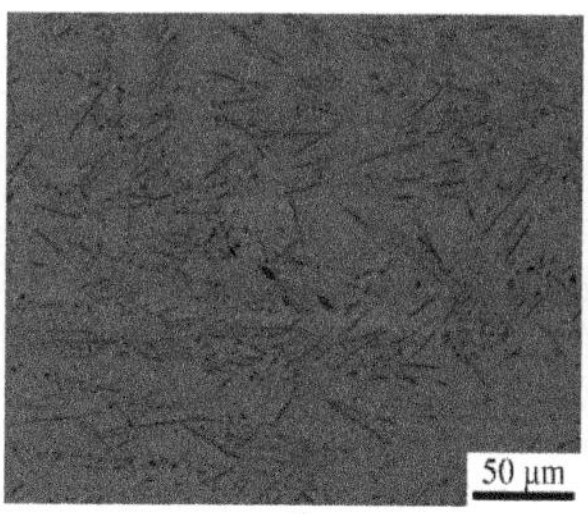

b

c

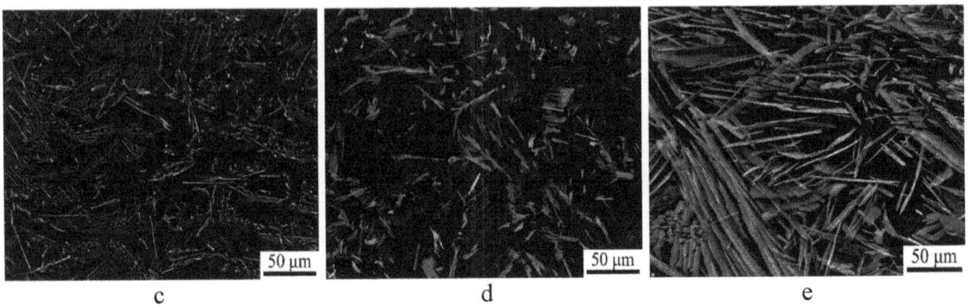

Fig. 1. Microstructure of the Alloy A (a, c), Alloy B (b, d) and Alloy C (c, e); SEM images of unetched (a-c) and etched (c-e) surfaces.

The amount of the borides significantly affected the mechanical behavior of the composites (Table 1). The non-reinforced Ti-6.5Al-2Zr-1Mo-1V alloy demonstrated yield strength, peak strength and compression ductility of 800 MPa, 1350 MPa and 15 %, respectively. The addition of borides resulted in 15-35% increase in strength without a visible drop in ductility. The Alloy C with the highest amount of the reinforcement attained yield strength of 1100 MPa, peak strength of 1670 MPa and compression ductility of 10 %.

Table 1. Mechanical properties in compression of the Ti-6.5Al-2Zr-1Mo-1V alloy and Ti-6.5Al-2Zr-1Mo-1V/TiB composites with different amounts of TiB_2 at 20 °C

Condition	Yield strength, MPa	Peak strength, MPa	ε, %
Base alloy	800	1350	18
Alloy A	920	1620	17
Alloy B	1020	1720	16
Alloy C	1100	1670	10

High-temperature deformation of MMCs in compression resulted (Fig. 2), after initial hardening transient, in continuous strengthening at 400, 600 °C or steady-state flow at the higher temperatures (i.e. at 800-950 °C). The observed softening can be ascribed to dynamic recrystallization/recovery processes typical of hot deformation of titanium alloys or titanium-based composites [7,13,14]. It can be noted that under compression at 400 °C the samples of all states (including non-reinforced Ti-6.5Al-2Zr-1Mo-1V alloy) fractured at already ~15-20 % of height reduction; at the same time, the values of yield strength varied from 470 MPa for the non-reinforced alloy to 700-750 MPa for the Alloys B and C. A similar trend in the strength properties was observed for the all studied higher deformation temperatures. However, the higher yield strength values of the Alloy C during deformation at 900 or 950 °C can be seen more obviously; the yield strengths of the Alloy C at 900 and 950 °C were found to be 110 and 70 MPa, respectively. It should be noted that these values was higher than the strength values of Ti/TiB and Ti-15Mo/TiB composites, under similar deformation conditions [13,14].

Materials Research Forum LLC
https://doi.org/10.21741/9781644902615-32

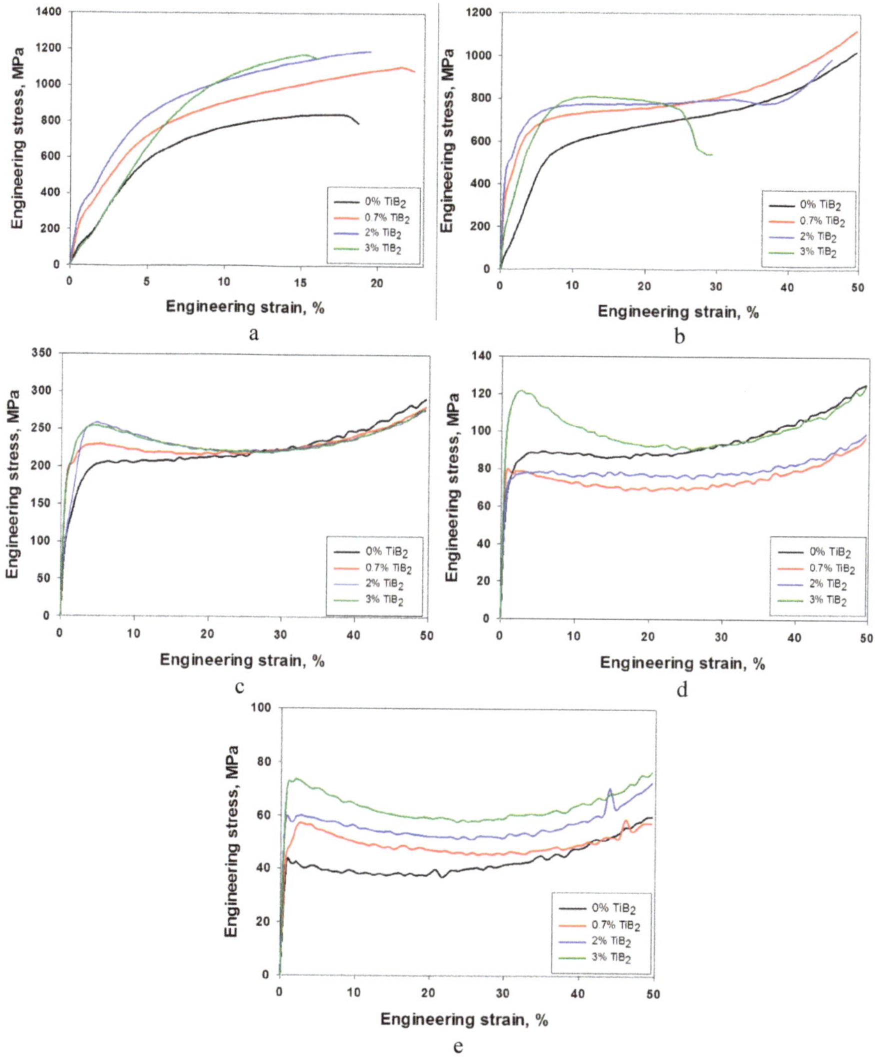

Fig.2. Mechanical behavior obtained during compression at 400 °C (a), 600 °C (b), 800 °C (c), 900 °C (d) and 950 °C (e) of the composites with different amounts of TiB.

Typical microstructures of the Alloy B compressed to 50 % at temperatures 800, 900 and 950 °C are shown in Fig. 3. Microstructure evolution during compression at elevated temperatures was associated with dynamic recovery/recrystallization and rearrangement of the borides along the metal flow direction and shortening of TiB fibers; the intensity of TiB shortening decreased with temperature. A change in the average length of the fibers leads to a corresponding decrease in the aspect ratio (which is an important factor influencing mechanical properties of the MMCs [7,13,15-17]) to ~5 at 800°C. At 950 °C the aspect ratio for the Alloy B was found to be ~10. The different shortening intensity is associated with much softer matrix at higher temperatures [13], and particularly at the β phase field temperatures (i.e. 950°C).

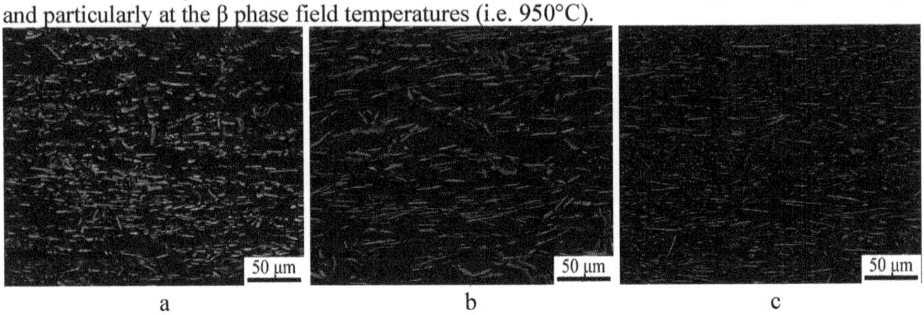

Fig. 3. SEM images of etched surfaces of the Alloy B strained at 800 °C (a), 900 °C (b) and 950 °C (c).

The results obtained clearly show a positive effect of borides on high-temperature strength of the Ti-6.5Al-2Zr-1Mo-1V alloy, in accord with some earlier results [18,19]. The alloy with the highest content of TiB (~10 vol. %) possesses significantly higher strength in comparison with the non-reinforced alloy in the whole studied temperature interval. It should be noted also that the composites did not show dramatic decrease in ductility even at room temperature. Some increase in ductility can be attained due to thermomechanical treatment via reducing the aspect ratio of the borides and "switching on" the Orowan mechanism [6,7,10]. Thus, further studies will be focused on the development of thermomechanical treatment of this composite in order to improve the strength-plasticity balance of the material.

Summary

The results of this study show a positive effect of the TiB fibers on strength of the Ti-6.5Al-2Zr-1Mo-1V alloy in the interval 20-950°C. The initial microstructure of the cast composites was composed of two-phase α+β matrix with embedded boride particles. The addition of borides resulted in 15-35% increase in strength without a considerable drop in ductility. Microstructure evolution during hot deformation was associated with dynamic recrystallization of the matrix and rearrangement/shortening of the TiB fibers.

Acknowledgements

The authors gratefully acknowledge the financial support from the Russian Science Foundation (Grant Number 23-49-00108). The authors are grateful to the personnel of the Joint Research Centre, Belgorod State University for their assistance with the instrumental analysis.

References

[1] C. Leyens, M. Peters, Titanium and Titanium Alloys: Fundamentals and Applications, Wiley-VCH: Weinheim, Germany, 2003. https://doi.org/10.1002/3527602119

[2] Q.J. Sun, X. Xie, Microstructure and mechanical properties of TA15 alloy after 433 thermo-mechanical processing, Mater. Sci. Eng. A 724 (2018) 493-501. https://doi.org/10.1016/j.msea.2018.03.109

[3] Z. Sun, H. Yang, Microstructure and mechanical properties of TA15 titanium alloy 436 under multi-step local loading forming, Mater. Sci. Eng. A 523(1-2) (2009) 184-192. https://doi.org/10.1016/j.msea.2009.05.058

[4] K.S. Ravi Chandran, K.B. Panda, S.S. Sahay, TiBw-reinforced Ti composites: Processing, properties, application, prospects, and research needs, JOM 56 (2004) 42-48. https://doi.org/10.1007/s11837-004-0127-1

[5] T. Godfrey, P. Goodwin, C. Ward-Close, Titanium particulate metal matrix composites - reinforcement, production methods, and mechanical properties. Adv. Eng. Mat. 2 (2000) 85-91. https://doi.org/10.1002/(SICI)1527-2648(200003)2:3<85::AID-ADEM85>3.0.CO;2-U

[6] K. Morsi, Review: Titanium-titanium boride composites, J. Mater. Sci. 54 (2019) 6753-6771. https://doi.org/10.1007/s10853-018-03283-w

[7] M. Ozerov, M. Klimova, V. Sokolovsky, N. Stepanov, A. Popov, M. Boldin, S. Zherebtsov, Evolution of microstructure and mechanical properties of Ti/TiB metal-matrix composite during isothermal multiaxial forging, J. Alloys Compd. 770 (2019) 840-848. https://doi.org/10.1016/j.jallcom.2018.08.215

[8] M. Ozerov, N. Stepanov, A. Kolesnikov, V. Sokolovsky, S. Zherebtsov, Brittle-to-ductile transition in a Ti-TiB metal-matrix composite, Mater. Lett. 187 (2017) 28-31. https://doi.org/10.1016/j.matlet.2016.10.060

[9] L. Huang, X. Cui, L. Geng, Y. Fu, Effects of rolling deformation on microstructure and mechanical properties of network structured TiBw/Ti composites, T NONFERR METAL SOC 22 (2012) 79-83. https://doi.org/10.1016/S1003-6326(12)61687-2

[10] J.H. Yang, S.L. Xiao, Y.Y. Chen, L.J. Xu, X.P. Wang, J. Tian, D.D. Zhang, Z.Z. Zheng, Microstructure evolution during forging deformation of (TiB+TiC+Y2O3)/α-Ti composite: DRX and globularization behavior, J. Alloys Compd. 827 (2020) 154170. https://doi.org/10.1016/j.jallcom.2020.154170

[11] R. Srinivasan, D. Miracle, S. Tamirisakandala, Direct rolling of as-cast Ti-6Al-4V modified with trace additions of boron, Mater. Sci. Eng., A 487 (1-2) (2008) 541-551. https://doi.org/10.1016/j.msea.2007.10.053

[12] L. Jia, X. Li, K. Kondoh, B. Chen, S.F. Li, J. Umeda, Z.L. Lu, Hybrid effect of TiCp and TiBw co-strengthening Ti matrix composites prepared by spark plasma sintering and hot extrusion, Mater. Char. 151 (2019) 6-14. https://doi.org/10.1016/j.matchar.2019.02.026

[13] M. Ozerov, M. Klimova, A. Kolesnikov, N. Stepanov, S. Zherebtsov, Deformation behavior and microstructure evolution of a Ti/TiB metal-matrix composite during high-temperature compression tests, Mater. Des. 112 (2016) 17-26. https://doi.org/10.1016/j.matdes.2016.09.051

[14] S. Zherebtsov, M. Ozerov, M. Klimova, D. Moskovskikh, N. Stepanov, G. Salishchev, Mechanical behavior and microstructure evolution of a Ti-15Mo/TiB titanium-matrix composite during hot deformation, Metals 9 (11) (2019) 1175. https://doi.org/10.3390/met9111175

[15] B. Chen, J. Shen, X. Ye, L. Jia, S. Li, J. Umeda, M. Takahashi, K. Kondoh, Length effect of carbon nanotubes on the strengthening mechanisms in metal matrix composites, Acta Mater. 140 (2017) 317-325. https://doi.org/10.1016/j.actamat.2017.08.048

[16] M.Y. Koo, J.S. Park, M.K. Park, K.T. Kim, S.H. Hong, Effect of aspect ratios of in situ formed TiB whiskers on the mechanical properties of TiBw/Ti-6Al-4V composites, Scr. Mater. 66 (2012) 487-490. https://doi.org/10.1016/j.scriptamat.2011.12.024

[17] S. Zherebtsov, M. Ozerov, E. Povolyaeva, V. Sokolovsky, N. Stepanov, D. Moskovskikh, G. Salishchev, Effect of hot rolling on the microstructure and mechanical properties of a Ti-15Mo/TiB metal-matrix composite, Metals 10 (2020) 40. https://doi.org/10.3390/met10010040

[18] H.T. Tsang, Effects of volume fraction of reinforcement on tensile and creep properties of in-situ TiB/Ti MMC, Scr. Mater. 37 (9) (1997) 1359-1365. https://doi.org/10.1016/S1359-6462(97)00251-0

[19] S. Wang, L. J. Huang, L. Geng, F. Scarpa, Y. Jiao, H. X. Peng, Significantly enhanced creep resistance of low volume fraction in-situ TiBw/Ti6Al4V composites by architecture network reinforcements, Sci. Rep. 7 (2017) 40823. https://doi.org/10.1038/srep40823

Materials Research Proceedings **32** (2023) 287-293 https://doi.org/10.21741/9781644902615-33

Investigation of the influence of kinematic asymmetry on the properties of laminated materials

BIRYUKOVA Olesya D.[1,a*], PESIN Alexander M.[1,b] and PUSTOVOYTOV Denis O.[1,c]

[1]Nosov Magnitogorsk State Technical University, Magnitogorsk 455000, Russia

[a]fimapatisonchik@inbox.ru, [b]pesin@bk.ru, [c]pustovoitov_den@mail.ru

Keywords: Aluminum Composites, Accumulative Roll Bonding, Kinematic Asymmetry, Laminated Materials, Tensile Strength, Yield Strength

Abstract. Trend of simultaneous improvement of such mechanical metal properties as strength, hardness and ductility remains popular. Asymmetric accumulative roll bonding is a method of severe plastic deformation that can create large shear deformations. These deformations in its turn are responsible for increase of mechanical properties. Initially computer modeling was carried out to study the influence of kinematic asymmetry parameters on the stress-strain state of laminated aluminum composites. For the study, such combinations of aluminum alloys as 5083/1070, 5083/2024, 6061/1070, 6061/2024 were considered. As a result of a numerical study, the conditions for obtaining high values of shear deformation, which affects the increase in mechanical properties in laminated aluminum composites, are determined. Experimental confirmation of the simulation results was carried out on a unique asymmetric rolling mill 400 installed in NMSTU. As the final stage of work, the development of rational technological schemes and regimes of cold and warm asymmetric accumulative roll bonding of laminated aluminum composites was carried out. The study showed a simultaneous increase in the strength and technological plasticity of received products, which differ in the ratio of total equivalent deformations and shear angles during at least 2 cycles of asymmetric accumulative roll bonding. It is shown that with an increase in value of the speed rolls ratio at the same inter-roll gap, the relative reduction increases, and the rolling force decreases by a factor of 2 (compared with these parameters at equal speeds of the work rolls). It is supposed to use such composites in space and automotive industries.

Introduction

For a lot of industrial application, including the automotive, aircraft and space industries, there is an acute problem of reducing the weight of the structures while maintaining or increasing the level of mechanical properties. For any metallic material, the fundamental mechanical characteristics include strength and ductility, which have a clearly expressed inversely proportional relationship. Very rarely, such materials simultaneously have high values of these properties. Materials processing technologies are being developed in such countries as Russia, China, Japan, South Korea, India, Germany etc. to solve this problem [1-6]. One of the ways to obtain new materials is asymmetric accumulative roll bonding. This method of rolling consists of several cycles with cutting and joining layers after each processing cycle [7-10]. Kinematic asymmetry is deliberately created; the work rolls speed ratio can be different. In Magnitogorsk (Russia) scientists of the "Zhilyaev Mechanics of Gradient Nanomaterials Laboratory" of Nosov Magnitogorsk State Technical University, investigate new fundamental effects of asymmetric deformation, which provide an increase in strength and plasticity [11-13]. It should be noted that the laboratory has unique scientific facility, an experimental reversible sheet rolling mill with an individual drive of work rolls. This equipment has no analogues in the world, according to such an important indicator as the work rolls speed ratio. This value can vary from 1 to 10 [14]. In the closest analogue, the South Korean mill, the maximum value of V_1/V_2 does not exceed 5.

Materials and Experimental Methods

To determine the technological parameters of the accumulative roll bonding process of laminated materials, samples of aluminum alloys were studied. The variation of laminated composites is presented in the form of the following combinations: 5083/1070, 5083/2024, 6061/1070, 6061/2024. On this issue, first of all, a computer simulation of asymmetric accumulative roll bonding process in the software package "Deform" and "QForm" was installed. Boundary and initial conditions for computer simulation of cold and warm accumulative roll bonding of laminated aluminum composites 5083/1070, 6061/1070, 5083/2024, 6061/2024 were established. The example of layer variations is presented on Fig. 1.

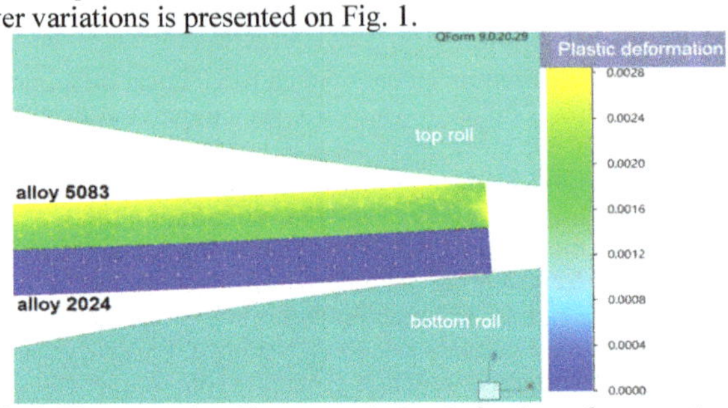

Fig.1. *The example of layers variation in laminated composite.*

It was found that it is rational to take the values of the friction coefficient not lower than 0.2 (for laminated aluminum composites 5083/1070, 6061/1070) and not higher than 0.3 (for laminated aluminum composites 5083/2024, 6061/2024).

The value of the work rolls speed ratio for the rectilinear movement of the metal must be increased with an increase in the thickness of the initial workpiece, reduction and strength of processed the material (in the range from 1 to 4). The distribution of layers in laminated aluminum composites along the length of the deformation zone depended on the value of the work rolls speed ratio (from 1 to 4), reductions (50–70%), as well as on the grades of aluminum alloys used (1070, 2024, 5083, 6061) and original total workpiece thicknesses (2 - 4 mm).

The most rational ratio of layer thicknesses in the initial workpiece is 2 to 1 (3 mm), at which the maximum values of the equivalent deformation and shear angle are achieved ($e > 2 \ldots 4$, $\varphi > 65 \ldots 70^{\circ}$). With an increase in the value of the work rolls speed ratio and equivalent deformation, the rolling force decreases (compared to the symmetrical case).

The next step was to verify technological regimes of asymmetric accumulative roll bonding of laminated aluminum composites 5083/1070, 6061/1070, 5083/2024, 6061/2024. The experimental plan included regimes with varying the type of laminated composite, the work rolls speed ratio (from 1 to 5), the position of the layers in the second rolling cycle, the type of heat treatment, and the value of the relative reduction (from 50 to 70%). Accumulative roll bonding regimes have been identified that made it possible to obtain high values of strength and technological plasticity, a high level of welding of composite layers, and a satisfactory surface quality. These regimes included cold and warm accumulative roll bonding, with two cycles of asymmetric deformation. Regimes including the value of the work rolls speed ratio 5 were excluded from the studies due to the melting of the metal in the deformation zone.

Cold asymmetric accumulative roll bonding was carried out:

1) without heat treatment;

2) using heat treatment (annealing at a temperature of 420° C for 90 - 120 minutes, followed by cooling in air).

Version 1 can be used when processing laminated aluminum composites 5083/1070 and 6061/1070, which consists of carrying out the initial stage of the first cycle with alloy 5083 or 6061 and the second stage of the first cycle with alloy 1070. The surface of the connection of the layers must be prepared (cleaning, degreasing). However, the second cycle of asymmetric accumulative roll bonding is not recommended to be carried out without heat treatment, because this can lead to delamination, as shown in Fig. 2.

According to version 2, annealing of alloys 1070, 2024, 5083 and 6061 was carried out at a temperature of 420°C for 90–120 minutes, followed by cooling in air. In this case, the appearance of irreversible defects, similar to the one shown in Fig. 2 is possible.

Fig.2. *Laminated aluminum composite 5083/2024 after cold accumulative roll bonding (cycle 2)*

Warm asymmetric accumulative roll bonding was carried out:

1) using preheating before rolling at a temperature of 420°C for 15 - 30 minutes;

2) using preheating before rolling at a temperature of 380°C for 5 to 30 minutes.

In both cases, the cooling of the metal occurred during the transport of the samples from the furnace to the rolling mill.

Version 1 was used in the processing of the laminated aluminum composites 5083/1070, 6061/1070 in the second cycle of asymmetric accumulative roll bonding, 5083/2024 and 6061/2024 in all processing cycles and was characterized by the formation of such defects as "transferring" and "wave" in all cases, as shown in Fig. 3.

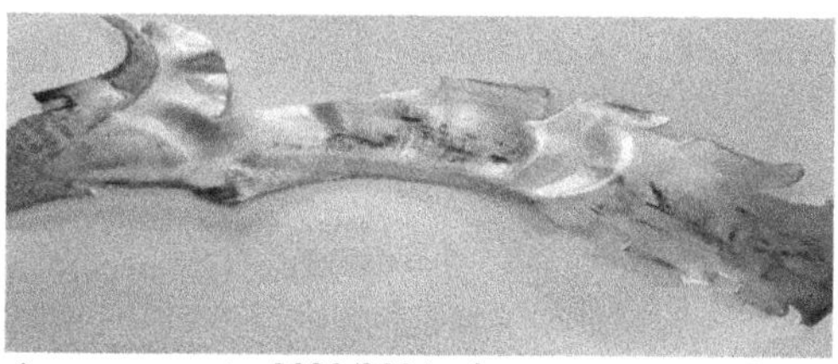

Fig.3. *Laminated aluminum composite 5083/2024 after warm accumulative roll bonding (cycle 2)*

Version 2 can be used for processing the laminated aluminum composites 5083/1070 and 6061/1070 in 1 and 2 cycles (depending on the arrangement of the layers); 5083/2024 and 6061/2024 in all processing cycles. The main results are presented in Fig. 4 (a – c).

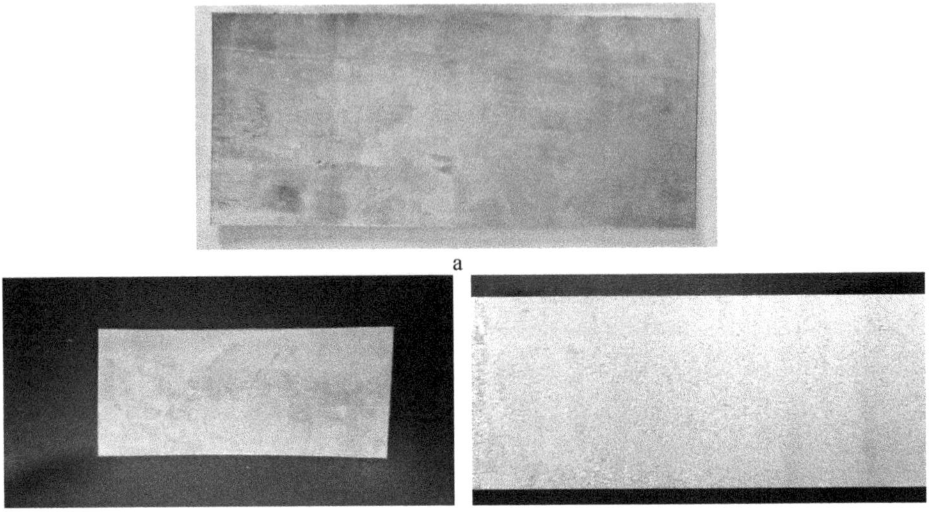

a

b c

Fig.4. *Laminated aluminum composite 5083/2024 (a), 6061/1070 (b), 6061/2024 (c) after warm accumulative roll bonding (cycle 2)*

The required flatness and surface quality can be obtained at the work rolls speed ratio of 10 rpm for the bottom roll and 2–6 rpm for the top roll. The reduction value ranged from 50% to 67%.

The results and discussion

In a number of cases, the displacement of the layers to each other in the deformation zone was observed. This led either to the absence of a connection between the alloys, or to partial welding and inversion of the front part of the workpiece according to such type of defect as "crescent". The reasons for the formation of these defects can be insufficient degrees of deformation, insufficient temperature in the deformation zone, incorrectly selected friction conditions and values of the work rolls speed ratio. One of the solutions of this problem can be the preliminary compounding of the laminated material surface in the first rolling cycle with the formation of a relief on the surface of the workpiece using relief rolls and subsequent smoothing of the workpiece in the next processing cycle.

An assessment was made of the influence of the work rolls speed ratio on the technological plasticity and rolling forces. The results for laminated aluminum composite 5083/2024 with an initial layer thickness of 1 mm (5083) and 2 mm (2024) and an established roll gap of 1.2 mm are presented in Table 1; for laminated aluminum composite 5083/1070 with an initial layer thickness of 1 mm (1070) and 2 mm (5083) and an established roll gap of 1.2 mm are presented in Table 2.

Table 1 - Influence of the values of the work rolls speed ratio on the technological plasticity and rolling forces of the laminated aluminum composite 5083/2024

The work rolls speed ratio, $\frac{rpm}{rpm}$	Final thickness of laminated aluminum composite 5083/2024, mm	Relative reduction, %	Rolling force, kN
10/10 (1)	1.90	37	1330
10/6 (1.7)	1.60	47	1080
10/5 (2)	1.53	49	990
10/4.5 (2.2)	1.47	51	940
10/4 (2.5)	1.35	55	900
10/3.5 (2.8)	1.25	58	790
10/3 (3.3)	1.20	60	700
10/2.5 (4)	1.00	67	600
10/2 (5)	melted		

Table 2 - Influence of the values of the work rolls speed ratio on the technological plasticity and rolling forces of the laminated aluminum composite 5083/1070

The work rolls speed ratio, $\frac{rpm}{rpm}$	Final thickness of laminated aluminum composite 5083/1070, mm	Relative reduction, %	Rolling force, kN
10/10 (1)	not welded		
10/8 (1.25)			
10/7 (1.4)	1.8	40	860
10/6 (1.7)	1.7	43	710
10/5 (2)	1.45	52	630
10/4 (2.5)	1.3	57	520
10/3 (3.3)	1.2	60	300
10/2 (5)	melted		

It is shown that with an increase in the work rolls speed ratio from 1 to 4 for laminated aluminum composite 5083/2024, the rolling force decreases with a simultaneous increase in the relative reduction. For laminated aluminum composite 5083/1070, the rolling force is also reduced and the reduction ratio increases with an increase in the rolls speed ratio from 1.4 to 3.3. Moreover some of samples or didn't weld either melted.

All samples after two processing cycles were tested on hardness, tensile strength, yield strength and relative elongation. The obtained characteristics were compared with the properties of materials currently used in the automotive and space industries. The values of plastic characteristics were approximately on the same level with the existing materials, and the strength values exceeded it by 15 - 20%, the hardness was 2 times higher.

From the obtained data it's obvious that in the first cycle of asymmetric accumulative roll bonding, a significant gradient is observed in laminated aluminum composites, which is characterized by hardness values in different layers (for example, HB = 109 units on the 5083 side and HB = 89 units on the 2024 side at the rolls speed ratio of 4 etc.). In the second cycle of deformation, in most cases, not only the hardness values decrease, but also the values of the first and second sides equalize relative to each other (for example, HB = 76 units on the 5083 side and

HB = 75 units on the 2024 side at a speed ratio of 4 and etc.). This is explained by the fact that during the deformation of the metal in the second cycle of asymmetric accumulative roll bonding, the characteristics of the laminated composite are averaged due to the sequential arrangement of the layers.

Both in the first and in the second cycle of asymmetric accumulative roll bonding, the values of yield strength and tensile strength decrease with an increase in the rolls speed ratio. The value of the relative elongation in the first cycle has low values (from 3% to 7%), in the second cycle with the rolls speed ratio of 2.5 - 4 it increases on average almost 2 times (up to 12% maximum), and with the rolls speed ratio 1 - 2.22 almost does not change. This is explained by the fact that with an increase in the rolls speed ratio, the temperature of the deformation heating of the sample in the deformation zone increases. Due to this, the technological plasticity of the metal also increases, which is expressed in a greater elongation (increase in drawing) of the finished laminated aluminum composite (in each subsequent experiment by 10–20% with an increase in the rolls speed ratio in the range of 1–4).

After mechanical testing of laminated aluminum composites, alloy 6061 was excluded from possible alloy for composite combination, because the values of strength and plastic characteristics did not meet the requirements of regulatory documentation. The value of the relative elongation in the second cycle has low values (from 1% to 3%), the yield strength varies from 184 to 192 MPa, and the tensile strength changes from 212 to 240 MPa. To obtain simultaneously high plastic and strength characteristics in these composites, it is necessary to use other regimes of asymmetric accumulative roll bonding.

Conclusion

Asymmetric accumulative roll bonding of laminated aluminum composites 5083/1070, 5083/2024, 6061/1070 and 6061/2024 was carried out on the asymmetric rolling mill 400 of the "Zhilyaev Mechanics of Gradient Nanomaterials Laboratory" of Nosov Magnitogorsk State Technical University. The regimes of the laminated aluminum composites processing are revealed, including such parameters as: type of material, total thickness of the sample and thickness of each layer, type of heat treatment, value of the rolls speed ratio, reductions, friction conditions, number of required cycles. For laminated aluminum composites 5083/1070, 6061/1070 and 5083/2024, 6061/2024, preheating is required before the asymmetric deformation process (rolling process must be warm). But laminated aluminum composites 6061/1070 and 6061/2024 should be excluded from the experiments due to the low mechanical properties.

The necessity of control of deformation heating in the deformation zone during rolling was revealed in order to avoid melting of the material or its sticking to the rolls (the rolls speed ratio should not exceed 4). It is important to observe the temperature limits for heating samples before rolling (in the case of warm rolling).

An increase in strength and ductility in the processing of laminated aluminum composites 5083/1070 and 5083/204 by the asymmetric accumulative roll bonding method was shown.

Acknowledgment

The study was supported by the grant of the Russian Science Foundation (No. 23-79-30015).

References

[1] Ma Y, Yang M, Yuan F, Wu XR. A review on heterogeneous nanostructures: a strategy for superior mechanical properties in metals, Metals. 9 (2019) 1-32. https://doi.org/10.3390/met9050598

[2] J. Jiang, Y. Ding, F. Zuo, A. Shan, Mechanical Properties and Microstructures of Ultrafine-Grained Pure Aluminum by Asymmetric Rolling. Scripta Materialia. 60 (2009) 905-908. https://doi.org/10.1016/j.scriptamat.2009.02.016

[3] Y.E. Beigelzimer, Simple shear of metals: what is it? Physics and tech. of high pres. 20 (2010) 40-52.

[4] A.M. Pesin, D.O. Pustovoito, O.D. Biryukova The effect of speed asymmetry on the strain state in aluminium bimetals during accumulative rolling, IOP Conf. Ser.: Mater. Sci. Eng. 447 (2018) 1-4. https://doi.org/10.1088/1757-899X/447/1/012066

[5] G.T. Sudha,B. Stalin,M. Ravichandran,M. Balasubramanian, Mechanical Properties, Characterization and Wear Behavior of Powder Metallurgy Composites - A Review. Materials Today: Proc. 22 (2020) 2582-2596. https://doi.org/10.1016/j.matpr.2020.03.389

[6] Z. Liao, M. Polyakov, O. Gavalda Diaz, D. Axinte, G. Mohanty, X. Maeder, J. Michler, M. Hardy. Grain refinement mechanism of nickel-based superalloy by severe plastic deformation - Mechanical machining case, Acta Materialia. 180 (2019) 2-14. https://doi.org/10.1016/j.actamat.2019.08.059

[7] M. Alizadeh, E. Salahinejad, Processing of ultrafine-grained aluminum by cross accumulative roll-bonding. Mat. Sc. and Eng. 595 (2014) 131-134. https://doi.org/10.1016/j.msea.2013.11.060

[8] C. Lu, H. Yu, K. Tieu, S. Hadi, A. Godbole, C. Kong. High strength and ductility ofultrathin laminate foils using accumulative roll bonding and asymmetric rolling. Metall.Mater. Trans. 46 (2015) 869-879. https://doi.org/10.1007/s11661-014-2640-3

[11] A. Pesin, D. Pustovoytov, Influence of process parameters on distribution of shear strain through sheet thickness in asymmetric rolling. Key Eng. Mat. 622 (2014) 929-935. https://doi.org/10.4028/www.scientific.net/KEM.622-623.929

[12] Q. Cui, K.Ohori, Grain refinement of high purity aluminum by asymmetric rolling, Mat.Sc. and Tech. 16 (2000) 1095- 1101. https://doi.org/10.1179/026708300101507019

[13] Y.H. Ji, J.J. Park, Development of Severe Plastic Deformation by Various Asymmetric Rolling Processes, Mat. Sc. and Eng. 499 (2009) 14-17. https://doi.org/10.1016/j.msea.2007.11.099

[14] Information on http://lmgn.magtu.ru/ru/oborudovanie.html

FEM simulation of dynamic recrystallization during asymmetric hot rolling of high-speed steel M2

Alexander Pesin[1,a], Denis Pustovoytov[1,b*], Ilya Pesin[1,c], Hailiang Yu[2,d], Puneet Tandon[3,e], Harshal Y Shahare [3,f], Abhay Kumar Dubey [3,g] and Leonid Nosov[1,h]

[1] Nosov Magnitogorsk State Technical University, Zhilyaev Laboratory of mechanics of gradient nanomaterials, 38 Lenin prospect, Magnitogorsk, 455000, Russian Federation

[2] Central South University, State Key Laboratory of High-Performance Complex Manufacturing, Light Alloys Research Institute, Changsha 410083, China

[3] PDPM Indian Institute of Information Technology, Design and Manufacturing, Jabalpur, India

[a]pesin@bk.ru, [b]pustovoitov_den@mail.ru, [c]ilya33392@bk.ru, [d]yuhailiang@csu.edu.cn, [e]ptandon@iiitdmj.ac.in, [f]1823604@iiitdmj.ac.in, [g]adubey@iiitdmj.ac.in, [h]nosov.leopold@yandex.ru

Keywords: High-Speed Steel, Dynamic Recrystallization, Asymmetric Hot Rolling, FEM, JMAK

Abstract. Dynamic recrystallization of high-speed steel (HSS) M2 occurs through a discontinuous mechanism, involving nucleation and growth of strain-free grains. The initiation and completion of discontinuous dynamic recrystallization (dDRX) must be carried out in a single rolling pass. The time for 99% dDRX depends on the critical strain, strain rate and equivalent strain, that generated in the workpiece material in a single rolling pass. The level of equivalent strain plays a key role in terms of the possibility to complete dDRX. High equivalent strain in conventional hot rolling for a single pass is limited by too high rolling forces. High equivalent strain at low rolling forces can be achieved by asymmetric hot rolling. Asymmetric rolling is a process based on purposefully created differences in the circumferential speeds of the work rolls. For such a process, a degree of asymmetry is defined by a speed ratio of the work rolls. The speed ratio is one of the most important parameters of the process, affecting the level of equivalent strain and the rolling force. In this work the single-pass asymmetric hot rolling process of HSS M2 at different temperatures in the range 850-1150 °C was simulated by FEM and JMAK model. The effects of speed ratio and thickness reduction on equivalent strain and rolling force were investigated. The effects of equivalent strain, strain rate and temperature on the recrystallized volume fraction and the average size of a dynamically recrystallized grains of HSS M2 are presented. Symmetric and asymmetric hot rolling processes are compared.

Introduction

The metal forming industry can use different methods for grain refinement in metals and alloys. One of these methods is based on grain refinement directly in the process of hot working due to the dynamic recrystallization (DRX). The DRX phenomena can occurring in different thermo-mechanical processing conditions for various metals and alloys. Several types of DRX are exist: discontinuous dynamic recrystallization (dDRX), continuous dynamic recrystallization (cDRX) and geometric dynamic recrystallization (gDRX). Different types of DRX are associated with different types of mechanisms of the formation of a new grain structure. Detailed characterization of these three types of DRX processes is given in the literature [1-3].

The dDRX is the most extensively studied recrystallization process. The dDRX occurs through a discontinuous mechanism, involving nucleation and growth of strain-free grains. dDRX operates in low to medium stacking fault energy materials at high temperatures ($T > T_m$) [1-3]. It is generally accepted that dDRX mainly initiates by strain induced bulging of original grain boundaries and a necklace structure which is composed of fine equiaxed recrystallized grains is formed. With regard to grain refinement, the important point is that dynamic recrystallization not only produces more refinement than the metadynamic recrystallization but also leads to considerably finer microstructures than conventional static recrystallization [1]. The dDRX has attracted more and more interests since it is the closest method to industrial practices.

The dDRX process initiates at the critical strain ε_c. The volume fraction X of dynamic recrystallization, can be calculated by JMAK model: $X = 1 - \exp(-\beta t^k)$, where β and k are coefficients and t is the time for dDRX to occur [4]. The time for 99% dDRX (when $X = 0.99$) strongly depends on strain ε and strain rate $\dot{\varepsilon}$. Large strain (e.g. $\varepsilon \geq 1$) are required for uniform grain refinement. The application of insufficient strain can lead to the presence of mixed grain sizes. Thus, strain is a key parameter that determines not only the initiating of dDRX, but also the possibility of its full completion. Significant grain refinement can be achieved at 99% dDRX due to large strain single-pass hot working at the lower temperature range.

The dDRX in hot working has been repeatedly studied for austenitic [5], microalloyed [6]. low-carbon [7] and high-speed tool steels (HSSs) [4,8]. HSSs represent an important class of materials for the manufacture of cutting tools. HSSs are used for most of the common types of cutting tools including lathe tools, drills, reamers, taps, milling cutters, end mills, hobs, saws, broaches, etc. The improvement of HSSs is carried out in the direction of finding the ways to improve microstructure and mechanical properties during hot working. Hot rolling is the main process for obtaining long semi-finished products (round bars, plates, sheets, strips, etc.) from HSSs. The conventional technologies include radial-shear hot rolling and plate/strip hot rolling. HSSs are difficult to deform, so conventional hot rolling is carried out with a large number of passes in order not to exceed the maximum allowable rolling forces. HSSs are usually hot rolled at high temperatures, e.g., 1100 °C to 1150 °C for the M2 grade, with the final rolling pass carried out at temperatures not less than 850 °C. In this range, their flow stress may reach 350…450 MPa, the same as the ultimate strength of ordinary steels at room temperature. Therefore, the ability to provide 99% dDRX in HSSs during single-pass hot rolling, especially at the lower temperature range is limited by too high rolling forces. Rolling force can be reduced due to the asymmetry of the hot rolling process by purposefully created difference in the circumferential speeds of the work rolls [9,10]. For such a process, a degree of asymmetry is defined by a speed ratio (SR) of the work rolls. The SR is one of the most important parameters of the process. Rolling with different circumferential speeds of the work rolls when both rolls are independently driven by two motors, is the most suitable way to implement asymmetric hot rolling in industry. However, studies of the effect of asymmetric hot rolling on the kinetics of dDRX of HSSs are yet unknown.

In this work the large strain single-pass asymmetric hot rolling process of HSS M2 at different temperatures in the range 850-1150 °C was simulated by FEM and JMAK model. The effects of speed ratio and thickness reduction on equivalent strain and rolling force were investigated. The effects of strain, strain rate and temperature on the recrystallized volume fraction and the average size of a dynamically recrystallized grains of HSS M2 are presented. Symmetric and asymmetric hot rolling processes are compared. The results of investigation can be used for design of effective technology of manufacturing HSS M2 with improved microstructure and mechanical properties.

Research Method

A coupled deformation and heat transfer simulation of the hot rolling was carried out using the commercial FEM code QForm 2D. Symmetric and asymmetric hot rolling processes were simulated and compared. Asymmetry was purposefully created by difference in the circumferential speeds of the work rolls, while the roll diameters were the same. The diameters of the work rolls were D = 340 mm, and the rolls were assumed as rigid. The angular speed of the upper (fast) work roll was 10 rpm in all simulation variants. Asymmetry was created by decreasing the angular speed of the lower (slow) work roll from 1.1 to 3.0 times with a step of 0.1. The initial thickness of the strip was 3.15 mm in all simulation variants. Single-pass rolling to the final thickness of 1.0 mm was performed. Thickness reduction corresponded to 68.25%. The modeling of the processes was performed at different initial temperatures: 1150, 1100, 1000, 900 °C with taking into account the increment of the metal's temperature due to the thermal effects of deformation and friction. Hot rolling was carried out without front and back tensions. High-speed tool steel AISI M2 was chosen as a material for the strip. AISI D2 from QForm 2D material library was chosen as a material for work rolls. The thermal constants of materials used in FEM simulations are shown in Table 1. Levanov's friction model (1) was used between rolls and strip. During hot rolling of steels friction factor m varies in the range of 0.7-1.0. In the present research the friction factor was assumed as constant in all simulation variants: $m = 0.85$.

$$\tau = mk\left(1 - e^{-1.25\frac{\sigma_n}{\sigma}}\right). \tag{1}$$

where τ is the frictional stress; m is the friction factor; k is the shear yield stress; σ_n is the normal contact pressure; σ is the flow stress.

Table 1. Properties of materials used in FEM simulation.

Material	Thermal conductivity, W/(m×K)	Heat capacity, J/(kg×K)	Heat transfer coefficient between work rolls and strip, W/(m^2×K)	Heat transfer coefficient between work rolls/strip and air, W/(m^2×K)	Environment temperature, °C
AISI M2	28.0	674	10000	30	20
AISI D2	21.5	460	10000	30	20

In order to accurately describe the flow behavior of HSS M2 in hot deformation, dDRX should be considered. Schematic representation of the stress-strain curve during dDRX is shown in Fig. 1. First, there is work hardening from σ_0 up to peak stress σ_p at peak strain ε_p then working softening up to steady stress σ_{ss} at steady strain ε_{ss} if no failure occurs.

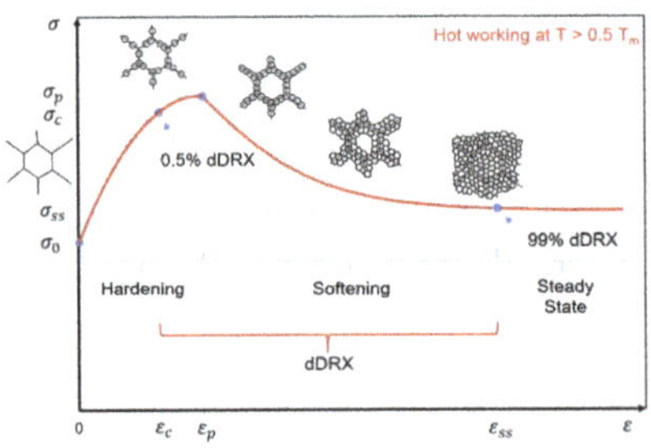

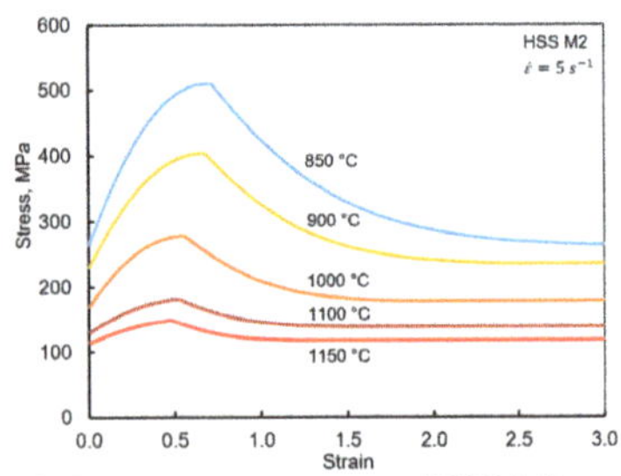

Fig. 1. Diagram of the stress-strain curve during dDRX. *Fig. 2. Stress-strain curves of HSS M2 used in FEM.*

Based on experimental values of σ_0, σ_c, σ_p, σ_{ss} and ε_c, ε_p, ε_{ss} from [4,8,11-13] the stress-strain curves for HSS M2 were approximated and used in FEM simulations. Equations (2)-(7) were also used (see below). Obtained stress-strain curves corresponded to the following ranges: $\varepsilon = 0\ldots3$, $\dot{\varepsilon} = 0.1\ldots50$ s^{-1}, $T = 850\ldots1150$ °C. As an example, stress-strain curves for different temperatures are shown in Fig. 2. A critical strain ε_c is necessary in order to initiate dynamic recrystallization. This occurs somewhat before the peak stress σ_p and peak strain ε_p of the stress-strain curve. Based on the experimental data from [11-13], the peak strain ε_p as a function of the Zener-Hollomon parameter Z was approximated:

$$\varepsilon_p = 0.0554 \cdot Z^{0.0515}. \tag{2}$$

$$Z = \dot{\varepsilon} exp\left(\frac{Q_{HW}}{RT}\right). \tag{3}$$

where $Q_{HW} = 455$ kJ/mol is the hot working activation energy [4], $R = 8.314$ J/mol/K is the universal gas constant, T is absolute temperature.

The relationship between critical strain ε_c and peak strain ε_p for HSS M2 was assumed as follows [4]: $\varepsilon_c = 0{,}65 \cdot \varepsilon_p$. The relationship between critical stress σ_c and peak stress σ_p was assumed as follows: $\sigma_c = 0{,}95 \cdot \sigma_p$.

The volume fraction X of dynamic recrystallized grain:

$$X = 1 - \exp\left(-\beta\left(\frac{\varepsilon - \varepsilon_c}{\dot{\varepsilon}}\right)^k\right). \tag{4}$$

Experimental coefficients β and k, depending on temperature T and strain rate $\dot{\varepsilon}$, for HSS M2 were taken and calculated from [4].

Taking into account the changing of volume fraction X of dynamic recrystallized grain the softening from the peak stress σ_p to the steady stress σ_{ss} was expressed:

$$\sigma_{p\_ss} = \sigma_p - \left(\sigma_p - \sigma_{ss}\right)\frac{(X - X_p)}{(X_{ss} - X_p)}. \tag{5}$$

where X_p and $X_{ss} = 0.99$ are volume fractions of dynamic recrystallized grain at peak strain ε_p and ε_{ss} respectively.

The time t_{ss} for 99% dDRX, depends on the steady state stress σ_{ss} in the following relationship [4]:

$$t_{ss} = A(\sinh \alpha\sigma_{ss})^n \exp\left(\frac{Q_{DRX}}{RT}\right). \tag{6}$$

where A is a material constant (Table 2), $n = -4.6$ is the stress exponent, $Q_{DRX} = 361$ kJ/mol is the steady state stress activation energy, $R = 8.314$ J/mol/K is the universal gas constant, T is absolute temperature, and $\alpha = 0.012$ MPa [4].

Table 2. Material constant A depending on temperature.

Temperature, °C	1150	1100	1000	900	850
$\log A$	-12.65	-12.49	-12.47	-12.22	-12.15

The time t_{ss} for 99% dDRX can also be calculated by the following:

$$t_{ss} = \frac{\varepsilon_{ss} - \varepsilon_c}{\dot{\varepsilon}}. \tag{7}$$

The steady state stress σ_{ss} was expressed from (6) with taking into account (7):

$$\sigma_{ss} = \frac{1}{\alpha}\sinh^{-1}\left[\left(\frac{1}{A}\left(\frac{\varepsilon_{ss}-\varepsilon_c}{\dot{\varepsilon}}\right)\exp\left(-\frac{Q_{DRX}}{RT}\right)\right)^{\frac{1}{n}}\right]. \quad (8)$$

The steady state strain ε_{ss} was expressed from (4) at $X = X_{ss} = 0.99$:

$$\varepsilon_{ss} = \varepsilon_c + \dot{\varepsilon}\left[-\frac{1}{\beta}\ln(1 - X_{ss})\right]^{\frac{1}{k}}. \quad (9)$$

The initial grain size D_0 was taken equal to 90 μm. The dynamic recrystallization grain size D_{dDRX} depends on the Zener-Hollomon parameter Z according to the following equation [4]:

$$D_{dDRX} = 2930 \cdot Z^{-0.15}. \quad (10)$$

Only dDRX was simulated. Static/metadynamic recrystallization and grain growth at high temperature were not simulated. The average grain size D_{avg} was calculated according to the "mixture" rule:

$$D_{avg} = X \cdot D_{dDRX} + (1 - X) \cdot D_0. \quad (11)$$

Simulation Results and Discussion

The dynamic recrystallization grain size D_{dDRX} decreases with decreasing temperature and increasing strain rate. Grain size up to ≈2 μm, ≈1.5 μm and ≈1.0 μm can be obtained, respectively, at temperatures of 1000 °C, 900 °C, 850 °C and a strain rate of 50 s^{-1} (Fig. 3). However, this grain refinement can be achieved in case when 99% dDRX has occurred due to a single-pass large strain ε, which is equal to or greater than the steady state strain ε_{ss}. Interrelation between steady state strain, strain rate and temperature for HSS M2 is shown in Fig. 4. For example, strain value ε must be at least 2.06 in order to provide 99% dDRX at 1000 °C and 20 s^{-1}. If strain rate increases up to 50 s^{-1}, then strain value ε must be at least 2.25 at the same temperature. At lower temperatures, strain should be even higher. At 900 °C strain value ε must be at least 2.82 in order to provide 99% dDRX at 50 s^{-1} (Fig. 4). The conventional hot rolling process cannot provide such large strain values in one pass, especially over the entire section of the strip.

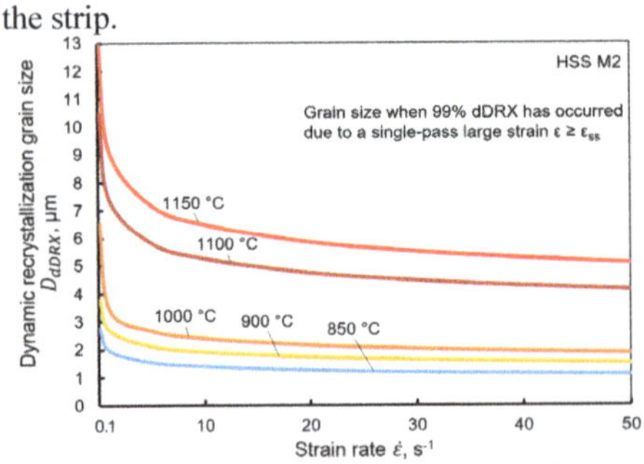

Fig. 3. Influence of strain rate and temperature on dynamic recrystallization grain size

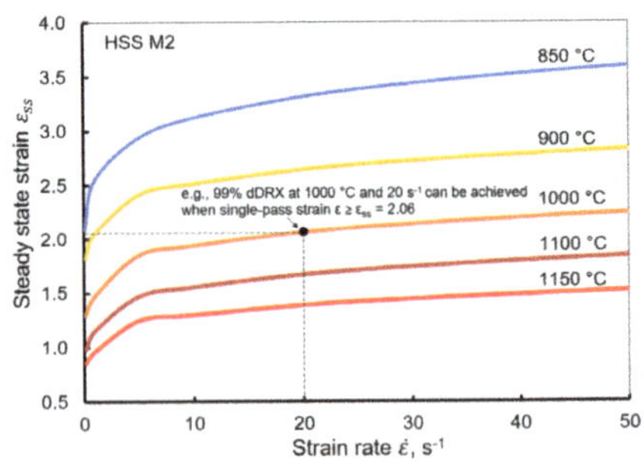

Fig. 4. Interrelation between steady state strain, strain rate and temperature for HSS M2.

Large strain can be achieved by single-pass asymmetric rolling. Work rolls speed ratio was varied from 1.0 (symmetric rolling) up to 1.1…3.0 with step 0.1 (asymmetric rolling). The initial temperature of the strip was also varied at the following values: 1150, 1100, 1000, 900 °C. A comparative analysis of many simulation variants showed that the finest average grain size in combination with a fairly uniform grain distribution over the strip thickness was achieved by asymmetric rolling with a work roll speed ratio of 1.5 at T = 1000 °C. Speed ratio of 1.5 means that the speed of upper work is 1.5 times faster than lower work roll. The average grain size of 2.9 µm in the center layer of the strip was achieved by asymmetric rolling (Fig. 5) in comparison with grain size of 14.2 µm obtained by symmetric rolling (Fig. 6). A decrease in the grain size in the central layer of the strip was achieved by almost 5 times.

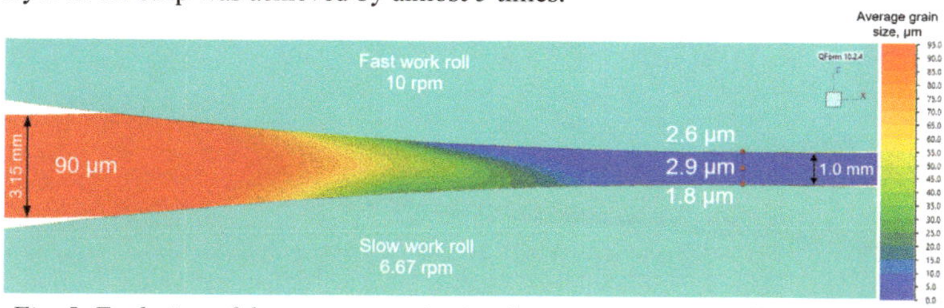

Fig. 5. Evolution of the average grain size during asymmetric rolling (SR = 1.5).

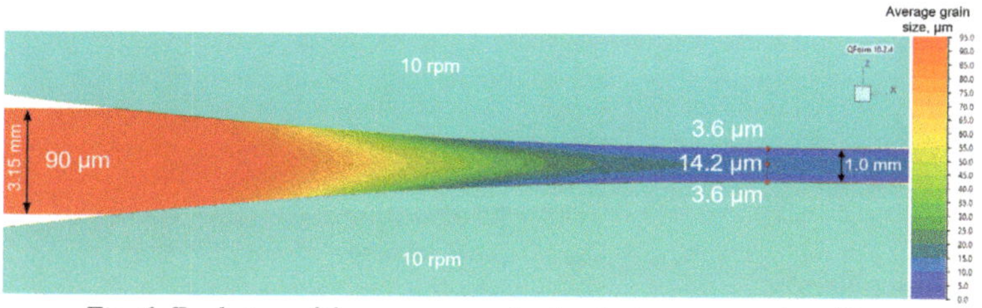

Fig. 6. Evolution of the average grain size during symmetric rolling.

At the same time, due to asymmetric rolling, a fairly uniform distribution of fine grain over the strip thickness was also achieved (Fig. 7). This is due to a more uniform distribution of strain over the thickness (Fig. 8).

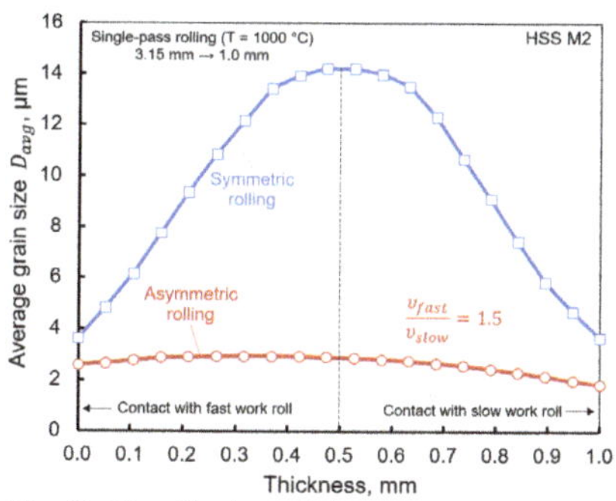

Fig. 7. Distribution of the average grain size over the strip thickness.

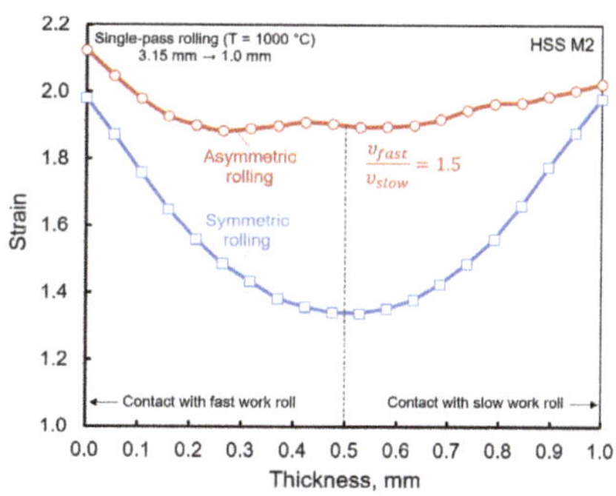

Fig. 8. Strain distribution over the strip thickness.

Temperature field during asymmetric rolling is shown in Fig. 9. The temperature in the deformation zone decreases from the initial 1000 °C to 860 °C at the contact with the slow work roll. The temperature difference at the contact with the upper and lower roll leads to an additional asymmetry factor. The lower contact temperature with the slow roll is due to the longer contact time. With symmetrical rolling, the temperature at the contact with the work rolls is also significantly reduced from 1000 °C to 890 °C (Fig. 10). Lower temperature and higher strain in the surface layers lead to grain refinement in both symmetric and asymmetric rolling. However, only asymmetric rolling makes it possible to refine the grain over the entire thickness of the strip due to the occurring of 100% dynamic recrystallization (Fig. 11). At the same time, the force during asymmetric rolling is significantly lower in comparison with symmetric rolling (Fig. 12).

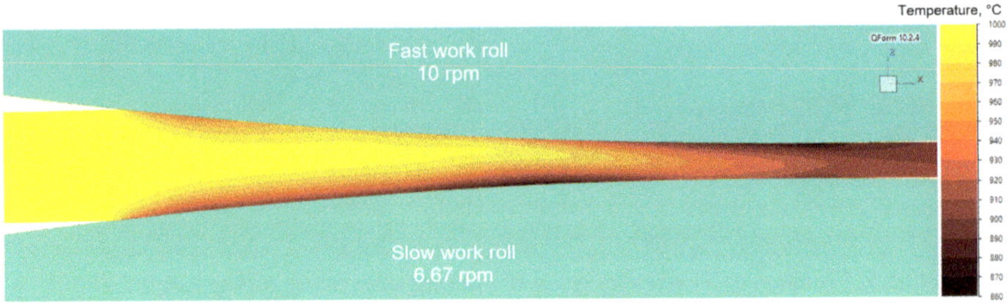

Fig. 9. Temperature field during asymmetric rolling.

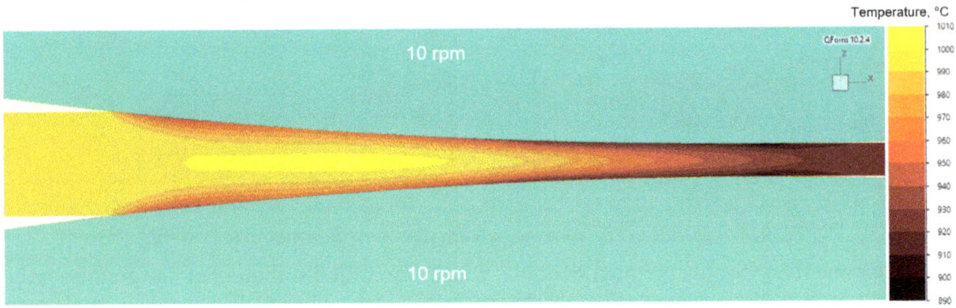

Fig. 10. Temperature field during symmetric rolling.

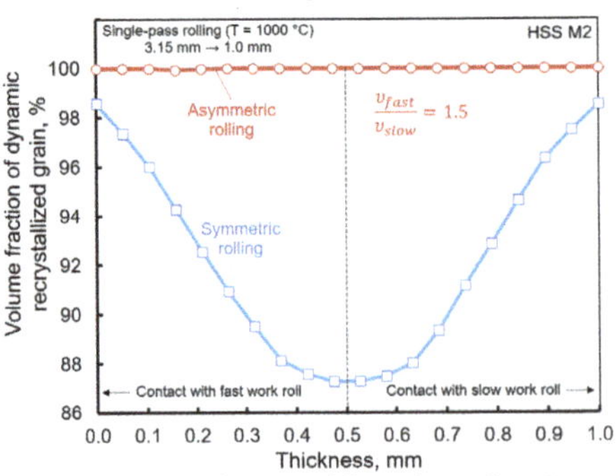

Fig. 11. Distribution of the volume fraction of dynamic recrystallized grain over the strip thickness.

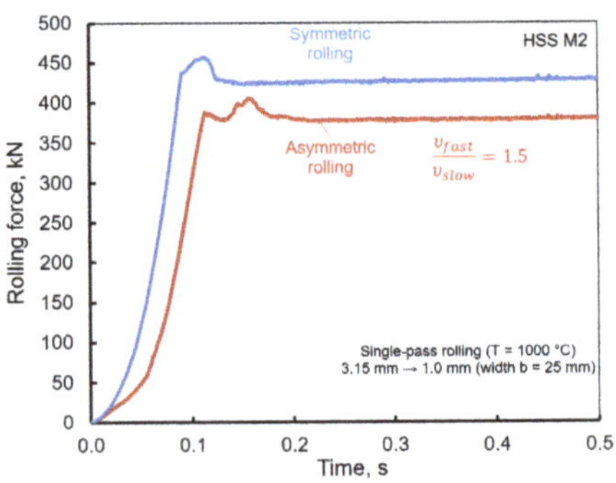

Fig. 12. Comparison of rolling forces during symmetric and asymmetric rolling.

Conclusion

Asymmetric hot rolling is a process which can provide the occurring of 99% dDRX in the difficult-to-deform high-speed steel M2 due to single-pass large strain at the lower temperature range. Large strain can be achieved by less rolling force in comparison with symmetric rolling.

In the present study the finest average grain size in combination with a fairly uniform grain distribution over the strip thickness was achieved by asymmetric rolling with a work roll speed ratio of 1.5 at T = 1000 °C. The average grain size of 2.9 μm in the center layer of the strip was achieved by asymmetric rolling in comparison with grain size of 14.2 μm obtained by symmetric rolling. A decrease in the grain size in the central layer of the strip was achieved by almost 5 times. However, experimental studies are required to confirm the results of FEM simulations.

The results of investigation can be used for design of effective technology of manufacturing HSS M2 with improved microstructure and mechanical properties. The main difficulty in the development of asymmetric hot rolling technology is to determine the optimal speed ratio of the work rolls.

Acknowledgements

The study was supported by the grant of the Russian Science Foundation (No. 23-79-30015).

References

[1] J.J. Jonas, Dynamic recrystallization - scientific curiosity or industrial tool? Mater. Sci. Eng. A 184 (1994) 155-165. https://doi.org/10.1016/0921-5093(94)91028-6

[2] K. Huang, R.E. Logé, A review of dynamic recrystallization phenomena in metallic materials, Materials & Design 111 (2016) 548-574. https://doi.org/10.1016/j.matdes.2016.09.012

[3] H.J. McQueen, Development of dynamic recrystallization theory, Mater. Sci. Eng. A 387-389 (2004) 203-208. https://doi.org/10.1016/j.msea.2004.01.064

[4] C.A.C. Imbert, H.J. McQueen, Dynamic recrystallization of A2 and M2 tool steels, Mater. Sci. Eng. A 313 (2001) 104-116. https://doi.org/10.1016/S0921-5093(01)00974-1

[5] A. Dehghan-Manshadi, M.R. Barnett, P.D. Hodgson, Hot deformation and recrystallization of austenitic stainless steel: part I. dynamic recrystallization, Metall. Mater. Trans. A. 39 (2008) 1359-1370. https://doi.org/10.1007/s11661-008-9512-7

[6] S.H. Cho, K.B. Kang, J.J. Jonas, Dynamic, static and metadynamic recrystallization of a Nb-microalloyed steel, ISIJ Int. 41 (2001) 63-69. https://doi.org/10.2355/isijinternational.41.63

[7] S.S. Dhinwal, C. Ernould, B. Beausir, Facilitating the occurrence of dynamic recrystallization in plain extra low-carbon steel by warm asymmetric rolling, Materials Characterization 189 (2022) 111942. https://doi.org/10.1016/j.matchar.2022.111942

[8] R. Milovic, D. Manojlovic, M. Andjelic, D. Drobnjak, Hot workability of M2 type high-speed steel, Steel research 63 (1992) 78-84. https://doi.org/10.1002/srin.199200474

[9] D. Pustovoytov, A. Pesin, P. Tandon, Asymmetric (hot, warm, cold, cryo) rolling of light alloys: a review, Metals 11 (2021) 956. https://doi.org/10.3390/met11060956

[10] G. Vincze, F.J.P. Simões, M.C. Butuc, Asymmetrical rolling of aluminum alloys and steels: a review, Metals (2020) 10, 1126. https://doi.org/10.3390/met10091126

[11] C. Imbert, N.D. Ryan, H.J. McQueen Hot workability of three grades of tool steel, Metallurgical Transactions A 15 (1984) 1855-1864. https://doi.org/10.1007/BF02664899

[12] C. Imbert, H.J. McQueen, Hot Ductility of Tool Steels, Canadian Metallurgical Quarterly 40 (2001). 235-244. https://doi.org/10.1179/000844301794388461

[13] C. Imbert, H.J. McQueen, Peak strength, strain hardening and dynamic restoration of A2 and M2 tool steels in hot deformation, Mater. Sci. Eng. A 313 (2001) 88-103. https://doi.org/10.1016/S0921-5093(01)00976-5

Superplasticity in Advanced Materials - ICSAM 2023 Materials Research Forum LLC
Materials Research Proceedings **32** (2023) 303-308 https://doi.org/10.21741/9781644902615-35

Improvement in strength and ductility of flame-retardant magnesium alloy through uniaxial hot pressing

Koichi Kitazono[1,a*] and Kotaro Wada[1,b]

[1] Graduate School of Systems Design, Tokyo Metropolitan University, Japan

[a] kitazono@tmu.ac.jp, [b] wada-kotaro@ed.tmu.ac.jp

Keywords: Magnesium Alloy, Heat Treatment, Tensile Test, Dynamic Recrystallization

Abstract. Flame-retardant magnesium alloys containing calcium element are attracting attention from aircraft and automotive industries because of their low density and high ignition temperature. Though thermomechanical treatments such as hot extrusion are effective to increase their mechanical properties, they cause an increase in cost. Present study focusses on commercial Mg-6Al-0.4Mn-2Ca and Mg-9Al-1Zn-2Ca alloys. Simple uniaxial hot pressing process is performed in these magnesium alloy plates. Tensile tests at room temperature revealed that the hot pressed Mg-6Al-0.4Mn-2Ca and Mg-9Al-1Zn-2Ca alloys have high tensile strength and elongation. Microstructural observation clarified that the improved mechanical properties were mainly due to dynamic recrystallization during the hot pressing. The present uniaxial hot pressing process has a potential as a simple thermomechanical treatment process for flame-retardant magnesium alloys.

Introduction

Many experimental studies have focused on the mechanical properties of lightweight magnesium alloys. For example, low temperature superplasticity in commercial Mg-Al-Zn alloys was induced by grain refinement through severe plastic deformation processes [1,2]. Recently, flame-retardant magnesium alloys have been developed by addition of calcium. Calcium element causes high strength and high heat resistance in Mg-Al alloys because of precipitation hardening by hard second phase (Al_2Ca) [3]. However, the low formability of Mg-Al-Ca alloys prevents the structural applications. Several thermomechanical treatment processes have been proposed in Mg-Al-Ca alloys. Hot extrusion is one of the effective process to increase both the strength and the ductility of Mg-Al-Ca alloys [4,5]. On the other hand, the hot extrusion process requires a large-scale equipment which increases the production cost.

The authors have proposed the simple uniaxial hot pressing process for enhanced formability of commercial magnesium alloys. The uniaxially hot pressed Mg-9Al-1Zn alloy showed high tensile strength as well as high elongation at room temperature [6]. In the present study, the hot pressing process is applied on Mg-Al-Ca alloys. The mechanical properties after the hot pressing process are evaluated by microstructural observation and tensile tests at room temperature.

Experimental Procedure

Hot extruded Mg-6Al-0.4Mn-2Ca alloy and cast Mg-9Al-1Zn-2Ca alloy were supplied by Fuji light metal Co., Ltd., Japan. Chemical compositions are shown in Table 1. Rectangular plates were cut from the starting materials. Uniaxial hot pressing was carried out using a Shimadzu CONCRETO 2000X compression testing machine. Specimens were compressed at 573 K, 623 K and 673 K in atmosphere. Boron nitride powder was used as a lubricant.

Table 1 Chemical compositions of as-extruded Mg-6Al-0.4Mn-2Ca alloy and as-cast Mg-9Al-1Zn-2Ca alloy (mass%).

Element	Al	Zn	Mn	Fe, Ni	Si	Cu	Ca	Mg
Mg-6Al-0.4Mn-2Ca	6.17	0.01	0.35	<0.001	0.03	0.001	2.04	Bal.
Mg-9Al-1Zn-2Ca	9.42	0.62	0.20	<0.001	0.04	0.001	2.16	Bal.

Microstructural observation was carried out using a JEOL JSM-IT800 field emission scanning electron microscope (FE-SEM). Crystal grains were evaluated by orientation imaging microscopy (OIM) analysis based on the electron back scattered diffraction (EBSD) patterns.

Tensile test specimens with different thicknesses were cut by electrical discharge machining. Tensile tests were performed using a Shimadzu Autograph AGX-50kVD universal testing machine at room temperature. Crosshead separation rate was fixed at 1 mm/min.

Results and Discussion

Inverse pole figure (IPF) maps of as-extruded Mg-6Al-0.4Mn-2Ca alloy and as-cast Mg-9Al-1Zn-2Ca alloy are shown in Fig. 1. Average crystal grain diameters of Mg-6Al-0.4Mn-2Ca alloy and Mg-9Al-1Zn-2Ca alloy were 8.3 μm and 1098 μm, respectively. In addition, the casting defects were observed in Mg-9Al-1Zn-2Ca alloy.

IPF maps of Mg-6Al-0.4Mn-2Ca alloy after the hot pressing at 573 K, 623 K and 673 K are shown in Fig. 2. Vertical directions are parallel to the compression direction. A part of area consisting of fine crystal grains were observed. This is due to the dynamic recrystallization during the hot pressing. However, the crystal grain sizes did not change significantly.

IPF maps of Mg-9Al-1Zn-2Ca alloy after the hot pressing at 573 K, 623 K and 673 K are shown in Fig. 3. Vertical directions are parallel to the compression direction. Area of dynamic recrystallization increased with increasing the temperature and the compressive strain. As a result, fine equiaxial grains were formed in specimens more than 50% compressive strain. In addition, the area of initial casting defects decreased after the hot pressing process.

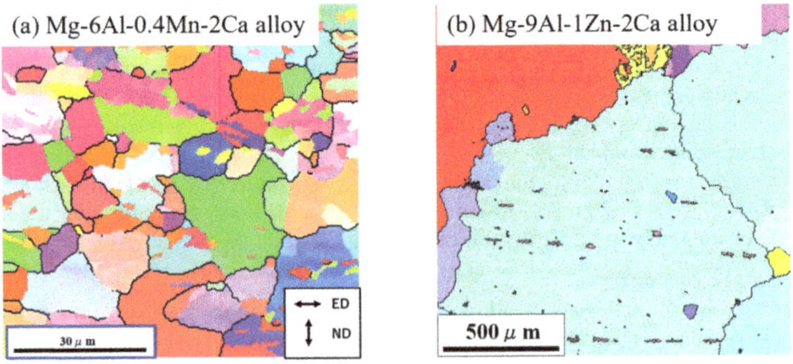

Fig. 1 IPF maps of (a) as-extruded Mg-6Al-0.4Mn-2Ca alloy and (b) as-cast Mg-9Al-1Zn-2Ca alloy.

Materials Research Forum LLC
https://doi.org/10.21741/9781644902615-35

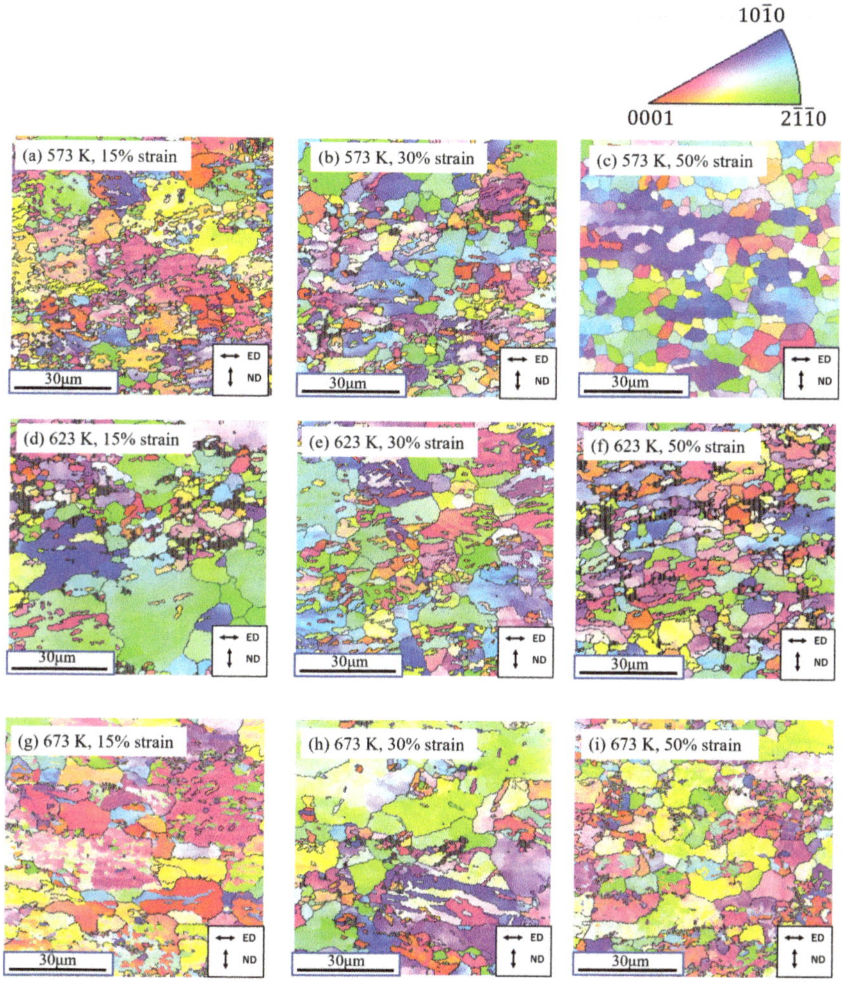

Fig. 2 IPF maps of Mg-6Al-0.4Mn-2Ca alloy. The specimens were hot pressed at 573 K to (a) 15%, (b) 30%, (c) 50% strains, at 623 K to (d) 15%, (e) 30%, (f) 50% strains, at 673 K to (g) 15%, (h) 30%, (i) 50% strains.

Materials Research Forum LLC
https://doi.org/10.21741/9781644902615-35

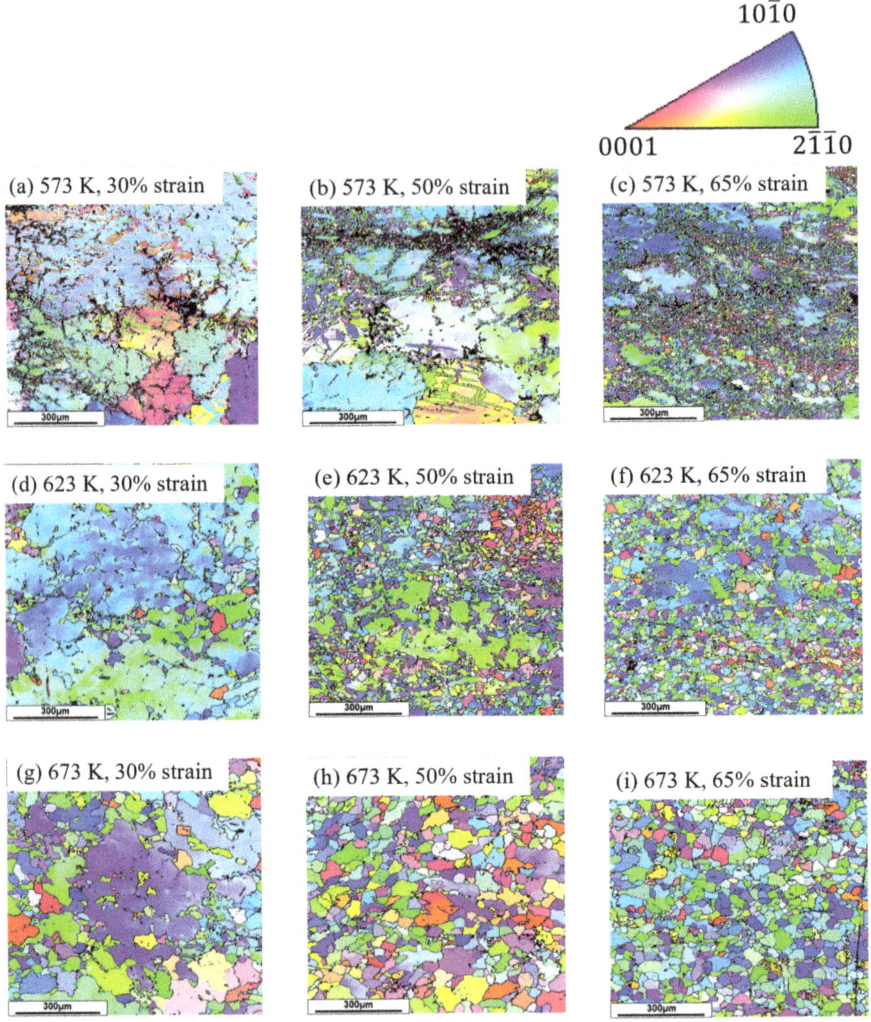

Fig. 3 IPF maps of Mg-9Al-1Zn-2Ca alloy. The specimens were hot pressed at 573 K to (a) 30%, (b) 50%, (c) 65% strains, at 623 K to (d) 30%, (e) 50%, (f) 65% strains, at 673 K to (g) 30%, (h) 50%, (i) 65% strains.

Results of tensile tests in Mg-6Al-0.4Mn-2Ca alloy are shown in Fig. 4. The 0.2% proof stress and the tensile strength were slightly increased after the hot pressing. The specimen compressed at 623 K to 15% strain showed the highest elongation of 16.4%. The enhancement of the mechanical properties was relatively low because of the initial fine equiaxial grains in as-extruded Mg-6Al-0.4Mn-2Ca alloy.

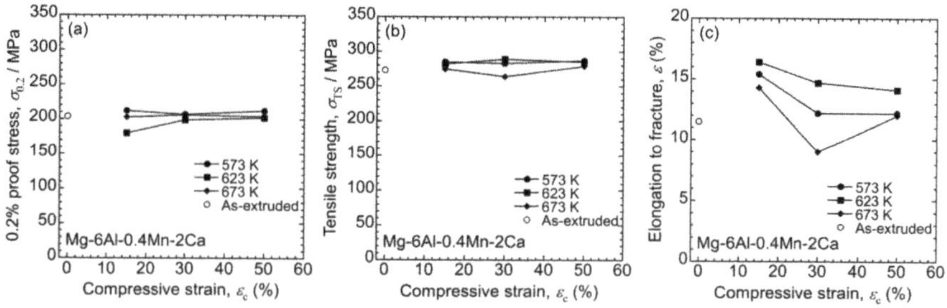

Fig. 4 (a) 0.2% proof stress, (b) tensile strength and (c) elongation to fracture in Mg-6Al-0.4Mn-2Ca alloys were plotted as a function of compressive strain.

Results of tensile tests in Mg-9Al-1Zn alloy are shown in Fig. 5. The 0.2% proof stress, the tensile strength and the elongation to fracture increased with increasing the compressive strain. The specimen compressed at 673 K to 65% strain showed the highest elongation of 8.1%. These results are due to the grain refinement induced by the dynamic recrystallization. Large crystal grains in as-cast Mg-9Al-1Zn-2Ca alloy were refined during the hot pressing process. In addition, casting defects decreased by the plastic deformation at high temperature. This is effective to increase the ductility.

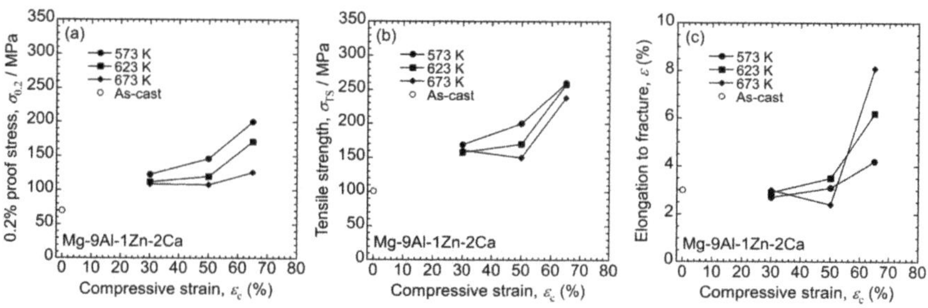

Fig. 5 (a) 0.2% proof stress, (b) tensile strength and (c) elongation to fracture in Mg-9Al-1Zn-2Ca alloys were plotted as a function of compressive strain.

Summary

Uniaxial hot pressing treatment was carried out to as-extruded Mg-6Al-0.4Mn-2Ca alloy and as-cast Mg-9Al-1Zn-2Ca alloys. During the high compression process, initial coarse grains were refined due to the dynamic recrystallization. Tensile tests revealed that the improvement in tensile strength and ductility of flame-retardant magnesium alloys. Especially, the mechanical properties of as-cast Mg-9Al-1Zn-2Ca alloy were significantly improved after the hot pressing process. Reduced casting defects is one reason of the improvements.

Acknowledgement

This work was supported in part by Light Metal Educational Foundation, Japan.

References

[1] M. Mabuchi, H. Iwasaki, K. Yanase, K. Higashi, Low temperature superplasticity in an AZ91 magnesium alloy processed by ECAE, Scr. Mater. **36** (1997) 681-686.

[2] M. Mabuchi, K. Ameyama, H. Iwasaki, K. Higashi, Low temperature superplasticity of AZ91 magnesium alloy with non-equilibrium grain boundaries, Acta Mater. **47** (1999) 2047-2057.

[3] Y. Kawamura, K. Ougi, S. Inoue, T. Kiguchi, M. Takafuji, H. Ihara, D. S. Shih, Advanced Mg-Al-Ca alloys with combined properties of high thermal conductivity, high mechanical strength and non-flammability, Mater. Trans. **63** (2022) 118-127.

[4] T. Homma, S. Hirawatari, H. Sunohara, S. Kamado, Room and elevated temperature mechanical properties in the as-extruded Mg-Al-Ca-Mn alloys, Mater. Sci. Eng. A **539** (2012) 163-169.

[5] X. Huang, Y. Chino, M. Yuasa, H. Ueda, M. Inoue, F. Kido, T. Matsumoto, Microstructure and mechanical properties of AZX912 magnesium alloy extruded at different temperatures, Mater. Sci. Eng. A **679** (2017) 162-171.

[6] N. Kitazono, D. Suzuki, R. Yamaguchi, K. Mitsuishi, K. Kitazono, Microstructural evolution through uniaxial hot pressing before age hardening of AZ91D alloy, Mater. Trans. **57** (2016) 1094-1100.

Experimental study of the effect of increasing technological plasticity during asymmetric rolling of aluminum alloys

Anna Kozhemyakina[1,a*], Alexander Pesin[1,b], Denis Pustovoytov[1,c], Leonid Nosov[1,d], Anna Baryshnikova[1,e], Natalia Lokotunina[1,f], and Dmitry Grachev[1,g]

[1] Nosov Magnitogorsk State Technical University, Laboratory of Mechanics of Gradient Nanomaterials named after A.P. Zhilyaev, 38 Lenin prospect, Magnitogorsk, 455000, Russian Federation

[a]kozhemiakina.a@yandex.ru, [b]pesin@bk.ru, [c]pustovoitov_den@mail.ru, [d]nosov.leopold@yandex.ru, [e]anyabar1999@mail.ru, [f]nml76@mail.ru, [g]omp_d.grachev@mail.ru

Keywords: Asymmetric Cold Rolling, High-Strength Steel Grades, Intermediate Annealing, Rolling Force, Laboratory and Industry Experiments

Abstract. In this paper the effect of asymmetric rolling on the possibility of increasing the technological plasticity of aluminum alloys was investigated. The experimental research was carried out on a laboratory asymmetric rolling mill with an individual drive of the work rolls with the possibility of creating a speed ratio from 1.0 to 5.0. It was shown that the increase of speed ratio of the work rolls from 1.0 to 5.0 significantly reduce the rolling force in comparison with symmetric rolling. Rolling force decreased in 1.9 times for alloy AD33 (AA6061), in 2.3 times for alloy AMg6, in 3.2 times for alloy D16 (AA2024). At the same time the technological plasticity was increased. Technological plasticity characterizes the ability of a material to undergo higher thickness reductions without fracture under certain conditions of stress, temperature, and strain rate. In asymmetric rolling the thickness reduction was increased from 48 to 87% for alloy D16, from 50 to 59% for alloy AMg6, and from 40 to 75% for alloy AD33 in comparison with symmetric rolling. In all cases the samples had initially room temperature and were subjected only to deformation heating and friction heating. Extremely high thickness reduction (87%) was achieved by a single pass asymmetric rolling (at speed ratio 5.0) for alloy D16. It was found that the ductility of the alloy D16 was 12.3% after asymmetric rolling with a thickness reduction of 87% and without the use of annealing. This was approximately 2 times higher than the initial ductility (6.2%) of the same alloy in the initial annealed state and much higher than ductility (0.3%) after symmetric rolling. New technological schemes of sheet rolling of aluminum alloys with high ductility and increased technological plasticity have been developed.

Introduction

Aluminum is one of main structural materials in the world, being second to only ferrous metals in production and consumption volumes. Now, Russia ranks as the world's second largest aluminum producer (after China) and the leading exporter of this metal [1].

One of main requirements for aluminum alloy sheets is technological plasticity or formability. Usually during rolling metals and alloys are hardened and, consequently, they show lower technological ductility [2]. Upon achievement of a definite level of deformation, technological plasticity is almost fully lost, and further deformation is not possible without additional heat treatment, followed, in turn, by complicating the process and reducing productivity, and introducing additional constraints of the process.

Since the 2010s, asymmetric rolling is seen in terms of forming the required properties of rolled products. Studies of Russian and foreign scientists [3-11] show that one of the most promising methods of obtaining the required properties of rolled products is an asymmetric thin sheet rolling process based on intentionally formed asymmetry due to the difference between peripheral speeds of work rolls.

However, the issues of increasing technological plasticity have not been properly studied; therefore, developing methods of improving technological plasticity of aluminum narrow strips during asymmetric rolling is still relevant.

The study is aimed at increasing technological plasticity of aluminum narrow strips by mismatching speeds of work rolls during asymmetric rolling.

Research methods and materials used. To carry out the studies, we selected aluminum alloys AMg6 and AD33, mostly used in the automotive industry, and alloy D16, highly used in all the industrial sectors, including aeronautical and transport engineering.

The experiments were conducted using asymmetric rolling of aluminum alloys, whose chemical composition is confirmed by the energy dispersive analysis:

- D16 (Si –0.26 %; Fe – 0.39 %; Cu – 4.00 %; Mn – 0.55 %; Mg – 1.48 %; Cr – 0.018 %; Zn – 0.16 %; Ti – 0.043 %; Ni – 0.030 %; B – 0.0018 %; Al – 93.09 %),
- AMg6 (Si – 0.17 %; Fe – 0.27 %; Cu – 0.045 %; Mn – 0.64 %; Mg – 6.15 %; Cr – 0.011 %; Zn – 0.027 %; Ti – 0.035 %; Ni – 0.008 %; Be – 0.0018 %; Al – 92.7 %),
- AS33 (Si – 0.58 %; Fe – 0.48 %; Cu – 0.22 %; Mn – 0.11 %; Mg – 0.80 %; Cr – 0.15 %; Zn – 0.080 %; Ti – 0.035 %; Be – 0.0014 %; Al – 97.45 %).

Research results. Basic data for simulating the process in DEFORM 2D/3D software: entry strip thickness h0 = 1.0-8.0 mm; initial temperature of the workpiece: 20 °C; work roll diameter R = 100-500 mm; friction is according to Coulomb's law; friction coefficient $\mu = 0.05$-0.4; deformation ratio (reduction) $\varepsilon = 10$-90 %; bottom roll peripheral speed $V_1 = 100$ mm/s; roll speed ratio is from 1 to 5.

It has been found that during asymmetric rolling ($V_1/V_2=5$) of specimens of aluminum alloy AMg6 at reduction $\varepsilon = 66$ %, rolling force decreases from 610.0 to 192.2 kN as compared with symmetric rolling; torque on the bottom roll increases by 1.4 times, from 3.8 to 5.6 kN·m, and torque on the top roll decreases by 1.3 times, to 2.9 kN·m. Rolling force during asymmetric rolling decreases by 2.9 (reduction is 64 %, $V_1/V_2=5$) and 3.5 times (reduction is 87 %, $V_1/V_2=5$) for alloys AD33 and D16, respectively.

To carry out the experimental study, we used laboratory and industrial rolling mill 400 at the Zhilyaev Laboratory of Mechanics of Gradient Nanomaterials. Key specifications of the mill are given in Table 1 [12,13].

Table 1 – Specifications of the pilot and experimental laboratory two-high rolling mill

Parameter	Value
Type	Reversing two-high rolling mill with individually-driven work rolls
Screw-down mechanism	Hydraulic
Work roll diameter	340 [mm]
Work roll body length	400 [mm]
Rolling force (max)	2500 [kN] (250 [tf])
Torque (max)	2·65 [kN·m]
Main drive power	2·75 [kW]
Thickness of rolled sheet workpieces	50 [mm] – 0.5 [mm]
Materials to be rolled	Al, Fe, Ti, and other metals and alloys, having σ_T up to 1500 [MPa]
Applied technologies	– symmetric rolling, – asymmetric rolling, – cryogenic rolling, – accumulative roll bonding.
Total equipment weight	20 [t]

The experiments on mill 400 proved validity of the designed computer models in terms of force values. The calculated rolling torques on every work roll deviated from the measured values by 15 % or less.

The experimental results of rolling narrow strips from D16 with various work roll speed ratios are given in Table 2.

Table 2 – Rolling parameters of narrow strips from aluminum alloy D16

Sample No.	Thickness, mm		Relative reduction, [%]	Force, [kN]	Roll speed ratio	Hardness, [HB 5/125]	Note
	Entry thickness	Exit thickness					
23.09.2021 –9	6.00	3.10	48	463.80	5.0/5.0	104	Fracture of the sample
23.09.2021 –19 (3)	6.00	2.60	58	181.40	8.0/4.0	118	Fracture of the sample
23.09.2021 –41	6.00	2.00	67	215.10	9.0/3.0	86	Fracture of the sample
23.09.2021 –61	6.00	1.90	68	188.60	8.0/2.0	88	Fracture of the sample
23.09.2021 –92	6.00	0.90	85	243.00	10.0/2.0	73	No fracture of the sample
23.09.2021 –94	6.00	0.65	89	230.00	10.0/1.5	68	No fracture of the sample
23.09.2021 –95	6.00	-	-	-	10.0/1.3	-	The sample melted

The conducted experiments showed an increase in technological plasticity of aluminum narrow strips during asymmetric rolling. So, during conventional symmetric rolling (V1/V2 = 5.0/5.0) of aluminum alloy D16, in a single pass the sample was broken, when we tried to roll it to a thickness of 3.1 mm (reduction was 48 %). By increasing the work roll speed ratio up to V1/V2 = 10.0/2.0, we produced the aluminum narrow strip with a final thickness of 0.8 mm, relative reduction per pass was 87 %. A further increase in the work roll speed ratio up to V1/V2 = 10.0/1.5 contributed to producing the aluminum narrow strip with a final thickness of 0.65 mm (reduction was 89 %). The sample melted at a roll speed ratio of ≈ 7.7 (V1/V2=10.0/1.3).

Increased technological plasticity, when introducing asymmetry, was also experimentally confirmed for the samples of aluminum alloys AMg6, from 50 to 59%, and for AD33, from 40 to 75%.

The studies performed showed a decrease in rolling force with an increase in the work roll speed ratio. So, when rolling the samples of aluminum alloy D16 at work roll speed ratio V1/V2 = 10.0/2.0, force decreased by over 3.2 times (from 463.8 kN to 144.7 kN) at the degree of reduction corresponding to the degree of reduction in the symmetric mode.

A similar character of force reduction during asymmetric rolling was found for the samples of aluminum alloys AMg6 and AD33. So, when rolling the samples of aluminum alloy AD33, force decreased by 1.9 times (from 353.9 kN to 184.8 kN) at a reduction of 40%; when rolling the samples of aluminum alloy AMg6, force decreased by 2.3 times (from 290.0 kN to 125.5 kN) at a reduction of 50%.

For the first time, as a result of the experiments on rolling the samples from aluminum alloy D16, it was found that an increase in the work roll speed ratio by 1 – 5 times entailed an increase in the relative elongation of the samples after fracture. So, after asymmetric rolling, relative elongation after fracture increased from 6.2 to 12.3% (Table 3), while, in the symmetric case, it decreased from 6.2 to 0.3.

Table 3 – Mechanical properties of the narrow strips from alloy D16 after rolling

Sample	Ultimate tensile strength σ_B, [MPa]	Yield stress $\sigma_{0.2}$, [MPa]	$\sigma_{0.2}/\sigma_B$	Relative elongation, δ [%]
The sample after asymmetric rolling without lubrication (V_1/V_2 = 10.0/2.0; relative reduction per pass is 87 %)	283.6±1.3	234.4±1.1	0.83	12.3±0.13
The sample after asymmetric rolling with lubrication (V_1/V_2 = 10.0/2.0; relative reduction per pass is 85 %)	295.9±1.1	255.7±1.2	0.86	10.3±0.10
The sample after symmetric rolling	262.9±1.0	249.8±1.2	0.95	0.3±0.04
Original sample	191.7±0.9	141.4±0.8	0.74	6.2±0.08

Materials Research Proceedings **32** (2023) 309-316
https://doi.org/10.21741/9781644902615-36

It should be noted that during asymmetric rolling $\sigma_{0.2}/\sigma_B$ decreases by 13 % (from 0.95 to 0.83) as compared with symmetric rolling.

Developing process flow charts to increase technological plasticity of aluminum alloys D16, AMg6 and AD33 during asymmetric rolling.

The experiments conducted with symmetric rolling of aluminum alloy D16 showed a relative reduction of up to 45 %, keeping the produced narrow strip unbroken. In view of this, producing narrow strips from aluminum alloy D16, 0.65 mm thick, in compliance with GOST 13726-97, by conventional symmetric rolling, required four stages of rolling and four stages of annealing (Fig. 1) [14].

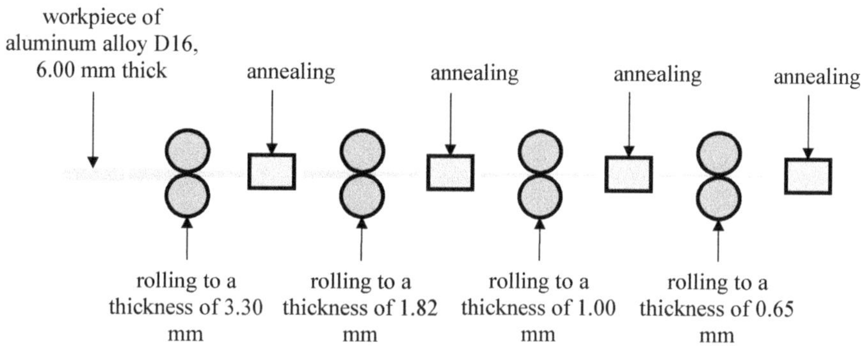

Fig. 1 – Process flow chart of symmetric rolling of the narrow strip from aluminum alloy D16

The above stated experimental data of rolling of the samples from aluminum alloy D16 with work roll speed ratio V_1/V_2 = 10.0/2.0 showed a potential increase in a single relative reduction up to 87 %, while keeping the narrow strip unbroken.

In this case, the conventional rolling process flow chart may be adjusted as stated in Fig.2.

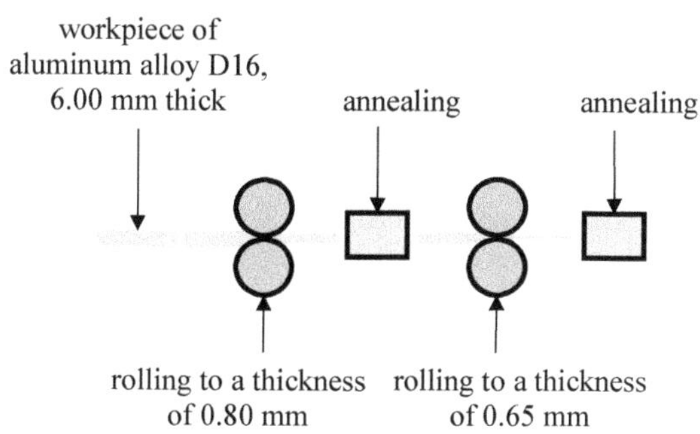

Fig. 2 – Process flow chart of asymmetric rolling of the narrow strip from aluminum alloy D16 with work roll speed ratio V_1/V_2 = 10.0/2.0

A further increase in the work roll speed ratio by 6.7 times (V_1/V_2 = 10.0/1.5) entail an increase in relative reduction per pass to 89 % (final thickness is 0.65 mm).

In this case, a process flow chart of asymmetric rolling of narrow strips from aluminum alloy D16 may be presented as in Fig. 3.

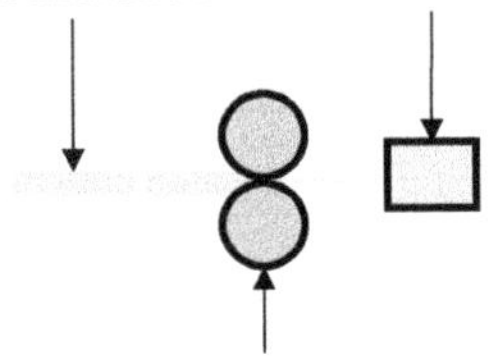

Fig. 3 – Process flow chart of asymmetric rolling of the narrow strip from aluminum alloy D16 with work roll speed ratio V_1/V_2 = 10.0/1.5

Similar process flow charts were developed for producing aluminum narrow strips from alloys AMg6 and AD33. The asymmetric rolling process flow chart for aluminum alloy AMg6 with work roll speed ratio V_1/V_2 = 10.0/2.0 contributes to excluding one rolling stage and annealing stage as compared with the conventional rolling sequence. Thus, by introducing asymmetry as a result of an increase in the work roll speed ratio up to 5, we can produce aluminum narrow strips from alloy AMg6, 0.78 mm thick, in a single pass.

The proposed process flow chart of rolling narrow strips from aluminum alloy AD33 at roll speed ratio V_1/V_2 = 8.0/2.0 shows that we can exclude three stages of rolling and three stages of annealing. Thus, the asymmetric rolling process flow chart for narrow strips from aluminum alloy AD33 at roll speed ratio V_1/V_2 = 8.0/2.0 can be used to produce aluminum narrow strips from alloy AD33, 0.50 mm thick, in a single pass.

Conclusion

The paper describes the approach developed at the Department of Materials Treatment Technologies at Nosov Magnitogorsk State Technical University and aimed at applying asymmetric rolling of metal narrow strips. In particular:

1) For the first time, it has been shown that an increase in the work roll speed ratio from 1.0 to 6.7 contributes to [15]:

- a significant decrease in rolling force as compared with symmetric rolling: by 1.9 times (for alloy AD33), by 2.3 times (for alloy AMg6), by 3.2 times (for alloy D16),
- higher technological plasticity (relative reduction increases without fracture of the samples) for D16 from 48 to 89 %, for AMg6 from 50 to 59 %, for AD33 from 40 to 75 %,
- increase in relative elongation of the samples to 12.3 % after asymmetric rolling of the narrow strip from alloy D16 as compared with 6.2 % in an annealed state.

2) The paper proposes new process flow charts for aluminum narrow strips of higher technological plasticity, excluding from one to three rolling stages and from one to three annealing stages.

Acknowledgment
The study was supported by a grant of the Russian Science Foundation (project No. 22-49-02041 dated 09.03.2022)

References

[1] Order of the Ministry of Industry and Trade of the Russian Federation No. 839 dated May 5, 2014 "On approval of the 2014 - 2020 and 2030 Strategy on Development of Ferrous Metallurgy in Russia and the 2014 - 2020 and 2030 Strategy on Development of Non-Ferrous Metallurgy in Russia". Information on https://www.garant.ru/products/ipo/prime/doc/70595824/#review

[2] V.N. Samokhvalov, Science of Metal Forming Processes, Samara University, Samara, 2019.

[3] V.M. Salganik, A.M. Pesin, Asymmetric Thin Sheet Rolling: Development of the Theory, Technologies and New Solutions, Moscow Institute of Steel and Alloys, Moscow, 1997.

[4] A.M. Pesin, D.O. Pustovoytov, O.D. Biryukova, A.E. Kozhemyakina, Asymmetric rolling of sheets and narrow strips: history and prospects for development, Bulletin of South Ural State University. Series: Metallurgy. 20 (3) (2020) 81-96.

[5] Y.H. Ji, J.J. Park, Development of severe plastic deformation by various asymmetric rolling processes, Materials Science and Engineering: A. 499 (1) (2009) 14-17. https://doi.org/10.1016/j.msea.2007.11.099

[6] Y.H. Ji, J.J. Park, W.J. Kim, Finite element analysis of severe deformation in Mg-3Al-1Zn sheets through differential-speed rolling with a high speed ratio, Materials Science and Engineering: A. 454 (2007) 570-574. https://doi.org/10.1016/j.msea.2006.11.076

[7] Q. Cui, K. Ohori, Grain refinement of high purity aluminum by asymmetric rolling, Materials Science and Technology. 16 (2000) 1095-1101. https://doi.org/10.1179/026708300101507019

[8] F.-Q. Zuo, J.-H. Jiang, A.-D. Shan, J.-M. Fang, X.-Yao Zhang, Shear deformation and grain refinement in pure Al by asymmetric rolling, Transactions of Nonferrous Metals Society of China. 18 (4) (2008) 774-777. https://doi.org/10.1016/S1003-6326(08)60133-8

[9] J. Jiang, Yi Ding, F. Zuo, A. Shan, Mechanical properties and microstructures of ultrafine-grained pure aluminum by asymmetric rolling, Scripta Materialia, 60 (10) (2009) 905-908. https://doi.org/10.1016/j.scriptamat.2009.02.016

[10] D.O. Pustovoytov, A.M. Pesin, A.A. Perekhozhikh, M.K. Sverdlik, Simulation of shear strain in a limit case of asymmetric thin sheet rolling, Vestnik of Nosov Magnitogorsk State Technical University, 1 (41) (2013) 65-68.

[11] D. Pustovoytov, A. Pesin, N. Lokotunina, A. Kozhemiakina, Influence of small microscopic grooves of work rolls on strain gradient induced in metal sheets during symmetric and asymmetric rolling, in: METAL 2019 - 28th International Conference on Metallurgy and Materials, Conference Proceedings, 2019, pp. 265-270. https://doi.org/10.37904/metal.2019.701

[12] Pustovoytov, D. Asymmetric (Hot, Warm,Cold, Cryo) Rolling of Light Alloys: A Review / D. Pustovoytov, A. Pesin, P. Tandon/ / Metals 2021. - 11 (956). - p. 1-46. https://doi.org/10.3390/met11060956

[13] O. Biryukova, A. Pesin, D. Pustovoytov, A. Kozhemiakina, L. Nosov, Obtaining laminated aluminum composites with a gradient structure based on asymmetric deformation, in: METAL 2021 - 30th Anniversary International Conference on Metallurgy and Materials, Conference Proceedings, 2021, pp. 496-501. https://doi.org/10.37904/metal.2021.4133

[14] A.M. Pesin, D.O. Pustovoytov, I.A. Pesin, A.E. Kozhemyakina, L.V. Nosov, A.I. Sverchkov, Developing asymmetric rolling process flow charts for aluminum narrow strips of higher strength and plasticity, Theory and Technology of Metallurgical Production, 2 (41) (2022) 32-42.

[15] Information on http://lmgn.magtu.ru/ru/oborudovanie.html

Posters

Microstructural and corrosion study of a "non-comercial" high manganese steel

Marcos Natan da Silva Lima[1,3,a], Mauro Andres Cerra Florez[1,b*], Jorge Luiz Cardoso[1,c], João Vitor da Silva Pinto[1], Lucas Moura Farias de Medeiros[1], Walney Silva Araújo[2,d], Hamilton Ferreira Gomes de Abreu[1,e]

[1]Materials Characterization Laboratory (LACAM), Department of Metallurgical and Materials Engineering, Federal University of Ceará, Campus do Pici, bloco 729, Fortaleza 60020-181, Ceará, Brazil

[2]Corrosion Research Laboratory (LACAM), Department of Metallurgical and Materials Engineering, Federal University of Ceará, Campus do Pici, bloco 729, Fortaleza 60020-181, Ceará, Brazil

[3]Processes for Conformation of Metallic Materials (PROCOMAME), Department Materials Science and Metallurgical Engineering, Universitat Politècnica de Catalunya: Barcelona, Catalunya, Bl. I, 08930, Spain

[a]marcos.natan.da.silva@upc.edu, [b]mauro.cerra@ufc.br, [c]jorgeluizjlc@gmail.com, [d]wsa@ufc.br, [e]hamilton@ufc.br

Keywords: High Mn Steel, Microstructural Characterization, Nacl Corrosion, Hot Rolling, E-Martensite

Abstract. High-Mn steels have great plasticity when subjected to deformation due to TWIP or TRIP effects. This work evaluated the microstructural evolution, the formation of the ε-Martensite phase taking into account the hot rolling of 80-60% and the solution annealing. Afterwards, microstructures were analyzed by SEM. Volume fraction of the Austenite and ε-Martensite phases were measured by EBSD technique. The steel obtained low energy levels of stacking fault, favoring the effect TRIP. Corrosion resistance in 0.1M NaCl solution was analyzed by open potential circuit and potentiodynamic polarization techniques. The analysis of the curves and the surface of the steel after the polarization tests showed that the steel with less strain had relatively nobler potential than the steel with more strain.

Introduction

High manganese steels are called cryogenic steels because they have high mechanical strength and good toughness at temperatures close to that of liquid Nitrogen. In applications at cryogenic temperatures, metals with FCC structure are widely used because they have good mechanical properties at low temperature, while metals with BCC structure show a brittle ductile transition [1]. FeMnC alloys with high levels of Manganese (20-30 wt.%) represent a very recent development of austenitic steels, which, through their differentiated mechanisms of deformation, in TWIP effect (Twinning Induced Plasticity) plastic deformation induced by twins and TRIP (Transformation Induced Plasticity) transformation induced by phase transformation, are of great scientific and economic relevance. Thus a work-scale study following the traditional routes of manufacture of austenitic steels (casting followed by hot mechanical conformation at temperatures above 1000 °C) is necessary, just as it has been practiced by several foreign research groups [2].

Two types of Martensite may occur in austenitic steels. The formation of ε-Martensite, whose crystalline structure is hexagonal compact (HC), and of α'-Martensite, of structure of body-centered cubic or body centered tetragonal (BCC / BCT) [3,4]. The ε-Martensite is still little explored in austenitic steels. The chemical composition with high manganese content and very low carbon content is very little reported in the literature, mainly by the non-formation of α'-Martensite phase, maintaining the tenacity and ductility properties of the steel. However, the results obtained so far are very promising for the industrial application in very low temperatures of the steel in question. For a research on how the TRIP effect occurs that has the consequence of the formation of ε-Martensite plates is of great relevance. The Stacking Fault Energy (SFE) increases with increasing Mn content, and the TWIP effect is well activated in the SFE range between (18-60 mJ/m^2), thus occurring to the formation of mechanical macros and to values of (SFE < 18 mJ/m^2) the martensitic transformation occurs in the microstructure of the material [4–6]. The ε-Martensite phase also forms at low temperatures in the absence of plastic deformation, whereas for the formation of α'- Martensite a more severe plastic deformation is required. As shown by some authors [5,7–10], the amount of ε-Martensite increases with deformation, reaches a maximum, in larger deformations decreases, while the amount of α'-Martensite increases progressively. Prior to the development of high strength low alloy steels or microalloyed steels (HSLA), carbon and manganese hardened steels thus had poor weldability. The carbon content of the steels produced was around ≅ 0.25%, and today the content has been reduced to around ≅ 0.05%. This reduction substantially improves the weldability of steel [11,12]. That is why the choice of such a low value of carbon in the composition of the steel under study.

In general, the corrosion behavior of high Mn steels in 3.5% NaCl solution is attributed to the high dissolution rate of Mn and Fe atoms in Chloride solutions [13,14]. The formation of unstable oxides in these types of steel allows Mn to be preferably dissolved at the oxide/electrolyte interface. There are some works on the electrochemical polarization behavior of Fe-Mn-Al alloys where some authors have reported that these steels do not passivate in aqueous solutions of 3.5% NaCl, 10% HCl, 0.01M H_2SO_4 or 10% HNO3 and are susceptible to pitting corrosion, but this depends on the alloying elements and their amounts [14-17]. It is also reported that increasing manganese content above 25%, decreases the corrosion resistance of the alloy in aqueous solutions [13-15], as well as adding aluminum to a 25% manganese steel has resulted in increased corrosion resistance in 1M Na_2SO_4, 50% HNO3 and 10-50% NaOH solutions, but this same steel showed no passivation in 10% HCl and 3.5% NaCl solutions [16]. Finally, it is mentioned that as the manganese content in high manganese steels increases, its corrosion resistance decreases [13-18], but the analysis of corrosion in Fe-C-Mn-Si steels is still little reported in the literature, especially for such low carbon values, hence the importance of the current research.

Materials and Methods

Material

The material used in this work was a steel model Fe-Mn-Si-C, manufactured by centrifugal casting, with chemical composition listed in Table 1. Chemical composition obtained in an optical emission spectrometer and compared to that provided by the manufacturer.

Table 1. Chemical composition (wt.%) of the cryogenic steel used in this work.

Mn	C	Si	Ni	Cr	Fe
27	0.027	1.0	0.19	0.41	Bal.

Characterization

The material was manufactured by centrifugal casting, aiming at the reduction of porosities and defects from the casting process. The SFE calculation was produced for the particular program in Microsoft Excel, fitting it to the alloy under study. The quantification of stacking fault energy was calculated by the proposed thermodynamic mathematical model and used by the author [19] for the Fe-Mn and FeMnC system [20]. The values of free energy, chemical composition, temperature and grain size ASTM of high steel Mn were used. The SFE calculation was carried out using a program specially in Microsoft Excel in order to adapt it to the alloy under study. The samples were hot rolled at 1200 °C, with five passes of the same reduction for deformed samples of 80-60%, with an initial thickness of 9.0 mm. The steel was solution annealed at 1100 °C for 60-90 minutes. After the solution annealing, the steel was cooled in water. After the thermal and mechanical treatments, the samples were prepared for the metallography and polished with alumina. Subsequently, electrolytic polishing was carried out for 20 min, with voltage parameters of 12-15 V and current ranging from 0-2 A. A chemical etching with HNO_3 + H_2O solution was used for a time of 1-5 s. For each condition, several micrographs were obtained in order to analyze the microstructure of Fe-Mn-Si-C steel as received and after deformation/solution annealed. The examined section corresponds to the thickness, plans of the rolling direction, DL. Microstructural analyzes were also carried out on the samples using a scanning electron microscope (SEM) (FE-SEM, Quanta 3D FEG, FEI) used in backscattering electron (BSE) mode. In this same equipment, maps were obtained from the Electron Backscatter Diffraction (EBSD) technique, using the software CHANNEL 5 for the acquisition of maps and quantification of phases. The maps were obtained with an average index of 80% of identified points.

Corrosion test

Corrosion tests were carried out using the Autolab potentiostat/galvanostat PGSTAT 302N controlled by the NOVA® program for data acquisition and treatment. The techniques of open circuit potential (OCP) and potentiodynamic polarization were used in a conventional three-electrode cell at room temperature and in triplicate. The reference electrode used was Ag/AgCl (saturated KCl), the counter electrode was a platinum plate with a rectangular geometrical area of approximately 150 mm^2, and the working electrodes were the samples of Fe27Mn1Si alloy (80-60%) deformed with an area of approximate 30 mm^2. Before the electrochemical measurements, the samples were ground and then washed with deionized water and ethanol. The OCP test was performed for 1 h. Potentiodynamic polarization curves were obtained in a potential range from -50 to +300 mV around the OCP. The scanning rate used was 1 mV/s and the measurements ended when 1 mA current was reached [14,15,17,18]. Due to the potential application of this steel in the construction of cryogenic storage tanks that are commonly located on ships or in areas close to the sea, the NaCl solution was chosen for conducting electrochemical tests. A 0.1M NaCl solution was used and the pH was measured before and after testing. After the corrosion tests, the corrosion products were analyzed with the help of the SEM coupled with an EDS detector.

Results and discussion

Stacking fault energy analysis

The performance fault energy quantification was balanced by the proposed mathematical model and used by Allain et al [21] for the Fe-Mn and FeMnC system. The following free energy components were used: apical temperature, temperature and ASTM grain size of high steel Mn. The addition of alloying elements such as Al and Ni increases SFE and tends to inhibit the martensitic transformation ($\gamma \rightarrow \varepsilon$). The addition of Si decreases SFE, thus favoring ε-Martensite formation during cooling and deformation. The temperature and grain size of the steel also influence the SFE values. The SFE (Γ) in a TRIP / TWIP steel can be estimated using the model proposed by Dumay et al, according to Eq. 1 and Eq. 2.

$$\Gamma = 2\rho\Delta G^{\gamma\rightarrow\varepsilon} + 2\sigma^{\gamma/\varepsilon} \quad (1)$$

$$\Delta G^{\gamma\to\varepsilon} = \Delta G^{\gamma\to\varepsilon}_{FeMnX} + x_c \Delta G^{\gamma\to\varepsilon}_{FeMnX/C} + \Delta G^{\gamma\to\varepsilon}_{mg} \quad (2)$$

The calculated SFE values for Fe-27Mn-1-Si-C steel at cryogenic temperatures (T = -196 ° C) were approximately 5.0 mJ/m^2 and at room temperature of 10.0 mJ/m^2.

Microstructural analysis

Fig. 1 shows the micrographics of the samples 80-60% deformed/solution annealed with water cooling. It can be observed that there was an increase of ε-martensite with water cooling. It is possible to observe the twins from the heat treatment (Annealing Twins) that differ from the twins originated from the high deformation. It is possible to affirm that it is common in some FCC metals (structure of the study material) the presence of twins from the heat treatment.

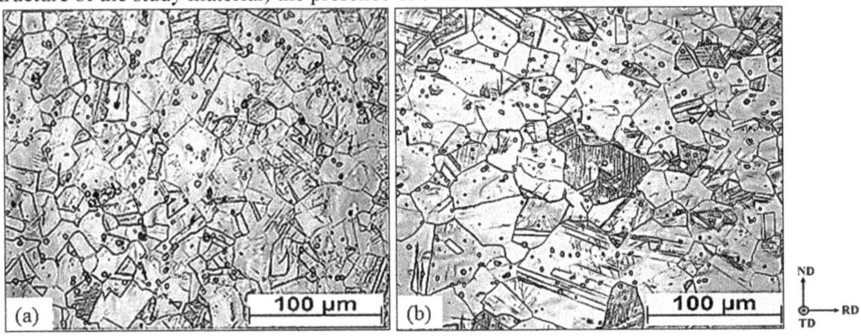

Fig.1. Micrographs of samples deformed with water cooling, 80% deformed (a) and 60% deformed (b).

Unlike the deformation mats, the annealing mats do not occur by shearing, but as irregularities in the growth of deformation-free grains from deformed grains. In the FCC metals, the contour of the blade is from the family (111), and a blade starts when a plan (111) fits the underlying plan in a local configuration of the type HC, rather than the continue FCC configuration.

Still in Fig. 1, microporosities can be observed. When the porosities appear with a white or light green region, it is because the alumina used during the polishing phase in the metallographic preparation covers the micropores. Stains from electrolytic polishing are also observed.

A high grain contour speed (in the fast cooling) favors the formation of annealing mats because the fault of growth in the heat treatment, which are responsible for the formation of mats, are then more frequent. Of course, a low SFE also favors the formation of heat treatment mats (Annealing Twins), since growth fault is easier to achieve [22,23].

After analysis by optical microscopy, grain sizes were calculated for all experimental conditions, as can be observed in Table 2. The method used was the measurement of the intercepts according to ASTM E112 – 13 [24]. The quantification of SFE was calculated by the proposed thermodynamic mathematical model and used by Allain et al [21] for the Fe-Mn and FeMnC system [20]. The values of free energy, chemical composition, temperature and grain size ASTM of high steel Mn were used. The SFE calculation was carried out using a program specially developed in Microsoft Excel, adapting it to the alloy under study.

Table 2. Grain size of the 80-60% deformed / solution annealed rolled samples with water cooling and the stacking fault energy for the Fe27Mn1Si alloy.

Condition	Cooling	Number of intercepts	Grain size (μm)	Error (+/-)
Deformed 80%	Water	183	51	0,2
Deformed 60%	Water	157	63	0.4
SFE (25°) ≅	10 mJ/m^2		-	-
SFE (-196°) ≅	5 mJ/m^2		-	-

In Fig. 2 it is possible to observe in detail the presence of ε-Martensite inside the grains, as well as the absence of it in others. In these cases, it can be explained by the absence of the preferential plan for this metastable phase to occur. There was no shear of the plan (111) - γ for the (100) - ε, for grains without ε-Martensite. In Fig. 2, the microstructure of the 80-60% hot-deformed sample can be observed. Still in Fig. 2, the image of a stunted grain with the presence of ε-Martensite in its interior can be better observed. Oxides can be observed in the sample microstructure, which in turn can be detrimental to the steel properties such as corrosion resistance and mechanical properties.

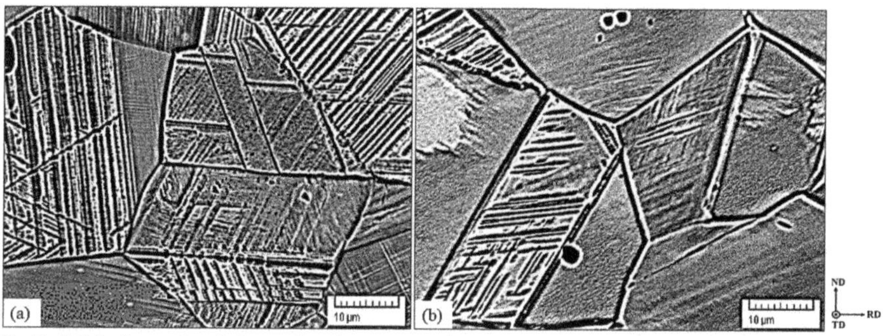

Fig.2. SEM of samples deformed with water cooling, 80% deformed (a) and 60% deformed (b).

The EBSD technique was an important evaluation for the identification and quantification of the phases present in the samples. In addition to the phase identification, it is possible to measure the phase quantity, and also plot the reverse pole figure map (IPF) for the analyzed micro-region, in addition to identify the preferred crystallographic orientation of each phase under study. Fig. 3 shows it that an increase of the martensitic phase occurred. This evidences the appearance of more epsilon martensite, even when the lamination is hot with subsequent cooling in water. For this and other reasons described in the text it is extremely important for research and industry to conduct research on the martensitic transformations of austenitic steels with high levels of Manganese. EBSD maps were also obtained with a high percentage of indexing, increasing the reliability of the technique.

In Fig. 3, one can see the austenitic and martensite grains. In Fig. 3a one can observe the IPF-map. In color, the grains of the austenite phase and in white the ε-martensite phase. In Fig. 3b one can observe the IPF-map, colored for the ε-martensite phase and in white, the grains of austenite phase. In Fig. 3c one can observe the identification of the phases, 90% austenite - and 10% ε-martensite.

Materials Research Proceedings **32** (2023) 319-329
https://doi.org/10.21741/9781644902615-37

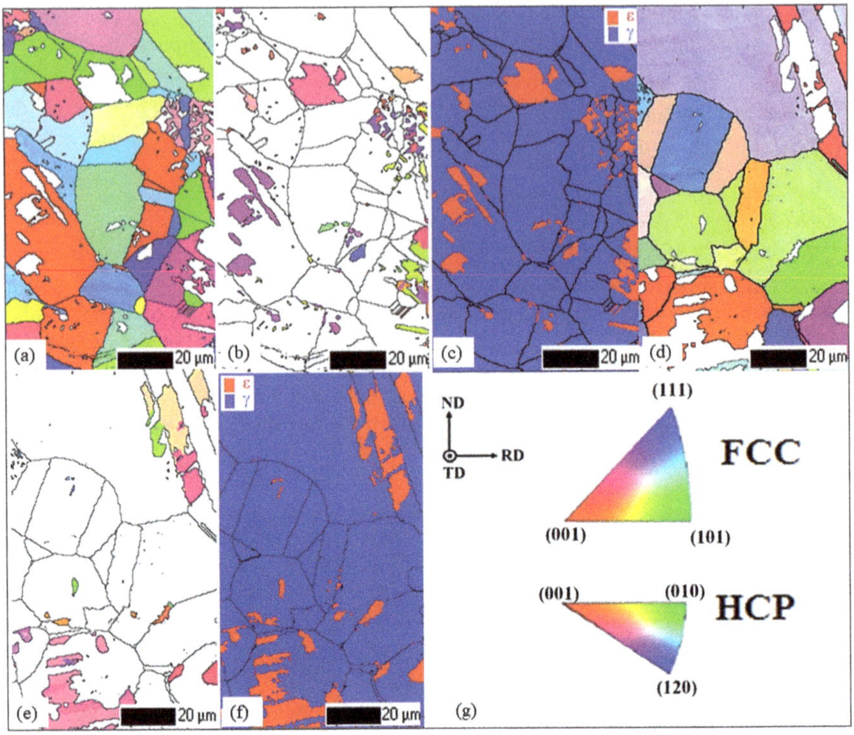

Fig.3. EBSD maps of samples deformed with water cooling. 80% (a) IPF-γ, (b) IPF-ε, (c) phase map, 60% (d) IPF-γ, (e) IPF-ε, (f) phase map, (g) corresponding standard IPF color triangle for γ and ε.

In Fig. 3d, the IPF-γ can be observed, in color the grains of austenite phase and the ε-Martensite phase in white. In Fig. 3e one can observe the map IPF-ε, in color the ε-Martensite phase and in white the grains of the austenite phase. In Fig. 3f one can observe the identification of phases, 89% austenite-γ and 11% ε-Martensite.

Corrosion analysis

The variation of OCP as a function of time for the two Mn high steel conditions evaluated in 0.1 M NaCl solution is shown in Fig. 4a. The curve presented for each condition of Fe27Mn1Si alloy was chosen as the most representative of the triplicate tests performed of each thermomechanical treatment condition in which the alloy was subjected. Fig. 4a shows that the steel with 60% deformation had slightly more noble behavior than the deformed 80% one. Although the values of OCP potentials are close, it is important to remember that it is the same steel, and that the difference in deformation could influence the electrochemical behavior, due to the TWIP phenomenon, more grain limits on the surface of the steel bring a greater area reactive susceptible to corrosion [14,15].

The 60% deformed sample had a decrease in potential in the first minutes of the test, and had variations until stabilized in 2000 s. Similarly, 80% deformed steel had variations with the difference that this steel was able to stabilize faster. This variation in potential is due to electrochemical interactions between the surface of the sample and the solution, that is, the formation of the double electrochemical layer until the steady state equilibrium is found [14,15].

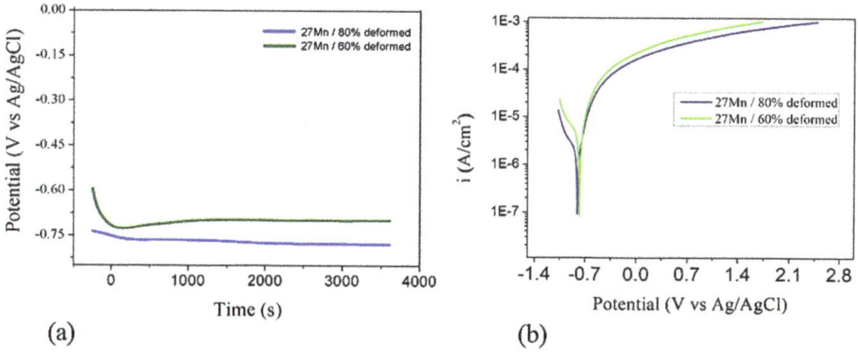

Fig.4. OCP (a) and anodic polarization curves (b) of samples deformed with water cooling (80-60%) deformed in 0.1M NaCl solution.

Fig. 4b shows the potentiodynamic polarization curves in 0.1 M NaCl solution for each Fe27Mn1Si alloy condition (80% and 60% deformed). It can be observed that high Mn steels do not exhibit passivating behavior. This is indicative of the low corrosion resistance of these steels in this medium, as also mentioned by Hyung et al [14-16]. Polarization curves show that the steels behaved similarly in both cathodic and anodic branches. There are slight variations in the current in the cathodic branch, but it is generally identified that the reaction is diffusion-controlled [14,15,25,27].

In the anodic branch, the behavior between them was similar, active dissolution of the material, as presented by Cerra et al [14,15]. More evident variations were found in the corrosion current (i_{corr}), it is possible to observe in the graph a lower corrosion rate in the deformed steel 60%. After the analysis with the Taffel curves, it was possible to obtain the icorr values, indicated in Table 3; these values confirm the less noble character of steel with 80% deformation. These differences found could be associated with the applied deformation (TWIP and TRIP effects) depending on the fact that it is the same steel (with the same chemical composition).

In our understanding, the differences in the results both in OCP and in the polarization tests could be associated with the deformation employed, thus, with the increased deformation, the corrosion resistance can decreases. This reasoning are due to the larger amount of grain boundaries per area (TWIP effect), to the amount of porosity and the greater presence of ε-martensite which contribute to higher surface reactivity and a faster surface oxidation [14,15,27,28], this was confirmed with the corrosion potential (icorr) shown in Table 3, showing a slightly less active surface (60% deformed steel) compared to 80% deformed steel.

Table 3. *Corrosion Potentials (Ecorr) of Fe27Mn1Si (80-60%) deformed in 0.1M NaCl solution.*

Corrosion parameters of the Fe27Mn1Si alloy		
	80% Deformed	60% Deformed
Ecorr (V)	-0.744	-0.776
Icorr (A/cm^2)	3.3E-5	1.8E-5

The measured pH of the 0.1 M NaCl solution before and after the tests was close to seven. The pH value of 7.0 was used to study and analyze the Pourbaix diagrams of the alloying elements of this steel. With the pH and the measured potentials, it was found that in steels with high manganese content, the elements Mn and Fe have a greater tendency to pass to solution as ions than to remain in the metal matrix, besides forming non-protective oxides on the surface in aqueous solutions [13-15,17,25,26,29].

In Fig. 5a, one can see an image obtained by SEM of the 80% deformed high manganese steel sample after potentiodynamic polarization tests. A layer of corrosion products can be observed across the sample surface.

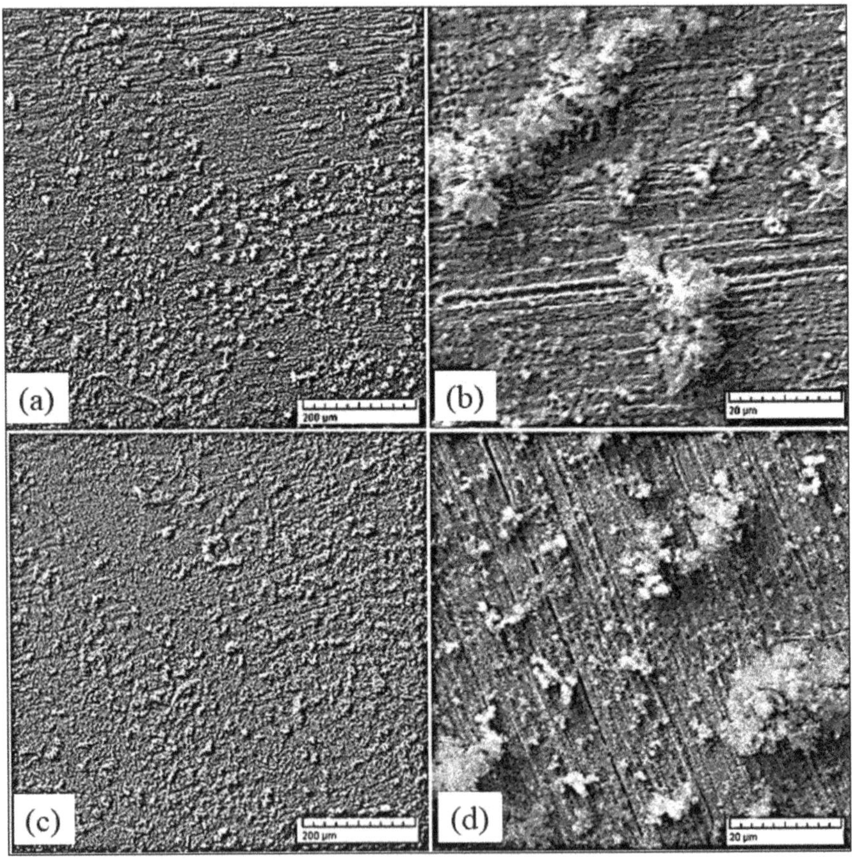

Fig.5. SEM of corrosion products in 0.1M aqueous NaCl solution of samples deformed with water cooling 80% deformed (a) and (b), 60% deformed (c) and (d).

In Fig. 5b, one can see an image obtained in SEM, where it is possible to identify the layer of corrosion products formed on the surface. These corrosion products (normally formed by the oxides γ-FeOOH (lepidocrocite) and α-FeOOH (goethite)) pass directly to the electrolyte, and this was observed by the yellow color (green rust) of the solution after the test as described by Reza et

Materials Research Forum LLC
https://doi.org/10.21741/9781644902615-37

al [26] and Cerra et al [14,15]. Still in Table 4, it can observe the result of the EDS punctual measurement of the corrosion products of high manganese steel for the conditions studied in this work. The composition was observed through the peaks of the elements Fe, Mn, Si and O (It is essential to indicate that oxygen was only taken into account for comparative purposes). Fig. 5c shows an image obtained by SEM of the sample of high manganese steel, after the potentiodynamic polarization tests. Corrosion products may be observed on the sample surface. Fig. 5d shows the corrosion products and the sanding marks in more detail. Comparing with the previous condition, a layer of corrosion products on the surface was also formed, and dissolved in the electrolyte, which was verified by the yellowish coloration of the solution after the polarization test. In Fig. 5d, one can see the result of the EDS spot measurement of $Fe_{27}MnSi$ alloy corrosion products.

Table 4. EDS of corrosion products in 0.1M solution of Fe27Mn1Si NaCl, 80% deformed and 60% deformed, respectively.

Elements	C Norm. [%wt] (80% def)	C Norm. [%wt] (60% def)
O	31.12	29.32
Si	1.45	1.38
Mn	13.20	13.58
Fe	54.23	55.72

Table 4 shows the quantification of the measurement for the elements made in the EDS analysis of the corrosion products mentioned above. Analyzing the EDS results made in the corrosion products, it was found that the layer contains iron, manganese, silicon and oxygen, which indicates that there are mainly oxides of these elements, as mentioned by Cerra et al [14,15].

Small peaks of the elements manganese and silicon were found, which also form oxides on the sample surface. It is important to mention that unlike Ni and Cr alloys, as mentioned earlier, the high manganese content negatively influences corrosion resistance. Manganese tends to pass into the solution as ions, leaving spaces in the matrix that are used by the electrolyte to corrode the steel [13-15,17,26,28].

Conclusions

Based on the experimental results obtained, it is possible to conclude that the FeMnSi steel presents well-activated transformation mechanisms (Austenite-γ → ε-Martensite), TRIP effect. The microstructure of the FeMnSi steel showed austenitic grains, microporosities from the smelting process. It also presented heat treatment twins, difference in grain size and percentage of phase before and after processing. The ε-martensite phase was found in greater proportions after deformation and heat treatment with cooling in water. It is possible to observe the nucleated martensitic phase within the grains due to the shear planes. This shear is due to the low stack failure energy and the thermal and mechanical stresses to which the alloy was subjected.

The phase transformation occurs from the shear of the FCC => HC structure planes and the high displacement density. Regarding the corrosion resistance of the high manganese steel under study, it did not show passivation due to the large amount of Mn in its composition, which in aqueous solutions, this element has an affinity to leave the metallic matrix and pass as ions, in addition to forming non-protective oxides on the surface of the metal sample.

The deformation of the steels influenced the reduction of the corrosion potential, due to the greater amount of grain contours per area, the amount of micropores and the greater presence of ε-martensite produced during the mechanical process.

The corrosion products formed on the surfaces of high manganese steels do not offer protection due to the presence of undesirable phases and high porosity, which allows electrolyte ions to reach the metal surface, accelerating the corrosion process of the steel.

Acknowledgments
The authors would like to thank Funcap (Cearense Foundation to Support Scientific and Technological Development) for the financial support for this research. The authors would like to thank the Central Analítica-UFC (funded by Finep-CT-INFRA, CAPES-Pró-Equipamentos, and MCTI-CNPq-SisNano2.0) for microscopy measurements, the LACAM-UFC and LPC-UFC Research Groups for technical support.

Declaration of competing interest
This work has no competing of interests or personal relationships that can influence it.

References

[1] Marquardt ED, Le JP, Radebaugh R. Cryogenic material properties database, in: Cryocoolers 11, Springer, 2002:681-687. https://doi.org/10.1007/0-306-47112-4_84

[2] Mi Z, Tang D, Jiang H, Dai Y, Li S. Effects of annealing temperature on the microstructure and properties of the 25Mn-3Si-3Al TWIP steel, Int J Miner. Metall Mater. 2009; 16:154-158. https://doi.org/10.1016/S1674-4799(09)60026-1

[3] Dafé SSF, Sicupira FL, Matos FCS, Cruz NS, Moreira DR, Santos DB. Effect of cooling rate on (ε, α') martensite formation in twinning/transformation-induced plasticity Fe-17Mn-0.06 C steel, Mater Res 2013; 16:1229-1236. https://doi.org/10.1590/S1516-14392013005000129

[4] Talonen J, Hänninen H, Nenonen P, Pape G. Effect of strain rate on the strain-induced γ→ α′-martensite transformation and mechanical properties of austenitic stainless steels, Metall Mater Trans. A. 2005;36: 421-432. https://doi.org/10.1007/s11661-005-0313-y

[5] Hamada AS, Karjalainen LP, Somani MC. The influence of aluminum on hot deformation behavior and tensile properties of high-Mn TWIP steels, Mater Sci Eng. A. 2007; 467:114-124. https://doi.org/10.1016/j.msea.2007.02.074

[6] Jung JE, Park J, Kim JS, Jeon JB, Kim SK, Chang YW. Temperature effect on twin formation kinetics and deformation behavior of Fe-18Mn-0.6 C TWIP steel, Met Mater Int. 2014;20: 27-34. https://doi.org/10.1007/s12540-014-1008-y

[7] Cina B. A transitional hcp phase in the γ→ α transformation in certain Fe-base alloys, Acta Metall. 1958; 6:748-762. https://doi.org/10.1016/0001-6160(58)90050-6

[8] B. Cina, Effect of cold work on the gamma-alpha transformation in some Fe-Ni-Cr alloys, J Iron Steel Inst. 1954; 177:406.

[9] Cullity BD, Stock SR. Elements of X-ray Diffraction, Prentice hall New Jersey, 2001.

[10] Pardal JM, Tavares SSM, Fonseca MPC, da Silva M.R, Neto JM, Abreu HFG. Influence of temperature and aging time on hardness and magnetic properties of the maraging steel grade 300, J Mater Sci. 2007; 42:2276-2281. https://doi.org/10.1007/s10853-006-1317-8

[11] Gorni A, da Silveira JD. Accelerated cooling of steel plates: the time has come, in: Quenching Cool. Residual Stress Distortion Control, ASTM International, 2010.

[12] Bai Y, Bai Q. Subsea pipelines and risers, Elsevier, 2005.

[13] Zhang YS, Zhu XM, Liu M, Che RX. Effects of anodic passivation on the constitution, stability and resistance to corrosion of passive film formed on an Fe-24Mn-4Al-5Cr alloy, Appl Surf. Sci. 2004; 222: 89-101. https://doi.org/10.1016/j.apsusc.2003.08.068

[14] Florez M.A.C., Lima, M. N. D. S., Araújo, W. S., & Silva, M. J. G. D. Characterization and Comparative Analysis of Corrosion Resistance of 4 High Manganese Steels Models in Aqueous Solution of NaCl. Mater Res. 2019, 22. https://doi.org/10.1590/1980-5373-mr-2019-0283

[15] Cerra Florez, M. A., Cardoso, J. L., Ferreira Gomes de Abreu, H., Araújo, W. S., & Gomes da Silva, M. J. Comparative study of corrosion resistance between four non-commercial high manganese steel models and 9% nickel steel in aqueous solution of H2SO4. Int J Mater Res, 2020, 111(8), 661-667. https://doi.org/10.3139/146.111925

[16] H.S. Seo, J.H. Park, J.Y. Park, J.G. Choi, J.K. Kim. Presented at Twenty-Fourth Int. Ocean Polar Eng. Conf., International Society of Offshore and Polar Engineers, 2014.

[17] Hamada AS, Karjalainen LP. Corrosion Behaviour of High-Mn TWIP Steels with Electroless Ni-P Coating, Open Corros. J. 2010;3:1-6. https://doi.org/10.2174/1876503301003010001

[18] Aperador WA, Ruiz JHB, Betancurt JD. Evaluation of erosion-corrosion resistance in Fe-Mn-Al austenitic steels, Mater Res. 2013;16:447-452. https://doi.org/10.1590/S1516-14392012005000179

[19] ASTM, Standard test methods and definitions for mechanical testing of steel products, ASTM A370. 2012.

[20] Dumay A, Chateau JP, Allain S, Migot S, Bouaziz O. Influence of addition elements on the stacking-fault energy and mechanical properties of an austenitic Fe-Mn-C steel, Mater Sci Eng A. 2008;483:184-187. https://doi.org/10.1016/j.msea.2006.12.170

[21] Allain S, Chateau JP, Bouaziz O, Migot S, Guelton N. Correlations between the calculated stacking fault energy and the plasticity mechanisms in Fe-Mn-C alloys, Mater Sci Eng A. 2004; 387:158-162. https://doi.org/10.1016/j.msea.2004.01.059

[22] Callister Jr WD, Rethwisch DG. Fundamentals of materials science and engineering: an integrated approach, John Wiley & Sons, 2012.

[23] Rath BB, Imam MA, Pande CS. Nucleation and growth of twin interfaces in FCC metals and alloys, Mater Phys Mech. 2000;1: 61-66.

[24] A. Standard, E112-13. Standard test methods for determining average grain size, ASTM International, West Conshohocken, (n.d.).2013.

[25] Opiela M, Grajcar A, Krukiewicz W. Corrosion behaviour of Fe-Mn-Si-Al austenitic steel in chloride solution, J Achiev Mater Manuf Eng.2009;33: 159-165.

[26] Razavi GR, Gholami H, Zirepour GR, Zamani D, Saboktakin M, Monajati H. Study corrosion of high-Mn steels with Mo in 3.5% NaCl solution, Conf Adv Mater Eng. IPCSIT, 2011. https://doi.org/10.4028/scientific5/AMR.457-458.334

[27] Lasek S, Mazancová E. Influence of thermal treatment on structure and corrosion properties of high manganese triplex steels, Metalurgija. 2013;52:441-444.

[28] Grajcar A, Kciuk M, Topolska S, Płachcińska A, Microstructure and corrosion behavior of hot-deformed and cold-strained high-Mn steels, J Mater Eng Perform.2016; 25: 2245-2254. https://doi.org/10.1007/s11665-016-2085-5

[29] Fajardo S, Llorente I, Jiménez JA, Bastidas JM, Bastidas DM. Effect of Mn additions on the corrosion behaviour of TWIP Fe-Mn-Al-Si austenitic steel in chloride solution, Corros Sci. 2019;154: 246-253. https://doi.org/10.1016/j.corsci.2019.04.026

Hot plastic behavior of an ultrafine-grained aluminum alloy fabricated by laser powder bed fusion and equal channel angular pressing

Jairo Alberto Muñoz[1,a*], Louis Huvelle[2,b], Enrique Manuel Huerta[1,c], and José María Cabrera[1,3,d]

[1]Department of Materials Science and Engineering EEBE, Universidad Politécnica de Catalunya, c/Eduard Maristany 10-14, 08019 Barcelona, Spain

[2]Ecole Européenne d'Ingénieurs en Génie des Matériaux EEIGM, Université de Lorraine, 6 Rue Bastien-Lepage, 54010 Nancy, France

[3]Fundació CIM-UPC, c/Llorens i Artigas 12, 08028 – Barcelona, Spain

[a]jairo.alberto.munoz@upc.edu, [b]louis.huvelle1@etu.univ-lorraine.fr, [c]enrique.manuel.huerta@upc.edu, [d]jose.maria.cabrera@upc.edu

Keywords: Additive Manufacturing (AM), Severe Plastic Deformation (SPD); Al Alloy, Plasticity

Abstract. This research presents the microstructural evolution and mechanical properties of a hypoeutectic AlSi11Cu alloy obtained through laser powder bed fusion (L-PBF) and subsequent severe plastic deformation. The as-built alloy demonstrated a structure formed by an Al matrix surrounded by a Si-enriched cellular network. Tensile tests indicated a yield strength of 350 MPa and elongation to fracture lower than 5% for the as-built material. After subjecting the alloy to severe plastic deformation (SPD) using equal channel angular pressing (ECAP), superior mechanical properties such as an elongation almost twice as high as the as-built condition (12% compared to 6%) and a high tensile yield strength (320 MPa) were observed. ECAP produced average grain size reduction from 10 μm in the as-built state to 1 μm after six ECAP passes. Microstructural analyses highlight in particular the grain refinement of the microstructure after the ECAP process, changing from a microstructure composed of columnar grains to a heterogeneous microstructure characterized by ultra-fine (grain sizes between 200 nm - 500 nm) and elongated grains (grains between 5 μm – 10 μm). This finding supposes a huge improvement for the mechanical performance of this alloy breaking the strength-ductility paradox. The superplastic properties of the materials were characterized, depending on the variation of the temperature and strain rate parameters. This showed that the superplastic behavior was favored by low strain rates (here 0.001 s^{-1}), and by increasing temperature (400 °C). Thus, elongations exceeding 70% were achieved with ECAP material.

Introduction

Throughout human history, great efforts have been made to develop structural materials with optimized mechanical properties for applications in different areas of industry, such as civil, medical, transportation, aerospace, and energy. Also, several manufacturing processes have been stablished amount them additive manufacturing (AM) and severe plastic deformation (SPD). AM also known as 3D printing processes have been attracting attention in recent years, because it could easily be adapted to generate alternative processing routes. AM is an industry that has existed for more than 30 years and has rapidly gained public and commercial interest due to its advantages in printing complex geometries without using assembly processes [1]. The development of these technologies is driven by a reduction in manufacturing cycle times, improved durability but also a reduction in manufacturing costs. These development paths have enabled additive manufacturing to find its way into many sectors such as aeronautics, aerospace, automotive and biomedicine [2].

Materials Research Forum LLC
https://doi.org/10.21741/9781644902615-38

According to ISO/ASTM 52900 [3], additive manufacturing is characterized as a process of assembling materials to make parts from 3D model data. Currently, the American Society for Testing and Materials (ASTM International) groups additive manufacturing into seven different families: Binder jetting, Directed energy deposition (DED), Material extrusion, Material jetting, Laser Powder Bed Fusion (L-PBF), sheet lamination, and vat photopolymerization. L-PBF is considered today as one of the most promising 3D printing techniques. This process, which takes place in a controlled atmosphere (e.g., argon) at high heating and cooling rates (e.g., 10^3 until 10^8 K s^{-1}), uses a high-energy laser beam to completely melt the metal powder [4]. This technique allows the production of very high-quality parts with excellent mechanical properties due to the heterogenous and complex microstructures [5]. However, the thermal gradients and the repetitive fusion of powder layers give rise to high residual stresses, lack of fusion, and porosity that affect funtional properties like mechanical strength and ductility [6,7].

On the other hand, metallic materials produced by severe plastic deformation (SPD) techniques possessing ultrafine grain structures have been shown to have unique mechanical, chemical and physical properties that can lead them to meet the demands of different industries. SPD processes are defined as metal forming processes in which a large deformation is introduced into a metal in order to produce metals with ultrafine grain sizes. The advantage of this type of grain refinement techniques is the attractive mechanical properties of the produced parts, such as high strength and a remarkable lightness, which are generally superior than their coarse-grained counterparts. As well as providing a significant improvement in material properties, SPD is a relatively simple process and there have been many ways in which the process has been improved to suit specific shapes and large parts [8,9]. Although ultrafine-grained (UFG) materials can exhibit a 2-3-fold increase in strength over material in the annealed state, it is also true that UFG materials exhibit mechanical inconvenience due to lack of strain hardening. This lack of hardening promotes limited homogeneous ductility. For this reason, the search for a high ductility combined with a high strength have been the objective of many investigations carried out during the last years [10,11]. In this context, there are several properties (e.g., microstructure and tensile behavior at high homologous temperatures), types of materials and new manufacturing routes such as the combination of AM and SPD that have not yet been studied and explored in great detail. Therefore, these topics represent a new field of study in search of materials with outstanding properties that require attention from the research community.

This research work brings together several well-known technologies: additive manufacturing, using the Laser Powder Bed Fusion (LPBF) method, which is a technology with great potential for large-scale use in industry, severe plastic deformation (SPD), using the Equal Channel Angular Pressing (ECAP) process, allowing materials to have novel microstructural and mechanical properties. Therefore, microstructural characterization using Electron Back Scattering Diffraction (EBSD) and tensile tests at room and hot temperatures for the as-built, and ECAP processed conditions were analyzed.

Materials and methods

The studied material is an almost eutectic aluminum-silicon alloy AlSi11Cu manufactured by L-PBF with the chemical composition indicated in Table 1. The metallic powders used to print the material were produced at the Volgograd Aluminium Plant (UC RUSAL), in Russia, using the molten metal spraying method. 20 mm diameter and 100 mm length rods were manufactured at the National University of Science and Technology (MISIS) in Moscow (Russia) using a selective laser melting (SLM) SLM280HL printer with the following operating conditions: scan speed of V = 1650 mm/s, layer thickness H = 50 μm, power of W = 370 W, and hatch spacing of 130 μm. Different samples for mechanical tests and microstructure characterization were cut-out of the rods using electrode discharge machining (EDM).

Table 1 Chemical composition of the studied alloy.

Element	Si	Cu	Mn	Mg	Ti	Zn	Fe	Pb	Ni	Ca	Al
Wt. %	10.6	0.78	0.54	0.47	0.25	0.21	0.03	0.03	0.01	0.01	Bal.

For the ECAP process an hydraulic press of 80 Tons was used. The ECAP die was designed with an inner angle of 120°, and the material was processed at 250 °C for six ECAP passes. This temperature was used because at room temperature the ECAP process produced cracks in the samples. Route B_C and Molybdenum disulfide (MoS_2) as lubricant were used to process the samples.

To characterize the alloy microstructure, a scanning electron microscope (SEM) Jeol, model JSM-7001F coupled with an EBSD detector operating at 15 eV and probe current of 14 A was used. Sample surfaces were prepared by mechanical polishing using carbon silicon sand papers and diamonds suspensions of 9 µm, 6 µm, 3 µm, and 1 µm. A final polishing step with colloidal silica of 0.03 µm particle size on a vibratory machine for 3 h was carried out to get a smooth surface free of scratches. Data was acquired with HKL Channel 5 software and analyzed with TSL-OIM 7.3b. Grains with less than 2 pixels were excluded from any statistical analysis. Optical microscopy (OM) and Kellers reagent were used to revel the melting pools before and after ECAP processing.

The mechanical properties of the specimens at room temperature were tested by means of uni-axial tensile tests using the Deben Microtest Tensile Stage machine. To measure the strain during the tensile tests, a so-called digital image correlation (DIC) non-contact technique was used. DIC analysis were carried out using a video correlation system equipped with one camera Balser of 5 MP and resolution of 2448 x 2048 at 75 fps. For a good illumination, one LED lightning grids was used. The deformation data was acquired using the Software VIC Gauge 2D for video extensometer purposes. The hot tensile tests were carried out using the Instron 5407 machine, coupled with a water-cooled infrared (IR) heater. Different temperatures and strain rates were analyzed.

Results and discussion

Fig. 1 gives the feedstock and as-built (AB) material characteristics using OM and SEM. Fig. 1a shows the powders spherical particles using SEM. With this image, some characteristics of the powders are obtained, such as the size of the particles and their size distribution, represented on Fig. 1b. Particles show an average size of 24.5 µm with a standard deviation of 8,8 µm. Some granulometric values are also calculated such as $d10$ = 13.6 µm, $d50$ = 23.6 µm, and $d90$ = 36.8 µm. This particle size distribution according to several researches is appropriate for a good powder flowability during the LPBF process [12]. Thus, powders can spread evenly on the powders bed, according to EPMA (European Powder Metallurgy Association (EPMA) and EuroAM (European Additive Manufacturing Group) [13]. Fig. 1c shows several layers of melting pools (MP) for the as-built material, which is the result of the selective fusion of powders during the L-PBF process. At high magnifications, SEM allows observing microstructural characteristics inside both the MP and the melting pool boundaries (MPB), as indicated in Fig. 1d and Fig. 1e. At this magnification, the structure of the alloy reveals an eutectic light-colored interconnected network (Si-rich zone) and a dark-colored cellular matrix (Al matrix). Inside the MPB, the interconnected network is expanded, elongated, unlike inside the MP, where the interconnected network is less expanded, showing a more equiaxed structure, as corroborated by Fig. 1f and Fig. 1g, respectively. This difference in structure may be due to melting/solidification/re-melting cycles between the printing layers, so some areas are subject to temperature gradients that reach higher values during printing. As MPBs are the boundary of two melting zones, they are more affected by this phenomenon.

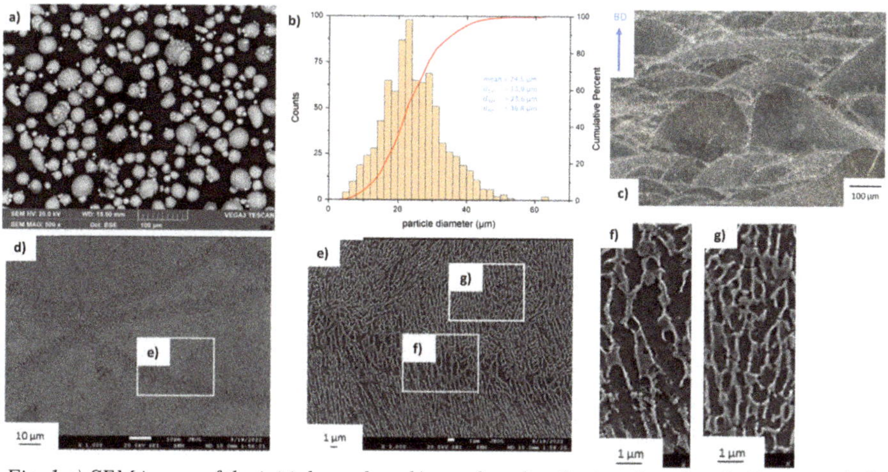

Fig. 1 a) SEM image of the initial powders, b) powders distribution size, c) as-built material, d) SEM image for the as-built material, e) SEM image close to one melting pool boundary, f) SEM image inside the melting pool boundary, and g) SEM image inside the melting pool.

Fig. 2a and Fig. 2b show the optical microstructure after SPD by 6 ECAP passes. Some differences on the optical microstructure are observed with respect to the AB condition. The ECAP process produced a shear effect on the optical microstructure, which is now characterized by elongated and rotated MPs. In this figure, significant microstructural changes can be observed. MPs are still visible and their height have been significantly reduced. Fig. 2c highlights that ECAP processing has resulted in significant microstructural changes. For example, the Si-rich network is destroyed after ECAP processing and the Al cellular structure is not as evident as in the AB condition, giving rise to a microstructure with a high density of particles.

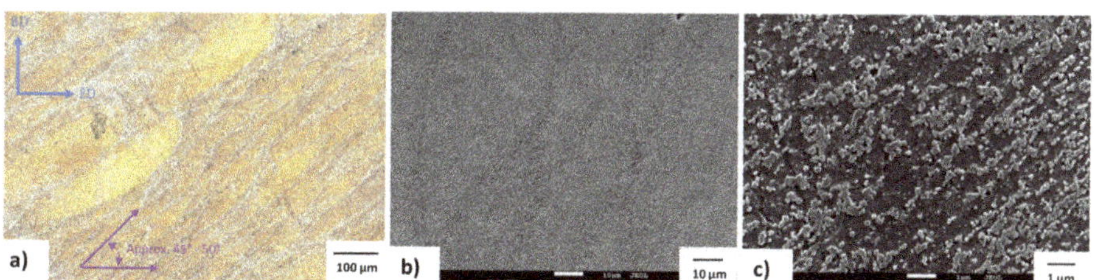

Fig. 2 a) Optical micrography for the ECAP processed material, b) SEM image after six ECAP passes, and c) cellular structure breakdown after six ECAP passes.

The microstructural characterization of the AB and ECAP materials can be followed by EBSD analyses, as shown in Fig. 3a and Fig. 3b, respectively. The low and high angle grain boundaries (LABG and HAGB) are also highlighted in white and black colors, respectively. The inverse pole figure (IPF) map of the AB material shows that the MPs present a morphology composed by columnar grains, while small and equiaxed grains are located around the MPBs. Thus, the grain size distribution for the AB material is represented by a single symmetric peak curve, suggesting an average grain size of ~10 µm. For the six ECAP passes material, different types of grains are visible on the IPF map of Fig. 3b. The grain size distribution on Fig. 3b suggests an average grain size of ~1 µm, which is 10 times smaller than the AB material. Although, there are still grains that have not been well refined, as the distribution shows the existence of some large grains that measure 20 µm approximatively. This behavior is coherent with the not-symmetric grain size distribution for the ECAP material concerning the like-Gauss function distributions observed on the AB condition, which corroborates a heterogeneous microstructure that mix ultrafine grains (UFGs) with coarse grains (CGs).

Fig. 4a and Fig. 4b show the tensile true and engineering stress-strain curves for all the studied conditions. The AB materials has a yield tensile strength (YTS) of 393 MPa, an ultimate tensile strength (UTS) of 565 MPa, a uniform elongation (Eu) and elongation at fracture (Ef) of 4.6% and 4.9%, respectively. It should be noted that the AB material has better mechanical properties than similar alloys produced by traditional forming processes (such as casting). Indeed, according to D. Song et al. such a cast alloy presents a YTS of 115 MPa, a UTS of 140 MPa, and elongation of 2,2% [14]. On the other hand, the ECAP processed material reached a YTS and UTS of 319 MPa and 494 MPa, respectively. The lower mechanical properties for the ECAP processed than the AB material can be related to the interconnected network breakdown, as shown in Fig. 2c. This behavior is related with the smooth strain hardening rate curve displayed on Fig. 4b for the ECAP processed material, while the AB condition presents a curve with multiple peaks leading to the strain localization according to the Considère criteria in Eq. (1) [15]. Thus, the interconnected network breakdown allows space for the dislocation motion and the new particles also generate a pinning effect maintaining a larger tensile ductility than the AB material.

$$\frac{d\sigma}{d\varepsilon} \geq 0. \qquad (1)$$

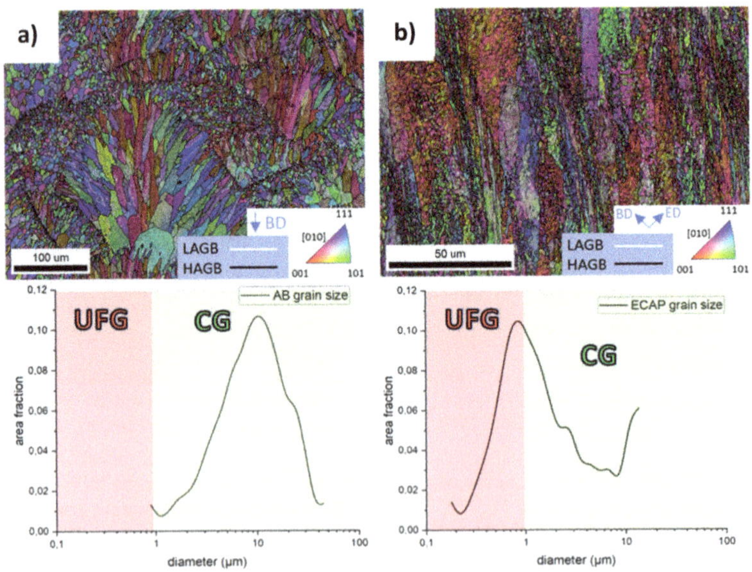

Fig. 3 EBSD analysis for the a) as-built and b) ECAP processed materials.

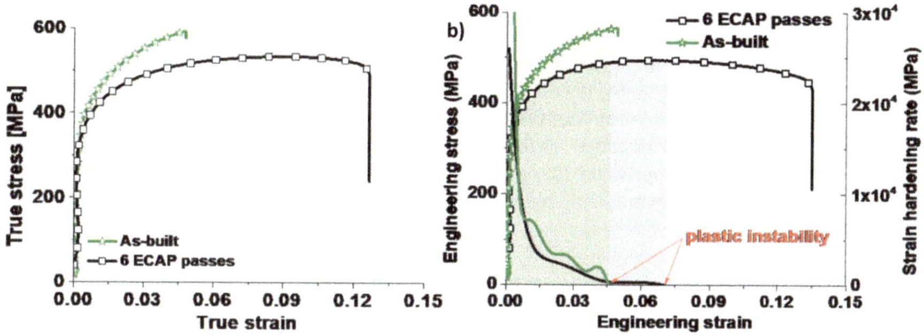

Fig. 4 a) True stress-strain curves and b) engineering stress-strain and strain hardening rate curves for the analyzed materials.

Fig. 5 indicates the hot tensile curves and some microstructures for different temperatures and strain rates. As the temperature increases the maximum stress decreases and elongation increases. This is well correlated with the grain size growth as the temperatures used correspond to homologues temperatures of $0.52T_M$, $0.61T_M$, and $0.69T_M$ the alloys melting temperature (T_M). Therefore, recovery and recrystallization phenomena occurred, as observed on the EBSD maps and tensile curves. At temperatures between 300 °C and 350 °C the EBSD maps on Fig. 5a and Fig. 5b show equiaxed grains with heterogeneous grain sizes, while at 400 °C a clear grain growth together with grains smaller than 1 μm are observed on Fig. 5c. Thus, the ECAP processed condition reached elongations around 70%, but did not show any superplastic behavior after 6 ECAP passes. This could be attributed to the heterogeneous microstructure, which was not fully UFG like the research work of Jiang Jing-hua et al. [16] on an as-cast AlSi11 alloy processed by sixteen ECAP passes with an average grain size of 300 nm. However, it can be stablished that the ECAP processing after the LPBF process is a good processing route to improve the strength-ductility of AM materials.

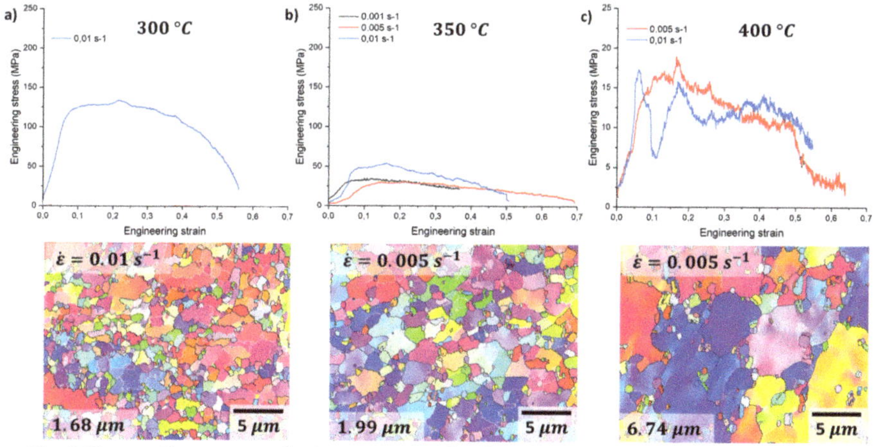

Fig. 5 Hot tensile curves and microstructures for the six ECAP passes material a different temperatures and strain rates, a) 300 °C, b) 350 °C, and c) 400 °C.

Summary
The ECAP processing of an AlSi11Cu alloy manufactured by L-PBF produced a heterogeneous microstructure with an excellent combination of strength-ductility. The ECAP process transformed the interconnected network into particles and reduced the melting pools heigh due to the compression and shear strain components during the extrusion process. These changes conducted to a better tensile plasticity because of the increased space for the dislocation motion and the pinning effect with the new particles. In terms of superplastic behavior, the ECAP processed alloy did not reach elongations beyond 70%, but the grain size showed a good thermal stability until 350 °C.

References

[1] W.J. Sames, F.A. List, S. Pannala, R.R. Dehoff, S.S. Babu, The metallurgy and processing science of metal additive manufacturing, Int. Mater. Rev. 61 (2016) 315–360. https://doi.org/10.1080/09506608.2015.1116649

[2] M. Pérez, D. Carou, E.M. Rubio, R. Teti, Current advances in additive manufacturing, Procedia CIRP. 88 (2020) 439–444. https://doi.org/https://doi.org/10.1016/j.procir.2020.05.076

[3] I. Astm, ASTM52900-15 standard terminology for additive manufacturing—general principles—terminology, 2015.

[4] J. Gunasekaran, P. Sevvel, I. John Solomon, Metallic materials fabrication by selective laser melting: A review, Mater. Today Proc. 37 (2021) 252–256. https://doi.org/https://doi.org/10.1016/j.matpr.2020.05.162

[5] J.A. Muñoz, S. Elizalde, A. Komissarov, J.M. Cabrera, Effect of heat treatments on the mechanical and microstructural behavior of a hypoeutectic Al alloy obtained by laser powder bed fusion, Mater. Sci. Eng. A. 857 (2022) 144091. https://doi.org/https://doi.org/10.1016/j.msea.2022.144091

[6] W. Chen, T. Voisin, Y. Zhang, J.-B. Florien, C.M. Spadaccini, D.L. McDowell, T. Zhu, Y.M. Wang, Microscale residual stresses in additively manufactured stainless steel, Nat. Commun. 10 (2019) 4338. https://doi.org/10.1038/s41467-019-12265-8.

[7] Y. Li, K. Zhou, P. Tan, S.B. Tor, C.K. Chua, K.F. Leong, Modeling temperature and residual stress fields in selective laser melting, Int. J. Mech. Sci. 136 (2018) 24–35. https://doi.org/https://doi.org/10.1016/j.ijmecsci.2017.12.001

[8] A. Azushima, R. Kopp, A. Korhonen, D.Y. Yang, F. Micari, G.D. Lahoti, P. Groche, J. Yanagimoto, N. Tsuji, A. Rosochowski, A. Yanagida, Severe plastic deformation (SPD) processes for metals, CIRP Ann. 57 (2008) 716–735. https://doi.org/https://doi.org/10.1016/j.cirp.2008.09.005

[9] K. Edalati, A. Bachmaier, V.A. Beloshenko, Y. Beygelzimer, V.D. Blank, W.J. Botta, K. Bryła, J. Čížek, S. Divinski, N.A. Enikeev, Y. Estrin, G. Faraji, R.B. Figueiredo, M. Fuji, T. Furuta, T. Grosdidier, J. Gubicza, A. Hohenwarter, Z. Horita, J. Huot, Y. Ikoma, M. Janeček, M. Kawasaki, P. Král, S. Kuramoto, T.G. Langdon, D.R. Leiva, V.I. Levitas, A. Mazilkin, M. Mito, H. Miyamoto, T. Nishizaki, R. Pippan, V. V Popov, E.N. Popova, G. Purcek, O. Renk, Á. Révész, X. Sauvage, V. Sklenicka, W. Skrotzki, B.B. Straumal, S. Suwas, L.S. Toth, N. Tsuji, R.Z. Valiev, G. Wilde, M.J. Zehetbauer, X. Zhu, Nanomaterials by severe plastic deformation: review of historical developments and recent advances, Mater. Res. Lett. 10 (2022) 163–256. https://doi.org/10.1080/21663831.2022.2029779

[10] X. Wu, Y. Zhu, Heterogeneous materials: a new class of materials with unprecedented mechanical properties, Mater. Res. Lett. 5 (2017) 527–532. https://doi.org/10.1080/21663831.2017.1343208

[11] Y. Zhu, X. Wu, Heterostructured materials, Prog. Mater. Sci. 131 (2023) 101019. https://doi.org/https://doi.org/10.1016/j.pmatsci.2022.101019

[12] Additive Manufacturing Overview, in: Addit. Manuf. Process., ASM International, 2020. https://doi.org/10.31399/asm.hb.v24.9781627082907

[13] Metal Powders key features, (2020). https://www.epma.com/additive-manufacturing (accessed April 29, 2023).

[14] D. Song, G. Wang, Z. Zhou, E.E. Klu, B. Gao, A. Ma, Y. Wu, J. Sun, J. Jiang, X. Ma, Developing a high-strength Al–11Si alloy with improved ductility by combining ECAP and cryorolling, Mater. Sci. Eng. A. 773 (2020) 138880. https://doi.org/10.1016/j.msea.2019.138880

[15] A. Considère, Memoire sur l'emploi du fer et de l'acier dans les constructions, Ann. Des Ponts Chaussées 1 Sem. 9 (1885) 574–775.

[16] J. Jiang, A. Ma, N. Saito, A. Watazu, P. Lin, Y. Nishida, Effect of microstructures on superplasticity of Al-11%Si alloy, Trans. Nonferrous Met. Soc. China. 17 (2007) 509–513. https://doi.org/https://doi.org/10.1016/S1003-6326(07)60124-1

Materials Research Proceedings **32** (2023) 338-354 https://doi.org/10.21741/9781644902615-39

Improved stochastic dissimilar diffusion bonding model with experimental validation

Bryan Ferguson[1,a], Neha Kulkarni[1,b], D.G. Sanders[1,c], Eric Bol[1,d]*, and M. Ramulu[1,e]*

[1]Department of Mechanical Engineering, Box 352600, University of Washington, Seattle, WA 98195, USA

[a] bjferg@uw.edu, [b] nehamech2004@gmail.com, [c] southernstargold@comcast.net, [d] boled84@gmail.com, [e] ramulum@uw.edu

Keywords: Diffusion Bonding, Voids, Superplastic Forming, Statistical Modeling

Abstract. Diffusion bonding is a solid-state welding operation that has seen wide spread use in the aerospace industry, especially in combination with superplastic forming. It combines two relatively flat, clean surfaces at high temperature to create a near flawless weld over a large surface area. Modelling of diffusion bonding has been challenging due in part because of the larger variations in voids formed from the mating surfaces. This paper attempts to compensate for that inadequacy by implementing a stochastic diffusion bonding model based with theoretical voids formed from interacting surfaces. The model uses a statistical version of Pilling's model in combination with surface roughness based initial conditions that estimates the possibilities of voids formed. The results of the model are compared with experimental results for three different alloys at a variety of process conditions along with presenting an investigation of the mechanics of the model for future improvements.

Introduction

Diffusion bonding is a solid-state welding operation that welds two mating surfaces together at nearly the structural properties of the parent materials. The process is commonly used in the aerospace industry in combination with superplastic forming which allows one to selectively weld and form multiple sheets to create complex stiffened structures for considerable weight and maintenance savings [1]. It also has the advantage of being able to weld different metals together and/or dissimilar alloys to produce new material systems. The bonding process requires excellent surface preparation and occurs at temperatures around 50%-70% of the material melting temperature. The primary defects in diffusion bonded parts are voids that are formed at the interface between the surfaces due to slight misfit of the surface topography. These voids are closed via plastic deformation, power law creep, and several high temperature diffusive processes [2].

Plastic deformation, creep, and diffusion are all dependent on the temperature, the applied pressure, the bonding time, and the material properties of the alloys used. Additionally, the size of the voids determines how much material will need to transfer to close them. This results in a large dependence on the surface geometry which has been the subject of several studies [3–5]. The purpose of this research is to both improve upon the existing modeling, as well as to demonstrate a more accurate way of accounting for the initial void geometry by providing more complex information about the surfaces and the way voids are formed. This model specifically deals with titanium alloys that are important to aerospace due to their extremely high strength to weight ratio and corrosion resistance. Titanium has good fracture toughness and forms a variety of alloys with varied properties. Both similar and dissimilar alloy combinations were modeled since it is often necessary to bond dissimilar titanium alloys to take advantage of differences in formability, strength, and corrosion resistance depending upon varied performance requirements.

Prior attempts at providing an accurate model that can predict diffusion bonding processing conditions date back to the 1970s. These models can be separated into groups that deal with void closure with pressure sintering models, and groups that deal with void closure solely using diffusion and superplastic collapse. The pressure sintering models include surface diffusion around the void and model creep with power law creep, while the alternative uses superplastic instead of creep and do not include any surface diffusion. The other significant difference between models is how they model surfaces and void geometry. Hamilton published the initial attempt to solve the problem in 1973 and modeled the voids as infinitely long pyramidal shapes and the closure process as solely superplastic collapse [6].

Derby and Wallach provided the next major step that used Hamilton's initial geometry and approach and then applied pressure sintering equations to incorporate diffusion. The closure of voids was based on the mechanisms of initial plastic collapse of the contacting points, surface and volume diffusion into the void from surface sources, surface and volume diffusion from the bonding interface, and power law creep collapse [7–9]. More models by Takahashi and Guo used lenticular voids with a combination of diffusion and power law creep collapse [10,11]. Hill et al. provided improvements to the Derby model using the same mechanisms, but deriving the equations for elliptical geometry [2].

Pilling et al. created several models starting in 1984 specifically for Ti-6Al-4V (Ti-64). These models were derived from research on void formation during superplasticity and models the bonding behavior as closure by a combination of superplastic collapse and stress-based diffusion. The models neglected the bonding mechanisms of surface diffusion included in the Derby and Hill models, and the initial model geometry was cylinders on the diffusion bonding plane as voids [12]. The second model developed by Pilling used the same mechanisms, but with cylinders oriented perpendicular to the diffusion bonding plane [13]. This model is the basis for the current stochastic diffusion bonding model. Salehi and Pilling looked into the effects of using the same equations, but with different void geometry, and rewrote Pilling's second model equations for oblate spheroids instead of vertical cylinders [14]. Orhan [15] and Ma [16] created more complex versions of Ti-64 models that coupled the two theories together. Li created the first version of the probabilistic model based on the Pilling 1988 model [17]. The stochastic formulation of Pilling's model was expanded for dissimilar alloys by Kulkarni [18].

The most important characteristic of these models is their accuracy and ability to predict the bonding processing parameters. While most of the models make some comparison with experimental data, the models present their results in different ways and are for different materials and processing conditions making a direct comparison between models challenging. The models dealing with titanium bonding, the subject of this study, compare with experimental results without micrographic correlation. They use results from structural shear tests and compare the shear strength in binary form (bonded or not bonded for a processing condition for above or below 95% shear strength) to the calculated bonding time. However, there is no set standard for structural testing the shear strength in diffusion bonding. Several papers that have discussed different methods have pointed to the discrepancies in the testing and lack of ability to distinguish between higher bonding percentages [19]. It is therefore debatable that the results were accurate. More importantly, these models are not modeling the shear strength or structural characteristics of the void but instead just modeling void closure. A summary of all the different models that are compared with Pilling's experimental data are presented in Figure 1.

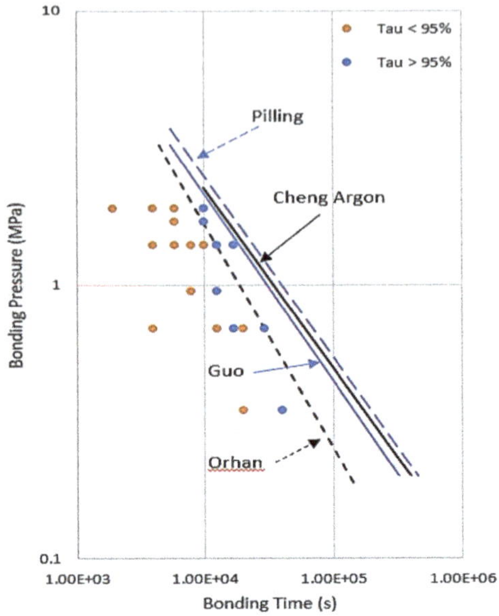

Figure 1: Bonding models compared with experimental results.

Experimental Methods

The model presented in this work uses a stochastic formulation, similar to Kulkarni et al., for dissimilar diffusion bonding with the inclusion of different initial conditions. Surface roughness profiles are interacted to determine the initial geometries of the voids followed by developing a statistical distribution based on them. This idea of overlapping the surfaces to determine the potential voids and then running statistics on them was pioneered by Kulkarni. The model in this paper provides a much more robust way of determining the potential voids.

The distribution of voids are applied to the equations derived by Pilling in a stochastic manner in the form of a Monte Carlo simulation. The surface roughness dataset is comprised of six different surface roughness profiles for each of the three materials for a total of 18 profiles. Three titanium alloy materials were involved which were TI-54M, Ti-64 standard grain, and Ti-6242S. Each material was prepared as it would be for diffusion bonding, and then surface profiles were measured using a Mahr XR20 MIT GD25 contact surface profilometer in random locations with three in the longitudinal direction and three in the transverse direction. Figure 2 displays the raw data for each profile which was aligned horizontally by subtracting a single linear fit of the data from the profile, not the series of linear fits commonly used in surface roughness measurements.

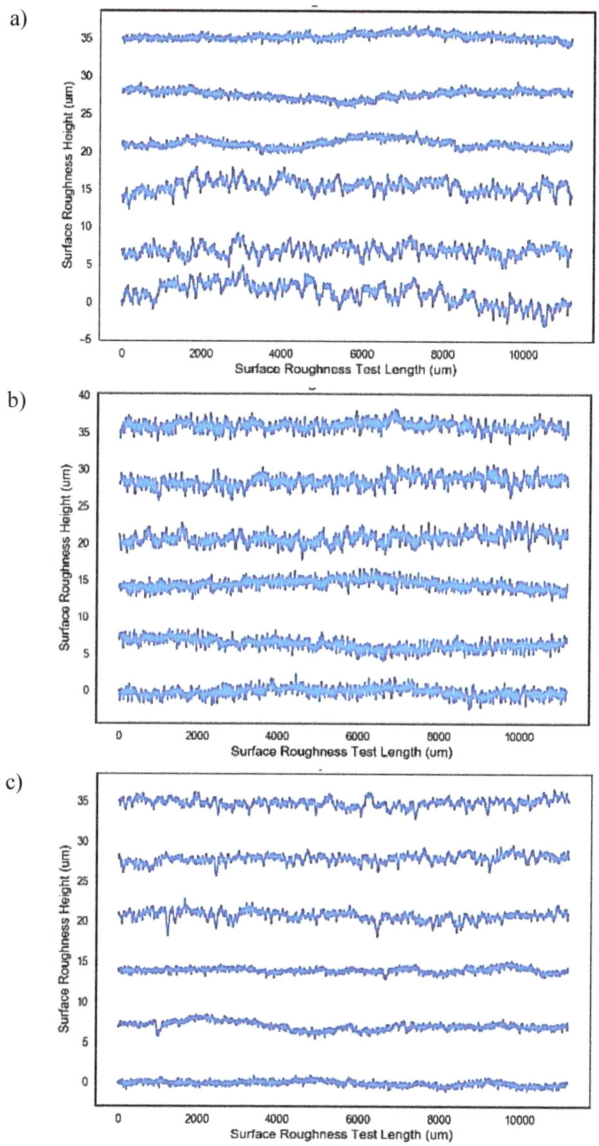

Figure 2: Surface roughness profiles from a) Ti54m, b) Ti6242S, and c) Ti64.

The initial conditions for the diffusion bonding model are based on the interaction of the two surfaces. The peaks of the surfaces contact and observe very high stresses causing them to go through an initial plastic deformation and flatten out. This occurs until the flow stress reaches the superplastic region. In the current model, the reduction in void height is assumed to be a simple overlap of the surface roughness profiles and that the initial plasticity of the peaks results in negligible closure of the voids. This initial collapse provides very little closure to begin with but serves to define the initial void geometry which has a large effect on the overall bonding time.

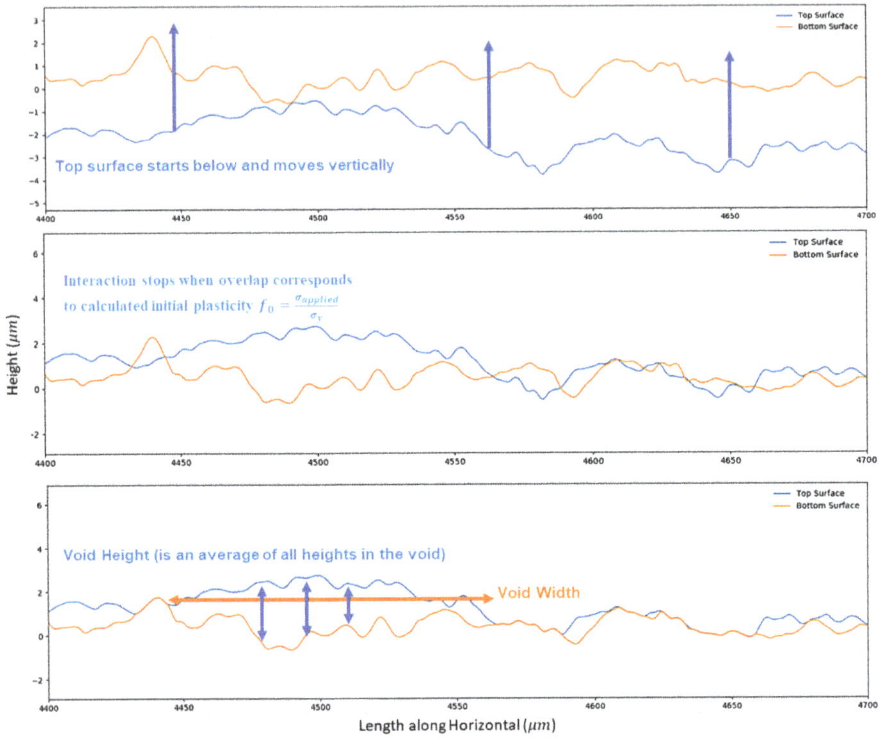

Figure 3: The surface profiles are overlapped to the point where they correspond. A calculated void is the average thickness over the specified width.

The surface profiles are in contact with each other until the overlap is at the same amount as specified in the Pilling model (the applied pressure over the yield strength of the material). For each of interactions between materials, one surface is translated relative to each. This is because there is no way of knowing exactly how the surface profiles will be mated in a production setting. This translation is repeated ten times in equal increments. Therefore, a total of 360 surface profile interactions per material combination are used to develop statistics on the void height and width. Figure 3 describes the algorithm procedure which entails flipping one of the profiles, translating it horizontally a given amount, overlapping it with the other profile until the contact stress would equal the yield strength, then determining the void height and width. The width of the void is the inside distance between each of the contact points. The height of the void is the average of the

 Materials Research Forum LLC
https://doi.org/10.21741/9781644902615-39

overlap height between the contact points. These are used as the basis for Pilling's geometry in the stochastic model shown in Figure 4.

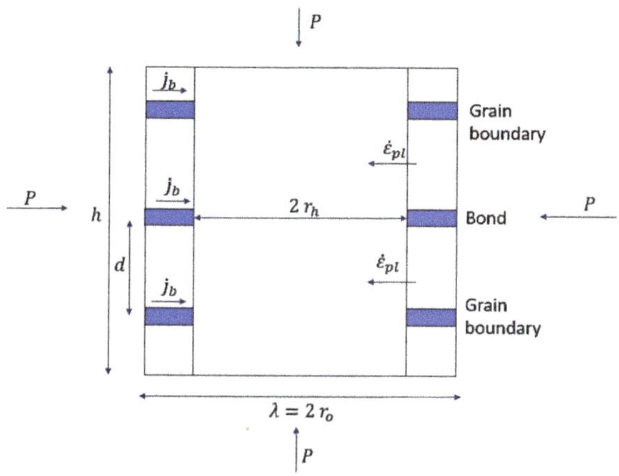

Figure 4: The Pilling void geometry is a revolved cylinder that has the axis of revolution along the bond plane normal vector.

The computed theoretical voids are tabulated and a distribution is fitted from the data. Many different statistical distributions were tried and the exponential distribution provided the best fit for both the height and the width of the voids. The void width had some significant outliers in the distribution where the surfaces randomly aligned to create an extremely large void. Void widths larger than the distribution fit maximum of 200 μm occurred around 7.5% of the time and were assumed to be negligible on the overall bonding performance. These large voids are most likely voids on the boundary of the surface profiles and would not occur in practice. An example of the distributions is presented in Figure 5.

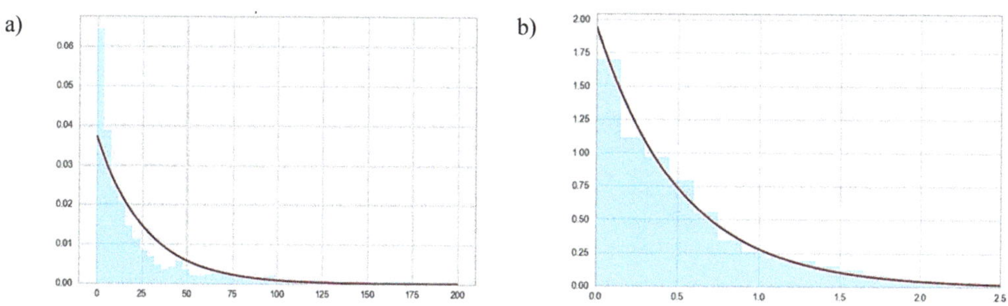

Figure 5: Example histogram of a) void width, and b) void height, with exponential distribution fits (Ti-54m/Ti-54m at low temperature and pressure).

The initial void geometry distributions are used as an input for a stochastic Pilling model. The primary equations were developed by Pilling and are for a specific void height and width. Using the probability distributions highlighted above, a random height and random width were selected. Then Pilling's equations were applied and the time required to completely bond was determined. The random selection of heights and widths and calculation of bonding was repeated 1000 times. Thus, a cumulative distribution was developed from the data to determine the percentage of randomly selected voids that would be closed after a given time for a set of processing conditions.

Pilling's equations are listed below in Eq. 1-7, and the explanation of terms along with the material properties are listed in Table 1. The initial bonding fraction, f, is determined by the applied stress divided by the high temperature yield strength of the material as shown in Eq. 1. The volume diffusion equation (Eq. 2), and the grain boundary diffusion equation (Eq. 3), were both derived from the stress dependent chemical potential due to the applied pressure. The plastic deformation from creep is modeled as deformation from an isostatic pressure with the creep equation (Eq. 4 and 5). The total change, bonded fraction per time, was integrated to determine the bonding time (Eq. 6). In a numerical implementation, the bonded fraction is incremented from the initial to complete closure while the height is kept uniform by applying the equations above. For each increment the change in time is computed. The bonding time for each random void is a summation of time increments. A flowchart of the simulation is provided in Figure 6.

$$f_0 = \frac{\sigma_{applied}}{\sigma_y}. \tag{1}$$

$$\left[\frac{df}{dt}\right]_{vol} = -\frac{4\Omega D_v}{kT}\left[\frac{f}{ln\left(\frac{1}{f}\right)-\frac{1-f}{2}}\right]\frac{1}{r_o^2}p. \tag{2}$$

$$\left[\frac{df}{dt}\right]_{gb} = -\frac{2\Omega D_{gb}\delta}{kT}\left[\frac{1}{ln\left(\frac{1}{f}\right)-\frac{1-f}{2}}\right]\frac{1}{r_o^2}p\left(\frac{1+\frac{h}{d}}{h}\right). \tag{3}$$

$$\dot{\varepsilon} = \frac{A\, exp^{(-Q_{sp}/RT)}\,\sigma^n}{T}. \tag{4}$$

$$\left[\frac{df}{dt}\right]_{plasticity} = \frac{-\dot{\varepsilon}}{\sigma_e}\left(3p\sqrt{f} - \frac{2\gamma\sqrt{f}}{r_0} + \frac{6\gamma}{r_0}\right). \tag{5}$$

$$t = \int_{f_0}^{f} \frac{-\sigma_e/\dot{\varepsilon}}{\left(-\frac{\dot{\varepsilon}}{\sigma_e}3p\sqrt{f} - \frac{2\gamma\sqrt{f}}{r_0} + \frac{6\gamma}{r_0}\right) + \frac{4\Omega D_v}{kT}\left[\frac{f}{\ln\left(\frac{1}{f}\right)-\frac{1-f}{2}}\right]\frac{1}{r_o^2}p + \frac{2\Omega D_{gb}\delta}{kT}\left[\frac{1}{\ln\left(\frac{1}{f}\right)-\frac{1-f}{2}}\right]\frac{1}{r_o^2}p\left(\frac{1+\frac{h}{d}}{h}\right)}\, df. \tag{6}$$

$$f(x;\mu,\lambda) = \left[\frac{\lambda}{2\pi x^3}\right] exp\frac{-\lambda(x-\mu)^2}{2\mu^2 x}, \text{ (for } x > 0, \mu > 0, \text{ and } \lambda > 0). \tag{7}$$

Table 1 – Material properties and variables for the Pilling equations (Pilling, 1988)

Properties	α	β
Atomic Volume, Ω (m^3)	1.76×10^{-29}	1.81×10^{-29}
Burgers vector, b (m)	2.9×10^{-10}	2.9×10^{-10}
Surface energy, γ (J m^2)	1.7	1.7
Grain boundary diffusivity, $D_{gb}\ \delta$	$3.6 \times 10^{-16} \times \exp(-97000/(R \times T))$	$5.4 \times 10^{-17} \times \exp(-153000/(R \times T))$
Grain boundary width, δ (m)	5.9×10^{-10}	5.72×10^{-10}
Volume boundary diffusivity, D_v	$8.6 \times 10^{-10} \times \exp(-150000/(R \times T))$	$1.9 \times 10^{-7} \times \exp(-153000/(R \times T))$
Shear modulus, G (MPa)	$4.3 \times 10^4 \times (1-((T-300)/T_m) \times 1.2)$	$2.05 \times 10^4 \times (1-((T-300)/T_m) \times 0.5)$
Creep constant, A	1.2×10^{-9}	1.2×10^{-9}
Strain rate creep constant, A_c	$A \times (D_{gb}\ \delta)_\alpha \times G_\alpha/(k \times T)$	$A \times (D_{gb}\ \delta)_\beta \times G_\beta/(k \times T)$
Gas constant, R (J/mol K)	8.314	
Botlzmann's constant, k (J/K)	1.38×10^{-23}	
Melting temperature, T_m (K)	1973	
Bonding temperature, T (K)	1046-1102	
Bonding pressure, P (MPa)	1-3	

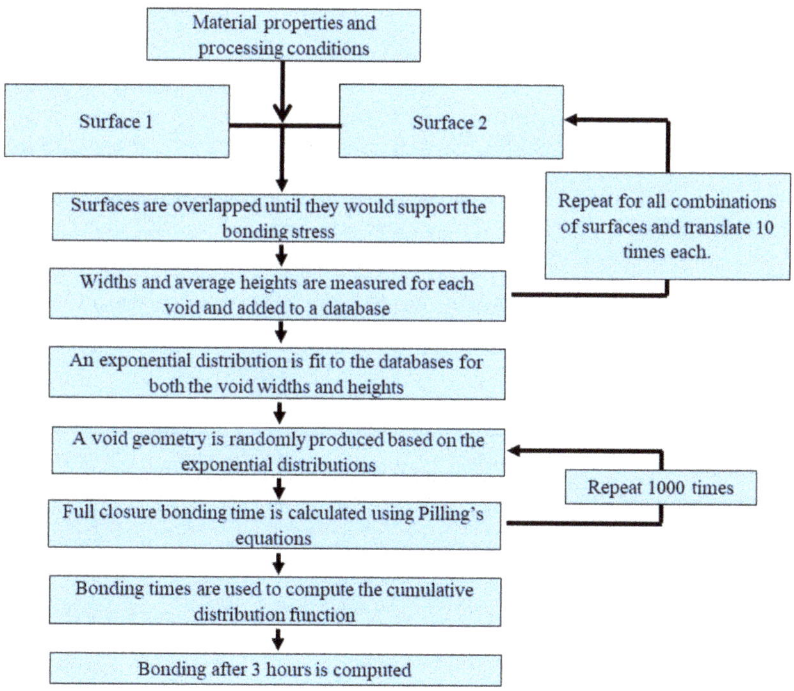

Figure 6: Program flowchart performed for each material combination and processing condition.

The simulation was compared with experimental data from TIMET, but the exact temperatures and pressure were not disclosed due to proprietary reasons. The pressures were labeled as low pressure (LP), intermediate pressure (IP), and high pressure (HP) while the temperatures were labeled as low temperature (LT), intermediate temperature (IT), and high temperature (HT). The grain sizes of the various materials are 2.4 μm for Ti-54M, 5.9 μm for Ti-64, and 4.1 μm for Ti-6242S [20]. The experimental results are summarized in Table 2 and are estimated from plots in the reference. The material properties for each of the calculations, shown in Table 1, is based on the rule of mixtures using the α/β ratio for each material. The α to β ratio depends heavily on the heat treatment of the alloys and is unknown for this data. The ratio is assumed to be 0.49 α for Ti-64 [21], 0.42 α for Ti-54m [22], and 0.33 α for Ti-6242 [23]. For every material/temperature/pressure combination the differences in α/β ratio generates a unique set of voids to be analyzed (54 unique sets).

Table 2 – Experimental results [20]

Materials	LT-LP	LT-IP	LT-HP	IT-LP	IT-IP	IT-HP	HT-LP	HT-IP	HT-HP
54m/54m	85%	95%	98%	100%	100%	100%	100%	100%	100%
64/64	51%	52%	53%	54%	81%	98%	91%	98%	99%
6242/6242	5%	8%	16%	37%	39%	49%	50%	59%	83%
54m/64	82%	95%	98%	100%	100%	100%	100%	100%	100%
54m/6242	44%	75%	81%	90%	95%	97%	98%	99%	100%
64/6242	17%	38%	47%	51%	60%	70%	70%	77%	97%

In addition to comparing work with published data, 9 conditions were simulated and experimentally verified in this study, which are shown in Table 3. Duplicate alloy combinations of Ti-6Al-4V fine grained (Ti-64FG to Ti-64FG), similar bonding of Ti-54M to Ti-54M), and dissimilar bonding with Ti-64FG to Ti-54M. Each condition was bonded at a moderate pressure for three hours. Each material combination was bonded at three temperatures, 0.530 T_m, 0.544 T_m, and 0.559 T_m, where T_m is the melting temperature. The details of the processing conditions are from [24].

Table 3 – Experimental data from specimens compared with predictions

Condition	Top Material	Bottom Material	Temperature	Experimental Bonded %	Calculated Bonded %
1	Ti-64 FG	Ti-64 FG	$0.530T_m$	64%	31%
2	Ti-64 FG	Ti-64 FG	$0.544T_m$	92%	83%
3	Ti-64 FG	Ti-64 FG	$0.559T_m$	100%	98%
4	Ti-54M	Ti-54M	$0.53T_m$	96%	24%
5	Ti-54M	Ti-54M	$0.544T_m$	100%	56%
6	Ti-54M	Ti-54M	$0.559T_m$	100%	100%
7	Ti-54M	Ti-64 FG	$0.530T_m$	95%	27%
8	Ti-54M	Ti-64 FG	$0.544T_m$	100%	73%
9	Ti-54M	Ti-64 FG	$0.559T_m$	100%	100%

Diffusion bonding specimens were created for each condition by taking two, clean and chemically milled, sheets of titanium and spot welding a sealing pattern around the perimeter. A gas tube was inserted and used to draw a vacuum between the sheets. Then the tube was welded shut leaving the vacuum between the sheets. The specimens were placed in a furnace at temperature and a superplastically expanding bladder was used to apply the bonding pressure. The specimens were cooled, sectioned and mounted, then ground using 200, 300, 400, 600, 800, and 1200 grit metallographic paper. Polishing was performed at 6 μm, 3 μm, and 1 μm diamond suspension, with a final polish of 0.05 μm alumina suspension with hydrogen peroxide. The specimens were etched for 10-15 seconds using Kroll's reagent, then imaged using an Olympus microscope at 500x. The bonding performance for each specimen was analyzed by looking at the percentage of bonded material along the bondline. The bonded fraction is the ratio of bonded material to total bond length along a simulated bondline, and the voids, and the bondline location, were determined by an image processing algorithm.

Results and Discussion

As discussed previously, none of the models provided a perfectly accurate fit to the experimental data and many of them cited specifically the lack of accurate surface geometry as a reason for this discrepancy. With the robust way of statistically incorporating the surface geometry presented in this paper, the surface effects should be well represented. However, as with the other bonding models, there were still some differences between the predicted bonding times and the experimental ones. The results of the simulation with respect to the experimental results is displayed in Figure 7. The high temperature – high pressure group was the most accurate of the processing conditions. However, some of the lower temperature and lower pressure groups had some significant inaccuracies. Figure 8 shows the average percentage discrepancy between experiments and calculations with respect to the temperature and pressure. The trend is fairly linear toward high pressure high temperature accuracy and low pressure low temperature inaccuracy.

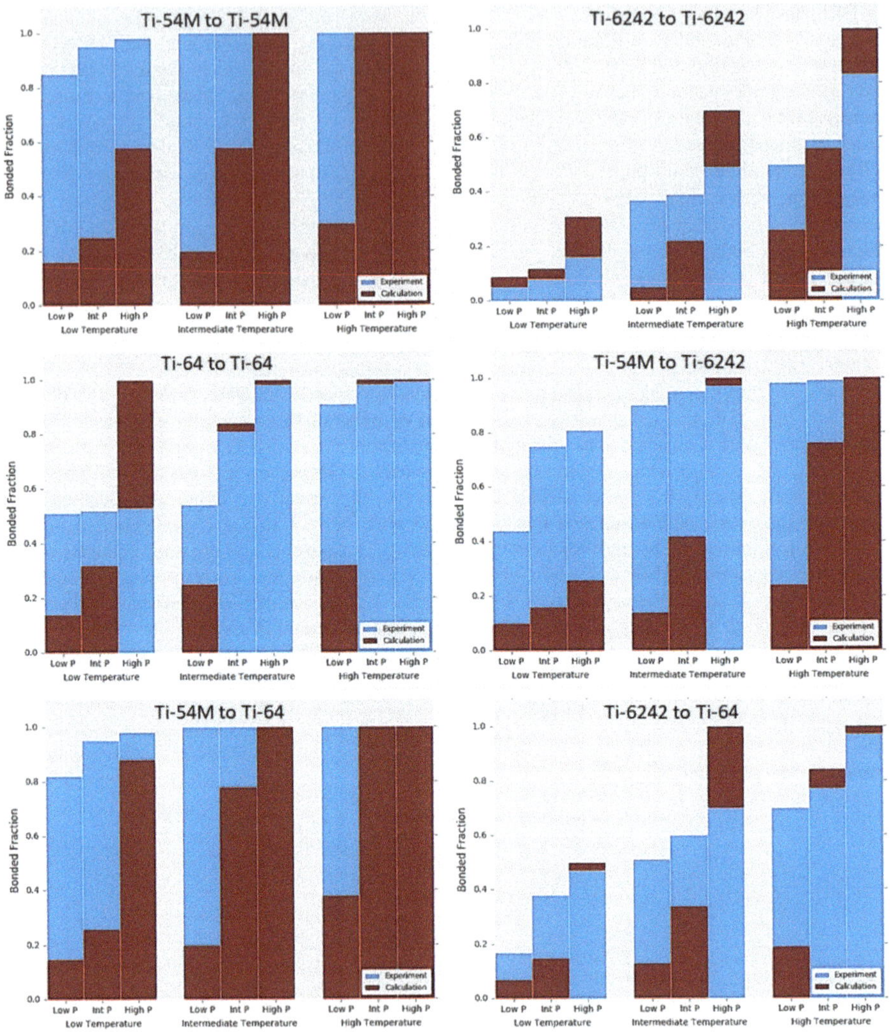

Figure 7: Program flowchart performed for each material combination and processing condition.

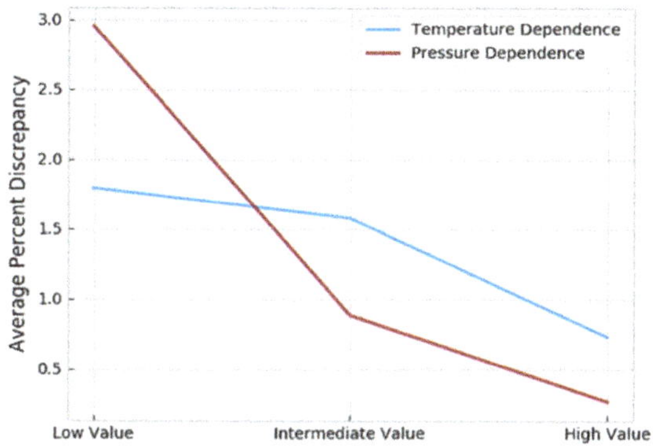

Figure 8: Average discrepancy between experimental data and calculated results by temperature and pressure. The high temperature and high pressure had the most accurate results.

While the theoretical aspects of a model determine its effectiveness, accurate material properties are also of paramount importance, and it is sometimes difficult to separate phenomenological issues with the theoretical model from discrepancies in the material properties. Specifically in this case, where the diffusion constants were based on the work of Ashby [24] which is a compilation of previous diffusion and deformation research from the 1960s-1980s, and it is meant for commercially pure titanium and not the alloys in this research. The material properties cited in previous titanium models separate the phases into α and β, and they have diffusion mobilities based on those two phases. Modern methods use the CALPHAD-type coupling of the thermodynamic and kinetic properties of a material to compute the diffusional mobilities making the diffusion of each alloying element a function of the Gibbs free energy and element concentrations [25,26]. Implementation of this would require more complex modeling of the different energies associated with the bonding process and is beyond the scope of this research. Therefore, some of the material properties used here and in other diffusion bonding research are most likely incorrect for these specific alloys.

To investigate how each material property is affecting the results of this simulation, a study was conducted. The effects of each parameter were calculated by using the base material properties in Table 1 along with the intermediate temperature and pressure and surface roughness constants for Ti-54M to Ti-54M at those processing conditions. By keeping everything else in the model constant and varying a single parameter, the relative change in complete bonding time is computed. For example, the temperature dependence in Figure 9 shows that if you reduce the temperature to 90% of the intermediate temperature, with all else being constant, the time required to obtain a fully bonded sample will increase by 7 times of what it would be at the intermediate temperature. The relative effects of each influential material property and processing condition is presented in a similar fashion.

a) b) c)

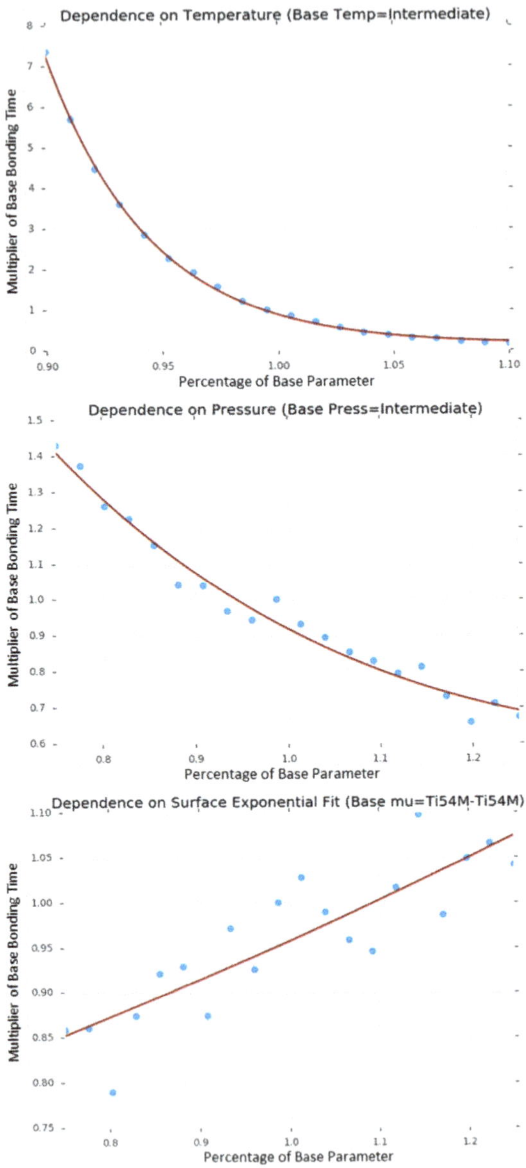

Figure 9: Dependencies on a) temperature, b) pressure, and c) surface parameter.

The primary processing conditions all have a strong effect on the bonding performance in the model and are in line with expectations. Most significant, is the temperature which softens the material for better closure during initial contact, increases the different diffusion constants, and increases the creep strain as the temperature increases providing a lower bonding time. A higher pressure provides more stress for power law creep and diffusion which reduces bonding time. As the surface parameter reduces, representing a distribution of smaller voids, the bonding time decreases while a larger surface parameter takes longer to close due to the larger voids. The effect of the surface is less extensive than the effects of temperature and pressure which is to be expected.

The material constants for the β phase are less conducive to bonding than the α phase. The amount of α and β will change depending on processing and heat treatment. Since the exact heat treatments were unknown for these materials, and the β% was unknown, the assumed α/β ratio could be a source of inaccuracy. For those papers that have investigated the rates of the different bonding mechanisms, creep has been the dominant bonding mechanism [2,10,13,15]. Pilling's model was no exception, and the bonding time was significantly affected by the creep material constants as shown in Figure 10. There is some debate as to whether the mechanism of plastic collapse is due to power-law creep (as in Derby and Hill's pressure sintering models), or if it is due to superplastic collapse (Hamilton and Pilling's models). While the equations are similar, the creep exponent is three times larger for power-law creep versus superplastic deformation. There are many additional constants that were checked but have no strong influence on the bonding time. Most notably is the grain size, which had no discernable influence, but is additionally incorporated into this version of Pilling's model and Hill's model.

A variation of parameters could be performed to determine exactly which combination of modified material properties allows this model to fit the experimental data accurately. However, this is unphysical and does not demonstrate the effectiveness of the model which is the point of this research. Because creep is the dominant closure mechanism, the creep exponent is the most influential material property, but the exact value of the exponent is unknown for these alloys, and there is some debate on the exponent value. Anyway, it is sufficient to modify this parameter alone in order to demonstrate how sensitive the results are to this material property. By lowering the creep exponent slightly more toward diffusive creep, Coble creep, and Nabarro-Herring creep, the accuracy of the model significantly increases especially in the lower temperature range. The changes in accuracy of the model simplify by a 0.025, 0.05, and 0.075 reduction in creep exponent which corresponds to an decrease in the average discrepancy between experiments and simulation from 28.32% to 23.0%, 19.5%, and 19.2% respectively. This demonstrates how sensitive these material properties can be and how measuring them on the specific alloys modeled would be necessary for an accurate diffusion bonding model.

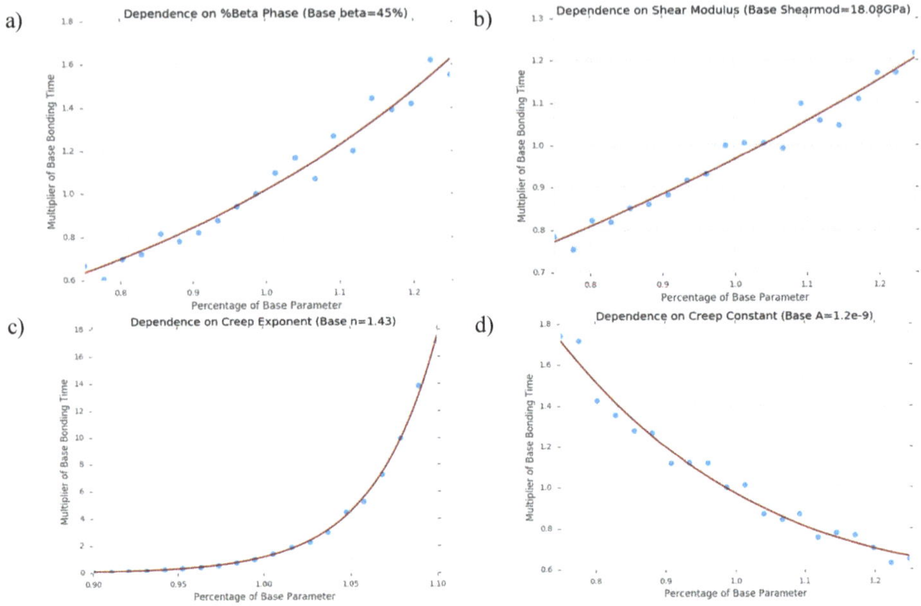

Figure 10: Dependence on a) alpha/beta ratio, b) shear modulus, c) creep exponent, and d) creep constant.

Summary

The stochastic Pilling model was computed with significantly more robust initial conditions computed through interactions of surface profiles. A database of potential voids was produced by overlapping several different surface roughness profiles in multiple ways to the overlap specified by the initial conditions of the diffusion bonding model. Pilling's model was then used to determine the statistical bonding percentage for the bonding time. While the model predicted the high temperature high pressure group very well, it was less accurate for other cases. An investigation into the effects of different material constants was performed, and comments were made on specific material properties that could cause issues. Further research could include simulations of different theoretical models in a similar fashion, implementation into three dimensions, and creation of an energy-based simulation with more modern material properties.

Acknowledgements

The authors sincerely thank The Boeing Company for funding the SPF/DB research at University of Washington. In addition, MR wants to acknowledge the Boeing-Pennell professorship funding and EB would like to acknowledge the Boeing Learning Together Program for educational funding.

References

[1] D. G. Sanders and M. Ramulu, "Examination of superplastic forming combined with diffusion bonding for titanium: Perspective from experience," J. Mater. Eng. Perform., vol. 13, no. 6, pp. 744-752, Dec. 2004. https://doi.org/10.1361/10599490421574

[2] A. Hill and E. Wallach, "Modelling solid-state diffusion bonding," Acta Metall., vol. 37, no. 9, pp. 2425-2437, 1989. https://doi.org/10.1016/0001-6160(89)90040-0

[3] M. F. Islam, J. Pilling, and N. Ridley, "Effect of surface finish and sheet thickness on isostatic diffusion bonding of superplastic Ti-6Al-4V," Mater. Sci. Technol., vol. 13, no. 12, pp. 1045-1050, Dec. 1997. https://doi.org/10.1179/mst.1997.13.12.1045

[4] H. Somekawa and K. Higashi, "The optimal surface roughness condition on diffusion bonding," Mater. Trans., vol. 44, no. 8, pp. 1640-1643, 2003. https://doi.org/10.2320/matertrans.44.1640

[5] A. Wang, O. Ohashi, and K. Ueno, "Effect of surface asperity on diffusion bonding," Mater. Trans., vol. 47, no. 1, pp. 179-184, 2006. https://doi.org/10.2320/matertrans.47.179

[6] C. H. Hamilton, "Pressure requirements for diffusion bonding titanium," Titan. Sci. Technol., pp. 625-648, 1973. https://doi.org/10.1007/978-1-4757-1346-6_46

[7] B. Derby and E. R. Wallach, "Diffusion bonding: development of theoretical model," Met. Sci., vol. 18, no. 9, pp. 427-431, 1984. https://doi.org/10.1179/030634584790419809

[8] B. Derby and E. R. Wallach, "Diffusion bonds in copper," J. Mater. Sci., vol. 19, no. 10, pp. 3140-3148, 1984. https://doi.org/10.1007/BF00549797

[9] B. Derby and E. R. Wallach, "Theoretical model for diffusion bonding," Met. Sci., vol. 16, no. 1, pp. 49-56, 1982. https://doi.org/10.1179/030634582790427028

[10] Z. X. Guo and N. Ridley, "Modelling of diffusion bonding of metals," Mater. Sci. Technol., vol. 3, no. 11, pp. 945-953, 1987. https://doi.org/10.1179/mst.1987.3.11.945

[11] Y. Takahashi and K. Inoue, "Recent void shrinkage models and their applicability to diffusion bonding," Mater. Sci. Technol., vol. 8, no. 11, pp. 953-964, 1992. https://doi.org/10.1179/mst.1992.8.11.953

[12] J. Pilling, D. W. Livesey, J. B. Hawkyard, and N. Ridley, "Solid state bonding in superplastic Ti-6Al-4V," Met. Sci., vol. 18, no. 3, pp. 117-122, 1984. https://doi.org/10.1179/msc.1984.18.3.117

[13] J. Pilling, "The kinetics of isostatic diffusion bonding in superplastic materials," Mater. Sci. Eng., vol. 100, pp. 137-144, 1988. https://doi.org/10.1016/0025-5416(88)90249-2

[14] M. T. Salehi, J. Pilling, N. Ridley, and D. L. Hamilton, "Isostatic diffusion bonding of superplastic Ti-6Al-4V," Mater. Sci. Eng. A, vol. 150, no. 1, pp. 1-6, 1992. https://doi.org/10.1016/0921-5093(90)90002-K

[15] N. Orhan, M. Aksoy, and M. Eroglu, "A new model for diffusion bonding and its application to duplex alloys," Mater. Sci. Eng. A, vol. 271, no. 1, pp. 458-468, 1999. https://doi.org/10.1016/S0921-5093(99)00315-9

[16] R. Ma, M. Li, H. Li, and W. Yu, "Modeling of void closure in diffusion bonding process based on dynamic conditions," Sci. China Technol. Sci., vol. 55, no. 9, pp. 2420-2431, Sep. 2012. https://doi.org/10.1007/s11431-012-4927-1

[17] S.-X. Li, S.-T. Tu, and F.-Z. Xuan, "A probabilistic model for prediction of bonding time in diffusion bonding," Mater. Sci. Eng. A, vol. 407, no. 1-2, pp. 250-255, Oct. 2005. https://doi.org/10.1016/j.msea.2005.07.003

[18] N. Kulkarni, M. Ramulu, and D. G. Sanders, "Modeling of Diffusion Bonding Time in Dissimilar Titanium Alloys: Preliminary Results," J. Manuf. Sci. Eng., vol. 138, no. 12, p. 121010, 2016. https://doi.org/10.1115/1.4034133

[19] F. A. Calvo, J. G. De Salazar, A. Urena, J. G. Carrion, and F. Perosanz, "Diffusion bonding of Ti-6Al-4V alloy at low temperature: metallurgical aspects," J. Mater. Sci., vol. 27, no. 2, pp. 391-398, 1992. https://doi.org/10.1007/BF00543928

[20] P. P. Gudipati and Y. Kosaka, "Diffusion Bonding of Similar and Dissimilar Titanium Alloys," presented at the Proceedings of the 13th World Conference on Titanium, 2016, pp. 1631-1636. https://doi.org/10.1002/9781119296126.ch273

[21] S. Jadhav, A. Powar, S. Patil, A. Supare, B. Farane, and R. Singh, "Effect of volume fraction of alpha and transformed beta on the high cycle fatigue properties of bimodal Ti6Al4V alloy," IOP Conf. Ser. Mater. Sci. Eng., vol. 201, p. 012035, May 2017. https://doi.org/10.1088/1757-899X/201/1/012035

[22] Y. Kosaka and P. GUDIPATI, "Method for the manufacture of alpha-beta ti-al-v-mo-fe alloy sheets," EP2721187 A1, 23-Apr-2014.

[23] M. E. Kassner, Y. Kosaka, and J. S. Hall, "Low-cycle dwell-time fatigue in Ti-6242," Metall. Mater. Trans. A, vol. 30, no. 9, pp. 2383-2389, 1999. https://doi.org/10.1007/s11661-999-0246-y

[24] P. M. Sargent and M. F. Ashby, "Deformation maps for titanium and zirconium," Scr. Metall., vol. 16, no. 12, pp. 1415-1422, 1982. https://doi.org/10.1016/0036-9748(82)90439-2

[25] L. Huang, Y. Cui, H. Chang, H. Zhong, J. Li, and L. Zhou, "Assessment of Atomic Mobilities for bcc Phase of Ti-Al-V System," J. Phase Equilibria Diffus., vol. 31, no. 2, pp. 135-143, Apr. 2010. https://doi.org/10.1007/s11669-009-9641-8

[26] Y. Liu, Y. Ge, D. Yu, T. Pan, and L. Zhang, "Assessment of the diffusional mobilities in bcc Ti-V alloys," J. Alloys Compd., vol. 470, no. 1, pp. 176-182, Feb. 2009. https://doi.org/10.1016/j.jallcom.2008.02.111

Development of asymmetric cold rolling technology of high-strength steel grades in order to exclude intermediate annealing operations

Alexander Pesin[1,a], Georgy Raab[1,b], Alexey Sverchkov[1,c], Denis Pustovoytov[1,d*], Gennady Kornilov[1,e], Alexey Bochkarev[1,f], Ilya Pesin[1,g] and Leonid Nosov[1,h]

[1] Nosov Magnitogorsk State Technical University, Zhilyaev Laboratory of mechanics of gradient nanomaterials, 38 Lenin prospect, Magnitogorsk, 455000, Russian Federation

[a]pesin@bk.ru, [b]giraab@mail.ru, [c]leshka-sverhc@rambler.ru, [d]pustovoitov_den@mail.ru, [e]korn_mgn@mail.ru, [f]analogsynth@mail.ru, [g]ilya33392@bk.ru, [h]nosov.leopold@yandex.ru

Keywords: Asymmetric Cold Rolling, High-Strength Steel Grades, Intermediate Annealing, Rolling Force, Electricity Consumption, Industry Experiment

Abstract. The technological process of manufacturing cold-rolled strip with a thickness of 0.5-1.5 mm from high-strength steel grades on cold rolling mills can consist of 2 or even 3 operating cycles "cold rolling - intermediate annealing". This is due to a significant hardening of such steels during cold rolling and, as a result, limitations on the maximum allowable rolling forces. Each additional operating cycle "cold rolling - intermediate annealing" significantly increases production costs and reduces the productivity of the technological process. One of the promising ways to reduce the rolling force is the asymmetric rolling. Rolling at different circumferential speeds of work rolls driven by two independent motors is the most suitable way to implement asymmetric rolling technology in industry. The paper presents data on the operation of the main electric drives of a five-stand industrial mill during symmetric and asymmetric cold rolling. Speed ratio of the work rolls was varied in the range from 1.0 to 1.5. Electrical and power parameters were calculated, measured and compared. Based on numerical simulation, laboratory and industry experiments it was shown that the asymmetric cold rolling makes it possible to reduce the rolling forces, increase the thickness reduction per pass and, as a result, obtain a thinner strip without the use of intermediate annealing.

Introduction

In the production of cold-rolled strip from high-strength steel grades with a final thickness of 0.5-1.5 mm on cold rolling mills, the technological process consists of 2 or even 3 operating cycles "cold rolling – intermediate annealing". This is due to the significant work hardening of such steels during cold rolling, and, as a result, the limitations of rolling forces. To be able to obtain a thin cold-rolled strip, intermediate annealing is required. Intermediate annealing is a recrystallization annealing at a temperature of 600 to 710 °C of work-hardened rolled metal in furnaces with a protective atmosphere. The duration of 1 intermediate annealing is 38-44 hours. Thus, each operating cycle "cold rolling – intermediate annealing" significantly increases production costs, reduces the productivity of the technological process.

One of the promising non-traditional ways to improve the mechanical properties, the accuracy of the geometric dimensions of the strips, as well as to reduce the rolling force is the speed asymmetry of the work rolls [1-3]. With a decrease in the thickness of sheets and strips, the effect of reducing the rolling force increases. Rolling with different circumferential speeds of the work rolls, when both rolls are driven by two independent motors, is the most suitable way to implement the asymmetric rolling technology in industry. Asymmetric rolling due to purposefully created differences in the circumferential speeds of the work rolls is also called "differential speed rolling". For such a process, a degree of asymmetry is defined by a ratio of circumferential speeds υ_1 and υ_2 of the work rolls:

$$S_R = \frac{v_1}{v_2}. \quad (1)$$

where S_R is the speed ratio and $v_1 > v_2$.

Previous extensive theoretical and experimental studies [1-11] of asymmetric rolling processes have revealed the positive aspects of speed asymmetry, which consist primarily in the possibility of a significant reduction in rolling force, reduction of elastic deformations of the stand, and an increase in thickness reductions per pass and the possibility of obtaining thinner strips compared to the conventional rolling process.

The mechanics of the asymmetric rolling process is based on the creation of additional shear deformations in the rolled metal. In contrast to the symmetric process, a shear deformation zone appears in the deformation zone during asymmetric rolling; a zone in which the forces of contact friction from the side of the fast (rotating at a higher speed) and slow (rotating at a lower speed) rolls are directed in opposite directions [11].

The practical application of asymmetric rolling is usually preceded by a series of preliminary calculations and experiments to determine the ultimate loads of mechanical and electrical equipment. The load on the transmission and drive of the fast roll noticeably increases, and the drive of the slow roll can switch to the generator (brake) mode. To eliminate the maximum unacceptable loads of electric drives, determining the extreme values of the asymmetry coefficient is an important task. The range of allowable values of speed asymmetry is determined by two restrictions: 1) the maximum allowable torque of the fast roll engine; 2) the minimum torque of the slow roll motor. The first condition excludes overloading and shutdown of the electric drive and the second one ensures that the slow roll engine is not allowed to switch to the generator (brake) mode. When choosing the asymmetry coefficient, it is necessary to take into account the factors affecting the quality of the strip surface and the stability of the rolling process. For example, the friction coefficient in the deformation zone and the temperature of the rolls with all the ensuing consequences depend on the amount and composition of the cutting fluid. The stability of the process is determined by the setting of local systems for controlling the speed of electric drives, the tension and thickness of the strip in the interstand gaps, and the degree of field weakening in systems with two-zone speed control.

The paper presents data on the operation of the main electric drives of a 5-stand industrial mill during symmetric and asymmetric cold rolling. Speed ratio of the work rolls was varied in the range from 1.0 to 1.5. Electrical, power and force parameters were calculated, measured and compared.

Research Method

Continuous 5-stand mill "630" (Fig. 1) is designed for cold rolling of narrow strip with a final thickness of 0.5-4.5 mm and a width of 250-500 mm from various steel grades. The individual electric drives of the work rolls are made according to the "thyristor converter – motor" system. Each of the work rolls is driven by a DC motor. Diameters of the work rolls and back-up rolls are 420 mm and 900 mm, respectively. The mill is equipped with an automatic tension and thickness control system. The main parameters of the electric drives (armature and excitation currents, voltage, and motor speed) were recorded using the "MONITOR" information electronic system. In addition, non-electrical parameters were recorded: the thickness and width of the strip, tension and rolling force. The experiment was carried out during cold rolling of high-strength steel (Table 1) at a rolling speed of 6 m/s.

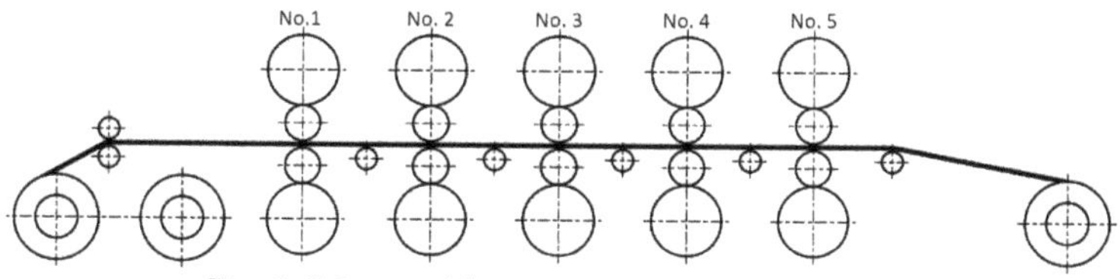

Fig. 1. Scheme of the continuous 5-stand mill "630"

The conventional technological process for the production of cold-rolled strip from high-strength steel includes the following operations: cold rolling (in 5 passes) from a thickness of 2.8 mm to 2.0 mm → intermediate annealing → cold rolling (in 5 passes) from a thickness of 2.0 mm to 1.3 mm. The task of the industrial experiment was to assess the technological feasibility of producing cold-rolled strip from high-strength steel according to the scheme: cold rolling (in 5 passes) from a thickness of 2.8 mm to 1.3 mm without intermediate annealing (Table 2).

Table 1. Chemical composition of high-strength steel.

Chemical composition, %														
C	Si	Mn	S	P	Cr	Ni	Cu	Al	Mo	V	Ti	Nb	N	B
0.67	0.27	0.97	0.0051	0.012	0.032	0.057	0.070	0.036	0.0060	0.0065	0.0050	0.0022	0.0054	0.0005

Table 2. Rolling pass schedule.

Stand No.	1	2	3	4	5
Entry thickness, mm	2.8	2.32	1.91	1.59	1.38
Exit thickness, mm	2.32	1.91	1.59	1.38	1.28
Thickness reduction, %	17.14	17.67	16.75	13.21	7.25

The industrial experiment included rolling in four modes: 1) symmetric rolling (mode 1); 2) asymmetric rolling with $S_R = 1.1$ (mode 2); 3) asymmetric rolling with $S_R = 1.2$ (mode 3) 4) asymmetric rolling with $S_R = 1.5$ (mode 4). In all cases, rolling in stands No. 1 and No. 5 was symmetrical. Asymmetry was created only in stands No. 2, No. 3 and No. 4. At the same time, the speed asymmetry was set to be the same in stands No. 2, No. 3, No. 4. With asymmetric rolling, the upper work roll had a higher speed than the lower one in all cases. For all variants of the experiment, the same rolling pass schedule was taken (Table 2).

The initial experimental data were oscillograms of electrical and power parameters exported from the "MONITOR" program in the form of data arrays. They are tabulated functions of time at a discrete point of 0.001 s. Their further processing was carried out in the MATLAB Simulink environment, which is well suited for working with large amounts of information. Due to the unequal operating time of the mill during symmetric and asymmetric rolling, the calculations were carried out for the steady state operation of electric drives. Determination of the required electrical and power parameters – power consumption, power consumption, torque and rolling power is made according to the well-known formulas given below.

The electrical power consumed by DC motor and the power consumption were calculated:

$$P_{el} = UI_a. \tag{2}$$

where U is motor voltage, V; I_a is armature current, A.

$$\Delta W_{el} = \int_{t_0}^{t} P_{el}(t)dt. \tag{3}$$

The calculation of the rolling torque and rolling power was carried out in accordance with the simplified scheme of the mechanical part of the electric drive (Fig. 2). This takes into account the efficiency of motors and gearboxes.

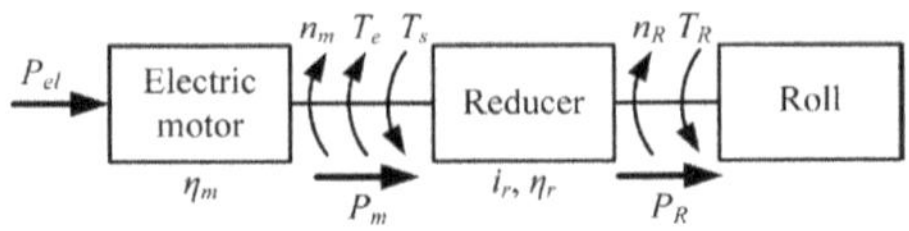

Fig. 2. Simplified scheme of the mechanical part of the electric drive.

The rolling torque was calculated as follows:

$$T_R = \frac{UI_a\eta_m\eta_r i_r}{\omega_m}. \quad (4)$$

where $\eta_m = 0.924$, $\eta_r = 0.95$ are motor efficiency and reducer efficiency; ω_m is angular speed of the motor, rad/s; i_r is gear ratio (i_r = 1.764; 1.474; 1.238; 1.088; 1,000 for stands No. 1-No. 5, respectively).

The rolling power was calculated as follows:

$$P_R = \frac{T_R\omega_m}{i_R}. \quad (5)$$

Equations (4) and (5) did not take into account friction torque and interstand tension. Average (X_{avg}) and root-mean-square (X_{rms}) values of the functions of electrical, power and force parameters (armature current, power, torque, etc.) were calculated according to instantaneous values of the $x(t)$ functions:

$$X_{avg} = \frac{1}{t-t_0}\int_{t_0}^{t} x(t)dt. \quad (6)$$

$$X_{rms} = \sqrt{\frac{1}{t-t_0}\int_{t_0}^{t} x^2(t)dt}. \quad (7)$$

The average and root-mean-square values were calculated in the steady state operation of the drive for the time interval $t - t_0$, which was 60 s. During the same time, the consumption of electricity was also determined. The block diagram of the model for processing experimental data, shown in Fig. 3 was implemented in MATLAB Simulink [12].

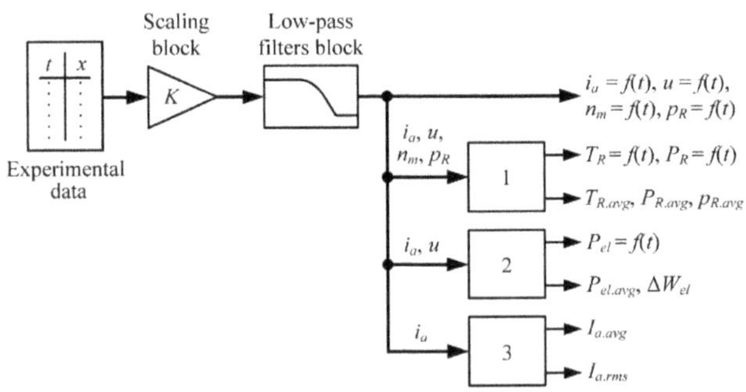

Fig. 3. Flow chart of the model for experimental data processing (where 1 is a power and force calculation block; 2 is an electric power and electricity consumption calculation block; 3 an average and root-mean-square armature current calculation block)

Simulation Results and Discussion

Symmetric rolling (mode 1) and asymmetric rolling with S_R = 1.1 and S_R = 1.2 (modes 2 and 3, respectively) were completed without any problems. The thickness of the strip along the entire

Materials Research Forum LLC
https://doi.org/10.21741/9781644902615-40

length and width was within the tolerance range. The rolling of the 4th mode with $S_R = 1.5$ could not be completed due to overheating of the rolls due to high frictional heating under conditions of insufficient cooling capacity of the emulsion.

The speed asymmetry caused a noticeable decrease in rolling forces in all stands, and this was accompanied by a decrease in the total rolling torque and power, with the exception of the first stand, where there was a slight increase in power. This was caused by a decrease in tension in the section "decoiler – stand No. 1" during asymmetric rolling. During asymmetric rolling, the armature current increased on the upper (faster) roll, and decreased on the lower (slower) roll. At the same time, the total load of the upper and lower rolls during asymmetric rolling was significantly reduced (at least 15%) compared to symmetrical rolling.

Oscillograms of the armature currents and rolling force of stand 2 during symmetric and asymmetric rolling in a steady-state operation of motors are given in Figs. 4-6. Comparing of the torques, power and force parameters for symmetric and asymmetric rolling are shown in Figs. 7-9.

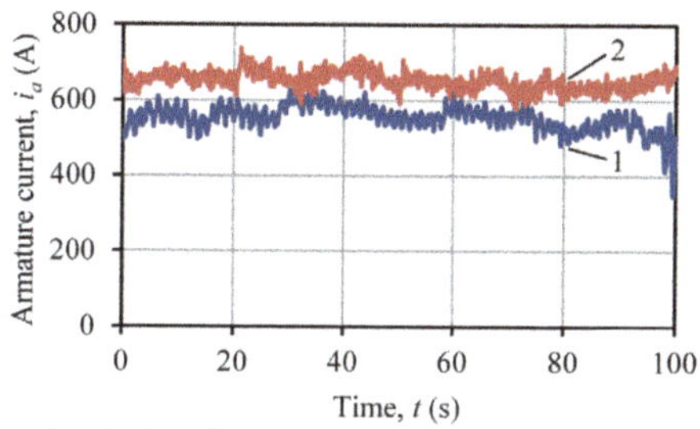

Fig. 4. Oscillograms of armature currents of motor of top roll of stand 2 (1 is symmetric rolling; 2 is asymmetric rolling).

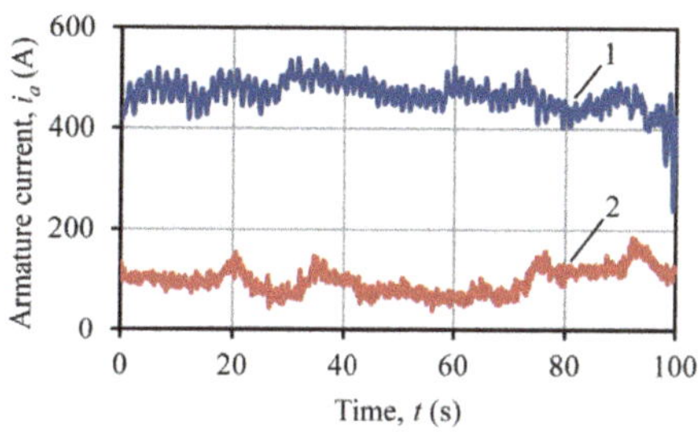

Fig. 5. Oscillograms of armature currents of motors of bottom roll of stand 2 (1 is symmetric rolling; 2 is asymmetric rolling).

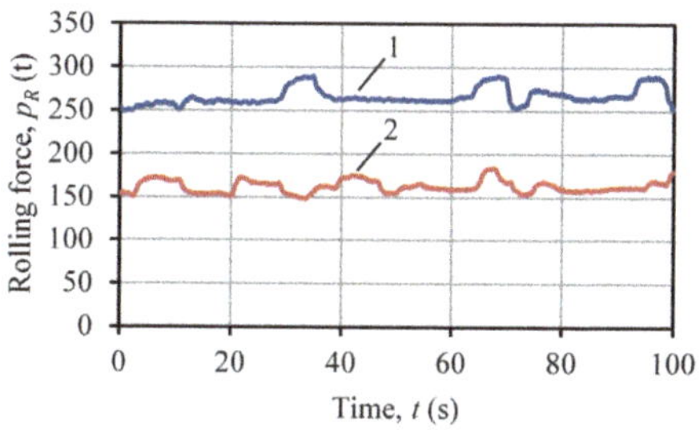

Fig. 6. Oscillograms of armature currents of rolling force in stand 2 (1 is symmetric rolling; 2 is asymmetric rolling).

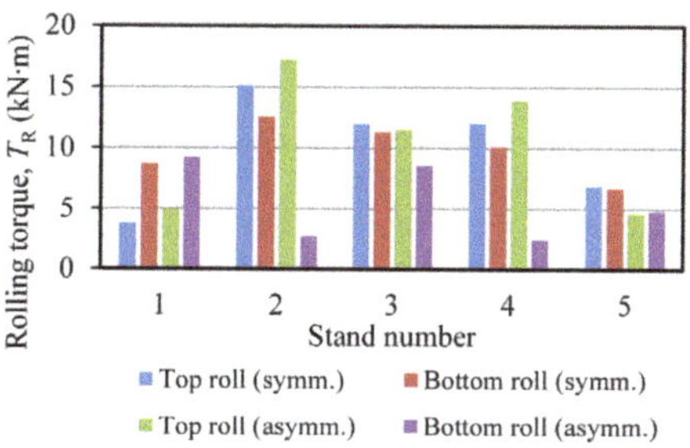

Fig. 7. Torques compared for symmetric and asymmetric rolling.

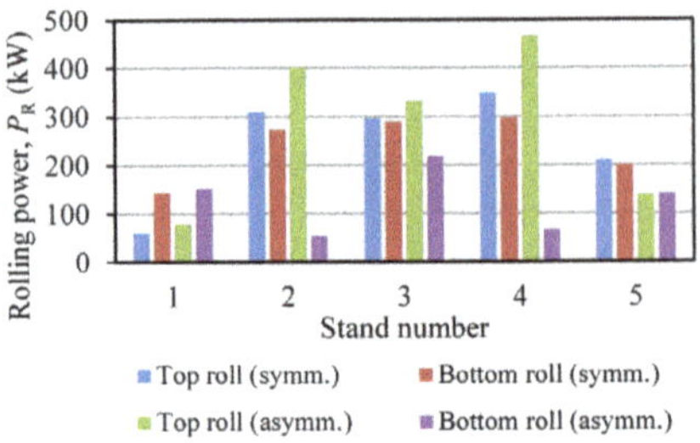

Fig. 8. Powers compared for symmetric and asymmetric rolling.

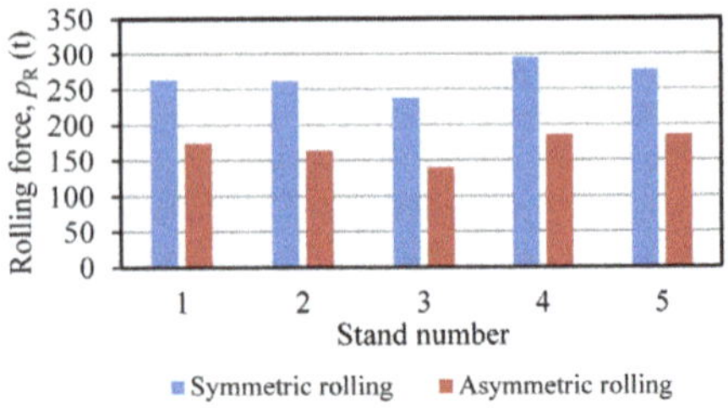

Fig. 9. Forces compared for symmetric and asymmetric rolling.

Table 3 shows a comparison of the experimental results presented for two cases – symmetric rolling (mode 1) and asymmetric rolling with $S_R = 1.2$ (mode 3). A comparison of the electricity consumption of the main drives showed that asymmetric rolling is a more efficient process compared to symmetric rolling. The difference was just over 15%.

Table 3. Comparison of electricity consumption.

Stand No.	1	2	3	4	5	$\sum \Delta W_{el}$
Rolling mode	Electricity consumption ΔW_{el}, kW·h					
Symmetric rolling	3.931	11.070	11.130	12.190	7.811	46.13
Asymmetric rolling ($S_R = 1.2$)	4.409	8.668	10.470	10.200	5.337	39.08
Difference, %	+10.8	-21.7	-5.93	-16.3	-31.7	-15.3

For asymmetric rolling, the following condition must be satisfied:

$$S_R = \frac{v_1}{v_2} < \frac{h_0}{h_1}. \tag{8}$$

where v_1 is the speed of the fast work roll; v_2 is the speed of the slow work roll; h_0 is the thickness before the rolling pass; h_1 is the thickness after the rolling pass.

That is, the speed ratio during asymmetric rolling should not be greater than the ratio of the input and output thicknesses of the strip in the corresponding pass.

The experiment on the implementation of the technology of asymmetric rolling in industrial conditions was carried out at a reduced speed (6 m/s), which differs significantly from the maximum. In normal speed mode, the motors operate with field weakening. The manual setting of the speed asymmetry, tested in this experiment, is planned to be implemented in the automatic mode for each stand individually, taking into account the rolling pass schedule and a steel grade.

The economic effect of using asymmetric rolling technology is to reduce the cost of additional cold rolling and intermediate annealing operations. In particular, when additional cold rolling operations are excluded, the costs for energy resources (electricity, technical water, compressed air), costs for auxiliary materials (liquid lubricant, inter-operational coil packaging), costs for rolls (transshipment, regrinding) are reduced. With the exclusion of additional intermediate annealing operations, the costs on process fuel (natural gas) and the costs on energy resources (electricity, pure nitrogen, hydrogen, industrial water) are reduced. In addition, the overall productivity of the technological process is significantly increased, primarily due to the elimination of intermediate annealing operations, the duration of which is 38...44 hours.

Conclusion

According to the results of an industrial experiment, the technical feasibility of implementing the technology of asymmetric rolling in a continuous 5-stand mill was confirmed. This technology can be used to reduce the number of operating cycles "cold rolling - intermediate annealing" in the production of a strip with a thickness of 0.5-1.5 mm from high-strength steel grades to reduce the cost of production and increase the productivity of the technological process.

A reduction in the total consumption of electricity by the main drives of 5 stands by 15% was established. Additional statistical analysis is required to identify patterns and confirm. After asymmetric rolling, there were no discontinuities, microcracks or microfractures in the strip material, and the mechanical properties met the requirements of consumers. With asymmetric rolling, the thickness of the strip along the entire length and width was within the tolerance field.

The results of investigation can be used for design of effective asymmetric cold rolling technology which makes it possible to reduce the rolling forces, increase the thickness reduction per pass and, as a result, obtain a thinner strip without the use of intermediate annealing.

Acknowledgements
The study was supported by the grant of the Russian Science Foundation (No. 23-79-30015).

References

[1] G. Vincze, F.J.P. Simões, M.C. Butuc Asymmetrical rolling of aluminum alloys and steels: a review, Metals (2020) 10(9):1126 https://doi.org/10.3390/met10091126

[2] D. Pustovoytov, A. Pesin, P. Tandon Asymmetric (hot, warm, cold, cryo) rolling of light alloys: a review, Metals (2021) 11(6):956 https://doi.org/10.3390/met11060956

[3] H. Yu, C. Lu, A.K. Tieu, H.J. Li, A. Godbole, S.H. Zhang, Special rolling techniques for improvement of mechanical properties of ultrafine-grained metal sheets: a review, Adv. Eng. Mat. 18 (2016) 754-769 https://doi.org/10.1002/adem.201500369

[4] S.S. Dhinwal, C. Ernould, B. Beausir, Facilitating the occurrence of dynamic recrystallization in plain extra low-carbon steel by warm asymmetric rolling, Materials Characterization 189 (2022) 111942 https://doi.org/10.1016/j.matchar.2022.111942

[5] J. Wang, X. Liu, X. Sun, Study on the relationship between asymmetrical rolling deformation zone configuration and rolling parameters, Int. J. Mech. Sci. 187 (2020) 105905 https://doi.org/10.1016/j.ijmecsci.2020.105905

[6] D. Konstantinov, D. Pustovoitov, A. Pesin Influence of microstructure on inhomogeneity of stress and strain in the deformation zone during asymmetric cold rolling of ferritic-pearlitic steels, Procedia Manufacturing 50 (2020) 514-519 https://doi.org/10.1016/j.promfg.2020.08.093

[7] A. Kawalek, The theoretical and experimental analysis of the effect of asymmetrical rolling on the value of unit pressure, J. Mater. Process. Technol. 157-158 (2004) 531-535 https://doi.org/10.1016/j.jmatprotec.2004.07.142

[8] H. Dyja, J. Markowski, D. Stoiniski, Asymmetry of the roll gap as a factor improving work of the hydraulic gauge control in the plate rolling mill, J. Mater. Process. Technol. 60 (1996) 73-80 https://doi.org/10.1016/0924-0136(96)02310-2

[9] M. Philipp, W. Schwenzfeier, F.D. Fischer, R. Wodlinger, C. Fischer, Front end bending in plate rolling influenced by circumferential speed mismatch and geometry, J. Mater. Process. Technol. 184 (2007) 224-232 https://doi.org/10.1016/j.jmatprotec.2006.11.027

[10] D. Kasai, A. Komori, A. Ishii, K. Yamada, S. Ogawa, Strip Warpage Behavior and Mechanism in Single Roll Driven Rolling, ISIJ Int. 56 (2016) 1815-1824 https://doi.org/10.2355/isijinternational.ISIJINT-2016-188

[11] Y. Tian, Y.C. Guo, Z.-D. Wang, G.-D. Wang, Analysis of rolling pressure in asymmetrical rolling process by slab method, J. Iron Steel Res. Int. 16 (2009) 22-26 https://doi.org/10.1016/S1006-706X(09)60055-8

[12] A.A. Bochkarev, G.P. Kornilov, Understanding the electric drive parameters of energy and forces during asymmetric rolling, Magnitogorsk rolling practice 2022. Magnitogorsk, NMSTU Publ. (2022) 48-49

Texture and strain rate sensitivity analysis of solid solution and precipitation hardening aluminum alloys processed by repetitive corrugation and straightening

Sergio Elizalde[1,2,a], Marco Ezequiel[1,3,b], Liliana Romero-Resendiz[1,4,c], Jose Maria Cabrera[2,5,d], Gonzalo González[1,e*]

[1] Instituto de Investigaciones en Materiales, Universidad Nacional Autónoma de México, Circuito Exterior S/N, Cd. Universitaria, A. P. 70-360, Coyoacán, C.P. 04510, Mexico.

[2] Departamento de Ciencia e Ingeniería de Materiales, EEBE - Universidad Politécnica de Cataluña, c/Eduard Maristany 10-14, 08019 Barcelona, Spain

[3] Univ. Lille, CNRS, INRAE, Centrale Lille, UMR 8207 - UMET - Unité Matériaux et Transformations, F-59000 Lille, France

[4] Facultad de Química, Departamento de Ingeniería Metalúrgica, Universidad Nacional Autónoma de México, México City, 04510, Mexico.

[5] Fundación CIM UPC, c/Llorens i Artigas 12, 08028 Barcelona, Spain

[a]sergio.elizalde@upc.edu, [b]ezequielam@univ-lille.fr, [c]liliana.rom7@comunidad.unam.mx, [d]jose.maria.cabrera@upc.edu, [e]joseggr@unam.mx

Keywords: Aluminum Alloys; Texture; Strain-Rate Sensitivity; Repetitive Corrugation and Straightening

Abstract. The potential of improving the mechanical strength by the RCS process is evaluated on the 5754, 6061, and 7075 aluminum alloys, which present different hardening mechanisms related to their respective alloying elements. This work compares the evolution of the texture and the mechanical properties of the different alloys through the RCS processing. The mechanical properties were evaluated by micro-hardness measurements, tensile tests at different temperatures, and strain rates to evaluate the strain-rate sensitivity. The results showed that after two RCS passes, the 6061 and 5754 alloys showed a relatively high strain-rate sensitivity at 300°C. In addition, an increment of 27%, 22%, 15% in hardness was obtained for the 5754, 6061 and 7075 alloys, respectively. Showing the potential of improvement in the mechanical resistance due to the different hardening mechanism. Furthermore, the crystallographic texture was characterized by the obtention of pole figures by X-ray diffraction and the calculation of their orientation distribution functions. The results showed the same trend in the three aluminum alloys, i.e., the initial texture components were conserved, but the texturized volume decreased.

Introduction

Aluminum alloys have excellent mechanical properties, low density, corrosion resistance, and recyclability, making them promising alternatives to some steel components in the automotive industry. Despite these advantages, their lower formability and significant spring back at room temperature constrain the application of aluminum alloys [1,2]. To overcome these disadvantages, recent studies have focused on reducing the grain size of these alloys to an ultrafine level to improve the mechanical properties of the alloys, as smaller grain sizes typically contribute to enhanced strength and hardness. Severe plastic deformation (SPD) techniques can significantly decrease the grain size and increase the mechanical properties; some examples are equal channel angular pressing (ECAP), high-pressure torsion (HPT), accumulative roll bonding (ARB), equal channel angular rolling (ECAR), constrained groove pressing (CGP), and repetitive corrugation and straightening (RCS) [3–6].

Most SPD techniques induce repetitive plastic deformation, overcoming standard deformation limits without significantly changing the initial shape. Since many modern engineering applications require metal sheets, SPD processes on such geometries are particularly interesting. ECAR, ARB, CGP, and RCS allow deforming materials in sheet format while maintaining the dimensions and shape of the material. Particularly, ARB is a simple and continuous process, but it requires special surface treatment to prevent bonding failure. In contrast, RCS requires no special surface treatment and offers more flexibility in improving mechanical properties by modifying die geometries and deformation routes. Exploring these techniques and processes presents a promising avenue for enhancing aluminum formability and mechanical properties for industrial applications [8].

Despite the development of these techniques, the lack of knowledge regarding post-SPD shaping hinders the broader commercialization of these SPD-processed materials [9,10]. Increasing the forming temperature is one proposed solution to enhance the post-SPD shaping of aluminum alloys. In this context, the strain rate sensitivity parameter becomes an essential indicator of the forming capacity, illustrating the ability of the material to sustain homogeneous deformation and delay fracture under the applied strain rate [1]. In addition, the grain size, notably when reduced to below 10 μm, significantly affects the hot forming processes, even being capable of inducing superplastic behavior and resulting in extraordinary homogeneous elongations [11].

On the other hand, the crystalline orientation distribution within a polycrystalline material, i.e., the crystallographic texture, significantly impacts mechanical properties such as anisotropy, formability, strain hardening, and post-forming surface roughness. In that sense, the material texture influences the rate and magnitude of strain hardening during deformation. Consequently, comprehending and controlling crystallographic texture is critical for optimizing metallic materials for specific manufacturing applications [12,13].

In aluminum alloys, crystallographic texture evolution typically manifests specific orientations during different manufacturing processes. For example, the texture generally evolves around the so-called copper, brass, and S orientations during the rolling process. In contrast, the annealing process often develops the Cube and Goss orientations. The impact of these texture components on the formability of aluminum alloys is a topic of ongoing debate in the scientific community. For example, some researchers argue that the cube texture component is beneficial in enhancing formability, while others contest that it might have a detrimental effect [14,15].

Aluminum alloys 5754, 6061, and 7075 have distinct hardening mechanisms due to their unique compositions and microstructural characteristics. The 5754 alloy is a non-heat treatable alloy with medium strength. Its hardening mechanism mainly relates to solid solution strengthening and strain hardening. Solid solution strengthening occurs due to the alloying elements (such as magnesium), which distort the lattice structure when added to the aluminum matrix, creating resistance to dislocation movement. This resistance enhances the strength of the material. Strain hardening, also known as work hardening, is the process by which the material becomes stronger as it is deformed, with dislocations in the crystal structure interacting and multiplying to create a harder and stronger material [16].

On the other hand, the 6061 and 7075 alloys are heat-treatable with medium to high strength. Their hardening mechanism relies on precipitation. This process involves three stages: solution heat treatment, quenching, and aging. Solution heat treatment dissolves the soluble alloying elements within the aluminum matrix, followed by rapid quenching to retain them in a super-saturated solid solution. During aging, these elements precipitate out of the solution, forming fine particles that hinder dislocation movement, increasing the alloy's strength and hardness. Therefore, the hardening mechanism is the primary difference between the studied alloys [16].

The objective of the present work is to compare the hardening evolution and hot formability of different aluminum alloys and the effect of the RCS on the crystallographic texture evolution through the process. Furthermore, the strain rate sensitivity is evaluated by hot tensile tests to characterize the RCS impact on the hot formability. Moreover, a quantitative analysis of the texture evolution is assessed by identifying the different texture components and calculating their respective volume fraction through the orientation distribution functions.

Methods

Commercially available 5754-H111, 6061-T6, and 7075-O aluminum sheets with 1 mm thickness were cut into squares of 120 x 120 mm. Table 1 summarizes the nominal chemical composition of these alloys. A pair of dies, fabricated from D2 tool steel and featuring a geometry based on two perpendicular extruded cuts with a sinusoidal profile, were used for the corrugation stages. A pair of analogous flat dies made of D2 steel were used for the straightening stages. The modified RCS process allows the alternation of the deformed zone with each pressing, and it consists of pressing the sheets and then rotating them 90° around the normal direction to generate alternated zone with low and high deformation levels. The semi-circular profile avoids stress concentrations in comparison to a V-groove profile, and it introduces less stress concentration points [17]. Every sequence of corrugation, straightening, 90° rotation, corrugation, and straightening is considered one RCS pass (Fig. 1).

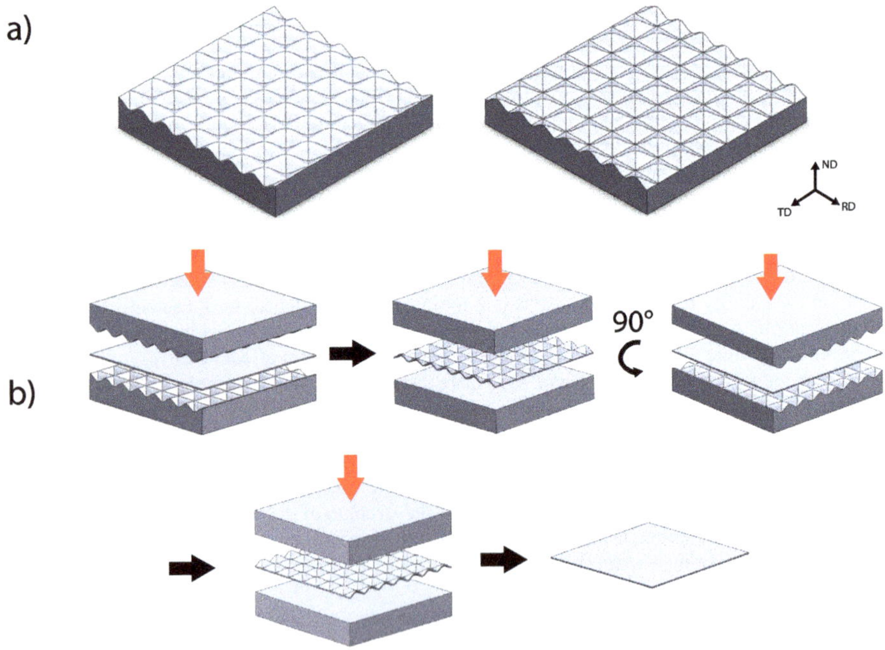

Figure 1 (a) Die profile. (b) One pass of RCS.

Table 1. Nominal composition of the aluminum alloys investigated in %wt [17–19].

Alloy	Si	Fe	Cu	Mn	Mg	Cr	Zn	Ti	Ni	Al
5754-H111	0.04	0.17	0.00	0.01	3.00	0.22	0.00	0.02	0.00	balance
6061-T6	0.77	0.24	0.16	0.03	1.03	0.07	0.03	0.02	0.00	balance
7075-O	0.00	0.3	1.8	0.00	2.2	0.28	5.3	0.06	0.01	balance

The hardening potential of the RCS process in the different aluminum alloys is compared based on the microhardness evolution. Therefore, Vickers microhardness measurements were taken in the deformed zone with a load of 100g ($HV_{0.1}$) with a holding time of 15 seconds. Before the hardness measurement, a representative part of a specimen was mechanically grounded and polished until mirror-finish for each composition.

Uniaxial tensile tests were conducted only for the 5754 and 6061 alloys. The tests were carried out on RCS samples subjected to one and two passes within the temperature range of 150 – 500 °C and strain rates between 3×10^{-5} and 3×10^{-2}. Figure 2 depicts the geometry of these samples, which possess a gauge length of 50 mm. Tensile tests were carried out on specimens after one and two RCS passes. These tests were performed using an INSTRON 5584 universal testing machine outfitted with a cylindrical furnace. The strain rate sensitivity coefficient (m) was derived following Equation (1).

$$\sigma = K\dot{\varepsilon}^m \tag{1}$$

where σ is the stress, K a material constant, and $\dot{\varepsilon}$ the strain rate. A pair of samples were examined for every distinct pairing of temperature and strain rate. Peak values recorded for tensile stress and associated strain rates were graphically represented on a logarithmic scale. Subsequently, a linear approximation of the data was undertaken. The slope of this plot reflects the strain rate sensitivity.

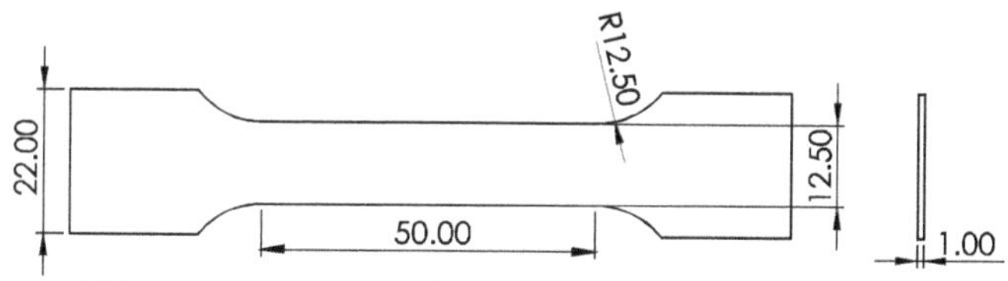

Figure 2 Dimensions of the sample employed for the hot tensile test in mm.

Finally, the crystallographic texture was determined utilizing a Rigaku Ultima IV diffractometer, operating with a Cu-Kα wavelength. The apparatus was arranged in 'in-plane' mode, with an angle range α spanning from 0.5 to 90° and β from 0 to 360°. Specimens were meticulously prepared to mirror-finish prior to the measurements. Orientation distribution functions (ODFs), typically used to represent the crystallographic orientation within Euler's space, were computed from the experimentally derived pole figures (111), (200), and (220). These ODFs were calculated using the MATLAB® toolbox MTEX software for samples in their initial condition and following one and two RCS cycles.

Materials Research Forum LLC
https://doi.org/10.21741/9781644902615-41

Results

Table 2 displays the average hardness values for the samples subjected to the RCS process up to two passes for the different aluminum alloys. These tests offer insight into the effects of the different hardening mechanisms on the RCS process.

Table 2. Microhardness values of the samples before and after being processed by RCS.

Average (HV0.1)	Initial condition	1P	2P
5754-H111	71.7	87.2	91.3
6061-T6	120	139.6	147.2
7075-O	66.2	76.5	75.4

The microhardness data demonstrate significant variations in hardness for different aluminum alloys subjected to the RCS process. In the case of the aluminum alloy 5754, a marked hardness increment of 22% was observed after one pass (1P), which further escalated to a 27% increase after two passes (2P). A similar trend was detected for aluminum alloy 6061-T6, albeit with slightly lower increments of 16.3% and 22% after the 1P and 2P conditions, respectively. Lastly, the hardness testing on aluminum alloy 7075-O revealed a comparable progression, registering an approximate increase of 15% in hardness for the 1P sample relative to its initial condition, and keep constant for further passes.

The most significant increase in microhardness is observed in the first RCS pass, a typical behavior typical of severe plastic deformation processes. However, in the 5754 alloy, the increase in the first pass is greater than that in the 6061 and 7075 alloys, which show a similar increase in the first pass. On the other hand, for the second RCS pass, the alloys whose primary hardening mechanism is precipitation tend to saturate more quickly. A non-significant hardness change was reported for RCS-processed alloy 7075 from the first to fourth RCS passes [19]. This result accords with the superior strain-hardening capacity of the 5754 alloy.

The increment in hardness might be attributable to the substantial dislocation density and grain boundaries induced by the RCS procedure. However, the nearly steady mechanical strength observed following the initial RCS pass could likely be ascribed to the concurrent influence of micro and ultrafine grains. The microscale grains might permit substantial deformations, thereby contributing to ductility, while the ultrafine grains could serve as barriers to the creation and sliding of dislocations, thereby enhancing the material's resistance.

Figure 3a and Figure 3b show the strain rate sensitivity coefficients at different temperatures for 6061 and 5754 samples undergoing one and two RCS cycles. It can be observed that both alloys exhibit relatively high values of the strain rate sensitivity 'm' coefficient at the tested conditions when compared with the typical values reported for aluminum alloys.

Materials Research Proceedings **32** (2023) 362-370
https://doi.org/10.21741/9781644902615-41

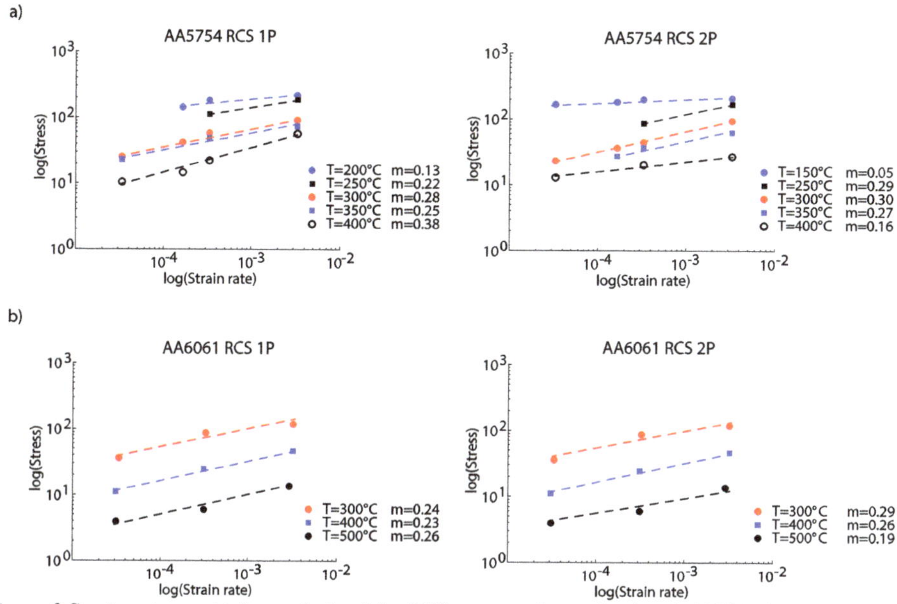

Figure 3 Strain rate sensitivity analysis of the RCS processed samples for a) AA5754 b) AA6061.

The highest 'm' value of 0.38 was achieved at 400 °C for the aluminum 5754 for one RCS pass, while the aluminum 6061 was at 300 °C for two RCS passes with a value of 0.29. Such value is similar to the exhibited by the 5754 in the same condition. On the other hand, the same trend can be observed in both alloys. For the one-pass condition, the 'm' values increase as the temperature rises, while this behavior is the opposite for two passes. They are showing the maximum value for that condition at 300 °C and getting lower as the temperature rises. The grain refinement obtained after the RCS processed samples could explain the high values for the strain rate sensitivity coefficients, since the grain size plays an important role in the deformation mechanisms at these temperatures. Furthermore, the inverse behavior in the two RCS passes could be because the recrystallization temperature decreased due to the deformation process, impacting the activation temperature associated with the deformation mechanism. This results in an enhanced capacity of uniform deformation at high temperatures given by the RCS process.

The quantitative analysis of the volume fraction of the different texture components that are typically developed in aluminum alloys was calculated from the ODF. Table 3 shows the result for the different alloys. Firstly, the aluminum alloy 5754-H111 exhibits a predominantly mixed texture, primarily composed of β-fiber and S-Cube fiber components. The β-fiber texture is defined by a series of <111> fiber textures that spread in the φ_2=45° section of an orientation distribution function (ODF). This texture component is typically observed in metals undergoing processes like rolling or drawing. On the other hand, a connecting fiber, S-Cube fiber, has been previously identified as a potential pathway for recrystallization leading to Cube. These observed texture components align well with the material's thermomechanical history, which underwent a process of rolling, leading to the generation of the S component, followed by an annealing process, resulting in the production of the S-Cube fiber. Throughout the process, the texture components remain consistent. However, there is a minor decrease in the intensity correlating with the number of RCS passes. This reduction in β-fiber, along with the lack of additional β-fiber generation during the process, could positively impact the material's forming capacity, given that previous studies suggest that the presence of β-fiber can be detrimental to the material's formability.

In the case of the aluminum alloy 6061-T6, the initial condition prominently exhibits a strong cube texture, characteristic of the annealed texture commonly found in aluminum alloys. This texture aligns well with the material's thermomechanical history because of the solid solution treatment performed previous to the aging needed to reach the T6 condition. As the number of RCS passes increases, the S component significantly increases. In addition, when subjected to two RCS cycles, a non-common texture component appears. This emergent texture component might be associated with the bending process inherent to the RCS process. Moreover, a substantial reduction in the intensity of the Cube texture component is evident. Literature suggests that this decrease in intensity relates to the rotation of some texture components induced by the bending state [20].

In the initial state of the 7075-O alloy, a prominent Cube orientation is discernible, likely attributable to the recrystallized grains formed during the annealing process. This texture aligns with the findings observed in the 6061 alloys. Interestingly, the Goss and P orientations also manifest in approximately equal measures. A significant observation is that as the intensity of the Cube texture dwindles, other texture components essentially sustain their presence. This pattern is evident across all three studied alloys, where the dominant texture component significantly decreases in intensity with each additional RCS pass. The marked decrease in Cube texture intensity concurrent with increasing applied deformation is consistent with numerous studies on rolled Al-based alloys.

Table 3. Quantitative analysis of the texture evolution through the RCS process.

Texture component	5754			6061			7075		
	0P	1P	2P	0P	1P	2P	0P	1P	4P
Goss	2.3	2.0	1.7	-	-	-	14.0	18.9	14.5
P	-	-	-	-	-	-	13.9	14.4	10.6
Brass	6.6	5.1	3.9	-	1.4	3.8	-	-	-
S	25.9	21.7	21.7	3.1	7.1	12.2	-	-	-
Cu	6.3	5.7	6.0	-	-	-	-	-	-
Cube	-	-	-	46.9	44.1	17.8	13.7	5.3	2.8
S-Cube connecting fiber	18.3	18.3	17.9	9.5	9.9	5.3	-	-	-

Conclusions

In conclusion, the Repetitive Corrugation and Straightening (RCS) process was applied to commercially available 5754, 6061, and 7075 aluminum alloys. The results displayed a significant hardening effect in the alloys, with distinct hardening mechanisms influencing the process. Microhardness tests revealed a significant increase in hardness, particularly for the 5754 alloy, which showed the greatest hardening potential. The rise in hardness is likely attributable to the high dislocation density and grain boundaries introduced by the RCS process.

Tensile tests conducted at different temperatures and strain rates showed 6061 and 5754 alloys exhibiting relatively high strain rate sensitivity 'm' coefficients. A peak 'm' value appears at specific temperatures for one and two RCS passes, providing further evidence of grain refinement's impact on the deformation mechanisms and the alloy's mechanical properties.

Crystallographic texture analysis showed that the RCS process had distinct effects on different alloy textures. In particular, the process significantly reduced the cube texture component in all the studied alloys, with the effect being most pronounced in the 7075-O alloy. The consistency in the other texture components suggests that the Goss and S texture components are able to remain during the RCS process.

In summary, the RCS process presents an effective approach to enhance the mechanical properties of the studied aluminum alloys. This study provides insight into the potential of the RCS process for industrial applications, such as automotive and aerospace, where enhanced mechanical properties are of significant importance. However, more comprehensive investigations considering more complex deformation processes and additional alloy types may be necessary to understand the RCS's potential better.

References

[1] Z. Chen, G. Fang, J.-Q. Zhao, Formability Evaluation of Aluminum Alloy 6061-T6 Sheet at Room and Elevated Temperatures, J. of Materi Eng and Perform. 26 (2017) 4626–4637. https://doi.org/10.1007/s11665-017-2895-0.

[2] W.S. Miller, L. Zhuang, J. Bottema, A.J. Wittebrood, P. De Smet, A. Haszler, A. Vieregge, Recent development in aluminium alloys for the automotive industry, Materials Science and Engineering: A. 280 (2000) 37–49. https://doi.org/10.1016/S0921-5093(99)00653-X.

[3] R.Z. Valiev, R.K. Islamgaliev, I.V. Alexandrov, Bulk nanostructured materials from severe plastic deformation, Progress in Materials Science. 45 (2000) 103–189. https://doi.org/10.1016/S0079-6425(99)00007-9.

[4] H. Yu, L. Su, C. Lu, K. Tieu, H. Li, J. Li, A. Godbole, C. Kong, Enhanced mechanical properties of ARB-processed aluminum alloy 6061 sheets by subsequent asymmetric cryorolling and ageing, Materials Science and Engineering: A. 674 (2016) 256–261. https://doi.org/10.1016/j.msea.2016.08.003.

[5] J. Huang, Y.T. Zhu, D.J. Alexander, X. Liao, T.C. Lowe, R.J. Asaro, Development of repetitive corrugation and straightening, Materials Science and Engineering: A. 371 (2004) 35–39. https://doi.org/10.1016/S0921-5093(03)00114-X.

[6] D.H. Shin, J.-J. Park, Y.-S. Kim, K.-T. Park, Constrained groove pressing and its application to grain refinement of aluminum, Materials Science and Engineering: A. 328 (2002) 98–103. https://doi.org/10.1016/S0921-5093(01)01665-3

[7] L. Romero-Resendiz, J.M. Cabrera, S. Elizalde, V. Amigó-Borrás, I.A. Figueroa, G. Gonzalez, Mechanical, stress corrosion cracking and crystallographic study on flat components processed by two combined severe plastic deformation techniques, Journal of Materials Research and Technology. 18 (2022) 1281–1294. https://doi.org/10.1016/j.jmrt.2022.03.010

[8] M. Ezequiel, I.A. Figueroa, S. Elizalde, J.M. Cabrera, C. Braham, L. Morin, G. Gonzalez, Numerical and experimental study of a 5754-aluminum alloy processed by heterogeneous repetitive corrugation and straightening, Journal of Materials Research and Technology. 9 (2020) 1941–1947. https://doi.org/10.1016/j.jmrt.2019.12.026

[9] A. Azushima, R. Kopp, A. Korhonen, D.Y. Yang, F. Micari, G.D. Lahoti, P. Groche, J. Yanagimoto, N. Tsuji, A. Rosochowski, A. Yanagida, Severe plastic deformation (SPD) processes for metals, CIRP Annals. 57 (2008) 716–735. https://doi.org/10.1016/j.cirp.2008.09.005.

[10] E. Bruder, Formability of Ultrafine Grained Metals Produced by Severe Plastic Deformation-An Overview, Adv. Eng. Mater. 21 (2019) 1800316. https://doi.org/10.1002/adem.201800316.

[11] A.A. Khamei, K. Dehghani, Effects of strain rate and temperature on hot tensile deformation of severe plastic deformed 6061 aluminum alloy, Materials Science and Engineering: A. 627 (2015) 1–9. https://doi.org/10.1016/j.msea.2014.12.081

[12] S. Suwas, R.K. Ray, Crystallographic Texture of Materials, Springer London, London, 2014. https://doi.org/10.1007/978-1-4471-6314-5

[13] J. Galán-López, L.A.I. Kestens, Optimization of Crystallographic Texture for Sheet-forming Applications Using Taylor-based Models, Metall Mater Trans A. 49 (2018) 5745–5762. https://doi.org/10.1007/s11661-018-4869-8

[14] K. Yoshida, T. Ishizaka, M. Kuroda, S. Ikawa, The effects of texture on formability of aluminum alloy sheets, Acta Materialia. 55 (2007) 4499–4506. https://doi.org/10.1016/j.actamat.2007.04.014

[15] F. Barlat, O. Richmond, Prediction of tricomponent plane stress yield surfaces and associated flow and failure behavior of strongly textured f.c.c. polycrystalline sheets, Materials Science and Engineering. 95 (1987) 15–29. https://doi.org/10.1016/0025-5416(87)90494-0

[16] K. Ma, H. Wen, T. Hu, T.D. Topping, D. Isheim, D.N. Seidman, E.J. Lavernia, J.M. Schoenung, Mechanical behavior and strengthening mechanisms in ultrafine grain precipitation-strengthened aluminum alloy, Acta Materialia. 62 (2014) 141–155. https://doi.org/10.1016/j.actamat.2013.09.042

[17] S. Elizalde, M. Ezequiel, I.A. Figueroa, J.M. Cabrera, C. Braham, G. Gonzalez, Microstructural Evolution and Mechanical Behavior of an Al-6061 Alloy Processed by Repetitive Corrugation and Straightening, Metals. 10 (2020) 489. https://doi.org/10.3390/met10040489

[18] M. Ezequiel, S. Elizalde, J.-M. Cabrera, J. Picas, I.A. Figueroa, I. Alfonso, G. Gonzalez, Formability of the 5754-Aluminum Alloy Deformed by a Modified Repetitive Corrugation and Straightening Process, Materials. 13 (2020) 633. https://doi.org/10.3390/ma13030633

[19] L. Romero-Resendiz, V. Amigó-Borrás, A. Vicente-Escuder, S. Elizalde, J.M. Cabrera, D. Pineda-Ruiz, I.A. Figueroa, G. Gonzalez, Effect of the microstructure generated by Repetitive Corrugation and Straightening (RCS) process on the mechanical properties and stress corrosion cracking of Al-7075 alloy, Journal of Materials Research and Technology. 15 (2021) 4564–4572. https://doi.org/10.1016/j.jmrt.2021.10.043

[20] W. Muhammad, U. Ali, A.P. Brahme, J. Kang, R.K. Mishra, K. Inal, Experimental analyses and numerical modeling of texture evolution and the development of surface roughness during bending of an extruded aluminum alloy using a multiscale modeling framework, International Journal of Plasticity. 117 (2019) 93–121. https://doi.org/10.1016/j.ijplas.2017.09.013

Materials Research Proceedings **32** (2023) 371-378 https://doi.org/10.21741/9781644902615-42

Exploring the effects of repetitive corrugation and strengthening on alloy A5083 using X-ray diffraction techniques

Carlos Alberto De la Trinidad[1,2a*], Ignacio A. Figueroa[1,b], José María Cabrera[2,3,c] and Gonzalo González[1,d*]

[1] Instituto de Investigaciones en Materiales, Universidad Nacional Autónoma de México, Circuito Exterior S/N, Cd. Universitaria, A.P. 70-360, Coyoacán, 04360, Ciudad de México, Mexico

[2] Departament de Ciència i Enginyeria del Materials, EEBE (Escola d'Enginyeria de Barcelona Est) - Universitat Politècnica de Catalunya, Av. Eduard Maristany, 16, 08019, Barcelona, Spain

[3] Fundació CIM UPC, c/Llorens I Artigas 12, 08028, Barcelona, Spain

[a]carlos.alberto.de.la.trinidad@upc.edu, [b]iafigueroa@unam.mx, [c]jose.maria.cabrera@upc.edu, [d]joseggr@unam.mx

Keywords: X-Ray Diffraction, Severe Plastic Deformation, Repetitive Corrugation and Straightening, Aluminium Alloy

Abstract. Aluminum alloys are highly versatile materials and their behavior can be improved by severe plastic deformation. X-Ray Diffraction (XRD) is a key tool for studying the microstructural changes and the evolution of the crystalline structure during this process. This combination of processing techniques helped the development of high-performance aluminum alloys for various industrial applications. The objective was to determine a heat treatment to promote a new crystallographic texture of AA5083 under a severe plastic deformation process. The determination of this thermal ratio, as well as the crystallographic characterization of the material, were studied through XRD, thermo-diffraction, Rietveld analysis, and pole figures (PF). The results showed that using a partial recrystallization heat treatment combined with the RCS process favors obtaining a characteristic recrystallization and deformation texture (*cube* and *brass* component). Furthermore, the evolution of diffraction peaks intensities at different temperatures confirm that recrystallization takes place during the process. Finally, the phases present in the A5083 alloy were determined by XRD.

Introduction

Aluminum alloys are widely used materials in various industries due to their excellent combination of properties, including high strength, lightness, and corrosion resistance. These alloys can further improve their characteristics through the process of severe plastic deformation, which allows them to obtain a refined microstructure and superior mechanical properties. The X-Ray diffraction technique plays a fundamental role in the study of the structural changes that occur during Severe Plastic Deformation (SPD) of aluminum alloys[1].

SPD involves applying large deformations to a material in one or several stages, generating a high density of crystalline defects, such as dislocations and subgrains. This process leads to significant changes in the microstructure of the material, including the formation of subgrains of reduced size, a higher dislocation density, and greater uniformity in the distribution of alloying elements[2]. These microstructural changes directly impact the mechanical properties of aluminum alloys, improving their strength and ductility.

XRD is an experimental technique used to analyze the crystalline structure of materials as well as phases of the sample. The diffraction pattern contains valuable information about the spatial arrangement of the atoms in the sample and the structural changes that occur during severe plastic deformation[3].

By studying aluminum alloys by XRD before and after severe plastic deformation, changes in the intensity and position of the diffraction peaks can be observed. These changes indicate alterations in the crystalline structure, such as the appearance of new phases, the formation of dislocations, and the generation of crystalline defects. In addition, XRD also makes it possible to determine important structural parameters, such as grain size and dislocation density[4].

The combination of severe plastic deformation and XRD provides a deep understanding of the microstructural mechanisms that govern the behavior of aluminum alloys during plastic deformation. This is essential for designing and developing alloys with improved properties for specific applications, such as the aerospace, automotive, and construction industries.

Material and methods

The starting material was a commercial AA5083 sheet of 1.2 cm in thickness. The nominal chemical composition is shown in Table 1.

Table 1. Composition of the studied AA5083 alloy.

Element	Si	Fe	Cu	Mn	Mg	Cr	Zn	Ti	Al
% wt	0.40	0.40	0.10	0.4	4.00	0.05	0.25	0.15	Balance

First, a stress relief heat treatment was carried out at 200°C for 1h and cooled in the air, to ensure homogeneity in the material to be worked on and eliminate any residual stress present. Following this, the material was rolled at 2.7 cm/min to reach a thickness of 1 mm and achieve the first crystallographic texture of the material.

As the first preparation step for any mechanical forming process, a stress-relieving heat treatment is performed to ensure homogeneity in the material to be worked on and eliminate any residual stress.

The second step is the most important for obtaining a crystallographic texture. In order to achive this, a matrix with a sinusoidal profile with an amplitude of 2mm and a period of 16mm was used. Samples were rotated 90° from the normal direction as shown in Fig. 1.

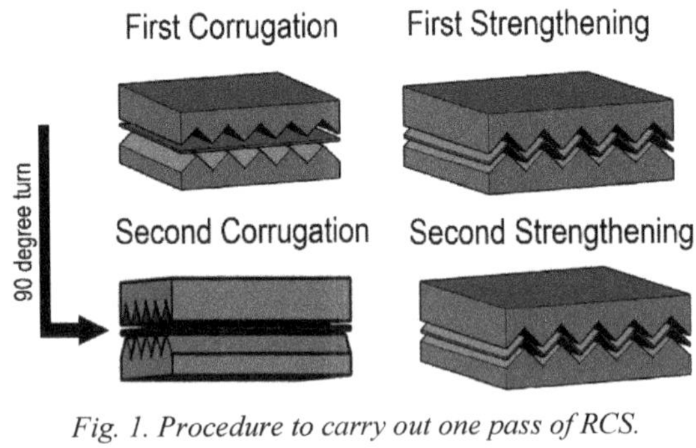

Fig. 1. Procedure to carry out one pass of RCS.

XRD, thermo-diffraction, and Rietveld analysis will be performed at each stage of the process.

Finally, it was determined a pole figure record using Rigaku X-ray diffraction equipment, observing the families of crystalline planes {200}, to validate the new crystallographic texture.

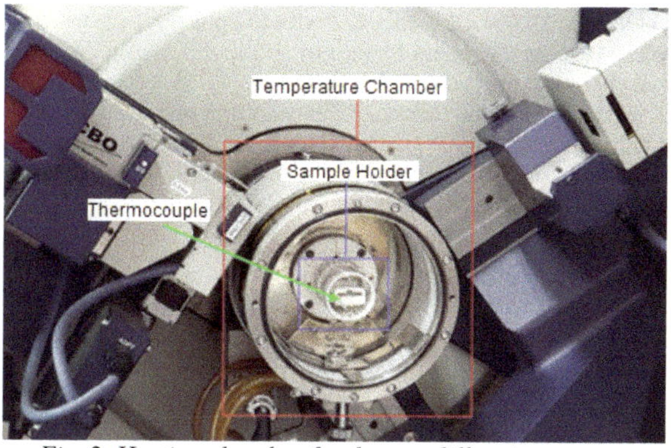

Fig. 2. Heating chamber for thermo-diffraction.

Results and discussion

XDR coupled with a heating chamber allows to identify the crystalline phases and the microstructural changes, indicating the texture of the material, based on the change in peaks width and their intensity.

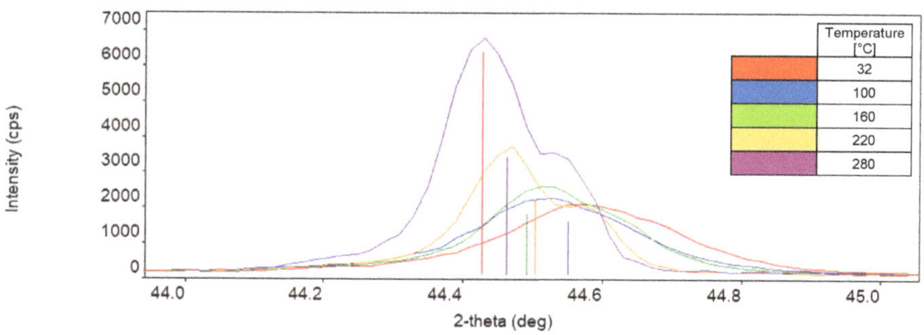

Fig. 3. Diffractograms of the peak (200) at different temperatures for the rolled A5083 alloy.

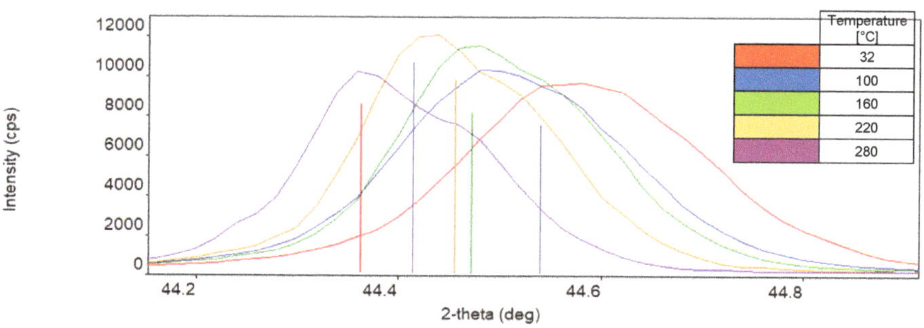

Fig. 4. Diffractograms of the peak (200) at different temperatures for the A5083 alloy subjected to 2 steps of the RCS process.

Materials Research Forum LLC
https://doi.org/10.21741/9781644902615-42

Figures 3 and 4 show the peaks associated with the plane (200), which was chosen for having a good balance between intensity and sensitivity to the effect of temperature. It is observed how the increase in temperature shifts the diffraction peaks to the left regardless of their previous state of deformation. This is due to the thermal expansion that occurs in the material, which causes an increase in the lattice parameter "a" [5]. In addition, in Fig. 3, an increase in the intensity of the peaks is observed, which indicates a texturing of the material. This texture needs a heat treatment before of the RCS process to favor the formation of a new texture.

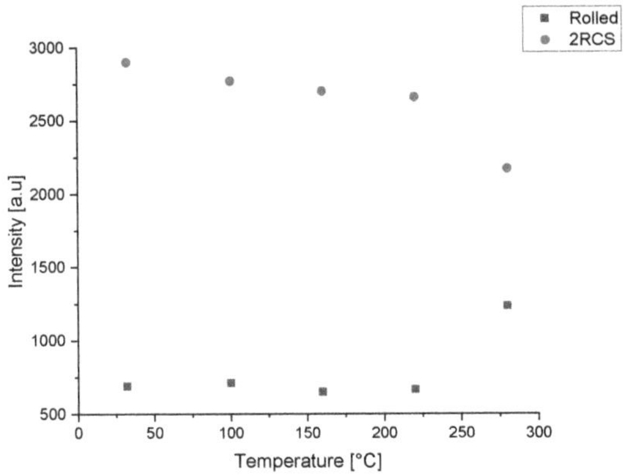

Fig. 5. Thermal profiles of the peak intensity (200), for the alloy A5083, subjected to the rolling process and two steps of RCS.

Fig. 5 shows in more detail the favoring of a crystallographic texture over the material, where there are marginal changes in the intensity of the peaks from 32 to 220°C; from 280°C it increases significantly, which is an indicator that from that temperature, a new preferential texture is favored[6].

The Crystallography Open Database (COD)[7] was used to identify the characteristic peaks of the aluminum phase, as well as the presence of precipitates. The data was processed with the PDXL2 software, showing the presence of Al, Al_3Mg_2, $MgZn_2$, y Mg_2Si.

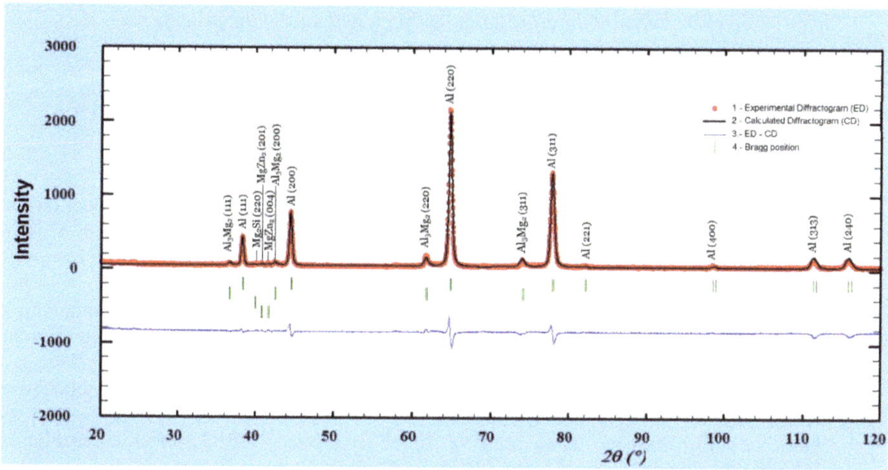

Fig. 6. X-ray diffractogram of the rolled A5083 alloy, refined with the Rietveld method.

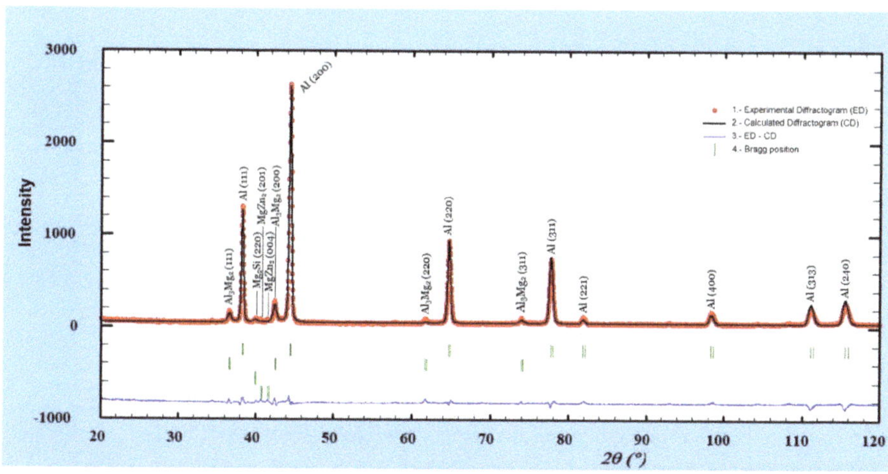

Fig. 7. X-ray diffractogram of the A5083 alloy subjected to two RCS steps, refined with the Rietveld method.

Figures 6 and 7 show the phases observed in the A5083 alloy, with a slight shift of the peaks corresponding to the (313) and (240) planes to the left.
With the help of the Rietveld refinement, the analysis of micro-deformations and reduction in grain size can be quantitatively calculated (see Table 2).

Table 2. Rietveld refinement for alloy A5083 subjected to the rolling process and 2RCS.

Rolled		2RCS	
Apparent Grain Size [A]	1334.91	Apparent Grain Size [A]	1499.64
Micro-deformations	1.79E-03	Micro-deformations	1.21E-03
Chi2	1.990	Chi2	0.890

Once the sample has been subjected to two RCS steps, the microstrains in the crystal lattice decrease due to the high amount of shear stresses to which the material is subjected. This is evidenced in the Rietveld refinement showed in Table 2.

With this information, a Partial Recrystallization Heat Treatment (PRHT) was carried out, which was adapted to promote the formation of a new crystallographic texture; the following thermal cycle was proposed:

- Heat the alloy from room temperature at a rate of 15°C/min until reaching a solubilization temperature of 400°C, and hold for 60min.
- Heat the alloy from the temperature of 400°C at a rate of 15°C/min until reaching a solubilization temperature of 500°C; hold for 30min.
- Cool in the air.

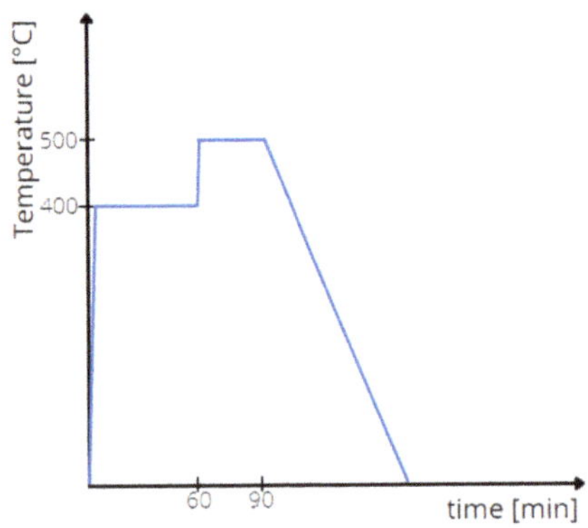

Fig. 8. Proposed Partial Recrystallization Heat Treatment (PRHT).

Finally, with an established thermal treatment, pole figures were made to observe the crystallographic texture changes qualitatively.

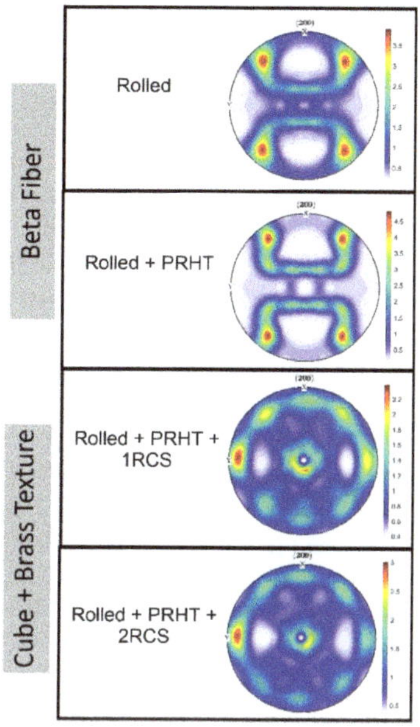

Fig. 9. Evolution of the Pole Figures of the planes (200), of the AA5083 subjected to a process to observe the microstructural evolution.

As shown in Fig. 9, introducing a suitable heat treatment (PRHT) between the rolling process and RCS can encourage the formation of a new crystallographic texture.

In this case, the starting point is a beta fiber (rolled), which is maintained during the PRHT, and when subjected to the RCS process, it changes to a component with a typical recrystallization texture (*cube*) coexisting with a typical component of shaping processes (*brass*).

The formation of textures in aluminum alloys produced by thermo-mechanical processes, including SPD processes, is a very striking and novel field of knowledge that is currently being studied[8]. It is intended to continue with the study of this work to finish explaining the phenomena of texturing that were produced.

Conclusions

- According to crystallographic analysis, a partial recrystallization heat treatment was determined to obtain a bimodal texture.
- The Al, Al_3Mg_2, $MgZn_2$, and Mg_2Si phases were identified with the help of XRD.
- A Rietveld analysis was quantified and determined the severity of the RCS process in the micro-deformations, as well as evidenced by the texturing of the material when subjected to this process.
- It was possible to obtain a characteristic crystallographic texture of recrystallization (*cube* component) with another characteristic of mechanical processes (*brass* component) for the A5083 alloy.

Acknowledgments

The authors would like to acknowledge CONACyT (Mexico) for the scholarship (No. 966568).

References

[1] E. Cerri and P. Leo, "Influence of severe plastic deformation on aging of Al-Mg-Si alloys," *Mater. Sci. Eng. A*, vol. 410–411, no. April, pp. 226–229, 2005. https://doi.org/10.1016/j.msea.2005.08.135

[2] I. Sabirov, M. Y. Murashkin, and R. Z. Valiev, "Nanostructured aluminium alloys produced by severe plastic deformation: New horizons in development," *Mater. Sci. Eng. A*, vol. 560, pp. 1–24, 2013. https://doi.org/10.1016/j.msea.2012.09.020

[3] J. Gubicza, N. Q. Chinh, G. Krállics, I. Schiller, and T. Ungár, "Microstructure of ultrafine-grained fcc metals produced by severe plastic deformation," *Curr. Appl. Phys.*, vol. 6, no. 2, pp. 194–199, 2006. https://doi.org/10.1016/j.cap.2005.07.039.

[4] F. Hadef and M. Ans, "X-ray analysis and Rietveld refinement of ball milled Fe50Al35Ni15 powder," *Surfaces and Interfaces*, vol. 26, no. June, p. 101303, 2021. https://doi.org/10.1016/j.surfin.2021.101303

[5] C. G. Pope, "X-ray diffraction and the bragg equation," *J. Chem. Educ.*, vol. 74, no. 1, pp. 129–131, 1997. https://doi.org/10.1021/ed074p129

[6] R. Song, D. Ponge, D. Raabe, and R. Kaspar, "Microstructure and crystallographic texture of an ultrafine grained C-Mn steel and their evolution during warm deformation and annealing," *Acta Mater.*, vol. 53, no. 3, pp. 845–858, 2005. https://doi.org/10.1016/j.actamat.2004.10.051

[7] Information on: "Crystallography Open Database." http://www.crystallography.net/cod/

[8] J. J. Sidor, R. H. Petrov, and L. A. I. Kestens, "Microstructural and texture changes in severely deformed aluminum alloys," *Mater. Charact.*, vol. 62, no. 2, pp. 228–236, 2011. https://doi.org/10.1016/j.matchar.2010.12.004

Keyword Index

5083 Aluminum Alloy 157

A320NEO 119
Accumulative Roll Bonding 287
Activation Energy 72
Additive Manufacturing (AM) 330
Airbus 119
Al Alloy 330
Alpha Case 55
Aluminium Alloy 371
Aluminium SPD Grain Refinement 189
Aluminum Alloys 173
Aluminum Alloys, Texture 362
Aluminum Composites 287
Aspect Ratio 280
As-Rolled 2A97 Al-Li Alloy 149
Asymmetric Cold Rolling 309, 355
Asymmetric Hot Rolling 294
AZ31 213

Boride Fibers 280

Complex Concentrated Alloys 244
Consolidation 213
Constitutive Equation 87, 244
Constitutive Parameters 264
Corrosion 197
Cup Forming 15

Deformation Behaviour 253
Deformation Mechanisms 227
Deformation 253
Diffraction Theory 25
Diffusion Bonding 64, 338
Dissimilar Material Joining 95
Double Gas Pressure 141
Duplex Stainless Steel 205, 220
Dynamic Grain Growth 181
Dynamic Recrystallization 25, 149, 294, 304
Dynamic Reverse Transformation 167

ECAP Misorientation Evolution 189
ECAP 197, 205
Electric Field Assisted 111
Electricity Consumption 355
Equal Channel Angular Pressing 3, 213

FEM 294
Finite Elements 127
Flow Stress 181
Forming 103
Four-Storey Structure 81
Free-Inflation Test 264
Friction Stir Forming 95
Friction Stir Processing (FSP) 173
Friction Stir Welding 95
Functional Properties 41

GBS Superplastic Constitutive Equation 189
Grain Boundary Sliding 167, 227
Grain Refinement 205
Grain Size 173

Heat Treatment 303
HF/SPF Simulations 119
HF/SPF Workshop 119
High Mn Steel 319
High Strength Steel 167
High-Entropy Alloy 244, 253
High-Pressure Sliding 15
High-Pressure Torsion 3, 25, 41, 72, 235
High-Speed Steel 294
High-Strength Steel Grades 309, 355

High-Strength Steel 274
High-Temperature Titanium Alloy 81
High-Temperature 25
Hot Forming 119
Hot Rolling 319
HPT 220
Hyperbolic Sine Model 264

Industrial Challenges 119
Industry Experiment 355
In-Situ 25
Intermediate Annealing 309, 355

JMAK 294

Kinematic Asymmetry 287

Laboratory and Industry Experiments 309
Laminated Materials 287
Laser-Scanning Confocal Microscopy 235
Lattice Structure 141
Logistics 119
Low-Temperature Superplasticity 167
LZ91 Mg-Li Alloy 111
Magnesium Alloy 213, 303
Martensite 274
Mechanical Behavior 280
Mechanical Interlock 95
Mechanical Properties 149, 197, 244, 280
Metal-Matrix Composite 280
Mg Alloy 197
Microstructural Characterization 319
Microstructural Evolution 181
Microstructure 64, 103, 173, 197, 280
Modelling 264

Nacl Corrosion 319
Nanostructure 72, 235
Neutron Diffraction 235
Neutron Scattering 25
New Misorientation Expression 189
Numerical Simulation 87

Opening and Closing Mechanism Hatch 157
Optical Fiber 95

Phase Transformation 25
Plastic Deformation 25
Plasticity 330
Production Workflow 119

Reciprocal Space 25
Repetitive Corrugation and Straightening 371
Repetitive Corrugation and Straightening
Retained Austenite 274
Rolling Force 355, 309

Secondary Phases 220
Severe Plastic Deformation (SPD) 3, 15, 41, 205, 213, 220, 330, 371
Sheet 64
Simulation 127
Solid-State Recycling 213
SP-700 95
Statistical Modeling 338
Strain Rate Sensitivity 41, 72, 103, 362
Superalloy 15
Superplastic Deformation Behavior 111
Superplastic Forming 55, 64, 81, 87, 119, 127, 141, 157, 253, 338
Superplastic Gas-Bulging 149
Superplasticity 15, 41, 72, 95, 103, 173, 181, 227

Surface Groove 141
Synchrotron Radiation 25

Technological Test 157
Tensile Strength 287
Tensile Test 303
Tensile 103
Thermo-Mechanical Processing 25
Ti2AlNb Alloy 87
Ti60, Diffusion Bonging 81
Ti6242 119
Ti-6Al-4V 55, 103, 119
Titanium Alloy 95, 181, 280
Titanium Aluminides 25
Titanium Matrix Composite 64
Titanium 55, 103, 119, 127
TRIP-Aided Steel 274
Tube High-Pressure Shearing 3

Ultrafine Grained 3, 72, 244

Voids 338

Wall Thickness 149
Warm Rolling 274

X-Ray Diffraction 235, 371

Yield Strength 287

E-Martensite 319

About the Editor

José-María Cabrera-Marrero, was born on December 8th, 1964 in Santa Cruz de Tenerife, Canary Islands, Spain. He holds a PhD in Industrial Engineering from the Technical University of Catalonia (UPC), having received for his doctoral thesis the extraordinary doctorate prize (1995) from the Higher Technical School of Industrial Engineers of Barcelona. Likewise, his final project was awarded the first prize (1992) by the College and Association of Industrial Engineers of Catalonia.

He has been the recipient of a research training grant (1992-1995) and a postdoctoral fellowship (1995-1996) from the Generalitat de Catalunya. He has carried out research stays at McGill University, Montreal (six months divided into two summers), at the National Polytechnic Institute of Mexico City (10 months between 1995 and 1996, where he obtained an appointment as a visiting professor), at the Federal University of Sao Carlos (Brazil) for 3 weeks in 2012, at the Materials Research Institute of the UNAM (Mexico) for two months (January-February 2017). He has made a sabbatical stay at the Institute of Research in Materials and Metallurgy of the Michoacan University of San Nicolás de Hidalgo, Morelia for 12 months (September 2019-August 2020), the latter financed by CONACyT.

He has been a professor of Metallurgy at the UPC since 1996 and is currently a University Professor ("Chair") in the Department of Materials Science and Engineering of the UPC, a position he has held since 2008. He was secretary of the department for 10 years, and deputy vicedirector of the same and responsible for the doctoral program for another 10 years. At the same time, he was responsible for 15 years of the metal forming processes area of the Manresa Technology Center (CTM) until March 2016. He is currently director of the CIM UPC Foundation (www.cimupc.org).

He lectures on Material Selection in Mechanical Design, Metallurgical Technology, Physical Metallurgy, Manufacture of Metals and Nanostructured Materials in the Faculty of Engineering of Barcelona East.

He has participated in more than 70 research projects (leading 50) both competitive public funding (responsible for five CICYT projects, a PETRI, an ALFA network, member of a CYTED project, leader of three European RCFS projects and participant in two others), and in projects with companies (including two large projects funded by the CENIT program). He has carried out a multitude of consultancies and analyses of metallurgical nature for both Spanish and foreign companies.

His area of knowledge is oriented to the characterization of the plastic behavior, mainly at high temperature, of metallic materials. It is basically in this area that Professor Cabrera has made his greatest contributions, especially by offering a universal constitutive equation that explains the hot flow curves of practically all types of metallic materials. This equation, of physical basis, allows to be introduced in process simulation software by numerical techniques allowing at the same time to make microstructural predictions. It has also been oriented lately to obtain ultrafine and nanometric grain structures in metallic materials by processes of severe plastic deformation (equal channel angular pressing and mechanical milling).

He has given lectures, talks and courses at universities, institutes and companies in France, England, Germany, Belgium, Greece, Russia, Mexico, Argentina, Colombia, Chile, Venezuela, Inida and Brazil.

The list of his publications can be found in:

https://scholar.google.co.uk/citations?user=N4HojSgAAAAJ&hl=es

www.ingramcontent.com/pod-product-compliance
Ingram Content Group UK Ltd.
Pitfield, Milton Keynes, MK11 3LW, UK
UKHW020659041026
15590UKWH00001B/353

9 781644 902608